Lecture Notes in Networks and Systems

Volume 338

The series "Lecture Notes in Networks and Systems" publishes the latest developments in Networks and Systems—quickly, informally and with high quality. Original research reported in proceedings and post-proceedings represents the core of LNNS.

Volumes published in LNNS embrace all aspects and subfields of, as well as new challenges in, Networks and Systems.

The series contains proceedings and edited volumes in systems and networks, spanning the areas of Cyber-Physical Systems, Autonomous Systems, Sensor Networks, Control Systems, Energy Systems, Automotive Systems, Biological Systems, Vehicular Networking and Connected Vehicles, Aerospace Systems, Automation, Manufacturing, Smart Grids, Nonlinear Systems, Power Systems, Robotics, Social Systems, Economic Systems and other. Of particular value to both the contributors and the readership are the short publication timeframe and the world-wide distribution and exposure which enable both a wide and rapid dissemination of research output.

The series covers the theory, applications, and perspectives on the state of the art and future developments relevant to systems and networks, decision making, control, complex processes and related areas, as embedded in the fields of interdisciplinary and applied sciences, engineering, computer science, physics, economics, social, and life sciences, as well as the paradigms and methodologies behind them.

Indexed by SCOPUS, INSPEC, WTI Frankfurt eG, zbMATH, SCImago.

All books published in the series are submitted for consideration in Web of Science.

For proposals from Asia please contact Aninda Bose (aninda.bose@springer.com).

More information about this series at https://link.springer.com/bookseries/15179

Krassimir T. Atanassov ·
Vassia Atanassova · Janusz Kacprzyk ·
Andrzej Kałuszko · Maciej Krawczak ·
Jan W. Owsiński · Sotir S. Sotirov ·
Evdokia Sotirova · Eulalia Szmidt ·
Sławomir Zadrożny
Editors

Uncertainty and Imprecision in Decision Making and Decision Support: New Advances, Challenges, and Perspectives

Selected papers from BOS-2020, held on December 14–15, 2020, and IWIFSGN-2020, held on December 10–11, 2020 in Warsaw, Poland

Springer

Editors
Krassimir T. Atanassov
Department Bioinformatics and Mathematical Modelling
Bulgarian Academy of Sciences, Institute of Biophysics and Biomedical Engineering
Sofia, Bulgaria

Vassia Atanassova
Institute of Biophysics and Biomedical
Bulgarian Academy of Sciences
Sofia, Bulgaria

Janusz Kacprzyk
Systems Research Institute
Polish Academy of Sciences
Warsaw, Poland

Andrzej Kałuszko
Systems Research Institute
Polish Academy of Sciences
Warsaw, Poland

Maciej Krawczak
WIT-Warsaw School of Information Technology
Warsaw, Poland

Jan W. Owsiński
Systems Research Institute
Polish Academy of Sciences
Warsaw, Poland

Sotir S. Sotirov
Faculty of Technical Sciences
"Prof. Assen Zlatarov" University
Burgas, Bulgaria

Evdokia Sotirova
Faculty of Natural Sciences
"Prof. Assen Zlatarov" University
Bourgas, Bulgaria

Eulalia Szmidt
Systems Research Institute
Polish Academy of Sciences
Warsaw, Poland

Sławomir Zadrożny
Systems Research Institute
Polish Academy of Sciences
Warsaw, Poland

ISSN 2367-3370 ISSN 2367-3389 (electronic)
Lecture Notes in Networks and Systems
ISBN 978-3-030-95928-9 ISBN 978-3-030-95929-6 (eBook)
https://doi.org/10.1007/978-3-030-95929-6

This Springer imprint is published by the registered company Springer Nature Switzerland AG
The registered company address is: Gewerbestrasse 11, 6330 Cham, Switzerland

Preface

This volume is composed of selected papers from two important conferences held in Warsaw, Poland. The first is the Sixteenth National Conference on Operational and Systems Research, BOS-2020, held on December 14–15, 2020, one of premiere conferences in the field of operational and systems research, not only in Poland, but also at the European level. The second is the Nineteenth International Workshop on Intuitionistic Fuzzy Sets and Generalized Nets, IWIFSGN-2020, held on December 10–11, 2020, in Warsaw, Poland, in turn—one of premiere conferences on fuzzy logic, notably on extensions of the traditional fuzzy sets, also comprising a considerable part on the generalized nets (GNs), an important extension of the traditional Petri nets. A joint publication of selected papers from BOS-2020 and IWIFSGN-2020 is well justified as, first, it follows a long tradition of such a joint organization, even collocation in some years. Secondly, from the substantial point of view, the scope of the BOS conferences covers all kinds of problems related to systems modeling, systems analysis, broadly perceived operational research, notably optimization, decision making and decision support, to just mention a few, often under uncertain and imprecise information. Therefore, in addition to traditional probabilistic and statistical tools and techniques, the use of fuzzy sets and their extensions is also relevant. The use of the intuitionistic fuzzy sets, which are at the core of the IWIFSGN conferences, may be here a very good example. This all has implied that the BOS and IWIFSGN conferences have always jointly been perfect venues for the exchange of ideas, cross-fertilization, and mutual inspiration.

The volume is composed of some parts that cover the main areas and challenges related to the above rationale and philosophy. Part I, "Advances in the processing of uncertain and imprecise information," contains papers that concern issues and problems of a more general interest, notably related to uncertainty and imprecision, with emphasis on data analysis and related aspects.

Jan G. Bazan, Urszula Bentkowska, and Marcin Mrukowicz ("Application of decision rules to discover knowledge for fitting aggregations to datasets") propose to use decision rules to determine the best interval-valued aggregation function in the classification algorithm, based on the dataset complexity measures. RSES toolset is employed. Examples of feature-based measures, dimensionality measures,

measures of separability of classes, and class imbalance measures are presented to show both the complexity of the classification problem in question and the effectiveness and efficiency of the method proposed.

Dawid Kosior and Barbara Pękala ("Influence of interval-valued measures on classification methods with missing values") extend the basic version of the *k*-nearest neighbor (*k*-NN) algorithm to include similarity and indistinguishability measures that reflect uncertainty. The measures used take into account the uncertainty in decision making concerning the classification process. The algorithm is adapted to the processing of attribute values stored as interval values and is tested on data with missing values using three approaches to determine the missing value of individual objects.

Przemysław Grzegorzewski and Maciej Romaniuk ("Bootstrap methods for fuzzy data") are concerned with the bootstrap, which is recognized as a universal method supporting statistical inference in situations where other approaches fail, also in the area of fuzzy data analysis. After a review of the bootstrap methods oriented at statistical reasoning with random fuzzy samples, a new flexible bootstrap approach designed for asymmetrical random fuzzy numbers is proposed. Finally, a comparison of the existing bootstrap approaches is shown to provide some hints for the practitioners.

Jarosław Stańczak and Jan W. Owsiński ("Evolutionary *k*-means clustering method with controlled number of clusters applied to determine the typology of Polish municipalities") are concerned with the development of a method of grouping data items in a number of categories, depending on the values of easy-to-interpret parameters, giving the ability to regulate to a certain degree the level of detail of data division (the number, size, and location of detected clusters). The problem's importance and difficulty have become the impetus to develop the evolutionary *k*-means method and to examine its properties in conjunction with the test data, referring to several important parameters of Polish municipalities. The method proposed is based on the widely known *k*-means approach which, in combination with the evolutionary algorithm and some easy-to-obtain information on grouped data, makes it possible to obtain an adaptive method of data clustering, in which the degree of detail in the distribution of data can be easily adjusted using one parameter, and, depending on the value of this parameter, the method produces more detailed or more coarse data clustering, involving more or less of the detected clusters. The new evolutionary *k*-means method is shown with the data considered and the obtained results, compared with the results generated by geographers.

Part II, "Aspects of uncertain and imprecise information," is mainly concerned with theoretical and analytic issues and problems related to the representation of imperfect, uncertain and imprecise, information, mainly dealing with some specialized representation of fuzzy sets and fuzzy numbers, intuitionistic fuzzy sets, also with interval-valued extensions.

Katarína Čunderlíková ("Conditional intuitionistic fuzzy mean value in connection with IF-probability") defines the notion of a conditional intuitionistic fuzzy mean value for the intuitionistic fuzzy observables using the intuitionistic fuzzy probability. The relation is shown between the conditional intuitionistic fuzzy mean value and the conditional intuitionistic fuzzy probability, related to the intuitionistic fuzzy probability. A martingale convergence theorem for the conditional intuitionistic fuzzy mean value is formulated in this case, too.

Krzysztof Piasecki and Anna Łyczkowska-Hanćkowiak ("Heikin-Ashi technique with use of oriented fuzzy numbers") are concerned with the Heikin-Ashi technique, which is known to be one of the methods to identify and follow trends using a price time series in financial markets. From the viewpoint of methodology, the Heikin-Ashi candles are determined as a transformation of the usual Japanese candles/candlesticks. Moreover, the Heikin-Ashi candles represent the former and the present quotation changes. This is a significant advantage of Heikin-Ashi candles over Japanese candles. On the other hand, the Japanese candles may be described by oriented fuzzy numbers (OFNs). The main goal of this paper is to present the Heikin-Ashi candles as a transformation of the Japanese candles, described by the OFNs. Additionally, the impact of the Heikin-Ashi technique on the imprecision that plagues the Japanese candles is studied. The results are illustrated on a case study.

Stoyan Poryazov, Velin Andonov, and Emiliya Saranova ("Three intuitionistic fuzzy estimations of uncertainty in service compositions") propose three estimations of uncertainty of the service of requests by service devices: intuitionistic fuzzy traffic characterization, intuitionistic fuzzy flow characterization, and intuitionistic fuzzy time characterization. All are defined as intuitionistic fuzzy pairs, and the estimation of uncertainty is done using the operations defined over the intuitionistic fuzzy pairs.

Krassimir Atanassov, Evgeniy Marinov, and Peter Vassilev ("An interval-valued intuitionistic fuzzy estimation of the area of 2D-figures based on Pick's formula") propose an interval-valued intuitionistic fuzzy estimation of the interior of a simple closed curve, which is based on a previous research, focused on an iterative procedure for the estimation of the area surrounded by a simple closed curve in the real 2D space. Pick's formula is employed for calculating the area surrounded by the special types of polygons.

Wojciech T. Dobrosielski ("The Golden Ratio of Area method based on fuzzy number area as a defuzzifier") presents a new method of defuzzification that may be implemented in fuzzy controllers. The method combines the golden ratio and the widely used center of gravity defuzzification and is termed the golden ratio of area, or GRoA, for short. To verify the efficiency of the GRoA method, the results are compared with those obtained by using other existing defuzzification methods. An example of control of a combustion process is considered. The results show that there are cases that significantly deviate from central values and should be eliminated in the control. Another conclusion is that the application of a pattern other than central value can help attain a surer selection of the defuzzification methods. The analysis shows that the GRoA method works properly and may be used in fuzzy control. Finally, an extension with the use of the ordered fuzzy numbers (OFNs) is advocated.

Part III, "Mathematical modeling of processes and systems," includes contributions related to various approaches, tools, and techniques, with their algorithmic implementations, for the mathematical and computer modeling of various processes and systems, in particular in telecommunication, transportation, process engineering, etc.

Krasiyan Andreev and Ivelina Vardeva ("Generalized net model of implementation of port knocking on RouterOS") deal with an important problem related to computer networks, more specifically the network security feature of port knocking as a security builder for MikroTik communication devices via the RouterOS operating system. The application of the generalized nets is proposed, with the index matrices for describing the states of the token.

Toncho Boyukov and Krassimir T. Atanassov ("Generalized nets as a tool for modeling of railway networks. Part 2") provide a new (second) example that shows that the generalized nets can be used as a tool for the modeling of railway networks. With respect to Part 1 of the paper, published elsewhere, in which a generalized net model of the railway network in Southern Bulgaria was shown, now the model is applied to the railway network in Northern Bulgaria.

Milan Guzan, Tibor Vince, Ján Molnár, Matej Bereš, and Branislav Sobota ("Acceleration of the calculation of boundary surface cross-sections") present a new approach to the acceleration of the calculation of the cross section through the boundary surface of Chua's circuit generating chaos. An optimal test cuboid lying on a chaotic attractor is designed, then a second one is added, symmetrically placed also on a chaotic attractor, and finally another, large cuboid, is added. The latter cuboid is located inside the chaotic attractor, i.e., within the regions of attractiveness, corresponding to the chaotic attractor in the individual cross sections of the boundary surface. This approach yields an almost double acceleration of the calculations.

Part IV, "Decision making and decision support systems," is concerned with various aspects and issues related to the broadly perceived decision making and decision support, in particular decision support systems, which are one of the core research areas in operations and systems research. The problems related to the uncertainty, imprecision, and incompleteness of information are crucial here too, and this is an important reason why the contributions to both BOS-2020 and IWIFSGN-2020 are complementary and cross-fertilizing each other.

Jerzy Józefczyk, Mirosław Ławrynowicz, and Grzegorz Filcek ("On problems and methods of coordinated scheduling and location") investigate a combined approach for job scheduling on parallel machines and the location of machines, referred to in the literature as ScheLoc. A general classification of such a two-part problem involving different solving methods is first presented. A sequential approach requires both parts' successive consideration, unlike the joint proposal with their simultaneous investigation. The uncertain version is also proposed to replace the two-part deterministic problem with a single counterpart with interval release dates of jobs. The new method is illustrated on a specific job scheduling and machine location subproblem.

Emil Hadzhikolev, Stanka Hadzhikoleva, Daniela Orozova, and Kostadin Yotov ("A comprehensive approach to assessing higher and lower order thinking skills") propose a formal model of a software process for multicomponent assessment. It aims at dividing the process into several well-defined steps, defining reusable abstract meta-models and assessments. The new model proposed derives final assessments using fuzzy sets.

Hristo Blidov and Lyubka Doukovska ("Evaluating the general claim process through temporal intuitionistic fuzzy pairs") provide a critical review and evaluation of the development of court proceedings in civil disputes or the so-called general claim process from the initiation of the case to its conclusion by the court of first instance. The use of the temporal intuitionistic fuzzy pairs is proposed.

Przemysław Juszczuk and Lech Kruś ("Investigating the efficiency of market indicators in trading systems") evaluate the usability of different market indicators and propose and analyze three methods that can be used to estimate the efficiency of market indicators for specific trading instruments under different measures of efficiency, taking into account the correctness of signals generated by the given indicator, their profitability, and the maximal drawdown concept. The new measures are formulated and analyzed, and implemented in an experimental trading system used on the Forex market.

Andrzej Łodziński ("Supporting the process of negotiations with the use of multi-criteria analysis") presents a method for supporting the process of negotiations with the use of multi-criteria analysis. Solutions sought are found by a procedure involving an interactive selection of certain proposals. The parties submit their proposals concerning the subjects of the negotiations. These proposals comprise the values of parameters of the multi-criteria optimization problem. The selection of solutions is accomplished by solving the optimization problem with parameters defining the aspiration levels of each party involved. Finally, the solutions obtained by the parties are evaluated.

Hubert Zarzycki, Dawid Ewald, Oskar Skubisz and Piotr Kardasz ("A comparative study of two nature-inspired algorithms for routing optimization") present and compare two algorithms inspired by nature, the genetic algorithm, and the ant colony optimization algorithm, applied to the problem of determining the optimal route. A comparison of results obtained is provided, with an analysis of advantages and disadvantages of both algorithms to help choose a proper algorithm for a particular task.

Sebastian Porębski ("Selection of *t*-norms for calculating belief measure and their influence on support decision with uncertainty") deals with information processing with belief function of the Dempster–Shafer theory (DST) generalized for a fuzzy environment. Within the domain of explainable artificial intelligence, different set theories are proposed that try to upgrade the system readability while keeping its reliability: fuzzy, intuitionistic, rough, soft, to just name a few. Usually, they are connected to particular knowledge-based inference mechanisms: probabilistic, evidential, possibilistic, etc. The work focuses on effective and verified decision support solution originating from the DST and making it possible to process uncertain knowledge and imprecise information. The study validates the use of the fuzzy inclusion measures, particularly exploring different triangular norm operations. A comparison of experimental differences reveals that the minimum operation can be outperformed by other *t*-norms, and their choice is highly problem dependent.

Dawid Ewald, Huber Zarzycki, and Jacek M. Czerniak ("Certain aspects of the OFNBee algorithm operation for different fuzzifiers") describe a new hybrid method for the bee optimization (metaheuristic) using the arithmetic of ordered fuzzy numbers (OFNs). The aim of this combination is to more realistically represent the decision-making mechanisms of bees, commonly referred to as waggle dance. The OFN notation makes it possible to more accurately transfer information between individual bees, which contributes to the improvement of the algorithm. Some defuzzification method is first proposed, using the golden ratio and Mandala factor functions. Results obtained by using the proposed methods are shown and analyzed.

Part V, "Engineering applications," contains important contributions, in which the approaches, tools, and techniques, as well as algorithms, in particular those presented in previous sections, are applied to the solution of important problems in the broadly perceived engineering applications. The papers are concerned with telecommunication, waste management, forest fires, etc.

Łukasz Apiecionek, Robert Palka, Mateusz Biedziak, Maciej Tomczak, Łukasz Blechman, and Dawid Ewald ("Hybrid system for intelligent waste measurement method") are concerned with waste management, which is more and more significant as a growing amount of waste is generated all over the world. Among many aspects, control and monitoring are here essential. National regulations on recycling levels, reuse, and recovery of various municipal waste fractions set higher targets for the recycling rates. This calls for an improvement of waste management. The only realistic and relatively inexpensive solution in this respect is based on a proper weighing and cataloguing of the types of the waste collected from a given location. A hybrid system for the intelligent waste measurement is proposed, and an analysis of promising results is shown.

Łukasz Apiecionek, Robert Palka, Mateusz Biedziak, Maciej Tomczak, Łukasz Blechman, and Dawid Ewald ("Method for multiplication transmission interfaces") deal with a problem related to mobile telecommunication. With a rapidly growing popularity of all kinds of mobile solutions, including smartphones, smartwatches, and tablets, the spectrum of their applications is growing, in particular due to smaller and smaller dimensions and an ever increasing performance. As many companies and institutions operate globally, often over large areas and in different locations, then it is important to provide a secure, and efficient and effective communication by using all available transmission interfaces. Since the equipment and solutions used, for instance, in security and military sectors are exposed to many dangers, then it is necessary that these devices meet required standards and be resistant to climate change and extremely difficult operating conditions. The use of the method for multiplication transmission interfaces is shown.

Veselina Bureva, Evdokia Sotirova, Peter Vassilev, Vassia Atanassova, Olympia Roeva, Krassimir Atanassov, and Hristo Tsakov ("Application of game method for modeling to locate a forest fire ignition point in the presence of wind") are concerned with the modeling of forest fires spread in the presence of wind, illustrated by a real example of the forest fires in the Haskovo region and Yambol region of Bulgaria. The game method for modeling (GMM) is used to determine the forest fire ignition point from the constructed map of fire damages and is applied to the map containing the initial configuration of the observed area to simulate the fire.

Wojciech T. Dobrosielski ("An alternative fuzzy logic inference method on the example of a PID controller") presents a different approach to the inference process of a fuzzy controller using the OFNs (ordered fuzzy numbers) and their related arithmetic operations. The solutions of inference processing are based on the Mamdani or Takagi–Sugeno architectures, which involve a rule base. A control problem is formulated involving the OFNs. The approach simplifies the controller, with a simpler mathematical model of the process, discussed in the paper. Possible solutions for a PID controller under the OFNs are discussed.

Dariusz Ruciński ("Data quotation period and the quality of ANN learning as a Polish Power Exchange model") deals with the impact of the length of data quotations on the Day Ahead Market and thus the number of training and testing pairs on the quality of the resulting artificial neural network model of the Polish Power Exchange. Emphasis is on the determination of the quality of artificial neural network reproduction as a multidimensional approximation of the Polish Power Exchange Day Ahead Market model. The data used cover the period of 2002–2019, and MATLAB and Simulink, with m-file library, are employed. The results obtained are promising.

Nikolay Ikonomov, Pencho Marinov, Peter Vassilev, Olympia Roeva, Dafina Zoteva, Vassia Atanassova, and Krassimir Atanassov ("3D software implementation of the game method for modeling of forest fires in MyGL software tool") show a preliminary version of the 3D visualization software tool My OpenGL Game of Life (MyGL) v. 2.0 for creating abstract game models of forest fires using concepts related to Conway's Game of Life. The OpenGL technology is used for the 3D visualization of the cells. The primary focus is on cube cells, while hexagonal prisms have been considered for backward compatibility. Three algorithms are currently implemented. The feasibility of the MyGL software is illustrated with an example of a forest fire scenario.

Marek Pecha ("Balancing predictive relevance of ligand biochemical activities") presents a technique for balancing predictive relevance models related to supervised modeling of ligand biochemical activities to biological targets. The non-calibrated models are trained by employing conventional supervised machine learning technique, more specifically the support vector machines (SVMs). Unfortunately, the SVMs have a serious drawback as they are sensitive to imbalanced datasets, outliers, and a high multi-collinearity among training samples. Thus, an additional calibration could be required for balancing a predictive relevance of models. As a technique for this balancing, Platt's scaling is proposed. The results obtained are shown on examples of single-target models trained on datasets exported from the ExCAPE database.

Danail Stratiev, Dafina Zoteva, Dicho Stratiev, and Krassimir Atanassov ("Modelling the process of production of automotive gasoline by the use of Generalized Nets") propose a generalized net model of the process of production of different grades of automotive gasoline. The model can be used for the synchronization and optimization of the process. For the simulation, the Generalized Net Integrated Development Environment (GN IDE), a platform-independent software for constructing and simulating generalized nets models, is used.

Part VI, "Social, medical and health care applications," gathers highly relevant applications of the concepts and algorithms, proposed in the previous parts of the book, to various problems that are crucial for social systems, biology, medicine, and health care. These fields are more and more important in the present world, plagued by many problems exemplified by aging of the society, the pandemic, and numerous other social problems.

Jan W. Owsiński, Cristian Ciurea, Florin Gh. Filip, and Jarosław Stańczak ("Students' forward thinking – an analysis of cultural divides through the 'reverse clustering'") present an analysis of data collected about the educational track of university students using the "reverse clustering" approach. The results are interpreted in the context of a crowdsourcing effort made by the authors and the respondents of the questionnaire in order to extract relevant data for the countries involved in the study. The results obtained and reported indicate that the cultural profiles of the respondents have a great influence regarding the responses associated with the educational and professional perspectives of the students involved.

Gabriela Hristova and Veselina Bureva ("Generalized net model of library activities using intuitionistic fuzzy estimations") discuss a generalized net (GN) model of the library processes. The proposed generalized net model of library can be used for monitoring the working processes. In the last step of the generalized net model, intuitionistic fuzzy evaluations for book and intuitionistic fuzzy estimations of the library are presented. These can help analyze trends in borrowing books from a library.

Lyudmila Todorova, Valentina Ignatova, Peter Vassilev, and Jivko Surchev ("Generalized net model of computer based registration and rehabilitation of cognitive impairments in multiple sclerosis") deal with a cognitive decline that occurs with a physical disability in approximately two-thirds of persons suffering from sclerosis multiplex (SM) and has a devastating influence on the quality of life. The modern immunomodulatory therapy has a satisfactory effect on the relapse rate and disease progression but does not reduce the degree of cognitive impairment. A generalized net model for the registration, assessment, and rehabilitation of cognitive impairment in persons with the SM is developed. The assessment of a cognitive status is done by using computer-based neuropsychological tests. The use of the InterCriteria Analysis to analyze the output of the model provides information about the type of impaired cognitive domains and the degree of cognitive deterioration. This allows the therapist to build an appropriate rehabilitation strategy for each individual patient. The system is implemented in a cloud and is planned to be provided as an IT service for neurological departments and clinics.

Valentin Vassilev, Evdokia Sotirova, and Krassimir Atanassov ("A generalized net model of the normal heart functioning") present a generalized net (GN) model of a normal heart functioning, which gives the possibilities of evaluating the heart's parameters and of analyzing its behavior and functioning. Moreover, some disorders of the cardiovascular system can also be dealt with.

Zhelyana Ivanova, Veselina Bureva, and Sotir Sotirov ("Generalized net model of biometric multifactor authentication system") propose a generalized net (GN) model of a biometric multifactor authentication system. The model uses biometric identification methods such as hand (palm) geometry, palm vein, and cardiac signal, acquired in the same time. The features are compared to the stored information for individuals. In the end, a rank-level fusion is applied to make an authentication decision.

We hope that the broad coverage of the volume, the inspiring and interesting contributions, included therein, which present both the state-of-the-art and original contributions, will be of much interest and use for a wide research community.

We wish to express our deep gratitude to the contributors for their great works, as well as to other participants of the BOS-2020 and IWIFSGN-2020 conferences, whose presence and active participation have helped make the conferences a success. Special thanks are due to anonymous peer reviewers whose deep and constructive remarks and suggestions have greatly helped to improve the quality and clarity of the contributions.

And last but not least, we wish to thank Dr. Tom Ditzinger, Dr. Leontina di Cecco, and Mr. Holger Schaepe for their dedication and help to implement and finish this important publication project on time, while maintaining the highest publication standards.

Krassimir T. Atanassov
Vassia Atanassova
Janusz Kacprzyk
Andrzej Kałuszko
Maciej Krawczak
Jan W. Owsiński
Sotir Sotirov
Evdokia Sotirova
Eulalia Szmidt
Sławomir Zadrożny

Contents

Mathematical Modeling of Processes and Systems

Decision Making and Decision Support Systems

Engineering Applications

Social, Medical and Health Care Applications

Advances in the Processing of Uncertain and Imprecise Information

Application of Decision Rules to Discover Knowledge for Fitting Aggregations to Datasets

Jan G. Bazan, Urszula Bentkowska(✉), and Marcin Mrukowicz

College of Natural Sciences, Institute of Computer Science, University of Rzeszów, Pigonia 1, 35-310 Rzeszów, Poland
{bazan,ududziak,mmrukowicz}@ur.edu.pl

Abstract. In this paper we propose to use decision rules in order to choose possibly the best interval-valued aggregation function in the classifying algorithm based on the dataset complexity measures. In discovering knowledge we apply the existing toolset RSES. We consider examples of feature-based measures, dimensionality measures, measures of separability of classes, and class imbalance measures to reflect the complexity of the classification problem.

1 Introduction

Aggregation functions [6] and interval-valued aggregation functions are widely discussed and successfully applied in diverse areas of applications. In this contribution we would like to propose the method of discovering knowledge about the applicability of interval-valued aggregation functions in classifying algorithms depending on the dataset complexity measures.

If there are many aggregations for the uncertainty intervals, the question arises how to select aggregations for specific datasets in such a way as to obtain the highest quality of the constructed classifier. It is mainly about obtaining a high AUC parameter. The obvious method of answering this question is to conduct experiments sequentially using all proposed aggregations for all datasets. In this way, we obtain full information about the quality of individual aggregations in the context of each dataset. Apart from the fact that this approach is not always possible, e.g. due to the complexity of calculations, it does not answer the question why for a given set one aggregation achieved a better result and the other a worse result.

In this work, we are interested in the problem of matching aggregation to a dataset based solely on the properties of that dataset. In particular, we limit this problem to the problem of explaining why for a fixed pair of aggregations and a fixed dataset, one aggregation is better than the other aggregation. This allows to select one of the two aggregations for a given dataset, which is more appropriate to construct the classifier for that dataset. Certainly, this problem can be generalized to more aggregation pairs, creating a tool for selecting aggregation for a given dataset from any number of them.

K. T. Atanassov et al. (Eds.): IWIFSGN 2020/BOS/SOR 2020, LNNS 338, pp. 3–14, 2022.
https://doi.org/10.1007/978-3-030-95929-6_1

We apply decision rules and the RSES tool [2,16] to discover knowledge useful for fitting aggregations to datasets. The conditional attributes of these rules are based on the dataset complexity measures which are useful in analyzing the difficulty of a classification problem by using descriptors extracted from a learning dataset [8]. In [11] it was proposed an automatic extraction method of the domains of competence for classifiers using dataset complexity measures. There are many other interesting papers discussing diverse dataset complexity measures and applying the extracted knowledge in classification tasks (cf. [10]).

The paper is organized as follows. In Sect. 2 we recall the notion of diverse types of interval-valued aggregation functions. In Sect. 3, we provide the information about data complexity measures. Next, details on implementation (Sect. 4) and application of RSES for fitting aggregation functions to datasets are discussed (Sect. 5). Finally, in Sect. 6 discussion on the results is presented.

2 Interval-Valued Aggregation Functions

Let us denote $L^I = \{[\underline{x}, \overline{x}] : \underline{x}, \overline{x} \in [0,1],\ \underline{x} \leqslant \overline{x}\}$. We may consider diverse comparability relations between intervals. Below we present some of them.

Well-known partial order for intervals is of the form (cf. [9])

$$[\underline{x}, \overline{x}] \preceq [\underline{y}, \overline{y}] \Leftrightarrow \underline{x} \leqslant \underline{y},\ \overline{x} \leqslant \overline{y}.$$

We will also use the following comparability relations on L^I (cf. [15]):

$$[\underline{x}, \overline{x}] \preceq_{\pi} [\underline{y}, \overline{y}] \Leftrightarrow \underline{x} \leqslant \overline{y},$$

$$[\underline{x}, \overline{x}] \preceq_{\nu} [\underline{y}, \overline{y}] \Leftrightarrow \overline{x} \leqslant \underline{y}.$$

The admissible linear orders were introduced in [7]. Namely, an order $\leq_{L^I}$ on L^I is called admissible if it is linear and satisfies that for all $x, y \in L^I$, such that $x \preceq y$, then $x \leq_{L^I} y$.

An aggregation function on the domain L^I fulfils some monotonicity condition (based on a chosen comparability relation between intervals) and boundary conditions. Namely,

Definition 1. An operation $\mathcal{A} : (L^I)^n \to L^I$ is called an aggregation function on L^I if it is increasing, i.e.,

$$\underset{\mathbf{x}_i, \mathbf{y}_i \in L^I}{\forall} \mathbf{x}_i \leqslant \mathbf{y}_i \Rightarrow \mathcal{A}(\mathbf{x}_1, \ldots, \mathbf{x}_n) \leqslant \mathcal{A}(\mathbf{y}_1, \ldots, \mathbf{y}_n)$$

and $\mathcal{A}(\underbrace{\mathbf{0}, ..., \mathbf{0}}_{n\times}) = \mathbf{0},\ \ \mathcal{A}(\underbrace{\mathbf{1}, ..., \mathbf{1}}_{n\times}) = \mathbf{1}$, where $\mathbf{0} = [0,0]$, $\mathbf{1} = [1,1]$ and $\leqslant$ is some comparability relation on L^I.

As a result we obtain diverse classes of interval-valued aggregations depending on the monotonicity conditions involved. They belong for example to the family of recently introduced pos-aggregations or nec-aggregations [3], where the comparability relation in the monotonicity condition is $\preceq_\pi$ or $\preceq_\nu$, respectively. These families of aggregations are not disjoint and moreover they may intersect with others families of interval-valued aggregations, for example with interval-valued aggregation functions with respect to $\preceq$ or linear orders (cf. [5]).

We present here only examples of aggregations applied in our classification algorithm and as a result these are aggregations which we would like to study in the context of fitting them to the datasets. Namely,

$$\mathcal{A}_1(\mathbf{x}_1, \mathbf{x}_2, ..., \mathbf{x}_n) = \left[\frac{\underline{x}_1 + \underline{x}_2 + ... + \underline{x}_n}{n}, \frac{\overline{x}_1 + \overline{x}_2 + ... + \overline{x}_n}{n}\right],$$

where $\mathcal{A}_1$ is an aggregation function with respect to $\preceq$ and it is a pos-aggregation function and a nec-aggregation function.

$$\mathcal{A}_2(\mathbf{x}_1, \mathbf{x}_2, ..., \mathbf{x}_n) = [\frac{\underline{x}_1 + \underline{x}_2 + ... + \underline{x}_n}{n}, \max(\frac{\underline{x}_1 + \overline{x}_2 + ... + \overline{x}_n}{n}, ..., \frac{\overline{x}_1 + \cdots + \overline{x}_{n-1} + \underline{x}_n}{n})],$$

where $\mathcal{A}_2$ is a nec-aggregation function and an aggregation function with respect to $\preceq$ but it is not a pos-aggregation function.

$$\mathcal{A}_3(\mathbf{x}_1, \mathbf{x}_2, ..., \mathbf{x}_n) = \left[\frac{\underline{x}_1 + ... + \underline{x}_n}{n}, \frac{\overline{x}_1^2 + ... + \overline{x}_n^2}{\overline{x}_1 + ... + \overline{x}_n}\right],$$

$$\mathcal{A}_4(\mathbf{x}_1, ..., \mathbf{x}_n) = \left[\frac{\underline{x}_1 + ... + \underline{x}_n}{n}, \frac{\overline{x}_1^3 + ... + \overline{x}_n^3}{\overline{x}_1^2 + ... + \overline{x}_n^2}\right].$$

$\mathcal{A}_3$ and $\mathcal{A}_4$ are pos-aggregation functions but they are neither aggregation functions with respect to $\preceq$ nor nec-aggregation functions, where it used the convention $\frac{0}{0} = 0$.

$$\mathcal{A}_5(\mathbf{x}_1, ..., \mathbf{x}_n) = \left[\sqrt{\frac{\underline{x}_1^2 + ... + \underline{x}_n^2}{n}}, \sqrt[3]{\frac{\overline{x}_1^3 + ... + \overline{x}_n^3}{n}}\right],$$

$$\mathcal{A}_6(\mathbf{x}_1, ..., \mathbf{x}_n) = \left[\sqrt[3]{\frac{\underline{x}_1^3 + ... + \underline{x}_n^3}{n}}, \sqrt[4]{\frac{\overline{x}_1^4 + ... + \overline{x}_n^4}{n}}\right],$$

where $\mathcal{A}_5$ and $\mathcal{A}_6$ are aggregation functions with respect to $\preceq$ and they are pos-aggregation functions but they are not nec-aggregation functions.

$$\mathcal{A}_7(\mathbf{x}_1, \mathbf{x}_2, \ldots, \mathbf{x}_n) = [\min(\frac{\overline{x}_1 + \underline{x}_2 + \cdots + \underline{x}_n}{n}, ..., \frac{\underline{x}_1 + \cdots + \underline{x}_{n-1} + \overline{x}_n}{n}), \frac{\overline{x}_1 + \overline{x}_2 + \cdots + \overline{x}_n}{n}].$$

$\mathcal{A}_7$ is a nec-aggregation function and an aggregation function with respect to $\preceq$ but it is not a pos-aggregation function.

$$\mathcal{A}_8(\mathbf{x}_1, ..., \mathbf{x}_n) = \left[\sqrt[n]{\underline{x}_1 \cdot ... \cdot \underline{x}_n}, \frac{\overline{x}_1^2 + ... + \overline{x}_n^2}{\overline{x}_1 + ... + \overline{x}_n}\right],$$

$$\mathcal{A}_9(\mathbf{x}_1, ..., \mathbf{x}_n) = \left[\sqrt{\frac{\underline{x}_1^2 + ... + \underline{x}_n^2}{n}}, \frac{\overline{x}_1^3 + ... + \overline{x}_n^3}{\overline{x}_1^2 + ... + \overline{x}_n^2} \right]$$

where for $\mathcal{A}_8$, $\mathcal{A}_9$ it is used the convention $\frac{0}{0} = 0$. $\mathcal{A}_8$ and $\mathcal{A}_9$ are pos-aggregation functions but they are neither aggregation functions with respect to $\preceq$ nor nec-aggregation functions.

$$\mathcal{A}_{10}(\mathbf{x}_1, ..., \mathbf{x}_n) = \left[\sqrt{\frac{\underline{x}_1^2 + ... + \underline{x}_n^2}{n}}, \sqrt{\frac{\overline{x}_1^2 + ... + \overline{x}_n^2}{n}} \right],$$

where $\mathcal{A}_{10}$ is an aggregation function with respect to $\preceq$, a pos-aggregation function and a nec-aggregation function. More information about the given examples of aggregations may be found in [5].

3 Dataset Complexity Measures

Dataset complexity measures were introduced in [8] and later discussed by many authors. We consider only a few examples of them and we adapt the conventions applied in [10]. Moreover, the chosen measures are representatives of existing families of measures.

F1 (maximum Fisher's discriminant ratio) is one of the feature-based measures which characterize how informative the available features are to separate the classes. If there is at least one very discriminative feature in the dataset, the problem can be considered simpler than if there is no such an attribute. All measures from this category require the features to have numerical values. Originally, F1 takes the value of the largest discriminant ratio among all the available features. In this contribution we take the inverse of the original F1 formulation into account (cf. [10]). As a result, the F1 values become bounded in the $(0, 1]$ interval and higher values indicate more complex problems (which means that no individual feature is able to discriminate the classes). Low values of the F1 measure indicate that there is at least one feature whose values show little overlap among the different classes. As a result, it indicates the existence of a feature for which a hyperplane perpendicular to its axis can separate the classes fairly. If the required hyperplane is oblique to the feature axes, F1 may not be able to reflect the underlying simplicity. It is worth to mention that F1 measure is most effective if the probability distributions of the classes are approximately normal.

T2 is one of the dimensionality measures. The measures from this category give an indicative of data sparsity. They are based on the dimensionality of the datasets, either original or reduced. The idea behind this measure is that it can be more difficult to extract good models from sparse datasets, due to the probable presence of regions of low density that will be arbitrarily classified. Originally, T2 divides the number of examples in the dataset by their dimensionality. In this paper we take the inverse of this formulation in order to obtain higher values for more complex datasets (cf. [10]). Therefore, lower T2 values indicate less sparsity and therefore simpler problems.

N1 is one of the measures of separability of classes. N1 presents fraction of borderline points (considering a minimum spanning tree). These type of measures analyze neighborhoods of the data points. N1 estimates the size and complexity of the required decision boundary through the identification of the critical points in the dataset, namely those very close to each other but belonging to different classes. Higher N1 values indicate the need for more complex boundaries to separate the classes or that there is a large amount of overlapping between the classes.

C1 and C2 are examples of the class imbalance measures (a large difference in the number of examples per class in the training dataset). If the differences are huge, most of the classification techniques tend to favor the majority class. If the problem has a high imbalance in the proportion of examples per class, it can be considered more complex than a problem for which the proportions are similar. C1 is measures entropy of class proportions. It will achieve minimum value for balanced problems (in which all proportions are equal). These can be considered simpler problems according to the class balance aspect. C2 measures imbalance ratio and it is a well known index computed for measuring class balance. Larger values of C2 are obtained for imbalanced problems.

4 Details of Implementation

As an exemplary algorithm we consider here the binary classifier F given in [4] to deal with missing values in datasets. However, to simplify the discussion we concentrate here only on the missing value level 0.0. In this algorithm interval-valued aggregation functions were applied. The experiments were made using stratified 10-fold cross validation. To evaluate the results AUC (area under the ROC curve) was reported (the experiments were repeated 10 times and the arithmetic mean and standard deviation of the results were determined).

We considered datasets from UC Irvine Machine Learning Repository [18]. Details of these datasets (number of objects and number of attributes) are given in Table 1. Moreover, for a few datasets we removed some number of objects from the chosen class (taking into account the decision of the classifier). As a result we obtained diverse dataset complexity measures in these two versions of datasets. Two versions of datasets are denoted by number 1 and 2. We used the implementation project 'The Data Complexity Measures' in Python [17] to determine the considered dataset complexity measures.

We have chosen the pair of aggregation functions $\mathcal{A}_1$ and $\mathcal{A}_8$ to present our method of fitting aggregation to a dataset. Aggregation $\mathcal{A}_1$ is the well-known interval-valued aggregation operator. Application of aggregation $\mathcal{A}_8$ in the classifier F (see [4]) often resulted in extreme values of AUC, i.e. the best or the worst ones. This is why we wanted to discover the dataset details that yield the adequate results. Moreover, $\mathcal{A}_8$ is an example of pos-aggregation functions, i.e. recently introduced type of interval-valued aggregation functions.

5 RSES for Aggregation Functions

Datasets used in classifiers may be presented by data tables where objects correspond to rows and attributes correspond to columns of the given data table. In this contribution we consider decision tables of the form $\mathbf{T} = (U, A, d)$ (cf. [14]), where A is the set of attributes, U is the set of objects, and d is a decision attribute (distinguished from the set A).

Now let us assume that we have a given collection of k datasets $D_1, ..., D_k$, calculated n properties of datasets $a_1, ..., a_n$ (we understand each such property as a function that computes a numerical value for each dataset), two fixed aggregations A and B and the classification results obtained by means of these aggregations for each of the datasets $D_1, ..., Dk$ ($AUCA$ and $AUCB$ values). Based on these data and parameters, we can create a rectangular table, the rows of which correspond to the datasets $D_1, .., D_k$, and the columns to the attributes representing the properties $a_1, ..., a_n$ and the classification results $AUCA$ and $AUCB$ obtained for aggregation A and B. Moreover, we can add a decision attribute to this table that has the following three values:

- the value 1 for a given row when $AUCA - AUCB > \delta$,
- the value 2 for a given row when $AUCB - AUCA > \delta$,
- value of 0 for a given row when $|AUCA - AUCB| \leqslant \delta$,

where δ is a fixed rational number (e.g., $\delta = 0.001$) which is used to compare the $AUCA$ and $AUCB$ values.

It is easy to notice that the decision attribute represents information for which datasets:

- aggregation A significantly exceeds aggregation B (value 1),
- aggregation B significantly exceeds aggregation A (value 2),
- aggregations A and B have the same efficiency in constructing the classifier.

Table 1 shows an example of this type of table constructed for the datasets considered in this paper and aggregation functions $\mathcal{A}_1$ and $\mathcal{A}_8$.

Table 1. Datasets with all corresponding attributes

Dataset	F1	N1	C1	C2	T2	NoAttr	Class0	Class1	NoObj	AUC A1	AUC A8	Decision
Biodeg 1	0.84	0.27	0.07	0.19	0.04	43	699	356	1055	0.90283	0.90226	0
Biodeg 2	0.87	0.06	0.75	0.91	0.058	43	699	30	729	0.87854	0.88092	2
Breast 1	0.37	0.07	0.04	0.12	0.019	11	212	357	569	0.99187	0.99213	0
Breast 2	0.62	0.06	0.45	0.72	0.045	11	212	30	242	0.98889	0.9873	1
German 1	0.87	0.46	0.11	0.27	0.025	25	700	300	1000	0.74402	0.73814	1
German 2	0.94	0.09	0.75	0.91	0.034	25	700	30	730	0.65857	0.64976	1
Ozone	0.98	0.03	0.87	0.96	0.042	74	1719	30	1749	0.72602	0.72602	0
Spam	0.95	0.02	0.91	0.97	0.02	59	2788	30	2818	0.7634417	0.7634425	0

In Table 1 the considered attributes regarding the datasets are: dataset complexity measures F1, N1, C1, C2, T2, moreover the number of attributes

(NoAttr), the number of objects in class 0 (Class 0), the number of objects in class 1 (Class 1), the number of objects (NoObj), the values $AUCA1$, $AUCA8$, and Decision.

From a data mining point of view, the above table is a decision table where the columns F1 to noObj are the conditional ones and Decision is the decision attribute ($AUCA1$ and $AUCA8$ should be removed because they were used to define the decision attribute so they are strongly correlated with it). In this way, we obtain a decision table given by Table 2.

Table 2. Datasets with the corresponding attributes except for AUC values

Dataset	F1	N1	C1	C2	T2	NoAttr	Class0	Class1	NoObj	Decision
Biodeg 1	0.84	0.27	0.07	0.19	0.04	43	699	356	1055	0
Biodeg 2	0.87	0.06	0.75	0.91	0.058	43	699	30	729	2
Breast 1	0.37	0.07	0.04	0.12	0.019	11	212	357	569	0
Breast 2	0.62	0.06	0.45	0.72	0.045	11	212	30	242	1
German 1	0.87	0.46	0.11	0.27	0.025	25	700	300	1000	1
German 2	0.94	0.09	0.75	0.91	0.034	25	700	30	730	1
Ozone	0.98	0.03	0.87	0.96	0.042	74	1719	30	1749	0
Spam	0.95	0.02	0.91	0.97	0.02	59	2788	30	2818	0

For such decision table, we could construct classifiers that can indicate which aggregation is better for a given dataset based on the properties of that dataset. However, in this paper we are interested in exploring the knowledge from the decision table to answer the question why one aggregation is better than another for a given dataset. As a result, we need a method of knowledge discovery from data. One of the best known methods of discovering knowledge from data is the method based on discovery of decision rules from data. In this paper, we assume that the decision rules are expressions of the type $(a_1 = v_1) \wedge (a_2 = v_2) \wedge ... \wedge (a_l = v_l) \Rightarrow (d = v_d)$ where a_i is the conditional attribute from the decision table and v_i is the value of this attribute, for $i = 1, ..., l$. Furthermore, d is the decision attribute and v_d is the value of the decision attribute. Each such rule consists of a predecessor (part of the rule before the implication sign $\Rightarrow$) and a successor (part of the rule after the sign $\Rightarrow$). The permise is a conjunction of expressions like $(a = v)$ called descriptors, where a is a conditional attribute and v is the value of this attribute. In the successor there is one descriptor which is built based on the decision attribute.

We say that a given *rule* r is *applicable* to some object u (a row in the decision table), or that an object u matches a rule r if the attribute values of this object satisfy the predecessor of the rule. We associate a number of numerical parameters with each rule, such as the *matching parameter* equal to the number of objects matching the rule's predecessor, *support* equal to the number of objects simultaneously matching the rule's predecessor and successor, and *confidence* equal to the support/matching quotient. These parameters help to determine the quality of the rule. For example, a rule with the value confidence equal to 1.0 is often called a *certain rule* in the context of the dataset for which it was generated.

Each rule carries some knowledge about the relationship in the dataset between the values of the selected conditional attributes and the value of the decision attribute. This knowledge allows us to conclude about specific objects, i.e. if the values of the conditional attributes match the predecessor, then with some confidence the value of the decision attribute has the value that appears in the successor.

Particularly important for practical applications are the so-called rule with a minimum number of descriptors in the rule's predecessor, because a small number of premises are enough to draw a conclusion based on the rule. In this work, to generate rules with the minimum number of descriptors, we use the RSES system, which generates all such rules for a given decision table [2,16]. This method is often called an exhaustive method. From the practical point of view, the method consists in applying an algorithm computing all reducts for each object individually, which results in obtaining decision rules with a minimal number of descriptors in relation to individual objects (see [1,13,14] for more details).

When the conditional attributes in the decision table for which we want to generate decision rules are numeric and they have potentially many values, often before generating the rules, such data must be discretized. The reason for this is that if the conditional attributes of a given decision table have a large number of values, the values of objects outside the decision table (called test objects) will often not match any decision rule. Meanwhile, we want to have the determined knowledge in the form of decision rules to be applicable not only to the objects from a given decision table, but also to other objects unknown during the induction of the rules. Therefore, for decision tables with numerical attributes, discretization methods are used, i.e. scaling the values of the attributes. Attribute value scaling is based on the fact that certain specific values called *cuts* are selected from the range of variability of the analyzed attributes, dividing the sets of attribute values into ranges, each of which (after adding a label) is a new attribute value. After applying such an operation, the number of attribute values rapidly decreases and the generated rules are more universal.

There are many discretization methods, but in this work we use two methods available in RSES as the global method and the local method. Generally speaking, these methods determine cuts that distinguish well between objects from different decision classes.

Both methods belong to a bottom-up approaches which add cuts for a given attribute one-by-one in subsequent iterations of algorithm. The difference between these two methods lies in the way in which the candidate for a new cut is evaluated. In the global method, we evaluate all objects in the data table at every step. In the local method, we only consider a part of objects that are related to the candidate cut, i.e., which have the value of the attribute considered currently in the same range as the cut candidate. More details about the method one may find in [1,12,16].

Below (cf. Figs. 1 and 2) we present Table 2 after the global and local discretization, respectively.

F1	N1	C1	C2	T2	NoAttr	Class0	Class1	NoObj	Decision
"(0.495,Inf)"	*	*	*	"(-Inf,0.05)"	"(34.0,Inf)"	*	*	*	0
"(0.495,Inf)"	*	*	*	"(0.05,Inf)"	"(34.0,Inf)"	*	*	*	2
"(-Inf,0.495)"	*	*	*	"(-Inf,0.05)"	"(-Inf,34.0)"	*	*	*	0
"(0.495,Inf)"	*	*	*	"(-Inf,0.05)"	"(-Inf,34.0)"	*	*	*	1
"(0.495,Inf)"	*	*	*	"(-Inf,0.05)"	"(-Inf,34.0)"	*	*	*	1
"(0.495,Inf)"	*	*	*	"(-Inf,0.05)"	"(-Inf,34.0)"	*	*	*	1
"(0.495,Inf)"	*	*	*	"(-Inf,0.05)"	"(34.0,Inf)"	*	*	*	0
"(0.495,Inf)"	*	*	*	"(-Inf,0.05)"	"(34.0,Inf)"	*	*	*	0

Fig. 1. Table 2 after the global discretization - screen view from the RSES toolset

In tables after dicretization in some columns may occur the signs *. It means that according to the given discretization method the attribute values are not necessary to distinguish between decision classes.

F1	N1	C1	C2	T2	NoAttr	Class0	Class1	NoObj	Decision
"(0.495,Inf)"	*	*	*	*	"(34.0,Inf)"	*	*	"(729.5,Inf)"	0
"(0.495,Inf)"	*	*	*	*	"(34.0,Inf)"	*	*	"(-Inf,729.5)"	2
"(-Inf,0.495)"	*	*	*	*	"(-Inf,34.0)"	*	*	"(-Inf,729.5)"	0
"(0.495,Inf)"	*	*	*	*	"(-Inf,34.0)"	*	*	"(-Inf,729.5)"	1
"(0.495,Inf)"	*	*	*	*	"(-Inf,34.0)"	*	*	"(729.5,Inf)"	1
"(0.495,Inf)"	*	*	*	*	"(-Inf,34.0)"	*	*	"(729.5,Inf)"	1
"(0.495,Inf)"	*	*	*	*	"(34.0,Inf)"	*	*	"(729.5,Inf)"	0
"(0.495,Inf)"	*	*	*	*	"(34.0,Inf)"	*	*	"(729.5,Inf)"	0

Fig. 2. Table 2 after the local discretization - screen view from the RSES toolset

6 Discussion on the Results

We generated the set of rules for both tables using the RSES toolset .

The set of rules after the global discretization:

- $(NoAttr = \text{“}(34.0, Inf)\text{”}) \wedge (NoObj = \text{“}(729.5, Inf)\text{”}) \Rightarrow (Decision = 0)$ 3
- $(F1 = \text{“}(0.495, Inf)\text{”}) \wedge (NoAttr = \text{“}(-Inf, 34.0)\text{”}) \Rightarrow (Decision = 1)$ 3
- $(NoAttr = \text{“}(-Inf, 34.0)\text{”}) \wedge (NoObj = \text{“}(729.5, Inf)\text{”}) \Rightarrow (Decision = 1)$ 2
- $(NoAttr = \text{“}(34.0, Inf)\text{”}) \wedge (NoObj = \text{“}(-Inf, 729.5)\text{”}) \Rightarrow (Decision = 2)$ 1
- $(F1 = \text{“}(-Inf, 0.495)\text{”}) \Rightarrow (Decision = 0)$ 1

The set of rules after the local discretization:

- $(T2 = \text{“}(-Inf, 0.05)\text{”}) \wedge (NoAttr = \text{“}(34.0, Inf)\text{”}) \Rightarrow (Decision = 0)$ 3
- $(F1 = \text{“}(0.495, Inf)\text{”}) \wedge (NoAttr = \text{“}(-Inf, 34.0)\text{”}) \Rightarrow (Decision = 1)$ 3
- $(T2 = \text{“}(0.05, Inf)\text{”}) \Rightarrow (Decision = 2)$ 1
- $(F1 = \text{“}(-Inf, 0.495)\text{”}) \Rightarrow (Decision = 0)$ 1

Information about the support of each rule was placed after the rule. Moreover, $-Inf$ denotes minus infinity and as a result $(-Inf, 0.05)$ is the interval of values smaller than 0.05. Similarly, Inf denotes plus infinity and as a result $(0.495, Inf)$ is the interval of values greater than 0.495.

Analyzing the obtained set of values we see that particularly interesting is the rule

$$(F1 = \text{``}(0.495, Inf)\text{''}) \wedge (NoAttr = \text{``}(-Inf, 34.0)\text{''}) \Rightarrow (Decision = 1)\ 3$$

which is repeated in both set of rules and moreover this rule has 3 supporting objects (all from class 1). It means that if $F1$ is greater than 0.495 and the number of attributes does not exceed 34, then aggregation function $\mathcal{A}_1$ obtains better classification results (AUC value) than aggregation function $\mathcal{A}_8$.

The rule

$$(F1 = \text{``}(-Inf, 0.495)\text{''}) \Rightarrow (Decision = 0)\ 1$$

says that if the value $F1$ is smaller than 0.495, then both aggregation functions $\mathcal{A}_1$ and $\mathcal{A}_8$ behave similarly.

The rule

$$(T2 = \text{``}(-Inf, 0.05)\text{''}) \wedge (NoAttr = \text{``}(34.0, Inf)\text{''}) \Rightarrow (Decision = 0)\ 3$$

says that if the value $T2$ is small and there is large number of attributes (over 34), then aggregation functions $\mathcal{A}_1$ and $\mathcal{A}_8$ behave similarly.

However, according to the rule

$$(T2 = \text{``}(0.05, Inf)\text{''}) \Rightarrow (Decision = 2)\ 1$$

if the value $T2$ is greater than 0.05, then the aggregation function $\mathcal{A}_8$ obtains better results than the aggregation function $\mathcal{A}_1$.

Generally speaking, after the analysis of the above rules it follows that dataset complexity measures $F1$ or $T2$ were important to distinguish between the usefulness of aggregation functions $\mathcal{A}_1$ and $\mathcal{A}_8$. If the value of $F1$ or $T2$ is below a determined cut (lower values of $F1$ or $T2$ indicate a lower complexity of the problem), then no matter which aggregation to use. It is intuitive and we have additionally determined adequate cut of the value of each measure. On the other hand, $\mathcal{A}_1$ (respectively $\mathcal{A}_8$) is better fitted to a dataset if the appropriate dataset complexity measure has the value higher than determined cut. In other words, each of the considered aggregations is better fitted to a specific dataset - $\mathcal{A}_1$ for datasets with higher $F1$ values and $\mathcal{A}_8$ for higher $T2$ values. These results are again intuitive. Namely, for less complex classification problems the choice of aggregation functions may be arbitrary while with more complex problems the choice of aggregation (to obtain better classification results) needs to be considered according to the discovered knowledge.

7 Conclusions

In this paper, we proposed the method of applying decision rules in the problem of fitting an aggregation function to the classification problems depending on the complexity of datasets. We limited ourselves to two examples of aggregation functions, a few datasets and a few dataset complexity measures just to demonstrate the suggested method. In future work, we plan to conduct a comprehensive analysis on the aggregation functions in the context of predicting the best choice based on the available knowledge of the complexity of the problem.

Acknowledgements. This work was partially supported by the Centre for Innovation and Transfer of Natural Sciences and Engineering Knowledge of University of Rzeszów, Poland, the project RPPK.01.03.00-18-001/10.

References

1. Bazan, J.G., Nguyen, H.S., Nguyen, S.H., Synak, P., Wróblewski, J.: Rough set algorithms in classification problem. In: Polkowski, L., Tsumoto, S., Lin, T.Y. (eds.) Rough Set Methods and Applications. Studies in Fuzziness and Soft Computing, vol. 56, 2000, pp. 49–88., Physica, Heidelberg (2000). https://doi.org/10.1007/978-3-7908-1840-6_3
2. Bazan, J.G., Szczuka, M.: The rough set exploration system. In: Peters, J.F., Skowron, A. (eds.) Transactions on Rough Sets III. LNCS, vol. 3400, issue number 3, pp. 37–56. Springer, Berlin (2005), Heidelberg. https://doi.org/10.1007/11427834_2
3. Bentkowska, U.: New types of aggregation functions for interval-valued fuzzy setting and preservation of pos-B and nec-B-transitivity in decision making problems. Inf. Sci. **424**, 385–399 (2018)
4. Bentkowska, U., Bazan, J.G., Rząsa, W., Zaręba, L.: Application of interval-valued aggregation to optimization problem of k-NN classifiers for missing values case. Inf. Sci. **486**, 434–449 (2019)
5. Bentkowska, U., Bazan, J.G., Mrukowicz, M., Zaręba, L., Molenda, P.: Multi-class classification problems for the k-NN algorithm in the case of missing values. In: 2020 IEEE International Conference on Fuzzy Systems (FUZZ-IEEE), Glasgow, UK, pp. 1–8 (2020). https://doi.org/10.1109/FUZZ48607.2020.9177592
6. Beliakov, G., Bustince, H., Calvo, T.: A practical Guide to Averaging Functions. Studies in Fuzziness and Soft Computing, vol. 329. Springer, Cham (2016). https://doi.org/10.1007/978-3-319-24753-3
7. Bustince, H., Fernandez, J., Kolesárová, A., Mesiar, R.: Generation of linear orders for intervals by means of aggregation functions. Fuzzy Sets Syst. **220**, 69–77 (2013)
8. Ho, T.K., Basu, M.: Complexity measures of supervised classification problems. IEEE Trans. Pattern Anal. Mach. Intell. **24**(3), 289–300 (2002)
9. Kulish, U.W., Miranker, W.L.: Computer Arithmetic in Theory and Practice. Academic Press, Cambridge (1981)
10. Lorena, A.C, Garcia, L.P.F., Lehmann, J., Souto, M.C.P., Ho, T.K.: How Complex is your classification problem? A survey on measuring classification complexity, Cornell University. https://arxiv.org/abs/1808.03591
11. Luengo, J., Herrera, F.: An automatic extraction method of the domains of competence for learning classifiers using data complexity measures. Knowl. Inf. Syst. **42**, 147–180 (2015). https://doi.org/10.1007/s10115-013-0700-4
12. Nguyen, H.S.: Approximate Boolean reasoning: foundations and applications in data Mining. In: Peters, J.F., Skowron, A. (eds.) Transactions on Rough Sets V. LNCS, vol. 4100, pp. 334–506. Springer, Heidelberg (2006). https://doi.org/10.1007/11847465_16

13. Pawlak, Z., Skowron, A.: A rough set approach for decision rules generation. In: 1993 ICS Research Report 23/93, Warsaw University of Technology and Proceedings of the IJCAI 1993 Workshop W12: The Management of Uncertainty in AI. France (1993)
14. Pawlak, Z., Skowron, A.: Rudiments of rough sets. Inf. Sci. **177**, 3–27 (2007)
15. Pękala, B., Bentkowska, U., De Baets, B.: On comparability relations in the class of interval-valued fuzzy relations. Tatra Mt. Math. Publ. **66**, 91–101 (2016)
16. RSES. Project. https://www.mimuw.edu.pl/~szczuka/rses/
17. Dataset complexity measures project. https://pypi.org/project/data-complexity/
18. UC Irvine Machine Learning Repository. http://archive.ics.uci.edu/ml/

Influence of Interval-Valued Measures on Classification Methods with Missing Values

Dawid Kosior[1] and Barbara Pękala[1,2(✉)]

[1] College of Natural Sciences, Institute of Computer Science, University of Rzeszów, Rzeszów, Poland
bpekala@ur.edu.pl

[2] Department of Artificial Intelligence, University of Information Technology and Management, Rzeszów, Poland

Abstract. In this paper, the basic version of the k-nearest neighbors (k-NN) algorithm has been extended to include similarity and indistinguishability measures that reflect uncertainty. The measures used take into account the uncertainty in the decision-making process when classifying objects. The algorithm has been adapted to the processing of attribute values stored as interval values. The algorithm was tested on data with missing values. Three approaches were tested to determine the missing value of individual objects.

Keywords: IV-kNN · Interval-valued fuzzy set · Aggregation function · Precedence indicator · Indistinguishability · Similarity measure

1 Introduction

Since the introduction of fuzzy sets by Zadeh [1], many new approaches and theories have been proposed for the study and modeling of uncertainty. Fuzzy sets with interval values (Zadeh and Sambuc 1975) or intuitionistic fuzzy sets (Atanassov 1986) have proved particularly useful for representing uncertainty and modeling knowledge with uncertainty (see [2] for more details). Besides, various applications of fuzzy sets with interval values have been successfully proposed to solve real-world problems including pattern recognition, medical diagnostics, and decision making. The above results are possible due to significant advances in the fuzzy set theory with interval values. Because many researchers have proposed different distances, measures of inclusion, equivalence measures, or similarity measures for fuzzy sets with interval values and investigated different types of relationships between them with reflecting dependences to interval aggregation operators [3–12].

The main theme of the article is the use of the impact of the lack of data and methods of their supplementation on diagnostic methods, in particular, based

K. T. Atanassov et al. (Eds.): IWIFSGN 2020/BOS/SOR 2020, LNNS 338, pp. 15–27, 2022.
https://doi.org/10.1007/978-3-030-95929-6_2

on the degree of inclusion and the measure of equivalence between interval values in the classification problem, especially in the k-nearest neighbors (k-NN) algorithm. Here we consider the concept of a fuzzy equivalence in an interval-valued fuzzy setting which may be treated as a special kind of lattice with its underlying interpretation and semantics. Interval-valued fuzzy sets represent uncertainty observed in many fields of applications. In the proposed definition of interval-valued equivalence, we use the usual axioms for an equivalence relation, i.e. reflexivity, symmetry, and transitivity. Moreover, we study a reference to the methods using the similarity measure. The fundamental novelty of the proposed approach is the use of the different measures between intervals and the width of these intervals representing both the uncertainty of information and the imprecision of the membership function.

The paper is organized as follows. In Sect. 2 basic information on interval-valued fuzzy sets are recalled. Inclusion, similarity, and indistinguishability degree measures for interval-valued fuzzy sets are presented in Sect. 3 Finally, the algorithm of application mentioned measures in k-NN classifiers by missing data problem is considered (Sect. 4).

2 Interval-Valued Fuzzy Set Theory

Firstly, we recall definition of an interval-valued fuzzy set and the classically applied order for this setting.

Definition 1 (cf. [13,14]). An interval-valued fuzzy set IVFS $\widetilde{A}$ in X is a mapping $\widetilde{A} : X \to L^I$ such that $\widetilde{A}(x) = [\underline{A}(x), \overline{A}(x)] \in L^I$ for $x \in X$, where

$$\widetilde{A} \cap \widetilde{B} = \{\langle x, [\min\{\underline{A}(x), \underline{B}(x)\}, \min\{\overline{A}(x), \overline{B}(x)\}]\rangle : x \in X\},$$

$$\widetilde{A} \cup \widetilde{B} = \{\langle x, [\max\{\underline{A}(x), \underline{B}(x)\}, \max\{\overline{A}(x), \overline{B}(x)\}]\rangle : x \in X\} \quad \textit{and}$$

$$L^I = \{[x_1, x_2] : \ x_1, x_2 \in [0, 1], \ x_1 \leq x_2\}.$$

The well-known classical monotonicity (partial order) for intervals is of the form

$$[x_1, y_1] \leq_2 [x_2, y_2] \Leftrightarrow x_1 \leq x_2, \ y_1 \leq y_2 \quad \textit{and}$$

$$[\underline{x}, \overline{x}] <_2 [\underline{y}, \overline{y}] \Leftrightarrow [\underline{x}, \overline{x}] \leq_2 [\underline{y}, \overline{y}] \ \textit{and} \ (\underline{x} < \underline{y} \ \textit{or} \ \overline{x} < \overline{y}).$$

The family of all interval-valued fuzzy sets on a given universe X with $\leq_2$ is partially ordered and moreover it is a lattice. The operations joint and meet are defined respectively

$$[\underline{x}, \overline{x}] \vee [\underline{y}, \overline{y}] = [\max(\underline{x}, \underline{y}), \max(\overline{x}, \overline{y})], \ [\underline{x}, \overline{x}] \wedge [\underline{y}, \overline{y}] = [\min(\underline{x}, \underline{y}), \min(\overline{x}, \overline{y})].$$

Note that the structure $(L^I, \vee, \wedge)$ is a complete lattice, with the partial order $\leq_2$, where $\mathbf{1} = [1, 1]$ and $\mathbf{0} = [0, 0]$ are the greatest and the smallest element of $(L^I, \leq_2)$, respectively.

Since in many real-life problems we need a linear order to be able to compare any two intervals, we are interested in extending the partial order $\leqslant_2$ to a linear one. The concept of, so called, admissible order would be of useful there.

Definition 2 [15]. An order $\leq$ in L^I is called admissible if it is linear and satisfies that for all $x, y \in L^I$, such that $x \leq_{L^I} y$, then $x \leq y$.

Simply, an order $\leq$ on L^I is admissible, if it is linear and refines the order $\leq_{L^I}$.

Proposition 1 [15]. *Let* $B_1, B_2 : [0,1]^2 \to [0,1]$ *be two continuous aggregation functions, such that, for all* $x = [\underline{x}, \overline{x}], y = [\underline{y}, \overline{y}] \in L^I$, *the equalities* $B_1(\underline{x}, \overline{x}) = B_1(\underline{y}, \overline{y})$ *and* $B_2(\underline{x}, \overline{x}) = B_2(\underline{y}, \overline{y})$ *hold if and only if* $x = y$. *If the order* $\leq_{B_{1,2}}$ *on* L^I *is defined by* $x \leq_{B_{1,2}} y$ *if and only if* $B_1(\underline{x}, \overline{x}) < B_1(\underline{y}, \overline{y})$ *or* $(B_1(\underline{x}, \overline{x}) = B_1(\underline{y}, \overline{y})$ *and* $B_2(\underline{x}, \overline{x}) \leq B_2(\underline{y}, \overline{y}))$, *then* $\leq_{B_{1,2}}$ *is an admissible order on* L^I.

Example 1 Admissible (linear) orders on L^I:
- Xu and Yager order: $[\underline{x}, \overline{x}] \leq_{XY} [\underline{y}, \overline{y}]$ if and only if

$$\underline{x} + \overline{x} < \underline{y} + \overline{y} \text{ or } (\overline{x} + \underline{x} = \overline{y} + \underline{y} \text{ and } \overline{x} - \underline{x} \leq \overline{y} - \underline{y}).$$

- The first lexicographical order (with respect to the first variable), $\leq_{lex1}$ defined as: $[\underline{x}, \overline{x}] \leq_{lex1} [\underline{y}, \overline{y}]$ if and only if

$$\underline{x} < \underline{y} \text{ or } (\underline{x} = \underline{y} \text{ and } \overline{x} \leq \overline{y}).$$

- The second lexicographical order (with respect to the second variable), $\leq_{lex2}$ defined as: $[\underline{x}, \overline{x}] \leq_{lex2} [\underline{y}, \overline{y}]$ if and only if

$$\overline{x} < \overline{y} \text{ or } (\overline{x} = \overline{y} \text{ and } \underline{x} \leq \underline{y}).$$

Remark 1. In the later part we will use the notation $\leq$ both for the partial or admissible linear order, with **0** and **1** as minimal and maximal element of L^I, respectively. Notation $\leq_2$ will be used while the results for the admissible linear orders will be used with the notation $\leq_{Adm}$.

Now we recall the concept of an aggregation function on L^I which is an important notion in many applications. We consider aggregation functions both with respect to $\leqslant_2$ and $\leqslant_{Adm}$.

Definition 3 [16,17]. An operation $\mathcal{A} : (L^I)^n \to L^I$ is called an interval-valued aggregation function if it is increasing with respect to the order $\leq$ (partial or total) and

$$\mathcal{A}(\underbrace{\mathbf{0}, ..., \mathbf{0}}_{n\times}) = \mathbf{0}, \quad \mathcal{A}(\underbrace{\mathbf{1}, ..., \mathbf{1}}_{n\times}) = \mathbf{1}.$$

The special case of interval-valued aggregation operation is a representable interval-valued aggregation function with respect to $\leq_2$.

Definition 4 [18,19]. An interval-valued aggregation function $\mathcal{A} : (L^I)^n \to L^I$ is called representable if there exist aggregation functions $A_1, A_2 : [0,1]^n \to [0,1]$ such that

$$\mathbf{A}(x_1, ..., x_n) = [A_1(\underline{x}_1, ...\underline{x}_n), A_2(\overline{x}_1, ...\overline{x}_n)], \quad x_1, ..., x_n \in L^I.$$

The next result shows the characterization of representable aggregation functions on L^I.

Theorem 1 [20]. *An operation $\mathcal{A} : (L^I)^n \to L^I$ is a representable interval-valued aggregation function with respect to $\leq_2$ if and only if there exist aggregation functions $A_1, A_2 : [0,1]^n \to [0,1]$ such that for all $x_1, ..., x_n \in L^I$ and $A_1 \leqslant A_2$*

$$\mathcal{A}(x_1, ..., x_n) = [A_1(\underline{x}_1, ...\underline{x}_n), A_2(\overline{x}_1, ...\overline{x}_n)]. \tag{1}$$

Example 2. Lattice operations $\wedge$ and $\vee$ on L^I are representable aggregation functions on L^I with $A_1 = A_2 = \min$ in the first case and $A_1 = A_2 = \max$ in the second one. It holds true with respect to the order $\leq_2$, but not $\leq_{Lex1}$, $\leq_{Lex2}$ or $\leq_{XY}$. Moreover, many other examples of representable aggregation functions with respect to $\leq_2$ may be considered, such as:

- the representable arithmetic mean $\mathcal{A}_{mean}([\underline{x}, \overline{x}], [\underline{y}, \overline{y}]) = [\frac{\underline{x}+\underline{y}}{2}, \frac{\overline{x}+\overline{y}}{2}]$,
- the representable mean-power mean $\mathcal{A}_{meanpow}([\underline{x}, \overline{x}], [\underline{y}, \overline{y}]) = [\frac{\underline{x}+\underline{y}}{2}, \sqrt{\frac{\overline{x}^2+\overline{y}^2}{2}}]$,
- the representable mean-max $\mathcal{A}_{meanmax}([\underline{x}, \overline{x}], [\underline{y}, \overline{y}]) = [\frac{\underline{x}+\underline{y}}{2}, \max \overline{x}, \overline{y}]$

for $[\underline{x}, \overline{x}], [\underline{y}, \overline{y}] \in L^I$.

3 Interval-Valued Fuzzy Measures

In this section, we recall three measures that play a crucial role in our studies, i.e. precedence indicator, similarity, and equivalence measures with respect to uncertainty.

3.1 Precedence Indicator

Before we get to the main issue that interests us, i.e. the similarity and equivalence measures we recall the notion of a precedence indicator that use linear or partial order and uncertainty measure [21]. Which pay attention to the width of the intervals, as this width is seen as a way to measure the uncertainty related to the interval, and which we denote by w, where
$w(\alpha) = \overline{\alpha} - \underline{\alpha}$ for $\alpha \in L^I$.

The precedence indicators can be used to build subsethood or inclusion measures which have been discussed in axiomatic or constructive ways. We remind that we use the notation $\leq$ both for linear and partial orders.

Definition 5 [21]. A function Prec: $(L^I)^2 \to L^I$ is said to be a **precedence indicator** if it satisfies the following conditions for any $a, b, c \in L^I$

P1 if $a = 1_{L^I}$ and $b = 0_{L^I}$, then $\text{Prec}(a, b) = 0_{L^I}$;
P2 if $a < b$, then $\text{Prec}(a, b) = 1_{L^I}$ for any $a, b \in L^I$;
P3 $\text{Prec}(a, a) = [1 - w(a), 1]$ for any $a \in L^I$;
P4 if $a \leq b \leq c$ and $w(a) = w(b) = w(c)$, then $\text{Prec}(c, a) \leq \text{Prec}(b, a)$ and $\text{Prec}(c, a) \leq \text{Prec}(c, b)$ for any $a, b, c \in L^I$.

Here is the example of the construction of precedence indicator.

Proposition 2 [21]. *The operation*

$$\mathrm{Prec}_w(a,b) = \begin{cases} 1_{L^I}, & a < b, \\ [1 - max(w(a), r(a,b)), 1 - r(a,b)], & else \end{cases}$$

is the precedence indicator with respect to $\leq$, *where* $r(a,b) = \max\{|\underline{a} - \underline{b}|, |\overline{a} - \overline{b}|\}$ *for* $a, b \in L^I$.

3.2 Similarity Measure

In this section, we recall a class of similarity measures between interval-valued fuzzy sets. To construct this interval-valued similarity, interval-valued aggregation functions and interval-valued inclusion measure which take into account the width of the intervals are needed [22].

Definition 6 [23]. Let $A_1 : [0,1]^n \to [0,1]$ be an aggregation function. Then operation $S : IVFS(X) \times IVFS(X) \to L^I$, which satisfies the following items:
(SIMV1) $S(A,B) = S(B,A)$ for $A, B \in IVFS(X)$;
(SIMV2) $S(A,A) = [1 - A_1(w_A(x_1), ..., w_A(x_n)), 1]$;
(SIMV3) $S(A,B) = 0_{L^I}$, if $\{A(x_i), B(x_i)\} = \{0_{L^I}, 1_{L^I}\}$;
(SIMV4) if $A \preceq B \preceq C$ and $w_A(x_i) = w_B(x_i) = w_C(x_i)$, then $S(A,C) \leq S(A,B)$ and $S(A,C) \leq S(B,C)$
is called a similarity measure for $i = 1, ..., n$, where

$$A \preceq B \Leftrightarrow a_i \leq b_i \text{ for each } i \text{ and } a_i = A(x_i),\ b_i = B(x_i).$$

Proposition 3. *Let* Prec *be a precedence indicator. If* $\mathcal{A} = [A_1, A_2]$, $\mathcal{B} = [B_1, B_2]$ *are representable interval-valued aggregation functions for which* A_1 *is as in previous definition and self-dual,* $\mathcal{B}$ *is symmetric with the neutral element* 1_{L^I} *and* B_1 *is idempotent aggregation function, then the operation* $S : IVFS(X) \times IVFS(X) \to L^I$:

$$S(A,B) = \mathcal{A}_{i=1}^n(\mathcal{B}(\mathrm{Prec}(A(x_i), B(x_i)), \mathrm{Prec}(B(x_i), A(x_i))))$$

is a similarity measure, where $w_A(x_i) = w_B(x_i)$.

Example 3. The operation $S : IVFS(X) \times IVFS(X) \to L^I$:

$$S(A,B) = \mathcal{A}_{i=1}^n(\mathrm{Prec}_w(A(x_i), B(x_i)) \wedge \mathrm{Prec}_w(B(x_i), A(x_i))))$$

is a similarity measure, where $\mathcal{A} \in \{\mathcal{A}_{mean}, \mathcal{A}_{meanpow}, \mathcal{A}_{meanmax}\}$.

3.3 Equivalence Measures

In this section, we discuss some of the new types of interval-valued equivalence, partially presented in [24], which pays attention to the width of the input intervals and is defined both for partial order or the linear one.

Definition 7. Let $\mathcal{B} : (L^I)^2 \to L^I$ be an IV aggregation (with respect to the order $\leq$). An operation $E : (L^I)^2 \to L^I$ is called a $\mathcal{B}$-equivalence measure if for $\alpha, \beta, \gamma \in L^I$ it holds:
(E1) $E(\alpha, \alpha) = [1 - w(\alpha), 1]$ (reflexivity);
(E2) $E(\alpha, \beta) = E(\beta, \alpha)$ (symmetry);
(E3) $\mathcal{B}(E(\alpha, \beta), E(\beta, \gamma)) \leq E(\alpha, \gamma)$ ($\mathcal{B}$-transitivity).

We use the following notation $EM = E : (L^I)^2 \to L^I$, where E is an interval-valued $\mathcal{B}$-equivalence measure.

In the following proposition we may observe an interesting connection between the mentioned precedence indicators and equivalence measures.

Proposition 4. *Let $\mathcal{A}$ be a symmetric, idempotent IV aggregation,* Prec *be $\mathcal{B}$-transitive precedence indicator, where $\mathcal{B}$ be a symmetric IV aggregation and $\mathcal{A} \gg \mathcal{B}$ (the dominance property and $\mathcal{B}$-transitivity are considered with the same order). Thus a function $E_{\mathcal{A}} : (L^I)^2 \to L^I$,*

$$E_{\mathcal{A}}(\alpha, \beta) = \mathcal{A}(\text{Prec}(\alpha, \beta), \text{Prec}(\beta, \alpha))$$

is a $\mathcal{B}$-equivalence measure.

3.4 Indistinguishability

Now we propose a class of a new type of measures, namely indistinguishability measures, between interval - valued fuzzy sets (IVFSs). The inspiration of this approach is the fact that we develop all the notions with respect to the total orders of intervals and that we consider the width of intervals. So the uncertainty of the output is strongly connected with the uncertainty of the inputs. For creating the new interval-valued indistinguishability, IV aggregation functions and IV equivalence measure which take into account the width of the intervals are needed.

Definition 8. Let A_1 be a n-argument aggregation function and $\mathcal{B}$ be an IV aggregation function. Then $\mathbf{I} : IVFS(X) \times IVFS(X) \to L^I$, which satisfies the following items:
(I1) $\mathbf{I}(A, A) = [1 - A_1(w_A(\alpha_1), ..., w_A(\alpha_n)), 1]$;
(I2) $\mathbf{I}(A, B) = \mathbf{I}(B, A)$;
(I3) $\mathcal{B}(\mathbf{I}(A, B), \mathbf{I}(B, C)) \leq \mathbf{I}(A, C)$ for $A, B, C \in IVFS(X)$
is called a $\mathcal{B}$-indistinguishability measure between interval-valued fuzzy sets.

Proposition 5. *Let for a representable IV aggregation function $\mathcal{A} = [A_1, A_2]$ A_1 be self-dual, $\mathcal{B}$ be an IV aggregation function and $\mathcal{A} \gg \mathcal{B}$ ($\mathcal{A}$, $\mathcal{B}$ and the dominance are in accordance with the order $\leq$). If $E \in EM$, then $\boldsymbol{I}_{\mathcal{A}} : IVFS(X) \times IVFS(X) \to L^I$,*

$$\boldsymbol{I}_{\mathcal{A}}(A, B) = \mathcal{A}(E(A(\alpha_1), B(\alpha_1)), ..., E(A(\alpha_n), B(\alpha_n)))$$

is a $\mathcal{B}$-indistinguishability measure for IVFSs.

Example 4. An operation

$$\mathbf{I}_{\mathcal{A}}(A,B) = \mathcal{A}_{mean}(E_{\mathcal{A}_{mean\,z}}(A(\alpha_1),B(\alpha_1)),...,E_{\mathcal{A}_{mean\,z}}(A(\alpha_n),B(\alpha_n)))$$

is a $T_{L_{IV}}$-indistinguishability measure for IVFSs, where

$$E_{\mathcal{A}_{mean\,z}}(\alpha,\beta) = \begin{cases} [1-w(\alpha),1], & \alpha=\beta, \\ [\frac{1}{2},\frac{1}{2}], & \alpha<\beta \; or \; \beta<\alpha, \\ \mathbf{0}, & otherwise \; (\alpha\|\beta) \end{cases}$$

is $T_{L_{IV}}$-equivalence measure with respect to $\leq_{L^I}$ created by $\mathcal{A} = \mathcal{A}_{mean}$ by $E_{\mathcal{A}_z}(\alpha,\beta) = \mathcal{A}(\mathrm{Prec}_z(\alpha,\beta),\mathrm{Prec}_z(\beta,\alpha))$, where

$$\mathrm{Prec}_z(\alpha,\beta) = \begin{cases} [1-w(\alpha),1], & \alpha=\beta, \\ \mathbf{1}, & \alpha<\beta, \\ \mathbf{0}, & otherwise \end{cases}$$

for $\alpha,\beta \in L^I$ ($\mathcal{A}_{mean} \gg T_{L_{IV}}$).

4 Application in K-NN Classifiers

The classification problem consists of determining the class (category) to which a new, previously unknown object should be assigned. Among the most commonly used techniques for classification can be distinguished k-Nearest Neighbors classifier (k-NN) [25]. In this paper, we present the modified interval-valued fuzzy classifier for supporting decision-making processes based on imprecise (uncertain) data. The main goal was to develop a comprehensive and effective the approach that enables the modeling and processing of input data, and then the presentation of results, to be done in a way that preserves the valuable information concerning the amount of uncertainty at each stage of the process. Especially, the main problem will be to compare the effectiveness of the method based on the measure of similarity taking the measure of indistinguishability and, moreover, the impact of the lack of data.

4.1 Data Set Description

The dataset is a Wisconsin (diagnostic) breast cancer dataset. This is one of the popular datasets from UCI Machine Learning Repository [26]. Data are from November 1995. Data containing information on 569 instances. Features are computed from a digitized image of a fine needle aspirate (FNA) of a breast mass. They describe the characteristics of the cell nuclei present in the image.

Ten real-valued features are computed for each cell nucleus:

- radius (mean of distances from center to points on the perimeter)
- texture (standard deviation of gray-scale values)
- perimeter
- area

- smoothness (local variation in radius lengths)
- compactness ($\frac{perimeter^2}{area-1.0}$)
- concavity (severity of concave portions of the contour)
- concave points (number of concave portions of the contour)
- symmetry
- fractal dimension (*coastline approximation* – 1)

The decision attribute stores information about the diagnosis: malignant or benign represented by values of 0 or 1. The dataset consists of 212 objects with malignant diagnosis and 357 objects with the benign diagnosis.

4.2 Proposed Method

This steps list interval-valued k-Nearest Neighbors classifier (IV-kNN) shows the main steps of the proposed classification method.

Algorithm
Step 1. Data double step fuzzyfication;
1.1. Fuzzyfication. The real data are normalized to $[0, 1]$ by classical formula based on minimal values m_i and maximal M_i for $i - th$ attribute with real value b_i

$$a_i = \frac{b_i - m_i}{M_i - m_i}.$$

1.2. 2-fuzzyfication. Interval-valued fuzzy set for each instance/object is built in the following way:

$$[x(1 - 0.25g(x)), x(1 - 0.25g(x)) + 0.25g(x)] \in L^I.$$

Note that the length of the intervals is equal to $g(x) = 2\min(x, 1 - x)$, where g is ignorance function [27].
Step 2. Filling the missing value with one of the proposed methods:

1. For a missing value, insert a [0,1] interval.

2. Search for the most similar k objects and select their value for the missing attribute and calculate for the missing value insert the minimum, maximum or average of the selected values on the missing attribute:

$$[min(a_1, ...a_k), max(a_1, ...a_k)], \quad [avg(a_1, ...a_k), avg(a_1, ...a_k)].$$

3. Find the most similar l objects and choose their value for missing attribute and calculate for missing value insert minimum, maximum or average of selected values on missing attribute. For training objects, similar objects from the same decision class are selected:

$$[min(a_1, ..., a_l), max(a_1, .., a_l)], \quad [avg(a_1, ...a_l), avg(a_1, ...a_l)].$$

Step 3. Import new object (without decision);
Step 4. For k = 1...n
Calculating similarities ($S_{\mathcal{M}}$) using similarity (S) or indistinguishability (I) measure for a new object;
Step 5. Ordered of a interval-valued fuzzy similarities;
Step 6. A selection of the k most similar neighbors;
Step 7. Aggregation of individual classes among k most like neighbors ($C_{\mathcal{A}}$);
Step 8. Class selection - selection of the largest interval for each class [28];
Step 9. Output: Class for the new object.

4.3 Results and Discussion

The algorithm testing procedure was as follows:

- The data was divided into a training part and a test part. The training part have 70% of all and testing part have 30% from all objects.
- The selection of objects for both parts was random and made ten times. For each of the ten datasets, we test algorithm with different parameters were checked and accuracy, sensitivity, specificity and precision were calculated.
- From the ten values obtained in this way for accuracy, sensitivity, specificity and precision their arithmetic mean was calculated and presented as the value obtained by the algorithm with particular parameters.
- The algorithm was tested on data with 5%, 25%, 50% of missing values.

We will focus on presenting the conclusions of the algorithm analysis in three aspects:

1. modification of the value of k;
2. the selection of aggregation in the construction of the similarity measure and indistinguishability measure;
3. checking the different methods of filling in missing values and different sizes of missing values in the data;
4. the selection of ordering relations ($\leq_2, \leq_{XY}, \leq_{Lex1}, \leq_{Lex2}$).

In Tables 1, 2 and 3 we present the results of our research. By MS we denote the percent of missing values in data. For two methods of supplementing the missing values, two options for determining the interval value were used. The designated value was to replace the missing one. The first option constructed the range where the lower bound is the minimum of the lower bounds of ranges and the upper bound is the maximum of the upper bounds values of the ranges it was much better in the tests than the approach based on the mean value of the bounds of the range. However, this does not mean that it would not be otherwise for different data. In Tables 2 and 3 we have only presented the results using the first option.

The tests showed that the methods of filling missing values studied by us by using our modified IV-kNN are characterized by a slower decrease in accuracy of the classification for the third method compare to the first and the second,

Table 1. Table for the first method of filling missing values.

k	Accuracy	Sensitivity	Specificity	Precision	$C_{\mathcal{A}}$	$S_{\mathcal{A}}$	Order	MS	$S_{\mathcal{M}}$
	0,877295213	0,956354306	0,833021922	0,90946816	$\mathcal{A}_{meanmax}$	$\mathcal{A}_{mean}$	$\leq_{lex2}$	5%	
1	0,855623164	0,952316431	0,785443313	0,888174475	$\mathcal{A}_{mean}$	$\mathcal{A}_{mean}$	$\leq_{lex2}$	25%	I
	0,847856945	0,950069216	0,779792944	0,881347978	$\mathcal{A}_{meanmax}$	$\mathcal{A}_{mean}$	$\leq_{lex2}$	50%	
	0,863293246	0,952064793	0,805510005	0,894909111	$\mathcal{A}_{meanmax}$	$\mathcal{A}_{meanmax}$	$\leq_{lex1}$	5%	
1	0,841757424	0,946752024	0,758375556	0,874332808	$\mathcal{A}_{meanmax}$	$\mathcal{A}_{meanpow}$	$\leq_{lex2}$	25%	S
	0,837122851	0,946521682	0,758785615	0,870113533	$\mathcal{A}_{meanmax}$	$\mathcal{A}_{meanpow}$	$\leq_{lex2}$	50%	
	0,86671996	0,957699335	0,806798803	0,896665004	$\mathcal{A}_{meanmax}$	$\mathcal{A}_{mean}$	$\leq_{lex2}$	5%	
3	0,846573346	0,953220951	0,761929092	0,877267935	$\mathcal{A}_{mean}$	$\mathcal{A}_{mean}$	$\leq_{lex2}$	25%	I
	0,83858117	0,953411195	0,752340814	0,86839648	$\mathcal{A}_{meanmax}$	$\mathcal{A}_{mean}$	$\leq_{lex2}$	50%	
	0,856294636	0,953290925	0,788916968	0,886932849	$\mathcal{A}_{meanmax}$	$\mathcal{A}_{meanpow}$	$\leq_{lex2}$	5%	
3	0,83483665	0,9471642	0,741429156	0,866483001	$\mathcal{A}_{meanmax}$	$\mathcal{A}_{meanpow}$	$\leq_{lex2}$	25%	S
	0,828317541	0,946351344	0,738225444	0,860137423	$\mathcal{A}_{meanmax}$	$\mathcal{A}_{meanpow}$	$\leq_{lex2}$	50%	
	0,860619786	0,956844717	0,793790423	0,890199054	$\mathcal{A}_{min}$	$\mathcal{A}_{mean}$	$\leq_{XY}$	5%	
5	0,836040603	0,950029196	0,741432911	0,867363151	$\mathcal{A}_{min}$	$\mathcal{A}_{mean}$	$\leq_{lex2}$	25%	I
	0,832745543	0,949909305	0,744427506	0,863729177	$\mathcal{A}_{min}$	$\mathcal{A}_{mean}$	$\leq_{XY}$	50%	
	0,861724323	0,956774869	0,79687821	0,890825269	$\mathcal{A}_{mean}$	$\mathcal{A}_{mean}$	$\leq_{XY}$	5%	
5	0,835598403	0,948806975	0,742392271	0,867691953	$\mathcal{A}_{mean}$	$\mathcal{A}_{mean}$	$\leq_{lex2}$	25%	S
	0,82592191	0,944076375	0,738009497	0,86004546	$\mathcal{A}_{mean}$	$\mathcal{A}_{meanmax}$	$\leq_{2}$	50%	

Table 2. Table for the second method of filling missing values.

k	Accuracy	Sensitivity	Specificity	Precision	$C_{\mathcal{A}}$	$S_{\mathcal{A}}$	Order	MS	$S_{\mathcal{M}}$
	0,875164803	0,955569769	0,829275549	0,907565691	$\mathcal{A}_{meanmax}$	$\mathcal{A}_{mean}$	$\leq_{lex2}$	5%	
1	0,862216862	0,951867453	0,801919554	0,895761817	$\mathcal{A}_{mean}$	$\mathcal{A}_{mean}$	$\leq_{lex2}$	25%	I
	0,858307794	0,954335521	0,79649695	0,889915539	$\mathcal{A}_{meanmax}$	$\mathcal{A}_{mean}$	$\leq_{lex2}$	50%	
	0,860383256	0,950900338	0,800522779	0,892336445	$\mathcal{A}_{meanmax}$	$\mathcal{A}_{meanpow}$	$\leq_{lex2}$	5%	
1	0,848219814	0,95087641	0,769143143	0,879745901	$\mathcal{A}_{mean}$	$\mathcal{A}_{meanmax}$	$\leq_{lex1}$	25%	S
	0,844206219	0,949049521	0,770200942	0,875996301	$\mathcal{A}_{meanmax}$	$\mathcal{A}_{meanpow}$	$\leq_{lex2}$	50%	
	0,86430934	0,956728007	0,802580878	0,894504398	$\mathcal{A}_{meanmax}$	$\mathcal{A}_{mean}$	$\leq_{lex2}$	5%	
3	0,852343376	0,952460141	0,777622187	0,883955726	$\mathcal{A}_{mean}$	$\mathcal{A}_{mean}$	$\leq_{lex2}$	25%	I
	0,850073994	0,955676891	0,774964368	0,879456496	$\mathcal{A}_{meanmax}$	$\mathcal{A}_{mean}$	$\leq_{lex2}$	50%	
	0,853131268	0,952682155	0,782661228	0,88381208	$\mathcal{A}_{meanmax}$	$\mathcal{A}_{meanpow}$	$\leq_{lex2}$	5%	
3	0,841698444	0,948807864	0,756996901	0,873633797	$\mathcal{A}_{mean}$	$\mathcal{A}_{meanpow}$	$\leq_{lex2}$	25%	S
	0,838060591	0,948662744	0,757630117	0,869702387	$\mathcal{A}_{meanmax}$	$\mathcal{A}_{meanpow}$	$\leq_{lex2}$	50%	
	0,858213327	0,953285198	0,794692982	0,889330988	$\mathcal{A}_{meanpow}$	$\mathcal{A}_{mean}$	$\leq_{XY}$	5%	
5	0,84107295	0,950628378	0,753018736	0,87232769	$\mathcal{A}_{meanpow}$	$\mathcal{A}_{mean}$	$\leq_{lex2}$	25%	I
	0,840267842	0,950526542	0,760773919	0,871823934	$\mathcal{A}_{min}$	$\mathcal{A}_{mean}$	$\leq_{XY}$	50%	
	0,853580999	0,952845669	0,784110124	0,885236216	$\mathcal{A}_{mean}$	$\mathcal{A}_{mean}$	$\leq_{lex2}$	5%	
5	0,84107295	0,950628378	0,753018736	0,87232769	$\mathcal{A}_{meanpow}$	$\mathcal{A}_{mean}$	$\leq_{lex2}$	25%	S
	0,842305378	0,951388131	0,764925667	0,873062717	$\mathcal{A}_{mean}$	$\mathcal{A}_{mean}$	$\leq_{XY}$	50%	

with an increase in the number of missing values in the data. It is good to see this difference between the results on the set with 25% and 50% of the missing values. The use of the method with indistinguishability measure gave slightly

Table 3. Table for the third method of filling missing values.

k	Accuracy	Sensitivity	Specificity	Precision	$C_{\mathcal{A}}$	$S_{\mathcal{A}}$	Order	MS	$S_{\mathcal{M}}$
	0,876393879	0,955600931	0,83176004	0,908778205	$\mathcal{A}_{meanmax}$	$\mathcal{A}_{mean}$	$\leq_{lex2}$	5%	
1	0,863174599	0,952951685	0,802535894	0,896140721	$\mathcal{A}_{mean}$	$\mathcal{A}_{mean}$	$\leq_{lex2}$	25%	I
	0,859022019	0,957291234	0,792617045	0,888400774	$\mathcal{A}_{meanmax}$	$\mathcal{A}_{mean}$	$\leq_{lex2}$	50%	
	0,861068713	0,951032444	0,801776804	0,89305433	$\mathcal{A}_{meanmax}$	$\mathcal{A}_{meanpow}$	$\leq_{lex2}$	5%	
1	0,849603856	0,952318698	0,770173382	0,880453221	$\mathcal{A}_{mean}$	$\mathcal{A}_{meanmax}$	$\leq_{lex1}$	25%	S
	0,845324056	0,953290377	0,76583975	0,874440253	$\mathcal{A}_{meanmax}$	$\mathcal{A}_{meanmax}$	$\leq_{lex1}$	50%	
	0,864983644	0,956773312	0,803794131	0,895031392	$\mathcal{A}_{meanmax}$	$\mathcal{A}_{mean}$	$\leq_{lex2}$	5%	
3	0,85441494	0,954623268	0,779027873	0,88489549	$\mathcal{A}_{mean}$	$\mathcal{A}_{mean}$	$\leq_{lex2}$	25%	I
	0,850464446	0,958875752	0,770235821	0,877601312	$\mathcal{A}_{meanmax}$	$\mathcal{A}_{mean}$	$\leq_{lex2}$	50%	
	0,853241271	0,952609413	0,783001422	0,883974259	$\mathcal{A}_{meanmax}$	$\mathcal{A}_{meanpow}$	$\leq_{lex2}$	5%	
3	0,843326753	0,950761731	0,757763476	0,874303137	$\mathcal{A}_{mean}$	$\mathcal{A}_{meanpow}$	$\leq_{lex2}$	25%	S
	0,83960183	0,953539166	0,753201546	0,868218926	$\mathcal{A}_{meanmax}$	$\mathcal{A}_{meanpow}$	$\leq_{lex2}$	50%	
	0,856801942	0,951099929	0,794039717	0,888744803	$\mathcal{A}_{meanpow}$	$\mathcal{A}_{mean}$	$\leq_{XY}$	5%	
5	0,845019703	0,952937731	0,758746393	0,875059964	$\mathcal{A}_{meanpow}$	$\mathcal{A}_{mean}$	$\leq_{lex2}$	25%	I
	0,83997864	0,955184783	0,751303913	0,86832831	$\mathcal{A}_{meanpow}$	$\mathcal{A}_{mean}$	$\leq_{lex2}$	50%	
	0,855589473	0,953187164	0,787761604	0,887139876	$\mathcal{A}_{meanpow}$	$\mathcal{A}_{mean}$	$\leq_{lex2}$	5%	
5	0,845019703	0,952937731	0,758746393	0,875059964	$\mathcal{A}_{meanpow}$	$\mathcal{A}_{mean}$	$\leq_{lex2}$	25%	S
	0,83997864	0,955184783	0,751303913	0,86832831	$\mathcal{A}_{meanpow}$	$\mathcal{A}_{mean}$	$\leq_{lex2}$	50%	

better classification results, but the method based on similarity measure should not be considered worse, because the differences are less than 2%, these are slight differences that may result from the specificity of the data processed.

5 Conclusions

We applied the similarity measure and indistinguishability measure reflected uncertainty in the decision-making algorithm based on the interval-valued k-NN method with different methods of filling in missing values and different sizes of missing values in the data. In the future, we will propose new methods of constructing similarity and indistinguishability, we will analyze the impact of using different aggregations, we will analyze another way to fill missing values, moreover, we will check whether and to what extent the conclusions obtained are affected by a small modification of the input parameters, e.g. the weak ignorance function or its parameters.

References

1. Zadeh, L.A.: Fuzzy sets. Inf. Contr. **8**, 338–353 (1965)
2. Bustince, H., et al.: A historical account of types of fuzzy sets and their relationships. IEEE Trans. Fuzzy Syst. **24**(1), 174–194 (2016)
3. Bodenhofer, U., De Baets, B., Fodor, J.: A compendium of fuzzy weak orders: representations and constructions. Fuzzy Sets Syst. **158**, 811–829 (2007)

4. Bosteels, K., Kerre, E.E.: On a reflexivity-preserving family of cardinality-based fuzzy comparison measures. Inf. Sci. **179**, 2342–2352 (2009)
5. Bustince, H., Barrenechea, E., Pagola, M.: Image thresholding using restricted equivalence functions and maximizing the measures of similarity. Fuzzy Sets Syst. **158**, 496–516 (2007)
6. Pękala, B.: Operations on interval matrices. In: Kryszkiewicz, M., Peters, J.F., Rybinski, H., Skowron, A. (eds.) Rough Sets and Intelligent Systems Paradigms. RSEISP 2007. Lecture Notes in Computer Science, vol. 4585, pp. 613-621. Springer, Heidelberg (2007). https://doi.org/10.1007/978-3-540-73451-2_64
7. De Baets, B., De Meyer, H., Naessens, H.: A class of rational cardinality-based similarity measures. J. Comp. Appl. Math. **132**, 51–69 (2001)
8. De Baets, B., Janssens, S., De Meyer, H.: On the transitivity of a parametric family of cardinality-based similarity measures. Internat. J. Appr. Reas. **50**, 104–116 (2009)
9. Jayaram, B., Mesiar, R.: I-Fuzzy equivalence relations and I-fuzzy partitions. Inf. Sci. **179**, 1278–1297 (2009)
10. Pękala, B.: Preservation of properties of interval-valued fuzzy relations. In: IFSA/EUSFLAT Conference, pp. 1206–1210 (2009)
11. Pękala, B., Balicki, K.: Interval-valued intuitionistic fuzzy sets and similarity measure. Iran. J. Fuzzy Syst. **14**(4), 87–98 (2017)
12. Palmeira, E.S., Bedregal, B., Bustince, H., Paternain, D., De Miguel, L.: Application of two different methods for extending lattice-valued restricted equivalence functions used for constructing similarity measures on L-fuzzy sets. Inf. Sci. **441**, 95–112 (2018)
13. Zadeh, L.A.: The concept of a linguistic variable and its application to approximate reasoning. Inf. Sci. **8**, 199–251 Part I, 301–357, Part II, Inf. Sci. **9**, 43–80, Part III (1975)
14. Sambuc, R.: Fonctions ϕ-floues: Application á l'aide au diagnostic en pathologie thyroidienne, Ph.D. Thesis, Université de Marseille, France (1975). in French
15. Bustince, H., Fernandez, J., Kolesárová, A., Mesiar, R.: Generation of linear orders for intervals by means of aggregation functions. Fuzzy Sets Syst. **220**, 69–77 (2013)
16. Zapata, H., et al.: Interval-valued implications and interval-valued strong equality index with admissible orders. Int. J. Approx. Reason. **88**, 91–109 (2017)
17. Komorníková, M., Mesiar, R.: Aggregation functions on bounded partially ordered sets and their classification. Fuzzy Sets Syst. **175**, 48–56 (2011)
18. Deschrijver, G., Kerre, E.E.: Implicators based on binary aggregation operators in interval-valued fuzzy set theory. Fuzzy Sets Syst. **153**(2), 229–248 (2005)
19. Deschrijver, G.: Quasi-arithmetic means and OWA functions in interval-valued and Atanassovs intuitionistic fuzzy set theory. In: EUSFLAT-LFA 2011, Aix-les-Bains, France, pp. 506–513 (2011)
20. Elkano, M., Sanz, J.A., Galar, M., Pękala, B., Bentkowska, U., Bustince, H.: Composition of interval-valued fuzzy relations using aggregation functions. Inf. Sci. **369**, 690–703 (2016)
21. Pękala, B., et al.: Interval subsethood measures with respect to uncertainty for interval-valued fuzzy setting. Int. J. Comput. Int. Syst. **3**, 167–177 (2020)
22. Pękala, B., Dyczkowski, K., Grzegorzewski, P., Bentkowska, U.: Inclusion and similarity measures for interval-valued fuzzy sets based on aggregation and uncertainty assessment. Inf. Sci. **547**, 1182–1200 (2021)
23. Bustince, H., Marco-Detchart, C., Fernandez, J., Wagner, C., Garibaldi, J.M., Takáč, Z.: Similarity between interval-valued fuzzy sets taking into account the width of the intervals and admissible orders. Fuzzy Sets Syst. **390**, 23–47 (2020)

24. Pękala, B., Bentkowska, U., Fernandez, J., Bustince, H.: Equivalence measures for Atanassov intuitionistic fuzzy setting used to algorithm of image processing. In: 2019 IEEE International Conference on Fuzzy Systems (FUZZ-IEEE), pp. 1–6 (2019)
25. Cover, T.M., Hart, P.E.: Nearest neighbor pattern classification. IEEE Trans. Inf. Theory **13**(1), 21–27 (1967)
26. https://archive.ics.uci.edu/ml/index.php
27. Barrenechea, E., Fernandez, J., Pagola, M., Chiclana, F., Bustince, H.: Construction of interval-valued fuzzy preference relations from ignorance functions and fuzzy preference relations: Application to decision making. Knowl.-Based Syst. **58**, 33–44 (2014)
28. Pękala, B., Rak, E., Kosior, D., Mrukowicz, M., Bazan, J.G.: Application of similarity measures with uncertainty in classification methods. In: 2020 IEEE International Conference on Fuzzy Systems (FUZZ-IEEE), pp. 1–8 (2020)

Bootstrap Methods for Fuzzy Data

Przemyslaw Grzegorzewski[1,2] and Maciej Romaniuk[1,3](✉)

[1] Systems Research Institute, Polish Academy of Sciences, Newelska 6, 01-447 Warsaw, Poland
{pgrzeg,mroman}@ibspan.waw.pl

[2] Faculty of Mathematics and Information Science, Warsaw University of Technology, Koszykowa 75, 00-662 Warsaw, Poland

[3] Warsaw School of Information Technology, Newelska 6, 01-447 Warsaw, Poland

Abstract. The bootstrap has been recognized as a universal method supporting statistical inference in situations where other approaches fail. In particular, the bootstrap appears also useful in fuzzy data analysis. In this paper, we make a review of the bootstrap methods oriented on statistical reasoning with random fuzzy samples. Moreover, we suggest a new flexible bootstrap approach designed for asymmetrical random fuzzy numbers. Finally, we compare the existing bootstrap approaches to provide some hints for practitioners.

Keywords: Bootstrap · Canonical representation · Fuzzy numbers · Fuzzy random variables · Hypothesis testing · Random fuzzy numbers · Resampling

1 Introduction

The rapid development of computer science and the accompanying progress in the computing power paved the way to various machine learning methods and their extraordinary successes in many fields. However, fast computers turned out to be a priceless help even in so traditional parts of statistical reasoning like estimation, confidence interval determination or hypotheses testing. It happened so partially thanks to the brilliant idea of Bradley Efron and his bootstrap [13]. Through the repeated sampling from available data new possibilities for statistical reasoning and data analysis are opening up for us. It is possible thanks to the key bootstrap idea that the available sample is to the generated bootstrap samples as the population is to the primary sample.

The main advantages of the bootstrap are its distribution-free nature that no distributional assumptions are required. Moreover, the bootstrap can provide quite accurate results even if the sample size is small or if the data are complex. Next, we may apply the bootstrap to statistics with sampling distributions that are difficult to derive even asymptotically. Finally, the bootstrap is quite general and simple to implement, although there are some cases when it fails.

Bearing in mind the advantages mentioned above, it should come as no surprise that the bootstrap has been also recognized as a very powerful tool in

K. T. Atanassov et al. (Eds.): IWIFSGN 2020/BOS/SOR 2020, LNNS 338, pp. 28–47, 2022.
https://doi.org/10.1007/978-3-030-95929-6_3

statistics with fuzzy data. This kind of data although so useful for modeling imprecise observations causes a lot of trouble to analysts. Indeed, problems with subtraction and division of fuzzy numbers, the lack of universally accepted total ranking between fuzzy numbers, the absence of suitable models for the distribution of fuzzy random variables or no Central Limit Theorems for direct use with fuzzy random variables make statistical inference based on fuzzy data quite complicated. In many situations, the straightforward generalization of the classical statistical methodology for fuzzy data appears either too difficult or even impossible. Therefore, to break through the problem some researchers proposed to use the bootstrap in a fuzzy environment [15,17,29]. Following these first papers the bootstrap was successfully applied in statistical tests with fuzzy data, classification, fuzzy rating in questionnaires, quality control, etc. (see, e.g., [6,27,28,32–34,41]).

A growing number of publications indicated a weak point of excessive repetitions in bootstrap samples. To increase the diversity of simulated results several improvements in bootstrap for fuzzy data were proposed [20–22,35,36].

In this paper, we review the suggested methods of more flexible bootstrap. Moreover, we introduce another method dedicated especially for the primary samples which consist mostly of asymmetrical fuzzy numbers. Then we provide the results of a simulation study-oriented on the comparison of available bootstrap methods.

2 Fuzzy Numbers and Their Characteristics

Starting with the seminal paper by L.A. Zadeh [42] fuzzy sets became the basic tool for modeling imprecise notions and objects. Consequently, fuzzy numbers became indispensable for modeling imprecise numerical results of the experiments in various fields.

Definition 1 (see [11]). A **fuzzy number** A is a fuzzy set in $\mathbb{R}$ which satisfies the following properties:

(i) A is normal (i.e. $\exists x \in \mathbb{R}$ such that $A(x) = 1$),
(ii) A is convex (i.e. $A(\lambda x_1 + (1-\lambda)x_2) \geqslant \min\{A(x_1), A(x_2)\}$, $\forall x_1, x_2 \in \mathbb{R}$ and $\forall \lambda \in [0,1]$),
(iii) A is upper semicontinuous in every $x_0 \in \mathbb{R}$ (i.e. $\forall \varepsilon > 0, \exists \delta > 0$ such that $A(x) - A(x_0) < \varepsilon$, whenever $|x - x_0| < \delta$),
(iv) $cl\{x \in \mathbb{R} : A(x) > 0\}$ is bounded, where cl denotes the closure operator.

A family of all fuzzy numbers will be denoted further on by $\mathbb{F}(\mathbb{R})$.
An α-cut of a fuzzy number A, where $\alpha \in [0,1]$, is defined by

$$A_\alpha = \begin{cases} \{x \in \mathbb{R} : A(x) \geqslant \alpha\} & \text{if } \alpha \in (0,1], \\ cl\{x \in \mathbb{R} : A(x) > 0\} & \text{if } \alpha = 0. \end{cases}$$

One can see easily that each α-cut $A(\alpha)$ of a fuzzy number A is a closed interval $A_\alpha = [A_L(\alpha), A_U(\alpha)]$. Thus each fuzzy number is completely characterized both by its membership function $A(x)$ and by a family of its α-cuts $\{A_\alpha\}_{\alpha\in[0,1]}$. Two α-cuts are of special interest: $A_0 = \text{supp}(A)$ called the **support** and $A_1 = \text{core}(A)$ known as the **core** of fuzzy number A, respectively.

Membership functions of fuzzy numbers may assume various shapes. However, the most often used fuzzy numbers are the **trapezoidal fuzzy numbers** (denoted further as TPFNs) with membership functions of the form

$$A(x) = \begin{cases} \frac{x-a_1}{a_2-a_1} & \text{if } a_1 < x \leqslant a_2, \\ 1 & \text{if } a_2 \leqslant x \leqslant a_3, \\ \frac{a_4-x}{a_4-a_3} & \text{if } a_3 \leqslant x < a_4, \\ 0 & \text{otherwise,} \end{cases} \tag{1}$$

where $a_1, a_2, a_3, a_4 \in \mathbb{R}$ such that $a_1 \leqslant a_2 \leqslant a_3 \leqslant a_4$. If $a_2 = a_3$ then A is said to be a **triangular fuzzy number**, while of $a_1 = a_2$ and $a_3 = a_4$ we have the so-called **rectangular fuzzy number** or **interval fuzzy number**.

If A is a fuzzy number (1) then four real values a_1, a_2, a_3, a_4 describe A completely. In particular, any α-cut is given by $A_\alpha = [a_1 + (a_2 - a_1)\cdot\alpha, a_4 - (a_4 - a_3)\cdot\alpha]$. Hence, $\text{supp}(A) = [a_1, a_4]$ and $\text{core}(A) = [a_2, a_3]$, which means that the borders of the support and core give the complete description of a trapezoidal fuzzy number.

However, sometimes another parametrization of a trapezoidal fuzzy number may be more useful. Let us define the following parameters

$$\begin{aligned} c &:= \frac{a_2 + a_3}{2}, \\ s &:= \frac{a_3 - a_2}{2}, \\ l &:= a_2 - a_1, \\ r &:= a_4 - a_3. \end{aligned} \tag{2}$$

Here c indicates the center of the core while s is equal to the core radius and hence describes the spread of the core. Moreover, l and r stand for the spread of the left and the right arm of the membership function $A(x)$, respectively. Obviously, $c \in \mathbb{R}$, while $s, l, r \geqslant 0$. Using this parametrization a trapezoidal fuzzy number A may be denoted as $A(c, s; l, r)$. Obviously, $A(c; l, r)$ stands for a triangular fuzzy number, since then $s = 0$.

To simplify an interpretation of fuzzy numbers Delgado et al. [10] suggested the so-called **canonical representation** of a fuzzy number, defined by only two main values which characterize two basic features of a fuzzy number: its general location and its global spread (or vagueness). The first characteristic responsible for the location of $A \in \mathbb{F}(\mathbb{R})$, called simply the **value**, is defined as follows

$$\text{Val}(A) = \int_0^1 [A_U(\alpha) + A_L(\alpha)]\cdot\alpha \, d\alpha, \tag{3}$$

whereas the second one, called the **ambiguity**, is given by

$$\text{Amb}(A) = \int_0^1 [A_U(\alpha) - A_L(\alpha)]\cdot\alpha \, d\alpha. \tag{4}$$

In many cases such a far-reaching reduction in the description of imprecise objects is fully enough. However, in some situations it would be useful to be able to say something about symmetry (or possible asymmetry) of a fuzzy number. One method to cope with it is to consider substitute the left-hand and right-hand ambiguity instead of (4). To be more specific let us define the **left-hand ambiguity** and **right-hand ambiguity** as follows

$$\mathrm{Amb}_L(A) = \int_0^1 [q - A_L(\alpha)] \cdot \alpha \, d\alpha, \tag{5}$$

$$\mathrm{Amb}_U(A) = \int_0^1 [A_U(\alpha) - q] \cdot \alpha \, d\alpha, \tag{6}$$

where $q = \frac{1}{2}[A_L(1) + A_U(1)]$ is the middle point of core(A).

It is easily seen that knowing both the left-hand and right-hand ambiguity allows to calculate (4). Indeed

$$\mathrm{Amb}(A) = \mathrm{Amb}_L(A) + \mathrm{Amb}_U(A).$$

But it seems that for non-symmetrical fuzzy numbers the three-element canonical representation $\big(\mathrm{Val}(A), \mathrm{Amb}_L(A), \mathrm{Amb}_U(A)\big)$ is more informative than the original two-element representation $\big(\mathrm{Val}(A), \mathrm{Amb}(A)\big)$.

In the case of trapezoidal fuzzy numbers by substituting (2) into (3) and (5)–(6) we obtain the following three-element canonical representation of a trapezoidal fuzzy number $A(c, s; l, r)$:

$$\mathrm{Val}(A) = c + \frac{r - l}{6}, \tag{7}$$

$$\mathrm{Amb}_L(A) = \frac{s}{2} + \frac{l}{6}, \tag{8}$$

$$\mathrm{Amb}_U(A) = \frac{s}{2} + \frac{r}{6}. \tag{9}$$

Let us also mention one more representation of fuzzy numbers, utilized in our experiments, given by the **expected value** of the fuzzy number [12,26] and its **width** [4] defined by

$$\mathrm{EV}(A) = \frac{1}{2}\left[\int_0^1 A_L(\alpha)\, d\alpha + \int_0^1 A_U(\alpha)\, d\alpha\right], \tag{10}$$

$$\mathrm{w}(A) = \int_0^1 (A_U(\alpha) - A_L(\alpha))\, d\alpha. \tag{11}$$

Given a trapezoidal fuzzy number $A(c, s; l, r)$ we obtain

$$\mathrm{EV}(A) = c + \frac{r - l}{4}, \tag{12}$$

$$\mathrm{w}(A) = s + \frac{r + l}{4}. \tag{13}$$

Finally, let us consider one more tool extremely useful in fuzzy data analysis, i.e. a distance between fuzzy numbers. This is not surprising because, obviously, the concept of distance is indispensable in many areas of life. However, in fuzzy environment it is exploited even more eagerly because of such problems like the lack of a satisfying subtraction operator between fuzzy numbers or the lack of a universally accepted total ranking between fuzzy numbers. Solutions used in these situations are based just on concept of distance (see, e.g., [3]).

Many distance functions have been proposed in $\mathbb{F}(\mathbb{R})$. A metric, perhaps most often used in statistical context, was proposed by Gil et al. [14] and by Trutschnig et al. [40]. Suppose λ is a normalized measure associated with a continuous distribution with support in $[0, 1]$ and let $\zeta > 0$ be a fixed constant. Then for any $A, B \in \mathbb{F}(\mathbb{R})$ we define a metric D_θ^λ as follows

$$D_\zeta^\lambda(A,B) = \Big(\int_0^1 \big[\big(\operatorname{mid} A(\alpha) - \operatorname{mid} B(\alpha) \big)^2 + \zeta \cdot \big(\operatorname{spr} A(\alpha) - \operatorname{spr} B(\alpha) \big)^2 \big] d\lambda(\alpha) \Big)^{1/2}, \tag{14}$$

where $\operatorname{mid} A(\alpha) = \frac{1}{2}(A_L(\alpha) + A_U(\alpha))$ and $\operatorname{spr} A(\alpha) = \frac{1}{2}(A_U(\alpha) - A_L(\alpha))$ denote the mid-point and radius of the α-cut $A(\alpha)$, respectively. Both λ and ζ are applied as some weighting: λ – to weight the influence of each α-cut, while ζ – to weight the impact of the distance between the mid-points of α-cuts in contrast to the distance between their spreads. In practice, the most common λ is the Lebesgue measure on $[0, 1]$, while the most popular choice of ζ is 1 or $\frac{1}{3}$.

For more details on fuzzy numbers, their characteristics, properties and approximations we refer the reader to [2].

3 Random Fuzzy Numbers

Fuzzy numbers discussed in previous section are effectively used for modeling imprecise outputs of experiments met in engineering, social sciences, economy, etc. However, to create the complete model of the process or phenomenon under study we also need some mathematical formalism to combine imprecision with coexisting randomness that appear almost always in the real-life. To cope with this problem Puri and Ralescu [31] introduced the concept of a **fuzzy random variable**, currently most often referred to as a **random fuzzy number**.

Definition 2. Given a probability space $(\Omega, \mathcal{A}, P)$, a mapping $X : \Omega \to \mathbb{F}(\mathbb{R})$ is said to be a **random fuzzy number** if for all $\alpha \in [0, 1]$ the α-level function is a compact random interval.

Alternatively, we say that X is a random fuzzy number if and only if X is a Borel measurable function w.r.t. the Borel σ-field generated by the topology induced by the metric (14).

To characterize location of a random fuzzy number Puri and Ralescu [31] defined the Aumann-type mean $\mathcal{E}(X) \in \mathbb{F}(\mathbb{R})$ such that for each $\alpha \in [0,1]$ the α-cut $\big(\mathcal{E}(X)\big)(\alpha)$ is equal to the Aumann integral of $X(\alpha)$, i.e.

$$\big(\mathcal{E}(X)\big)(\alpha) = \big[\mathbb{E}(X_L(\alpha)),\, \mathbb{E}X_U(\alpha)\big]. \tag{15}$$

Dispersion of a random fuzzy number X can be measured by the D_θ^λ-Fréchet-type variance $\mathcal{V}(X)$ defined as follows

$$\mathcal{V}(X) = \mathbb{E}\Big(\big[D_\theta^\lambda(X, \mathcal{E}(X))\big]^2\Big).$$

It is worth noting that so defined variance $\mathcal{V}(X)$ of a random fuzzy number X is a nonnegative real value contrary to the mean $\mathcal{E}(X)$ which is a fuzzy number.

Similarly, as in the crisp case, a vector $\mathbb{X} = (X_1, \ldots, X_n)$ of random fuzzy numbers will be called a fuzzy random sample. Usually, unless stated otherwise, we will assume that a fuzzy random sample consists of i.i.d. (i.e. independent and identically distributed) random fuzzy numbers.

Random fuzzy numbers preserve some properties similar to those known for the real-valued random variables. On the other hand, there are no suitable models for the distribution of random fuzzy numbers. Moreover, there are not yet Central Limit Theorems for random fuzzy numbers which can be applied directly in statistical reasoning. Fortunately, Efron's brilliant idea of the bootstrap method [13] appeared also very fruitful in statistical inference with fuzzy data and opened the door for generalizations of such statistical techniques like hypotheses tests or classifiers (see, e.g. [6,15–17,23,29,34]).

4 Resampling Algorithms

4.1 General Remarks

Suppose a fuzzy random sample $\mathbb{X} = (X_1, \ldots, X_n)$ stands for the mathematical model of the experiment under study and let $\mathbb{x} = (x_1, \ldots, x_n)$ denote its actual realization, i.e. each $x_i \in \mathbb{F}(\mathbb{R})$, where $i = 1, \ldots, n$ is an imprecise output of X_i. Further on this very sample, contrary to new samples generated with various bootstrap methods, will be called the **primary sample** or the **initial sample**. On the other hand, bootstrap samples generated from the primary sample will be called as **secondary samples**.

Following Efron's [13] idea the secondary samples are drawing randomly with replacement from the primary sample, as it is shown in Fig. 1. Since the bootstrap is very generous we may generate as many secondary samples as we need, say b bootstrap samples, and then the i-th element of the j-th secondary sample is denoted by x^*_{ij}. Obviously, $x^*_{ij} \in \{x_1, \ldots, x_n\}$ for any $i = 1, \ldots, n$ and $j = 1, \ldots, b$.

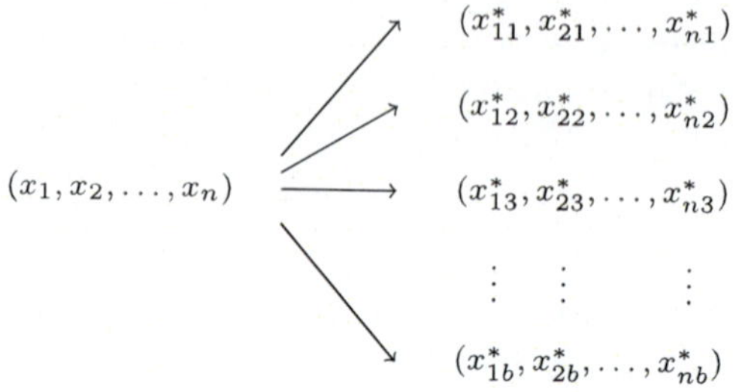

Fig. 1. A typical bootstrap scheme.

Since secondary samples are drawing with replacement from the primary one, most of the bootstrap samples contain repeated values. Moreover, if the size of the primary sample is small then all secondary samples consist of only a few distinct values. This is a serious disadvantage especially if the unknown population distribution is continuous. To deal with this trouble several modifications of the standard Efron's idea were proposed, e.g., the balanced bootstrap by Davison et al. [8] or Graham et al. [18] or the so-called smoothed bootstrap by Silverman and Young [38], Hall et al. [25] or De Angelis and Young [9].

Most of the papers where bootstrap was introduced to statistics with fuzzy data (like [15,17,29]) suffer from the same inconvenience of repetitive observations. This is the reason why research has been undertaken (see, [20–22,35,36]) into more flexible bootstrap methods for fuzzy data that would prevent duplication of observations. In this context, two main approaches can be found.

4.2 Bootstrap on Characteristic Points

The first one is restricted to trapezoidal fuzzy numbers. We start from the decomposition of each $x_i \in \mathbb{F}(\mathbb{R})$ into its "characteristic points". Following (2) we consider the left bound of the core $lc_i = c_i - s_i$, the core width $wc_i = 2s_i$, and the spreads l_i and r_i of the left and right arm, respectively. Thus given a primary sample $x_1, \ldots, x_n$ we obtain the following four sets: $\mathbf{LC} = \{lc_1, \ldots, lc_n,\}$, $\mathbf{WC} = \{wc_1, \ldots, wc_n,\}$, $\mathbf{L} = \{l_1, \ldots, l_n,\}$, $\mathbf{R} = \{r_1, \ldots, r_n,\}$.

Next, following the **d-method** [35,36] we draw randomly four numbers lc^*, wc^*, l^*, r^* from the respective sets $\mathbf{LC}, \mathbf{WC}, \mathbf{L}, \mathbf{R}$ which constitute a bootstrapped element $x^* \in \mathbb{F}(\mathbb{R})$. The procedure of drawing such fours independently with replacement is repeated n times to obtain a secondary sample $x_1^*, \ldots, x_n^*$.

This approach can be also applied to generate bootstrap fuzzy samples of interval-valued fuzzy numbers (see [37]). Moreover, it might be modified into the so-called **w-method** working as follows. Let $y_1 < y_2 < \ldots < y_m$ be a strictly increasing sequence of m values and let

$$we(x) = \frac{1}{2m}\delta_x(y_1) + \frac{1}{m}we_{1,2}(x) + \ldots + \frac{1}{m}we_{m-1,m}(x) + \frac{1}{2m}\delta_x(y_m),$$

where $\delta_x(.)$ is the Dirac measure and

$$we_{l-1,l}(x) = \frac{1}{y_l - y_{l-1}} \cdot \mathbb{1}(x \in [y_{l-1}, y_l]).$$

This way we obtain a mixed discrete-uniform distribution $we(x)$ which is applied in the w-method, contrary to the d-method, where the distribution with equal probabilities on a discrete set is considered (see [35,36]). Since each of the sets $\mathbf{LC}, \mathbf{WC}, \mathbf{L}, \mathbf{R}$ is ordered, we can generate lc^*, wc^*, l^*, r^* with respect to the density $w(x)$ to construct x_i^*. As noted in [36] by drawing elements with respect to $w(x)$ the secondary samples are more diversified than those obtained with the d-method. Indeed, now we can obtain values from the whole interval $[y_1, y_m]$.

4.3 Methods Preserving the Canonical Representation

The second approach is typical of several so-called "flexible" bootstrap methods (see [20–22]) aimed to increase the diversity of simulated fuzzy outcomes. Their key idea is to generate fuzzy numbers which possibly differ from the original ones but preserve the canonical representation [10] of fuzzy observations (i.e. the value and the ambiguity) or some other critical characteristics, like the expected value and the width of fuzzy numbers which belong to the primary sample.

This group of methods is not restricted to any particular type of fuzzy numbers in the primary sample. However, the secondary samples obtained as their outputs consist of trapezoidal fuzzy numbers.

One may ask whether such outputs are not too restrictive in general. However, many researchers noticed that trapezoidal fuzzy numbers are most common in applications since they are easy to handle and have a natural interpretation [2,30]. Moreover, even if the original data set consists of fuzzy numbers which are not trapezoidal, one often approximates them by such fuzzy numbers before further processing. A collection of such approximation algorithms can be found in [2]. Moreover, an approximation algorithm that preserves the value and the ambiguity of the original fuzzy number is given in [1], while several algorithms for trapezoidal approximations of fuzzy numbers preserving the expected interval are proposed in [19].

Given a primary sample $x_1, \ldots, x_n \in \mathbb{F}(\mathbb{R})$ which is a realization of a fuzzy random sample $X_1, \ldots, X_n$, we compute the value (8) and the ambiguity (9) of each observation which yields in the following set of pairs

$$\{(\mathrm{Val}(x_1), \mathrm{Amb}(x_1)), \ldots, (\mathrm{Val}(x_n), \mathrm{Amb}(x_n))\}. \tag{16}$$

The main idea of the **VA-method** is to create a secondary sample by generating randomly pairs from (16). Obviously, the value and the ambiguity characterize nicely a fuzzy number but do not identify it completely. Thus by imposing some restrictions on the generated fuzzy numbers we have still some room for the choice of their membership functions.

After drawing randomly a pair from (16) and denoting it by $(\mathrm{Val}^*, \mathrm{Amb}^*)$ we generate two random values s^* and c^* from the uniform distributions on some intervals (restricted by some constrains on (Val*, Amb*)) and next we calculate l^* and r^* by following formulas

$$\begin{aligned} s_i^* &\sim U\big[0, \mathrm{Amb}_i^*\big], \\ c_i^* &\sim U\big[\mathrm{Val}_i^* - \mathrm{Amb}_i^* + s_i^*, \mathrm{Val}_i^* + \mathrm{Amb}_i^* - s_i^*\big], \\ l_i^* &:= 3\big[\mathrm{Amb}_i^* - \mathrm{Val}_j^* + c_i^* - s_i^*\big], \\ r_i^* &:= 3\big[\mathrm{Amb}_i^* + \mathrm{Val}_j^* - c_i^* - s_i^*\big]. \end{aligned} \tag{17}$$

The **EW-method** (see [22]) resembles the VA-method but, instead of the value and the ambiguity, the expected value and the width of a fuzzy number are preserved now. Thus, given a primary sample $x_1, \ldots, x_n \in \mathbb{F}(\mathbb{R})$ we determine the expected value (12) and the width (13) of each fuzzy observation

$$\big\{\big(\mathrm{EV}(x_1), \mathrm{w}(x_1)\big), \ldots, \big(\mathrm{EV}(x_n), \mathrm{w}(x_n)\big)\big\}. \tag{18}$$

Then, after drawing randomly a pair (EV*, w*) from (18) we calculate

$$\begin{aligned} s_i^* &\sim U\big[0, \mathrm{w}^*\big], \\ c_i^* &\sim U\big[\mathrm{EV}^* - \mathrm{w}^* + s_i^*, \mathrm{EV}^* + \mathrm{w}^* - s_i^*\big], \\ l_i^* &:= 2\big[\mathrm{w}^* - \mathrm{EV}^* + c_i^* - s_i^*\big], \\ r_i^* &:= 2\big[\mathrm{w}^* + \mathrm{EV}^* - c_i^* - s_i^*\big]. \end{aligned} \tag{19}$$

for each $x_i^* = x_i^*(c_i^*, s_i^*; l_i^*, r_i^*)$, $i = 1, \ldots, n$, member of the secondary sample $x_1^*, \ldots, x_n^*$.

A general idea of the **VAF-method** (see [21]) is similar to the VA-method but now three parameters (the value, the ambiguity and the fuzziness) are calculated for each element x_i of the primary sample. Then a triple (Val*, Amb*, Fuzz*) is drawn randomly from the respective set

$$\big\{\big(\mathrm{Val}(x_1), \mathrm{Amb}(x_1), \mathrm{Fuzz}(x_1)\big), \ldots, \big(\mathrm{Val}(x_n), \mathrm{Amb}(x_n), \mathrm{Fuzz}(x_n)\big)\big\}. \tag{20}$$

Next, we reconstruct a trapezoidal fuzzy number $x_i^* = x_i^*(c_i^*, s_i^*; l_i^*, r_i^*)$, for each $i = 1, \ldots, n$, from the following formulas

$$\begin{aligned} s_i^* &:= \mathrm{Amb}_i^* - \tfrac{2}{3}\,\mathrm{Fuzz}_i^*, \\ c_i^* &\sim U\big[\mathrm{Val}_i^* - \tfrac{2}{3}\,\mathrm{Fuzz}_i^*, \mathrm{Val}_i^* + \tfrac{2}{3}\,\mathrm{Fuzz}_i^*\big], \\ l_i^* &:= 3\big[\mathrm{Amb}_i^* - \mathrm{Val}_i^* + c_i^* - s_i^*\big], \\ r_i^* &:= 3\big[\mathrm{Amb}_i^* + \mathrm{Val}_i^* - c_i^* - s_i^*\big]. \end{aligned} \tag{21}$$

Note that in the VAF-method we have less "degrees of freedom" in generating bootstrap samples since only s^* is now generated from the uniform distribution on some interval while c^*, l^* and r^* are directly calculated from (21). Actually, it is not surprising because here the extended three-parameter canonical representation of a fuzzy number is preserved. For more details on the VAF-method, we refer the reader to [21].

Below we propose a new method that actually belongs to the discussed second approach and hence should be placed in this very section. However, we put it in a separate section not only because we want to emphasize its novelty, but since it will be completed with the derivation of necessary formulas and with an algorithm ready for practical use.

4.4 VAA Algorithm

Given a primary sample $x_1, \ldots, x_n \in \mathbb{F}(\mathbb{R})$ let us compute the value of each fuzzy number (3), its left-hand ambiguity (5) and its right-hand ambiguity (6). This way we obtain the following set

$$\big\{(\mathrm{Val}(x_1), \mathrm{Amb}_L(x_1), \mathrm{Amb}_U(x_1)), \ldots, \\ (\mathrm{Val}(x_n), \mathrm{Amb}_L(x_n), \mathrm{Amb}_U(x_n))\big\}. \quad (22)$$

The main idea of the proposed new bootstrap technique is to create a secondary sample by generating randomly fuzzy observations not necessarily identical to those from the primary sample but preserving some of their crucial characteristics which deliver information on the primary location and possible asymmetry of the initial observations. Obviously, these three parameters do not identify a fuzzy number completely but leave some room on the generated membership functions of fuzzy numbers from the secondary sample.

Suppose our secondary sample $x_1^*, \ldots, x_n^*$ consists of trapezoidal fuzzy numbers $x_i^* = x_i^*(c_i^*, s_i^*; l_i^*, r_i^*)$, $i = 1, \ldots, n$. Thus, after drawing randomly a triple $(\mathrm{Val}^*, \mathrm{Amb}_L^*, \mathrm{Amb}_U^*)$ from (22) we can try to reconstruct values of the parameters $c_i^*, s_i^*, l_i^*, r_i^*$ characterizing x_i^*. Keeping in mind all necessary formulas related to this parametrization, i.e. (2), and combining them with equations which define the value, the left-hand ambiguity and the right-hand ambiguity, i.e. (7), (8) and (9), respectively, we obtain

$$\begin{aligned} 6c + r - l &= 6\,\mathrm{Val}^*, \\ 3s + l &= 6\,\mathrm{Amb}_L^*, \\ 3s + r &= 6\,\mathrm{Amb}_U^*, \end{aligned}$$

where $s, r, l \geqslant 0$, which is equivalent to

$$\begin{aligned} c &= \mathrm{Val}^* + \mathrm{Amb}_U^* - \mathrm{Amb}_U^L, \\ l &= 6\,\mathrm{Amb}_L^* - 3s, \\ r &= 6\,\mathrm{Amb}_U^* - 3s. \end{aligned} \quad (23)$$

Since $r, l \geqslant 0$ we have the following constrains: $s \leqslant 2\mathrm{Amb}_L^*$ and $s \leqslant 2\mathrm{Amb}_U^*$. Keeping in mind that $s \geqslant 0$ we finally obtain

$$0 \leqslant s \leqslant 2 \min\left\{\mathrm{Amb}_L^*, \mathrm{Amb}_U^*\right\}. \quad (24)$$

Summing up the aforementioned considerations as well as restrictions specified in (23) and (24), we obtain the following bootstrap algorithm for generating a secondary sample of trapezoidal fuzzy numbers.

Algorithm 1

Require: Fuzzy sample $x_1, \ldots, x_n \in \mathbb{F}(\mathbb{R})$

Ensure: A flexible bootstrap sample

for $i = 1$ to n **do**

  Compute $\mathrm{Val}(x_i), \mathrm{Amb}_L(x_i), \mathrm{Amb}_U(x_i)$

end for

for $i = 1$ to n **do**

  Generate randomly (with equal probabilities) a triple $(\mathrm{Val}^*, \mathrm{Amb}_L^*, \mathrm{Amb}_U^*)$ from

$$\{(\mathrm{Val}(x_1), \mathrm{Amb}_L(x_1), \mathrm{Amb}_U(x_1)), \ldots, (\mathrm{Val}(x_n), \mathrm{Amb}_L(x_n), \mathrm{Amb}_U(x_n))\}$$

  $c^* \leftarrow \mathrm{Val}^* + \mathrm{Amb}_U^* - \mathrm{Amb}_L^*$

  Generate s^* from the uniform distribution on the interval

$$[0, 2\min\{\mathrm{Amb}_L^*, \mathrm{Amb}_U^*\}]$$

  $l^* \leftarrow 6\mathrm{Amb}_L^* - 3s^*$

  $r^* \leftarrow 6\mathrm{Amb}_U^* - 3s^*$

  $x_i^* \leftarrow x_i^*(c^*, s^*; l^*, r^*)$

end for

Using Algorithm 1 we receive the bootstrap secondary sample $x_1^*, \ldots, x_n^*$ of fuzzy trapezoidal numbers, where $x_i^* = x_i^*(c_i^*, s_i^*; l_i^*, r_i^*)$, $i = 1, \ldots, n$. Usually, we repeat this algorithm as many times as required for the particular application, e.g. about 200 times for computing the standard error but not less than 1000 times in the case of a confidence interval determination or in hypothesis testing.

5 Case Study

Resampling techniques are commonly used to solve practical problems (see, e.g., [27,32,41]). Therefore, we start a comparison of the introduced bootstrap algorithms for fuzzy data with the Gamonedo cheese (a kind of blue cheese produced in Asturias, Spain) quality assessment. Some of the testers expressed their subjective opinions on the cheese quality using trapezoidal fuzzy numbers (TPFNs). Data containing their overall impressions about the cheese are given in [32,34].

Suppose we want to check if there is a significant difference between typical opinions of the two experts (i.e. Expert A and C in [32,34]). Thus we consider the following hypotheses testing problem

$$H_0 : \mathcal{E}(X) = \mathcal{E}(Y) \quad \text{vs.} \quad H_1 : \mathcal{E}(X) \neq \mathcal{E}(Y), \tag{25}$$

where $\mathcal{E}(X), \mathcal{E}(Y)$ correspond to the Aumman-type means (15) of the first and second population, respectively. We apply the statistical test for fuzzy data (denoted further on as the C2-test) discussed in Colubi [5] and Lubiano et al. [27].

All of the resampling methods discussed in this contribution as well as the classical bootstrap were used to estimate p-values for the various number of bootstrap replications (i.e. b equal to 100, 200 or 1000). To eliminate random fluctuations each simulation run was repeated 10000 times and the final p-value was calculated by averaging the outputs. The results are given in Table 1. Assuming the typical significant level (e.g. 0.05) the null hypothesis was never rejected. However, the classical bootstrap gave the most restrictive results (i.e. the respective p-values are the lowest), the d-method and w-method were the second ones, while the VA-method appeared as the most "flexible" one (i.e. its p-values are the biggest ones).

Table 1. Empirical C2-test size $\hat{\alpha}$ for real-life data.

b	100	200	1000
VA	0.9373	0.9371	0.9370
EW	0.9124	0.912	0.9121
d	0.8374	0.8372	0.8371
Bootstrap	0.8274	0.8275	0.8276
VAF	0.8629	0.8624	0.8630
w	0.8369	0.8364	0.8368
VAA	0.8931	0.8925	0.8959

Next we compared the resampling methods using four different primary samples with the small number of specially prepared TPFNs: 10 TPFNs with longer right spreads (Set 1), 5 TPFNs with longer right spreads and 5 TPFNs with longer left spreads (Set 2), 5 TPFNs with longer right spreads, the same number of TPFNs with longer left spreads and symmetrical ones (Set 3) and 10 TPFNs with longer right spreads (Set 4). To verify hypotheses (25) a special procedure was applied in which the initial set was compared with the same one under increasing shift ϵ (see also [22]).

The empirical C2-test sizes (i.e. a percentage of the rejections obtained in simulations when H_0 holds) for different ϵ obtained for Set 1 are presented in Table 2. As in Sect. 5 the number of the bootstrap replications b was set to 100, 200 or 1000 and then each simulation run was repeated 10000 times. In general, for small shift (like $\epsilon = 0.5$) the obtained empirical test sizes are very similar, although for the classical bootstrap they are smallest (similarly for the d- and the w-methods), while for the VAA-method they are biggest. On the contrary, for larger shifts (like $\epsilon = 0.75$ or $\epsilon = 0.8$) the empirical sizes are the biggest for the classical bootstrap and the lowest for the VA- and the EW-methods.

Table 2. Empirical C2-test sizes.

$\epsilon = 0.5$			
b	100	200	1000
VA	0.2559	0.2557	0.2557
EW	0.2560	0.2565	0.2567
d	0.2482	0.2481	0.2483
Bootstrap	0.2487	0.2489	0.2489
VAF	0.2518	0.2518	0.2516
w	0.2484	0.2477	0.2481
VAA	0.2629	0.2621	0.2624
$\epsilon = 0.75$			
b	100	200	1000
VA	0.0431	0.0430	0.0430
EW	0.0366	0.0363	0.0362
d	0.0793	0.0789	0.0791
Bootstrap	0.0868	0.0872	0.0870
VAF	0.0645	0.0648	0.0647
w	0.0789	0.0790	0.0794
VAA	0.0565	0.0565	0.0566
$\epsilon = 0.8$			
b	100	200	1000
VA	0.0286	0.0284	0.0284
EW	0.0231	0.0228	0.0228
d	0.0618	0.0616	0.0617
Bootstrap	0.0691	0.0693	0.0693
VAF	0.0478	0.0483	0.0480
w	0.0615	0.0617	0.0619
VAA	0.0397	0.0396	0.0396

6 Simulation Study

6.1 Applied Models of Fuzzy Numbers

Now we turn to a more detailed comparison of the considered resampling methods based on primary samples of various types of trapezoidal fuzzy numbers (TPFNs). To obtain random fuzzy samples we generated independent tuples consisting of four reals corresponding to (c, s, l, r) (see, e.g., [21,39]). Each element of a tuple was generated independently from the specified distribution. Therefore, choosing a distribution together with its parameters we obtain various types of fuzzy numbers, summarized in Table 3. Most of them are widely used in the literature (see, e.g., [20–22,28,35,36]) to model the primary fuzzy samples in numerical simulations.

Table 3. Types of the simulated fuzzy numbers.

FN's type	c	s	l	r
$\mathbb{F}^T_{\mathrm{N,E,U,U}}$	N(0, 4)	Exp(4)	U(0, 0.2)	U(0, 0.6)
$\mathbb{F}^T_{\Gamma,\mathrm{U,E,E}}$	$\Gamma(2,2)$	U(0, 0.2)	Exp(4)	Exp(2)
$\mathbb{F}^T_{\mathrm{B,Ucon}}$	$\beta(1,1)$	U (conditional)		
$\mathbb{F}^T_{\mathrm{B,Ucon'}}$	$\beta(5,5)$	U (conditional)		

The descriptions of the simulated fuzzy number types given in Table 3 are straightforward. For instance, $\mathbb{F}^T_{\mathrm{N,E,U,U}}$ stands for a trapezoidal fuzzy number with its center generated from the normal distribution (with the mean 0 and the standard deviation 4), the radius of its core is generated from the exponential distribution (with the parameter 4) and spreads of the left and the right arms generated by two independent uniform distributions (on intervals $(0, 0.2)$ and $0, 0.6$, respectively). The last types, $\mathbb{F}^T_{\mathrm{B,Ucon}}$ and $\mathbb{F}^T_{\mathrm{B,Ucon'}}$, represent a more complex fuzzy number related to the beta distribution describing centers and a few different conditional uniform distributions for s, l, r (see [28] for the details).

6.2 Standard Error Estimation

Resampling methods can be compared using the standard error estimation, which is a widely known problem in the literature. Let $X_1, \ldots, X_n$ denote a random sample from the distribution F_θ, where $\theta \in \Theta$ is an unknown parameter, $\widehat{\theta}$ stands for its estimator and $\mathrm{SE}(\theta) = \sqrt{\mathrm{Var}\widehat{\theta}}$ is the standard error of $\widehat{\theta}$. In many situations $\mathrm{SE}(\theta)$ cannot be calculated directly and the bootstrap is used to overcome the problem.

Let us estimate the standard error (related to the Fréchet-type variance (16)) of $\widehat{\theta}$ as follows

$$\widehat{\mathrm{SE}}(\widehat{\theta}) = \sqrt{\frac{1}{b-1}\sum_{i=1}^{b}\left[D^{\lambda}_{\zeta}(\widehat{\theta}^{*i}, \bar{\theta}^{*})\right]^2}, \tag{26}$$

where D^{λ}_{ζ} is the distance (14) with $\zeta = 1$ and λ denoting the Lebesgue measure and where $\widehat{\theta}^{*i}$ is the estimator based on the i-th bootstrap replication, and $\bar{\theta}^{*} = \sum_{i=1}^{b} \widehat{\theta}^{*i}$. In the following we apply $g(\theta) = \mathcal{E}(X)$, so $\widehat{\theta}^{*i}$ is given by the averaged fuzzy number for the i-th bootstrap replication.

We compared all considered resampling methods and the classical bootstrap random fuzzy samples of different sizes ($n = 5, 10, 30, 100$), numbers of the bootstrap replications ($b = 100, 200$), and the types of TPFNs were generated. To minimize random fluctuations each experiment was repeated 100000 times. Some results related to $\mathbb{F}^T_{\mathrm{N,E,U,U}}$ model can be found in Table 4, where the lowest values of the simulated standard errors are marked in boldface. As it is seen the w-method is usually the best approach for small sample sizes, while the VAF-method is the best for moderate and big samples. In general (when considering various models), the VAF-method and the EW-method behave in a rather stable manner, while the behavior of the VAA-method depends strongly on the type of data.

Table 4. Empirical standard errors for $\mathbb{F}^T_{\mathrm{N,E,U,U}}$.

n	5	10	30	100
b	100			
VA	0.7676	0.5943	0.3619	0.2015
EW	0.7609	0.5895	0.3588	0.2002
d	0.7709	0.5975	0.3641	0.2027
Bootstrap	0.7600	0.5887	0.3583	0.1997
VAF	0.7589	0.5878	**0.3579**	**0.1996**
w	**0.7326**	**0.5853**	0.3621	0.2025
VAA	0.7645	0.5936	0.3610	0.2012
b	200			
VA	0.7693	0.5949	0.362448	0.2017
EW	0.7608	0.5890	0.359303	0.2004
d	0.7741	0.5969	0.36426	0.2029
Bootstrap	0.7592	0.5889	0.35863	0.1999
VAF	0.7579	0.5886	**0.3586**	**0.1999**
w	**0.7320**	**0.5856**	0.3627	0.2029
VAA	0.7662	0.5931	0.3614	0.2029

6.3 Empirical Size of a Test

We also compared the considered resampling methods using another benchmark, i.e. a difference between the empirical test size and the nominal significance level (usually 5% significance was assumed). In this study, we focused our attention on the C2-test. Small and medium initial sample sizes ($n = 5, 10, 30, 100$) and the different numbers of bootstrap replications ($b = 100, 200$) were checked. As previously, the whole resampling procedure was iterated 100000 times for each numerical experiment.

Some results (for $\mathbb{F}^T_{\mathrm{N,E,U,U}}$ model) are presented in Table 5. In each case, the empirical test size closest to the nominal value of 0.05 is marked in boldface. Again the w-method is usually the best method; the VAF-method and the VA-method behave also nice; the EW-method behaves in a relatively stable manner.

Table 5. Empirical C2-test size $\hat{\alpha}$ for $\mathbb{F}^T_{\mathrm{N,E,U,U}}$.

n	5	10	30	100
b	100			
VA	**0.0509**	0.0594	0.0611	0.0627
EW	0.0466	0.0568	0.0590	0.0612
d	0.0438	0.0559	0.0585	0.0597
Bootstrap	0.0427	0.0559	0.0587	0.0593
VAF	0.0437	**0.0552**	0.0594	**0.0595**
w	0.0440	0.0556	**0.0576**	0.0611
VAA	0.0478	0.0576	0.0615	0.0609
b	200			
VA	**0.0455**	0.0537	0.0558	0.0581
EW	0.0413	0.0508	0.0535	0.0564
d	0.0392	0.0508	0.0547	0.0557
Bootstrap	0.0387	0.0512	**0.0542**	0.0552
VAF	0.0387	0.0512	0.05477	0.0547
w	0.0395	**0.0501**	0.0557	**0.054**
VAA	0.0423	0.05335	0.0565	0.0609

6.4 Power Analysis

To deepen our analysis, we provide a power study of the C2-test using different primary samples. We considered the null hypothesis H_0 as in (25) and examined the number of rejections under increasing shift $\epsilon = 0.05, 0.1, 0.15, 0.2, 0.25, 0.5, 075$ in the second primary fuzzy sample. Both samples of X and Y were generated using TPFNs types described in Table 3, assuming different sample sizes n and bootstrap repetitions b. Each experiment was repeated 100000 times.

Some results (for $\mathbb{F}^T_{\mathrm{N,E,U,U}}$ model) and $b = 100$ are shown in Table 6, where the highest power is marked in boldface. Here the VA-method was the overall winner with the highest percentage of rejections in all testing conditions. In general (i.e. including other models) this method was usually one of the best, especially for smaller sample sizes or a lower number of bootstrap replications (e.g. $b = 100$). The VAA-method or the EW-method were usually just after VA-method; the d-method, w- and the VAA-method were neither too good nor too bad, but the classical bootstrap (as compared with other resampling methods) was usually the worse method in hypotheses testing. These conclusions support the results from Sect. 5.

Table 6. C2-test power analysis for $\mathbb{F}^{T}_{\mathrm{N,E,U,U}}$.

n	5	10	30	100
ϵ	0.05			
VA	**0.0512**	**0.0593**	**0.0626**	**0.0667**
EW	0.0469	0.0569	0.0601	0.0649
d	0.0437	0.0561	0.0594	0.0636
Bootstrap	0.0428	0.0561	0.0598	0.0631
VAF	0.0439	0.0554	0.0603	0.0626
w	0.0444	0.0560	0.0594	0.0648
VAA	0.0475	0.0579	0.0617	0.0642
ϵ	0.1			
VA	**0.0515**	**0.0603**	**0.0658**	**0.0786**
EW	0.0470	0.0576	0.0639	0.0765
d	0.0439	0.0569	0.0631	0.0756
Bootstrap	0.0432	0.0573	0.0626	0.0746
VAF	0.0440	0.0564	0.0629	0.0735
w	0.0449	0.0569	0.0625	0.0758
VAA	0.0478	0.0585	0.0649	0.0758
ϵ	0.15			
VA	**0.0519**	**0.0619**	**0.0712**	**0.0980**
EW	0.0479	0.0592	0.0695	0.0957
d	0.0442	0.0582	0.0692	0.0943
Bootstrap	0.0436	0.0589	0.0679	0.0929
VAF	0.0444	0.0581	0.0683	0.0936
w	0.0456	0.0587	0.0688	0.0955
VAA	0.0482	0.0606	0.0706	0.0957
ϵ	0.2			
VA	**0.0527**	**0.0642**	**0.0796**	**0.1259**
EW	0.0485	0.0613	0.0772	0.1236
d	0.0453	0.0602	0.0764	0.1226
Bootstrap	0.0443	0.0612	0.0748	0.1208
VAF	0.0455	0.0607	0.0760	0.1213
w	0.0465	0.0611	0.0768	0.1226
VAA	0.0491	0.0633	0.0780	0.1242

7 Conclusions

For over 40 years the bootstrap is effectively applied in many fields of statistical analyses wherever traditional methods and techniques fail. Recently, the bootstrap is also used in fuzzy data analysis. A review of the bootstrap methods provided in this paper may help the users in choosing the optimal bootstrap method to suit their needs.

The reader certainly noticed that in this contribution we limited ourselves only to the analysis of the ontic fuzzy data (see [7]). Actually, this type of fuzzy data dominates in the statistical literature. However, the epistemic fuzzy data are also very common in applications and hence appropriately adapted bootstrap methods are of interest. To handle such observations a new bootstrap technique was proposed in [24]. The results given there, although preliminary, indicate that the epistemic bootstrap might be useful in different situations which is a good prognostic for further research.

References

1. Ban, A.I., Brândaş, A., Coroianu, L., Negruţiu, C., Nica, O.: Approximations of fuzzy numbers by trapezoidal fuzzy numbers preserving the ambiguity and value. Comput. Math. Appl. **61**, 1379–1401 (2011)
2. Ban, A.I., Coroianu, L., Grzegorzewski, P.: Fuzzy Numbers: Approximations, Ranking and Applications. Polish Academy of Sciences, Warsaw (2015)
3. Blanco-Fernández, A., et al.: A distance-based statistic analysis of fuzzy number-valued data (with rejoinder). Int. J. Approx. Reason. **55**(1487–1501), 1601–1605 (2014)
4. Chanas, S.: On the interval approximation of a fuzzy number. Fuzzy Sets Syst. **122**, 353–356 (2001)
5. Colubi, A.: Statistical inference about the means of fuzzy random variables: applications to the analysis of fuzzy-and real-valued data. Fuzzy Sets Syst. **160**(3), 344–356 (2009)
6. Colubi, A., Fernández-García, C., Gil, M.A.: Simulation of random fuzzy variables: an empirical approach to statistical/probabilistic studies with fuzzy experimental data. IEEE Trans. Fuzzy Syst. **10**(3), 384–390 (2002)
7. Couso, I., Dubois, D.: Statistical reasoning with set-valued information: ontic vs. epistemic views. Int. J. Approx. Reason. **55**, 1502–1518 (2014)
8. Davison, A.C., Hinkley, D.V., Schechtman, E.: Efficient bootstrap simulation. Biometrika **73**, 555–566 (1986)
9. De Angelis, D., Young, G.A.: Smoothing the bootstrap. Int. Stat. Rev. **60**, 45–56 (1992)
10. Delgado, M., Vila, M.A., Voxman, W.: On a canonical representation of a fuzzy number. Fuzzy Sets Syst. **93**, 125–135 (1998)
11. Dubois, D., Prade, H.: Operations on fuzzy numbers. Int. J. Syst. Sci. **9**, 613–626 (1978)
12. Dubois, D., Prade, H.: The mean value of a fuzzy number. Fuzzy Sets Syst. **24**, 279–300 (1987)

13. Efron, B.: Bootstrap methods: Another look at the jackknife. Ann. Stat. **7**, 1–26 (1979)
14. Gil, M.A., Lubiano, M.A., Montenegro, M., López, M.T.: Least squares fitting of an affine function and strength of association for interval-valued data. Metrika **56**, 97–111 (2002). https://doi.org/10.1007/s001840100160
15. Gil, M.A., Montenegro, M., González-Rodríguez, G., Colubi, A., Casals, M.R.: Bootstrap approach to the multi-sample test of means with imprecise data. Comput. Stat. Data Anal. **51**, 148–162 (2006)
16. González-Rodríguez, G., Colubi, A., Trutschnig, W.: Simulation of fuzzy random variables. Inf. Sci. **179**, 642–653 (2009)
17. González-Rodríguez, G., Montenegro, M., Colubi, A., Gil, M.A.: Bootstrap techniques and fuzzy random variables: synergy in hypothesis testing with fuzzy data. Fuzzy Sets Syst. **157**(19), 2608–2613 (2006)
18. Graham, R.L., Hinkley, D.V., John, P.W.M., Shi, S.: Balanced design of bootstrap simulations. J. Roy. Stat. Soc. Ser. B **52**, 185–202 (1990)
19. Grzegorzewski, P.: Trapezoidal approximations of fuzzy numbers preserving the expected interval - algorithms and properties. Fuzzy Sets Syst. **159**, 1354–1364 (2008)
20. Grzegorzewski, P., Hryniewicz, O., Romaniuk, M.: Flexible bootstrap based on the canonical representation of fuzzy numbers. In: Proceedings of EUSFLAT 2019. Atlantis Press (2019)
21. Grzegorzewski, P., Hryniewicz, O., Romaniuk, M.: Flexible bootstrap for fuzzy data based on the canonical representation. Int. J. Comput. Intell. Syst. **13**, 1650–1662 (2020)
22. Grzegorzewski, P., Hryniewicz, O., Romaniuk, M.: Flexible resampling for fuzzy data. Int. J. Appl. Math. Comput. Sci. **30**, 281–297 (2020)
23. Grzegorzewski, P., Ramos-Guajardo, A.B.: Similarity based one-sided tests for the expected value and interval data. In: Alonso, J.M., Bustince, H., Reformat, M. (eds.) Proceedings of the 2015 Conference of the International Fuzzy Systems Association and the European Society for Fuzzy Logic and Technology, pp. 960–966. Atlantis Press (2015)
24. Grzegorzewski, P., Romaniuk, M.: Epistemic bootstrap for fuzzy data (2021, Submitted)
25. Hall, P., DiCiccio, T.J., Romano, J.P.: On smoothing and the bootstrap. Ann. Stat. **17**, 692–704 (1989)
26. Heilpern, S.: The expected value of a fuzzy number. Fuzzy Sets Syst. **47**, 81–86 (1992)
27. Lubiano, M.A., Montenegro, M., Sinova, B., de Sáa, S.R., Gil, M.A.: Hypothesis testing for means in connection with fuzzy rating scale-based data: algorithms and applications. Eur. J. Oper. Res. **251**, 918–929 (2016)
28. Lubiano, M.A., Salas, A., Carleos, C., de Sáa, S.R., Gil, M.A.: Hypothesis testing-based comparative analysis between rating scales for intrinsically imprecise data. Int. J. Approx. Reason. **88**, 128–147 (2017)
29. Montenegro, M., Colubi, A., Casals, M.R., Gil, M.A.: Asymptotic and bootstrap techniques for testing the expected value of a fuzzy random variable. Metrika **59**(1), 31–49 (2004). https://doi.org/10.1007/s001840300270
30. Pedrycz, W.: Why triangular membership functions? Fuzzy Sets Syst. **64**(1), 21–30 (1994)
31. Puri, M.L., Ralescu, D.A.: Fuzzy random variables. J. Math. Anal. Appl. **114**, 409–422 (1986)

32. Ramos-Guajardo, A.B., Blanco-Fernández, Á., González-Rodríguez, G.: Applying statistical methods with imprecise data to quality control in cheese manufacturing. In: Grzegorzewski, P., Kochanski, A., Kacprzyk, J. (eds.) Soft Modeling in Industrial Manufacturing. SSDC, vol. 183, pp. 127–147. Springer, Cham (2019). https://doi.org/10.1007/978-3-030-03201-2_8
33. Ramos-Guajardo, A.B., Grzegorzewski, P.: Distance-based linear discriminant analysis for interval-valued data. Inf. Sci. **372**, 591–607 (2016)
34. Ramos-Guajardo, A.B., Lubiano, M.A.: K-sample tests for equality of variances of random fuzzy sets. Comput. Stat. Data Anal. **56**(4), 956–966 (2012)
35. Romaniuk, M.: On some applications of simulations in estimation of maintenance costs and in statistical tests for fuzzy settings. In: Steland, A., Rafajłowicz, E., Okhrin, O. (eds.) SMSA 2019. SPMS, vol. 294, pp. 437–448. Springer, Cham (2019). https://doi.org/10.1007/978-3-030-28665-1_33
36. Romaniuk, M., Hryniewicz, O.: Interval-based, nonparametric approach for resampling of fuzzy numbers. Soft. Comput. **23**, 5883–5903 (2019). https://doi.org/10.1007/s00500-018-3251-5
37. Romaniuk, M., Hryniewicz, O.: Discrete and smoothed resampling methods for interval-valued fuzzy numbers. IEEE Trans. Fuzzy Syst. **29**(3), 599–611 (2021)
38. Silverman, B.W., Young, G.A.: The bootstrap: to smooth or not to smooth? Biometrika **74**, 469–479 (1987)
39. Sinova, B., Gil, M.A., Colubi, A., Van Aelst, S.: The median of a random fuzzy number. The 1-norm distance approach. Fuzzy Sets Syst. **200**, 99–115 (2012)
40. Trutschnig, W., González-Rodríguez, G., Colubi, A., Gil, M.A.: A new family of metrics for compact, convex (fuzzy) sets based on a generalized concept of mid and spread. Inf. Sci. **179**, 3964–3972 (2009)
41. Wang, D., Hryniewicz, O.: A fuzzy nonparametric Shewhart chart based on the bootstrap approach. Int. J. Appl. Math. Comput. Sci. **25**(2), 389–401 (2015)
42. Zadeh, L.A.: Fuzzy sets. Inf. Control **8**, 338–353 (1965)

Aspects of Uncertain and Imprecise Information

Conditional Intuitionistic Fuzzy Mean Value in Connection with IF-Probability

Katarína Čunderlíková(✉)

Mathematical Institute, Slovak Academy of Sciences,
Štefánikova 49, 814 73 Bratislava, Slovakia
cunderlikova.lendelova@gmail.com, cunderlikova@savbb.sk

Abstract. The aim of this paper is to define a notion of a conditional intuitionistic fuzzy mean value for intuitionistic fuzzy observables. In this case we use an intuitionistic fuzzy probability. We show the relation between a conditional intuitionistic fuzzy mean value and conditional intuitionistic fuzzy probability related to the intuitionistic fuzzy probability. We formulate the martingale convergence theorem for a conditional intuitionistic fuzzy mean value in this case, too.

1 Introduction

In [5,6] we studied a conditional intuitionistic fuzzy mean value in connection with intuitionistic fuzzy state **m** and we proved the martingale convergence theorem for this case. In some cases we need to work with a intuitionistic fuzzy probability, so it is suitable to study a conditional intuitionistic fuzzy mean value in this connection. P. Grzegorzewski and M. Mrówka defined the probability on the family of intuitionistic fuzzy events

$$\mathcal{N} = \{(\mu_A, \nu_A)\ ;\ \mu_A + \nu_A \leq 1\},$$

where μ_A, ν_A are $\mathcal{S}$-measurable, as a mapping $\mathcal{P}$ from the family $\mathcal{N}$ to the set of all compact intervals in R by the formula

$$\mathcal{P}((\mu_A, \nu_A)) = \left[\int_\Omega \mu_A\ dP, 1 - \int_\Omega \nu_A\ dP\right],$$

where $(\Omega, \mathcal{S}, P)$ is probability space (see [7]). This intuitionistic fuzzy probability was axiomatically characterized by B. Riečan (see [9]). Since the intuitionistic fuzzy probability $\mathcal{P}$ can be decomposed to two intuitionistic fuzzy states **m** (see [10,13]), then we can use the results from papers [5,6] and we can defined a conditional intuitionistic fuzzy mean value in this sence. In this paper we show a relation between a conditional intuitionistic fuzzy mean value and a conditional intuitionistic fuzzy probability related to intuitionistic fuzzy probability. We formulate a version of a martingale convergence theorem for a conditional intuitionistic fuzzy mean value related to the intuitionistic fuzzy probability.

Remark that in a whole text we use a notation "IF" for short a phrase "intuitionistic fuzzy".

K. T. Atanassov et al. (Eds.): IWIFSGN 2020/BOS/SOR 2020, LNNS 338, pp. 51–59, 2022.
https://doi.org/10.1007/978-3-030-95929-6_4

2 Basic Notions

In this section we introduce the basic notions from IF-probability theory, see [1–3,12–14].

Definition 1. Let Ω be a nonempty set. An IF-set $\mathbf{A}$ on Ω is a pair (μ_A, ν_A) of mappings $\mu_A, \nu_A : \Omega \to [0,1]$ such that $\mu_A + \nu_A \leq 1_\Omega$.

Definition 2. Start with a measurable space $(\Omega, \mathcal{S})$. Hence $\mathcal{S}$ is a σ-algebra of subsets of Ω. An IF-event is called an IF-set $\mathbf{A} = (\mu_A, \nu_A)$ such that $\mu_A, \nu_A : \Omega \to [0,1]$ are $\mathcal{S}$-measurable.

The family of all IF-events on $(\Omega, \mathcal{S})$ will be denoted by $\mathcal{F}$, $\mu_A : \Omega \longrightarrow [0,1]$ will be called the membership function, $\nu_A : \Omega \longrightarrow [0,1]$ be called the non-membership function.

If $\mathbf{A} = (\mu_A, \nu_A) \in \mathcal{F}$, $\mathbf{B} = (\mu_B, \nu_B) \in \mathcal{F}$, then we define the Lukasiewicz binary operations $\oplus, \odot$ on $\mathcal{F}$ by

$$\mathbf{A} \oplus \mathbf{B} = ((\mu_A + \mu_B) \wedge 1_\Omega, (\nu_A + \nu_B - 1) \vee 0_\Omega)),$$
$$\mathbf{A} \odot \mathbf{B} = ((\mu_A + \mu_B - 1) \vee 0_\Omega, (\nu_A + \nu_B) \wedge 1_\Omega))$$

and the partial ordering is given by

$$\mathbf{A} \leq \mathbf{B} \Longleftrightarrow \mu_A \leq \mu_B, \nu_A \geq \nu_B.$$

Consider a probability space $(\Omega, \mathcal{S}, P)$. Then in [7] the IF-probability $\mathcal{P}(\mathbf{A})$ of an IF-event $\mathbf{A} = (\mu_A, \nu_A) \in \mathcal{F}$ has been defined as a compact interval by the equality

$$\mathcal{P}(\mathbf{A}) = \left[\int_\Omega \mu_A \, dP, 1 - \int_\Omega \nu_A \, dP \right].$$

Let $\mathcal{J}$ be the family of all compact intervals. Then the mapping $\mathcal{P} : \mathcal{F} \to \mathcal{J}$ can be defined axiomatically similarly as in [9].

Definition 3. Let $\mathcal{F}$ be the family of all IF-events in Ω. A mapping $\mathcal{P} : \mathcal{F} \to \mathcal{J}$ is called an IF-probability, if the following conditions hold:

(i) $\mathcal{P}((1_\Omega, 0_\Omega)) = [1,1]$, $\mathcal{P}((0_\Omega, 1_\Omega)) = [0,0]$;
(ii) if $\mathbf{A} \odot \mathbf{B} = (0_\Omega, 1_\Omega)$, then $\mathcal{P}(\mathbf{A} \oplus \mathbf{B}) = \mathcal{P}(\mathbf{A}) + \mathcal{P}(\mathbf{B})$;
(iii) if $\mathbf{A}_n \nearrow \mathbf{A}$, then $\mathcal{P}(\mathbf{A}_n) \nearrow \mathcal{P}(\mathbf{A})$.
(Recall that $[\alpha_n, \beta_n] \nearrow [\alpha, \beta]$ means that $\alpha_n \nearrow \alpha$, $\beta_n \nearrow \beta$, but $\mathbf{A}_n = (\mu_{A_n}, \nu_{A_n}) \nearrow \mathbf{A} = (\mu_A, \nu_A)$ means $\mu_{A_n} \nearrow \mu_A$, $\nu_{A_n} \searrow \nu_A$.)

IF-probability $\mathcal{P}$ is called **separating**, if

$$\mathcal{P}((\mu_A, \nu_A)) = [\mathcal{P}^\flat(\mu_A), 1 - \mathcal{P}^\sharp(\nu_A)],$$

where the functions $\mathcal{P}^\flat, \mathcal{P}^\sharp : \mathcal{T} \to [0,1]$ are probabilities.

Of course, each $\mathcal{P}(\mathbf{A})$ is an interval, denote it by $\mathcal{P}(\mathbf{A}) = [\mathcal{P}^\flat(\mathbf{A}), \mathcal{P}^\sharp(\mathbf{A})]$. By this way we obtain two functions

$$\mathcal{P}^\flat : \mathcal{F} \to [0,1], \mathcal{P}^\sharp : \mathcal{F} \to [0,1]$$

and some properties of $\mathcal{P}$ can be characterized by some properties of $\mathcal{P}^\flat, \mathcal{P}^\sharp$, see [10].

Theorem 1. *Let $\mathcal{P} : \mathcal{F} \to \mathcal{J}$ and $\mathcal{P}(\mathbf{A}) = [\mathcal{P}^\flat(\mathbf{A}), \mathcal{P}^\sharp(\mathbf{A})]$ for each $\mathbf{A} \in \mathcal{F}$. Then $\mathcal{P}$ is an IF-probability if and only if $\mathcal{P}^\flat$ and $\mathcal{P}^\sharp$ are IF-states.*

Proof. In [10] *Theorem 2.3.* □

Recall that by an **intuitionistic fuzzy state (*IF*-state) m** we understand each mapping $\mathbf{m} : \mathcal{F} \to [0,1]$ which satisfies the following conditions (see [11]):

(i) $\mathbf{m}((1_\Omega, 0_\Omega)) = 1$, $\mathbf{m}((0_\Omega, 1_\Omega)) = 0$;
(ii) if $\mathbf{A} \odot \mathbf{B} = (0_\Omega, 1_\Omega)$ and $\mathbf{A}, \mathbf{B} \in \mathcal{F}$, then $\mathbf{m}(\mathbf{A} \oplus \mathbf{B}) = \mathbf{m}(\mathbf{A}) + \mathbf{m}(\mathbf{B})$;
(iii) if $\mathbf{A}_n \nearrow \mathbf{A}$ (i.e. $\mu_{A_n} \nearrow \mu_A$, $\nu_{A_n} \searrow \nu_A$), then $\mathbf{m}(\mathbf{A}_n) \nearrow \mathbf{m}(\mathbf{A})$.

Now we explain the notion of an observable. Let $\mathcal{J}$ be the family of all intervals in R of the form

$$[a, b) = \{x \in R : a \leq x < b\}.$$

Then the σ-algebra $\sigma(\mathcal{J})$ is denoted by $\mathcal{B}(R)$ and it is called the σ-algebra of Borel sets, its elements are called Borel sets (see [15]).

Definition 4. By an *IF*-observable on $\mathcal{F}$ we understand each mapping $x : \mathcal{B}(R) \to \mathcal{F}$ satisfying the following conditions:

(i) $x(R) = (1_\Omega, 0_\Omega)$, $x(\emptyset) = (0_\Omega, 1_\Omega)$;
(ii) if $A \cap B = \emptyset$ and $A, B \in \mathcal{B}(R)$, then $x(A) \odot x(B) = (0_\Omega, 1_\Omega)$ and $x(A \cup B) = x(A) \oplus x(B)$;
(iii) if $A_n \nearrow A$ and $A_n, A \in \mathcal{B}(R)$, $n \in N$, then $x(A_n) \nearrow x(A)$.

In [8] we introduced the notion of product operation on the family of *IF*-events $\mathcal{F}$ and showed an example of this operation.

Definition 5. We say that a binary operation $\cdot$ on $\mathcal{F}$ is product if it satisfying the following conditions:

(i) $(1_\Omega, 0_\Omega) \cdot (a_1, a_2) = (a_1, a_2)$ for each $(a_1, a_2) \in \mathcal{F}$;
(ii) the operation $\cdot$ is commutative and associative;
(iii) if $(a_1, a_2) \odot (b_1, b_2) = (0_\Omega, 1_\Omega)$ and $(a_1, a_2), (b_1, b_2) \in \mathcal{F}$, then

$$(c_1, c_2) \cdot \big((a_1, a_2) \oplus (b_1, b_2)\big) = \big((c_1, c_2) \cdot (a_1, a_2)\big) \oplus \big((c_1, c_2) \cdot (b_1, b_2)\big)$$

and

$$\big((c_1, c_2) \cdot (a_1, a_2)\big) \odot \big((c_1, c_2) \cdot (b_1, b_2)\big) = (0_\Omega, 1_\Omega)$$

for each $(c_1, c_2) \in \mathcal{F}$;

(iv) if $(a_{1n}, a_{2n}) \searrow (0_\Omega, 1_\Omega)$, $(b_{1n}, b_{2n}) \searrow (0_\Omega, 1_\Omega)$ and $(a_{1n}, a_{2n}), (b_{1n}, b_{2n}) \in \mathcal{F}$, then $(a_{1n}, a_{2n}) \cdot (b_{1n}, b_{2n}) \searrow (0_\Omega, 1_\Omega)$.

In the following theorem is the example of product operation for IF-events.

Theorem 2. *The operation $\cdot$ defined by*

$$(x_1, y_1) \cdot (x_2, y_2) = (x_1 \cdot x_2, y_1 + y_2 - y_1 \cdot y_2)$$

for each $(x_1, y_1), (x_2, y_2) \in \mathcal{F}$ is product operation on $\mathcal{F}$.

Proof. In [8] *Theorem 1.* □

In [12] B. Riečan defined the notion of a joint IF-observable and he proved its existence.

Definition 6. Let $x, y : \mathcal{B}(R) \to \mathcal{F}$ be two IF-observables. The joint IF-observable of the IF-observables x, y is a mapping $h : \mathcal{B}(R^2) \to \mathcal{F}$ satisfying the following conditions:

(i) $h(R^2) = (1_\Omega, 0_\Omega)$, $h(\emptyset) = (0_\Omega, 1_\Omega)$;
(ii) if $A, B \in \mathcal{B}(R^2)$ and $A \cap B = \emptyset$, then $h(A \cup B) = h(A) \oplus h(B)$ and $h(A) \odot h(B) = (0_\Omega, 1_\Omega)$;
(iii) if $A, A_1, \ldots \in \mathcal{B}(R^2)$ and $A_n \nearrow A$, then $h(A_n) \nearrow h(A)$;
(iv) $h(C \times D) = x(C) \cdot y(D)$ for each $C, D \in \mathcal{B}(R)$.

Theorem 3. *For each two IF-observables $x, y : \mathcal{B}(R) \to \mathcal{F}$ there exists their joint IF-observable.*

Proof. In [12] *Theorem 3.3.* □

3 Conditional Intuitionistic Fuzzy Mean Value by Using IF-State

In this section we recall the notions associated with a conditional intuitionistic fuzzy mean value and an IF-state. In paper [5] we formulated the notion of indefinite integral.

Definition 7. If $x, y : \mathcal{B}(R) \to R$ are the IF-observables, such that y is integrable, (i.e. there exists $\mathbf{E}(y) = \int_R t \; d\mathbf{m}_y(t)$), then the indefinite integral is defined by the formula

$$\mathbf{E}(y_{x(B)}) = \int_R t \; d\mathbf{m}_{y_{x(B)}}(t)$$

for fixed $B \in \mathcal{B}(R)$.

There $y_{x(B)} : \mathcal{B}(R) \to \mathcal{F}$ is IF-observable given by the formula

$$y_{x(B)}(D) = \begin{cases} (0_\Omega, 1_\Omega), & \text{if } D = \emptyset \\ y(D \setminus \{0\}) \cdot x(B), & \text{if } D \neq R, D \in \mathcal{B}(R) \\ (1_\Omega, 0_\Omega), & \text{if } D = R \end{cases}$$

for fixed $B \in \mathcal{B}(R)$. The IF-observable $y_{x(B)}$ is integrable, if the IF-observable y is integrable (see [5]). Now we can introduce a notion of conditional intuitionistic fuzzy mean value (expectation).

Definition 8. If $x, y : \mathcal{B}(R) \to R$ are the IF-observables, y is integrable, then the conditional IF-mean value (expectation) $\mathbf{E}(y \mid x) = f$ is the Borel measurable function such that

$$\mathbf{E}\big(y_{x(B)}\big) = \int_B \mathbf{E}(y \mid x)\, d\mathbf{m}_x$$

for each $B \in \mathcal{B}(R)$.

There the measure $\mathbf{E}\big(y_{x(B)}\big)$ is a finite generalized measure and the existence of a Borel measurable function f flow from the Radom-Nikodym theorem (see [5]). We know a connection between a conditional IF-mean value and a conditional IF-probability associated with IF-state, too (see [5]).

Remark 1. Take $\mathbf{A} \in \mathcal{F}$ and define the IF-observable $y_\mathbf{A} : \mathcal{B}(R) \to \mathcal{F}$ by

$$y_\mathbf{A}(B) = \begin{cases} (0_\Omega, 1_\Omega), & \text{if } B = \emptyset \\ \mathbf{A}, & \text{if } B = \{1\} \\ y_\mathbf{A}(B \cap \{1\}), & \text{if } B \neq \emptyset, B \neq R, B \in \mathcal{B}(R) \\ (1_\Omega, 0_\Omega), & \text{if } B = R \end{cases}$$

Then $\mathbf{E}(y_\mathbf{A} \mid x) = \mathbf{p}(\mathbf{A} \mid x)$ holds $\mathbf{m}_x$ - almost everywhere.

There $\mathbf{p}(\mathbf{A} \mid x)$ is the **conditional intuitionistic fuzzy probability** introduced by B. Riečan in [14] as a Borel measurable function f (i.e. $B \in \mathcal{B}(R) \Longrightarrow f^{-1}(B) \in \mathcal{B}(R)$) such that

$$\int_B \mathbf{p}(\mathbf{A} \mid x)\, d\mathbf{m}_x = \mathbf{m}\big(\mathbf{A} \cdot x(B)\big)$$

for each $B \in \mathcal{B}(R)$.

Theorem 4 (Martingale convergence theorem). *Let $g_n : R \to R$ be the Borel measurable functions, $g_n^{-1}(\mathcal{B}(R)) \subset g_{n+1}^{-1}(\mathcal{B}(R))$ $(n = 1, 2, \ldots)$, $\bigcup_{n=1}^{\infty} g_n^{-1}(\mathcal{B}(R)) = \mathcal{B}(R)$. Let $x : \mathcal{B}(R) \to \mathcal{F}$ be an IF-observable, $x_n = x \circ g_n^{-1}$, $\mathbf{m}_x \ll \mathbf{m}_{x_n}$ $(n = 1, 2, \ldots)$. Let $y : \mathcal{B}(R) \to \mathcal{F}$ be an integrable IF-observable. Then*

$$\mathbf{E}(y \mid x_n) \to \mathbf{E}(y \mid x)$$

$\mathbf{m}_x$- *almost everywhere.*

Proof. In [6] *Theorem 4.2.* □

4 Conditional Intuitionistic Fuzzy Mean Value by Using IF-Probability

In this section we formulate a conditional IF-mean value using IF-probability. Such an IF-probability consists of two IF-states, we could use the results from Sect. 3.

Definition 9. Let $x, y : \mathcal{B}(R) \to R$ be the IF-observables, y be integrable and $\mathcal{P} = [\mathcal{P}^\flat, \mathcal{P}^\sharp]$ be an IF-probability. Then the function $\mathbf{E}(y \mid x) : \mathcal{B}(R) \to \mathcal{J}$ is the conditional IF-mean value, if there exist the Borel measurable functions $\mathbf{E}^\flat(y \mid x), \mathbf{E}^\sharp(y \mid x) : R \to R$ such that

$$\mathbf{E}^\flat\left(y_{x(B)}\right) = \int_B \mathbf{E}^\flat(y \mid x)\, d\mathcal{P}_x^\flat$$

$$\mathbf{E}^\sharp\left(y_{x(B)}\right) = \int_B \mathbf{E}^\sharp(y \mid x)\, d\mathcal{P}_x^\sharp$$

for each $B \in \mathcal{B}(R)$. There

$$\mathbf{E}(y \mid x) = \left[\mathbf{E}^\flat(y \mid x), \mathbf{E}^\sharp(y \mid x)\right].$$

Theorem 5. *Let $x, y : \mathcal{B}(R) \to R$ be the IF-observables, y be integrable and $\mathcal{P} = [\mathcal{P}^\flat, \mathcal{P}^\sharp]$ be an IF-probability. The function $\mathbf{E}(y \mid x) : \mathcal{B}(R) \to \mathcal{J}$ is the conditional IF-mean value with respect to the IF-probability $\mathcal{P}$, if and only if the function $\mathbf{E}^\flat, \mathbf{E}^\sharp : R \to R$ are the conditional IF-mean values with respect to the IF- states $\mathcal{P}^\flat, \mathcal{P}^\sharp$ respectively.*

Proof. It is consequent on *Definition* 9 and *Definition* 8. □

In [4] we introduced a notion of a conditional intuitionistic fuzzy probability with respect to the IF-probability $\mathcal{P}$ as a function $\mathbf{p}(\mathbf{A} \mid y) : \mathcal{B}(R) \to \mathcal{J}$ such that there exist the Borel measurable functions $\mathbf{p}^\flat(\mathbf{A} \mid y), \mathbf{p}^\sharp(\mathbf{A} \mid y) : R \to R$ with a property

$$\left[\int_B \mathbf{p}^\flat(\mathbf{A} \mid y)\, d\mathcal{P}_y^\flat, \int_B \mathbf{p}^\sharp(\mathbf{A} \mid y)\, d\mathcal{P}_y^\sharp\right] = \mathcal{P}(\mathbf{A} \cdot y(B))$$

for each $B \in \mathcal{B}(R)$. There $\mathbf{p}(\mathbf{A} \mid y) = \left[\mathbf{p}^\flat(\mathbf{A} \mid y), \mathbf{p}^\sharp(\mathbf{A} \mid y)\right]$. Now we show its connection with a conditional IF-mean value from *Definition* 9.

Remark 2. Take $\mathbf{A} \in \mathcal{F}$ and define the IF-observable $y_\mathbf{A} : \mathcal{B}(R) \to \mathcal{F}$ by

$$y_\mathbf{A}(B) = \begin{cases} (0_\Omega, 1_\Omega), & \text{if } B = \emptyset \\ \mathbf{A}, & \text{if } B = \{1\} \\ y_\mathbf{A}(B \cap \{1\}), & \text{if } B \neq \emptyset, B \neq R, B \in \mathcal{B}(R) \\ (1_\Omega, 0_\Omega), & \text{if } B = R \end{cases}$$

Then the equality $\mathbf{E}(y_{\mathbf{A}} \mid x) = \mathbf{p}(\mathbf{A} \mid x)$ holds $\mathcal{P}_x$ - almost everywhere.

Using *Remark* 1 and *Definition* 9 we obtain that

$$\mathbf{E}(y_{\mathbf{A}} \mid x) = \left[\mathbf{E}^{\flat}(y_{\mathbf{A}} \mid x), \mathbf{E}^{\sharp}(y_{\mathbf{A}} \mid x)\right]$$

and

$$\mathbf{E}^{\flat}(y_{\mathbf{A}} \mid x) = \mathbf{p}^{\flat}(\mathbf{A} \mid x) \text{ holds } \mathcal{P}_x^{\flat} \text{ - almost everywhere,}$$
$$\mathbf{E}^{\sharp}(y_{\mathbf{A}} \mid x) = \mathbf{p}^{\sharp}(\mathbf{A} \mid x) \text{ holds } \mathcal{P}_x^{\sharp} \text{ - almost everywhere.}$$

Hence

$$\mathbf{E}(y_{\mathbf{A}} \mid x) = \left[\mathbf{E}^{\flat}(y_{\mathbf{A}} \mid x), \mathbf{E}^{\sharp}(y_{\mathbf{A}} \mid x)\right] = \left[\mathbf{p}^{\flat}(\mathbf{A} \mid x), \mathbf{p}^{\sharp}(\mathbf{A} \mid x)\right] = \mathbf{p}(\mathbf{A} \mid x)$$

holds $\mathcal{P}_x$ - almost everywhere.

Theorem 6 (Martingale convergence theorem). *Let $g_n : R \to R$ be the Borel measurable functions, $g_n^{-1}(\mathcal{B}(R)) \subset g_{n+1}^{-1}(\mathcal{B}(R))$ $(n = 1, 2, \ldots)$, $\bigcup_{n=1}^{\infty} g_n^{-1}(\mathcal{B}(R)) = \mathcal{B}(R)$. Let $x : \mathcal{B}(R) \to \mathcal{F}$ be an IF-observable, $x_n = x \circ g_n^{-1}$. Let $\mathcal{P} = [\mathcal{P}^{\flat}, \mathcal{P}^{\sharp}]$ be an IF-probability, $\mathcal{P}_x^{\flat} \ll \mathcal{P}_{x_n}^{\flat}$, $\mathcal{P}_x^{\sharp} \ll \mathcal{P}_{x_n}^{\sharp}$ $(n = 1, 2, \ldots)$. Let $y : \mathcal{B}(R) \to \mathcal{F}$ be an integrable IF-observable. Then*

$$\mathbf{E}(y \mid x_n) \to \mathbf{E}(y \mid x)$$

$\mathcal{P}_x$- almost everywhere.

Proof. By *Definition* 9 and *Theorem* 5 we have

$$\mathbf{E}(y \mid x_n) = \left[\mathbf{E}^{\flat}(y \mid x_n), \mathbf{E}^{\sharp}(y \mid x_n)\right],$$
$$\mathbf{E}(y \mid x) = \left[\mathbf{E}^{\flat}(y \mid x), \mathbf{E}^{\sharp}(y \mid x)\right],$$

where $\mathbf{E}^{\flat}(y \mid x_n)$, $\mathbf{E}^{\flat}(y \mid x)$ are the conditional IF-mean values with respect to the IF-state $\mathcal{P}^{\flat}$ and $\mathbf{E}^{\sharp}(y \mid x_n)$, $\mathbf{E}^{\sharp}(y \mid x)$ are the conditional IF-mean values with respect to the IF-state $\mathcal{P}^{\sharp}$.

Using *Theorem* 4 we obtain

$$\mathbf{E}^{\flat}(y \mid x_n) \to \mathbf{E}^{\flat}(y \mid x) \quad \mathcal{P}_x^{\flat} \text{ - almost everywhere,}$$
$$\mathbf{E}^{\sharp}(y \mid x_n) \to \mathbf{E}^{\sharp}(y \mid x) \quad \mathcal{P}_x^{\sharp} \text{ - almost everywhere,}$$

Hence

$$\mathbf{E}(y \mid x_n) = \left[\mathbf{E}^{\flat}(y \mid x_n), \mathbf{E}^{\sharp}(y \mid x_n)\right] \to \left[\mathbf{E}^{\flat}(y \mid x), \mathbf{E}^{\sharp}(y \mid x)\right] = \mathbf{E}(y \mid x)$$

$\mathcal{P}_x$- almost everywhere. □

5 Conclusion

The paper is concerned in a probability theory on intuitionistic fuzzy sets. We formulated the notion of conditional intuitionistic mean value in a case of an intuitionistic fuzzy probability. We proved the martingale convergence theorem for a conditional intuitionistic fuzzy mean value related to the intuitionistic fuzzy probability. We show the connection between a conditional intuitionistic fuzzy mean value and a conditional intuitionistic fuzzy probability in an IF-probability case.

References

1. Atanassov, K.T.: Intuitionistic Fuzzy Sets: Theory and Applications. Studies in Fuzziness and Soft Computing, Physica-Verlag, Heidelberg (1999). https://doi.org/10.1007/978-3-7908-1870-3
2. Atanassov, K.T.: On Intuitionistic Fuzzy Sets. Studies in Fuzziness and Soft Computing, Springer, Berlin (2012). https://doi.org/10.1007/978-3-642-29127-2
3. Atanassov, K.T.: Intuitionistic fuzzy sets. Int. Bioautom. **20**, S1–S6 (2016). Reprint. VII ITKR Session, Sofia, 20–23 June 1983 (Deposed in Centr. Sci.-Techn. Library of the Bulg. Acad. of Sci., 1697/84) (in Bulgarian)
4. Čunderlíková, K.: Conditional intuitionistic fuzzy probability and martingale convergence theorem using IF-probability. Notes Intuitionistic Fuzzy Sets **26**(3), 13–21 (2020)
5. Čunderlíková, K.: Conditional intuitionistic fuzzy mean value. Axioms **10**(2), 97, 1–8 (2021)
6. Čunderlíková, K.: Martingale convergence theorem for the conditional intuitionistic fuzzy mean value. Notes on Intuitionistic Fuzzy Sets **27**(2), 94–102 (2021)
7. Grzegorzewski, P., Mrówka, E.: Probability of intuitionistic fuzzy events. In: Grzegorzewski, P., Hryniewicz, O., Gil, M.Á. (eds.) Soft Methods in Probability, Statistics and Data Analysis. Advances in Intelligent and Soft Computing, vol. 16, pp. 105–115. Physica, Heidelberg (2002). https://doi.org/10.1007/978-3-7908-1773-7_8
8. Lendelová, K.: Conditional IF-probability. In: Lawry, J., et al. (eds.) Soft Methods for Integrated Uncertainty Modelling. Advances in Soft Computing, vol. 37, pp. 275–283. Springer, Heidelberg (2006). https://doi.org/10.1007/3-540-34777-1_33
9. Riečan, B.: A descriptive definition of the probability on intuitionistic fuzzy sets. In: Wagenecht, M., Hampet, R. (eds.) EUSFLAT 2003, pp. 263–266. Zittau Görlitz University of Applied Sciences (2003)
10. Riečan, B.: On the probability on IF-sets and MV-algebras. In: Notes on IFS, vol. 11, issue number 6, 21–25 (2005)
11. Riečan, B.: On a problem of Radko Mesiar: general form of IF-probabilities. Fuzzy Sets Syst. **152**, 1485–1490 (2006)
12. Riečan B.: On the probability and random variables on IF events. In: Ruan, D., et al. (eds.) Applied Artifical Intelligence. In: Proceedings of the 7th FLINS Conference, Genova, pp. 138–145 (2006)
13. Riečan, B.: Probability Theory on Intuitionistic Fuzzy Events, Algebraic and Proof-Theoretic Aspects of Non-classical logics: Papers in Honour of Daniele Mundici's 60th Birthday. LNCS, vol. 4460. Springer, Heidelberg (2007). https://doi.org/10.1007/978-3-540-75939-3. Probability theory on intuitionistic fuzzy events

14. Riečan, B.: Analysis of fuzzy logic models. In: Koleshko, V. (eds.) Intelligent Systems, INTECH, pp. 219–244 (2012)
15. Riečan, B., Neubrunn, T.: Integral, Measure, and Ordering. Kluwer Academic Publishers, Dordrecht and Ister Science, Bratislava (1997)

Heikin-Ashi Technique with Use of Oriented Fuzzy Numbers

Krzysztof Piasecki and Anna Łyczkowska-Hanćkowiak(✉)

Institute of Economics and Finance, WSB University in Poznań, 61-895 Poznań, Poland
{krzysztof.piasecki,anna.lyczkowska-hanckowiak}@wsb.poznan.pl

Abstract. A candlestick is a style of chart describing a volatility of exchange quotation. The most well-known kind of candlesticks are Japanese Candlesticks based on Japanese Candles which represent current changes in quotations. The Heikin-Ashi technique is also known as one of the methods to identify and follow the trends using a price time series in financial markets. From the point-view of methodology, Heikin-Ashi candles are determined as a transformation of usual Japanese candles. Moreover, Heikin-Ashi Candles represent former and present quotation changes. This is a significant advantage of Heikin-Ashi candles over Japanese Candles. On the other hand, any Japanese Candles may be described by oriented fuzzy numbers. The main goal of this paper is to present any Heikin-Ashi candle as transformation of Japanese candles described by oriented fuzzy number. Additionally, we study the impact of Heikin-Ashi technique on imprecision burdening Japanese candles. All considerations are illustrated by numerical case study.

Keywords: Japanese Candle · Oriented fuzzy number · Heikin-Ashi technique

1 Introduction

The widely accepted model of an imprecise value is a fuzzy number (FN) [2]. The notion of an ordered FN is introduced by Kosiński in [8–10] as an extension of the concept of FN. For formal reasons, Kosiński's theory was revised [17] in such a way that the revised definition of ordered FNs fully corresponds to the intuitive Kosiński's definition. We call the ordered FN defined by Kosiński as "Kosiński's number" [18, 21, 22]. The ordered FN defined in a revised way is called an oriented FN (OFN) [18]. In this way, in the concept of ordered FNs, we distinguished two of its types: Kosiński's numbers and OFNs.

A Japanese Candlestick (JCT) is a style of financial chart used to describe variability of financial assets' exchange quotations [16]. Each JCT chart is given a time series of Japanese Candles (JCs). Any JCT is a very convenient tool for a synthetic record of high-frequency time series of financial data [1, 3, 4, 12, 14, 23]. The Heikin-Ashi technique is such a JC transformation that eliminates irregularities from a JCT chart [24]. This way, we obtain a better picture of trends and consolidations.

K. T. Atanassov et al. (Eds.): IWIFSGN 2020/BOS/SOR 2020, LNNS 338, pp. 60–71, 2022.
https://doi.org/10.1007/978-3-030-95929-6_5

If we use JCs for the financial analysis, then we need a linguistic evaluation of their attributes. With the essence of things, each such evaluation is imprecise. Then, these linguistic characteristics may be transformed with the use of fuzzy set theory. For this reason, JCs are usually represented by trapezoidal FNs [6, 11, 15].

For any security, each quantified JC may be used as such an approximation of its market price that it is additionally equipped with information about the price trend. Therefore, JCs are also described by ordered FNs. The Kosiński's numbers are applied for representing JCs in [5, 13]. A description of JCs with the use of OFNs is given in [19].

The main aim of this article is to apply the Heikin-Ashi technique for JCs described by OFNs. Our paper is organized as follows. In Sect. 2, we present the necessary information about OFNs. Section 3 contains the main information about the candlesticks drawn by the investors and conventional Heikin-Ashi technique. Section 4 is the main part of our paper. There we discuss representation of Heikin-Ashi Candles by OFNs. Section 5 contains final conclusions.

2 Oriented Fuzzy Numbers - Selected Definitions

The symbol $\mathcal{F}(\mathbb{R})$ denotes the family of all fuzzy sets in the real line $\mathbb{R}$. Our considerations we will restrict to the following kind of OFNs.

Definition 1 [17] **:** For any monotonic sequence $(a, b, c, d) \subset \mathbb{R}$, the trapezoidal OFN (TrOFN) $\overleftrightarrow{Tr}(a, b, c, d) = \overleftrightarrow{\mathcal{T}}$ is defined as the pair of orientation $\overrightarrow{a, d} = (a, d)$ and fuzzy set $\mathcal{T} \in \mathcal{F}(\mathbb{R})$ determined by its membership function $\mu_{Tr}(\cdot|a, b, c, d) \in [0; 1]^{\mathbb{R}}$ given by the identity.

$$\mu_{Tr}(x|a, b, c, d) = \begin{cases} 0, & x \notin [a, d] \equiv [d, a], \\ \frac{x-a}{b-a}, & x \in [a, b[\equiv]b, a] \\ 1, & x \in [b, c] \equiv [c, b] \\ \frac{x-d}{c-d}, & x \in]c, d] \equiv [d, c[. \end{cases} \tag{1}$$

The space of all TrOFNs we denote by the symbol $\mathbb{K}_{Tr}$.

After Klir [7], we distinguish the ambiguity and indistinctness as a components of imprecision. The ambiguity of any OFN is understood as a lack of a clear recommendation between one value among various others. In [20] it is shown that TrOFN ambiguity may be evaluated with use of the energy measure $d \in \left[\mathbb{R}_0^+\right]^{\mathbb{K}_{Tr}}$ given by the identity

$$d\left(\overleftrightarrow{Tr}(a, b, c, d)\right) = \frac{1}{2}|d + c - b - a| = \frac{1}{2}(|d - a| + |c - b|). \tag{2}$$

Indistinctness of any OFN is understood as a lack of explicit distinction between recommended and not recommended values. In [20] it is shown that TrOFN indistinctness may be evaluated with the use of the entropy measure $e \in \left[\mathbb{R}_0^+\right]^{\mathbb{K}_{Tr}}$ given by the identity.

$$e\left(\overleftrightarrow{Tr}(a, b, c, d)\right) = \frac{1}{4}|d - c + b - a| = \frac{1}{4}(|d - a| - |c - b|). \tag{3}$$

3 Heikin-Ashi Technique

Japanese Candlestick (JCT) is a kind of financial chart used to describe the volatility of the security quotation [16]. The basic elements of JCT are single Japanese Candles (JCs) describing the volatility of quotations within a given time range.

The JC notion is a special case of Generalized Candle (GC). Any GCs are composed of:

- The body determined as a rectangle between the open price P_o and the closed price P_c,
- The upper wick determined as the line between the body and the high price

$$P_h \geq \max\{P_o, P_c\} \tag{4}$$

- The lower wick determined as the line between the body and the low price

$$P_l \leq \min\{P_o, P_c\}. \tag{5}$$

The high price P_h and the low price P_l together are called extreme prices. Each GC is usually described by the ordered quadruple (P_o, P_c, P_h, P_l) containing reported prices. The GC represented by the quadruple (P_o, P_c, P_h, P_l) we will denote as follows

$$\widetilde{\mathcal{GC}} = \widetilde{\mathcal{GC}}(P_o, P_c, P_h, P_l). \tag{6}$$

A body illustrates the trend in trades. We can note the following cases here:

$$P_o < P_c, \tag{7}$$

$$P_o > P_c, \tag{8}$$

$$P_o = P_c. \tag{9}$$

If the condition (7) is met, then the candle body is white. Then, the GC is called a White Candle or a White Spinning. Any White Candle may be considered as a bullish signal, i.e., a forecast of an increase in quotations. If the condition (8) is satisfied, then the candle body is black. Then, the GC is called a Black Candle or a Black Spinning. Any Black Candle may be considered as a bearish signal, i.e., a forecast of a decline in quotations. The general case of a White Candle and a Black Candle are presented in Fig. 1.

Any GC need not have either a body or a wick. If a GC fulfils the condition (9) then it does not have a body. Such a JC is called a Doji. Individual types of candles differ among one another in a way the quadruple (P_o, P_c, P_h, P_l) is determined.

Let a fixed security be given as $\mathcal{Y}$. In the observed time interval $\Theta = [0, T]$, the volatility of the security $\mathcal{Y}$ quotations is described by the function $Q : [0, T] \to \mathbb{R}^+$. In a time range Θ we distinguish a series of subranges $\{\Theta_i\}_{i=1}^n$ satisfying the conditions

$$\forall_{i=1,2,\ldots,n}: \Theta_i = [s_i, t_i] \subset \Theta, \tag{10}$$

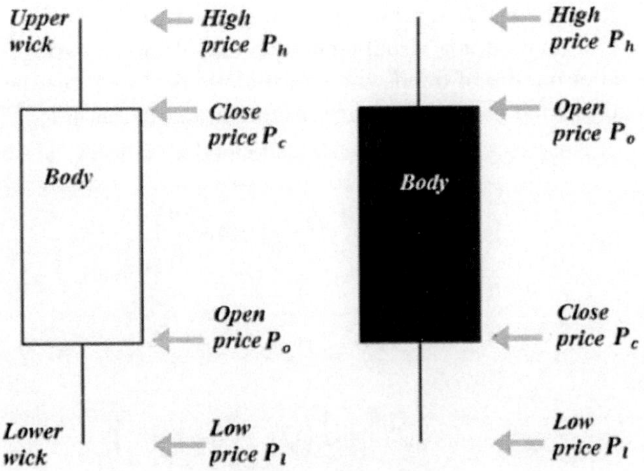

Fig. 1. Generalized candles.

$$\exists_{\Delta t>0}, \forall_{i=1,2,\ldots,n}: t_i - s_i = \Delta t, \tag{11}$$

$$\forall_{i=2,3,\ldots,n}: t_{i-1} \le s_i. \tag{12}$$

In each time range Θ_i $(i = 1, 2, \ldots, n)$ the volatility of quotations $\mathcal{Y}$ is determined by JC

$$\mathcal{J}^{(i)} = \widetilde{\mathcal{GC}}\left(P_o^{(i)}, P_c^{(i)}, P_h^{(i)}, P_l^{(i)}\right), \tag{13}$$

where the quadruple $\left(P_o^{(i)}, P_c^{(i)}, P_h^{(i)}, P_l^{(i)}\right)$ represents four important pieces of information about $\mathcal{Y}$ quotations:

- the open price:

$$P_o^{(i)} = Q(s_i), \tag{14}$$

- the close price:

$$P_c^{(i)} = Q(t_i), \tag{15}$$

- the high price:

$$P_h^{(i)} = \max\{Q(t): t \in \Theta_i\}, \tag{16}$$

- the low price:

$$P_l^{(i)} = \min\{Q(t): t \in \Theta_i\}. \tag{17}$$

This way we determine JCT represented by the time series $\{\mathcal{J}^{(i)}\}_{i=1}^{n}$ of JCs.

The Heikin-Ashi method is a visual technique that eliminates irregularities from a JCT, offering a better picture of trends and consolidations [24]. From the formal point-view, Heikin-Ashi method is based on transforming a time series $\{\mathcal{J}^{(i)}\}_{i=1}^{n}$ into a time series $\{\mathcal{H}^{(i)}\}_{i=1}^{n}$ of Heiken-Ashi Candles (HCs) defined as follows

$$\mathcal{H}^{(i)} = \widetilde{\mathcal{GC}}\left(Q_o^{(i)}, Q_c^{(i)}, Q_h^{(i)}, Q_l^{(i)}\right), \tag{18}$$

where

$$Q_o^{(1)} = P_o^{(1)}, \tag{19}$$

$$\forall_{i=2,3,\ldots,n}:\ Q_o^{(i)} = \frac{1}{2}\left(Q_o^{(i-1)} + Q_c^{(i-1)}\right), \tag{20}$$

$$\forall_{i=1,2,\ldots,n}:\ Q_c^{(i)} = \frac{1}{4}\left(P_o^{(i)} + P_c^{(i)} + P_h^{(i)} + P_l^{(i)}\right), \tag{21}$$

$$\forall_{i=1,2,\ldots,n}:\ Q_l^{(i)} = \min\left\{P_l^{(i)}, Q_o^{(i)}, Q_c^{(i)}\right\} = \min\left\{P_l^{(i)}, Q_o^{(i)}\right\}, \tag{22}$$

$$\forall_{i=1,2,\ldots,n}:\ Q_h^{(i)} = \max\left\{P_h^{(i)}, Q_o^{(i)}, Q_c^{(i)}\right\} = \max\left\{P_l^{(i)}, Q_o^{(i)}\right\}, \tag{23}$$

The above presented dependency (20) explicitly binds a current condition of HC with the quotations observed in previous periods. The experience gathered by the investors prove the practical use of this graphical way of presenting the quotation changes.

Example 1: In time period of 03 February 2020 and 31 March 2020 the price of stock of GDRPROJECT on Warsaw Stock Exchange was observed. I following days the fluctuation of the price was recorded by the means of JCs presented in the left column of Fig. 1. According to stock exchange tradition:

- green defines bullish signals (white candles),
- red colour defines bearish signals (black candles).

Next, using the dependencies (19–23) HCs were determined – presented in the right column of Fig. 1. The comparison of representation clearly shows that the time series of following HCs defines market trends in more stable way. IT enable the use of more stable investment strategies.

Table 1. Quotations of GDPROJECT – comparison of candlestick representation

Japanese Candles	Heikin-Ashi Candles
$\overleftrightarrow{\mathcal{GC}}$(280.60, 293.50, 279.50, 295.50)	$\overleftrightarrow{\mathcal{GC}}$(280.60, 286.78, 279.50, 295.50)
$\overleftrightarrow{\mathcal{GC}}$(293.50, 305.00, 290.30, 305.00)	$\overleftrightarrow{\mathcal{GC}}$(283.69, 298.45, 283.69, 305.00)
$\overleftrightarrow{\mathcal{GC}}$(303.00, 308.00, 302.30, 320.60)	$\overleftrightarrow{\mathcal{GC}}$(291.07, 308.48, 291.07, 320.60)
$\overleftrightarrow{\mathcal{GC}}$(309.50, 313.70,305.70, 314.50)	$\overleftrightarrow{\mathcal{GC}}$(299.77, 310.85,299.77, 314.50)
$\overleftrightarrow{\mathcal{GC}}$(313.30, 308.80,307.40,316.80)	$\overleftrightarrow{\mathcal{GC}}$(305.31, 311.58,305.31,316.80)
$\overleftrightarrow{\mathcal{GC}}$(310.90,310.70,310.00,313.40)	$\overleftrightarrow{\mathcal{GC}}$(308.44,311.25,308.44,313.40)
$\overleftrightarrow{\mathcal{GC}}$(313.00,310.50,310.50,317.00)	$\overleftrightarrow{\mathcal{GC}}$(309.85,312.75,309.85,317.00)
$\overleftrightarrow{\mathcal{GC}}$(311.00,313.20, 308.40,315.80)	$\overleftrightarrow{\mathcal{GC}}$(311.30,312.10, 308.40,315.80)
$\overleftrightarrow{\mathcal{GC}}$(313.10, 318.70,312.20, 318.70)	$\overleftrightarrow{\mathcal{GC}}$(311.70, 315.68,311.70, 318.70)
$\overleftrightarrow{\mathcal{GC}}$(319.30, 326.00,317.50, 326.80)	$\overleftrightarrow{\mathcal{GC}}$(313.69, 322.40,313.69, 326.80)
$\overleftrightarrow{\mathcal{GC}}$(327.60,326.00,323.60, 336.00)	$\overleftrightarrow{\mathcal{GC}}$(318.04,328.30,318.04, 336.00)
$\overleftrightarrow{\mathcal{GC}}$(325.40, 329.00, 323.00,330.80)	$\overleftrightarrow{\mathcal{GC}}$(323.17, 327.05, 323.00,330.80)
$\overleftrightarrow{\mathcal{GC}}$(330.00,330.80, 325.70, 331.90)	$\overleftrightarrow{\mathcal{GC}}$(325.11,329.60, 325.11, 331.90)
$\overleftrightarrow{\mathcal{GC}}$(330.60, 328.50, 326.60,331.90)	$\overleftrightarrow{\mathcal{GC}}$(327.36, 329.40, 326.60,331.90)
$\overleftrightarrow{\mathcal{GC}}$(328.00, 331.00, 326.60, 334.00)	$\overleftrightarrow{\mathcal{GC}}$(328.00, 331.00, 326.60, 334.00)
$\overleftrightarrow{\mathcal{GC}}$(327.60, 320.20, 319.00, 330.00)	$\overleftrightarrow{\mathcal{GC}}$(329.14, 324.20, 319.00, 330.00)
$\overleftrightarrow{\mathcal{GC}}$(316.70, 307.30, 305.30, 321.70)	$\overleftrightarrow{\mathcal{GC}}$(326.67, 312.63, 305.30, 326.67)
$\overleftrightarrow{\mathcal{GC}}$(301.60, 307.80, 297.00, 316.60)	$\overleftrightarrow{\mathcal{GC}}$(319.65, 305.75, 297.00, 319.65)
$\overleftrightarrow{\mathcal{GC}}$(308.00, 292.50, 290.70, 311.00)	$\overleftrightarrow{\mathcal{GC}}$(312.70, 300.55, 290.70, 312.70)
$\overleftrightarrow{\mathcal{GC}}$(276.00, 278.50, 261.60, 286.00)	$\overleftrightarrow{\mathcal{GC}}$(306.62, 275.53, 261.60, 306.62)
$\overleftrightarrow{\mathcal{GC}}$(287.90, 284.40, 272.00, 298.00)	$\overleftrightarrow{\mathcal{GC}}$(291.07, 285.58, 272.00, 298.00)
$\overleftrightarrow{\mathcal{GC}}$(295.00, 305.80, 290.20, 306.80)	$\overleftrightarrow{\mathcal{GC}}$(288.32, 299.48, 288.32, 306.80)
$\overleftrightarrow{\mathcal{GC}}$(300.30, 295.60292.50, 304.40)	$\overleftrightarrow{\mathcal{GC}}$(293.90, 298.20, 292.50, 304.40)
$\overleftrightarrow{\mathcal{GC}}$(303.00, 297.60,294.20, 305.50)	$\overleftrightarrow{\mathcal{GC}}$(296.05, 300.08,294.20, 305.50)
$\overleftrightarrow{\mathcal{GC}}$(294.60, 293.20, 283.00, 294.90)	$\overleftrightarrow{\mathcal{GC}}$(298.06, 291.43, 283.00, 298.06)
$\overleftrightarrow{\mathcal{GC}}$(266.00, 270.00, 259.50, 284.00)	$\overleftrightarrow{\mathcal{GC}}$(294.74, 269.88, 259.50, 294.74)
$\overleftrightarrow{\mathcal{GC}}$(275.00, 270.30, 268.00, 280.00)	$\overleftrightarrow{\mathcal{GC}}$(282.31, 273.33, 268.00, 280.00)
$\overleftrightarrow{\mathcal{GC}}$(271.00, 251.00, 242.00, 271.00)	$\overleftrightarrow{\mathcal{GC}}$(277.82, 258.75, 242.00, 277.82)
$\overleftrightarrow{\mathcal{GC}}$(228.60, 216.10, 205.20, 235.00)	$\overleftrightarrow{\mathcal{GC}}$(268.28, 221.08, 205.20, 268.28)
$\overleftrightarrow{\mathcal{GC}}$(230.00, 243.90,223.80, 253.00)	$\overleftrightarrow{\mathcal{GC}}$(244.68, 237.68,223.80, 253.00)
$\overleftrightarrow{\mathcal{GC}}$(235.90, 252.00, 222.00, 272.00)	$\overleftrightarrow{\mathcal{GC}}$(241.18, 245.48, 222.00, 272.00)
$\overleftrightarrow{\mathcal{GC}}$(262.00, 268.60, 255.30, 272.80)	$\overleftrightarrow{\mathcal{GC}}$(243.33, 264.65, 243.33, 272.80)
$\overleftrightarrow{\mathcal{GC}}$(269.50, 258.30, 255.30, 281.20)	$\overleftrightarrow{\mathcal{GC}}$(253.99, 266.08, 253.99, 281.20)
$\overleftrightarrow{\mathcal{GC}}$(258.30, 273.50, 256.50, 273.50)	$\overleftrightarrow{\mathcal{GC}}$(260.03, 265.45, 256.50, 273.50)
$\overleftrightarrow{\mathcal{GC}}$(281.00, 266.50, 256.40,283.30)	$\overleftrightarrow{\mathcal{GC}}$(262.74, 271.80, 256.40,283.30)
$\overleftrightarrow{\mathcal{GC}}$(257.10, 265.20,250.00, 272.00)	$\overleftrightarrow{\mathcal{GC}}$(267.27, 261.08,250.00, 272.00)
$\overleftrightarrow{\mathcal{GC}}$(277.10, 276.60, 268.30, 280.60)	$\overleftrightarrow{\mathcal{GC}}$(264.17, 275.65, 264.17, 280.60)
$\overleftrightarrow{\mathcal{GC}}$(284.00, 277.60, 272.00, 287.70)	$\overleftrightarrow{\mathcal{GC}}$(269.92, 280.33, 269.33, 287.70)
$\overleftrightarrow{\mathcal{GC}}$(275.90, 281.10, 273.50, 281.80)	$\overleftrightarrow{\mathcal{GC}}$(275.12, 278.08, 273.50, 281.80)
$\overleftrightarrow{\mathcal{GC}}$(282.50, 279.00, 274.60, 283.80)	$\overleftrightarrow{\mathcal{GC}}$(276.60, 279.98, 274.60, 283.80)
$\overleftrightarrow{\mathcal{GC}}$(280.00, 280.90, 273.20, 280.90)	$\overleftrightarrow{\mathcal{GC}}$(278.29, 278.75, 273.20, 208.90)
$\overleftrightarrow{\mathcal{GC}}$(283.30, 289.10,281.20, 289.90)	$\overleftrightarrow{\mathcal{GC}}$(278.52, 285.88,278.52, 289.90)

4 Representation of Heikin-Ashi Candles by Trapezoidal Oriented Fuzzy Numbers

Each candlestick representation is an imprecise reflection of the quotations volatility. It complies with the nature of things. It causes multiple proposals of representing JCs by FNs [5, 6, 11, 13, 15]. In [19] the method of representation of any JC by TrOFN is presented. The proposed method can be used directly to deteine TrOFN representing any GC.

Let us consider any GC $\widetilde{\mathcal{GC}}(P_o.P_c.P_h.P_l)$, where the quadruple $(P_o.P_c.P_h.P_l)$ satisfies conditions (4) and (5). If, along with the latter, the condition (9) is not fulfilled, then the candlestick represented by TrOFN

$$\overleftrightarrow{\mathcal{GC}}(P_o.P_c.P_h.P_l) = \overleftrightarrow{Tr}\left(P_b.P_o.P_c.P_f\right). \tag{24}$$

where:

- the back price $P_b \in \{P_l.Ph\}$ is the extreme price fulfilling the condition

$$|P_b - P_o| \leq |P_b - P_c|. \tag{25}$$

- the face price $P_f \in \{P_l.P_h\}$ is the extreme price fulfilling the condition

$$\left|P_f - P_c\right| \leq \left|P_f - P_o\right|. \tag{26}$$

In this way, for the White Candle, the back price. P_b is equal to the low price P_l. and for the Black Candle, the back price, P_b is equal to the high price P_h. Moreover, for the White Candle, the face price P_f is equal to the high price P_h and for the Black Candle the face price P_f is equal to the low price P_l. With this reinterpretation of extreme prices, we get a modernized model of GCs which has a body. Then, any White Candle is represented by positively oriented TrOFN. In this representation any Black Candle is described by negatively oriented TrOFN. The general cases of a White Candle and a Black Candle are presented in Fig. 2.

Now. we consider the case (9) when the considered GC is a Doji. Then we additionally take into account that GC $\widetilde{\mathcal{GC}}(P_o.P_o.P_h.P_l)$ describes the volatility of quotations of the security $\mathcal{Y}$ in a given time range $\hat{\theta}$. Then, each observed value of quotations $\overline{\overline{Q}} \in Q\left(\hat{\theta}\right)$ is assigned the last moment $\tau\left(\overline{\overline{Q}}.\hat{\theta}\right)$ of its observation. This moment is determined by the identity:

$$\tau\left(\overline{\overline{Q}}.\hat{\theta}\right) = \max\left\{t \in \hat{\theta} : Q(t) = \overline{\overline{Q}}\right\}. \tag{27}$$

In this way, we interpret extreme prices as an earlier or later one. Then, the back price, P_b and face price P_f are defined as follows:

$$P_b = \begin{cases} P_l. & \tau\left(P_l.\hat{\theta}\right) \leq \tau\left(P_h.\hat{\theta}\right). \\ P_h. & \tau\left(P_l.\hat{\theta}\right) > \tau\left(P_h.\hat{\theta}\right). \end{cases} \tag{28}$$

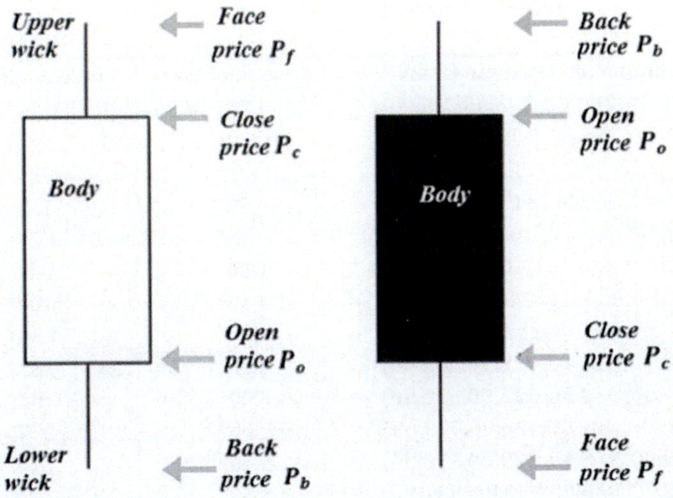

Fig. 2. Generalized candles—modernized model. Source: Own elaboration.

$$P_f = \begin{cases} P_h. \ \tau\left(P_l.\hat{\theta}\right) \le \tau\left(P_h.\hat{\theta}\right). \\ P_l. \ \tau\left(P_l.\hat{\theta}\right) > \tau\left(P_h.\hat{\theta}\right). \end{cases} \tag{29}$$

Due to this reinterpretation of extreme prices also any Doji can be represented by TrOFN (24).

The above presented method can be used to represent JCs, as well as, HCs. In that case we will obtain the possibility to implement the energy measure (2) and the entropy measure (3) to assess the imprecision of information on quotations volatility. It will also enable to examine the impact of Heikin-Ashi technique on the imprecision of candlesticks.

Example 2: All candlesticks describing the volatility of quotations of GDPROJECT in Example 1, can be noted by TrOFNs. We use dependencies (24–29). Representation of used JCs is presented in Tables 2. In Table 3 the representation of used HCs is presented. For each candlestick representation, the measures of energy and entropy, accordingly with the use of (2) and (3), which are also presented in Table 2 and 3. In each day of quotations the calculated levels of energy and entropy measures were compared. It was due to the search of more precise candlesticks. Lower levels of energy and entropy measures were desired. More preferable values were marked with the * symbol. This comparison allowed for the following conclusions:

- in most cases the increase of ambiguity resulted in the decrease of indistinctness,
- in most cases the increase of indistinctness resulted in the decrease of ambiguity,
- in some cases, JCs were unequivocally more accurate than HCs,
- no HC, unequivocally more accurate than JC, was found.

Table 2. Representation of Japanese Candles

Representation of Japanese Candles	Energy measure	Entropy measure
$\overleftrightarrow{Tr}$(279.50, 280.60, 293.50, 295.50)	13.45	0.28*
$\overleftrightarrow{Tr}$(290.30, 293.50, 305.00, 305.00)	13.10*	0.80*
$\overleftrightarrow{Tr}$(302.30, 303.00, 308.00, 320.60)	11.65*	3.33
$\overleftrightarrow{Tr}$(305.70, 309.50, 313.70, 314.50)	6.50*	1.15
$\overleftrightarrow{Tr}$(316.80, 313.30, 308.80,307.40)	6.95*	1.23*
$\overleftrightarrow{Tr}$(313.40, 310.90,310.70,310.00)	1.80*	0.80
$\overleftrightarrow{Tr}$(317.00, 313.00,310.50,310.50,)	4.50*	1.00*
$\overleftrightarrow{Tr}$(308.40, 311.00,313.20, 315.80)	4.80	1.30*
$\overleftrightarrow{Tr}$(312.20, 313.10, 318.70,318.70)	6.05	0.23*
$\overleftrightarrow{Tr}$(317.50, 319.30, 326.00,326.80)	8.00*	0.65*
$\overleftrightarrow{Tr}$(336.00, 327.60,326.00,323.60)	7.00*	2.70
$\overleftrightarrow{Tr}$(323.00, 325.40, 329.00,330.80)	5.70*	1.05
$\overleftrightarrow{Tr}$(325.70, 330.00,330.80, 331.90)	3.50*	1.35
$\overleftrightarrow{Tr}$(331.90, 330.60, 328.50, 326.60)	3.70	0.80*
$\overleftrightarrow{Tr}$(326.60, 328.00, 331.00, 334.00)	5.20*	1.10
$\overleftrightarrow{Tr}$(330.00, 327.60, 320.20, 319.00)	9.20	0.90*
$\overleftrightarrow{Tr}$(321.70, 316.70, 307.30, 305.30)	12.65*	1.88
$\overleftrightarrow{Tr}$(297.00, 301.60, 307.80, 316.60)	12.90	3.35
$\overleftrightarrow{Tr}$(311.00, 308.00, 292.50, 290.70)	17.90	1.20*
$\overleftrightarrow{Tr}$(261.00, 276.00, 278.50, 286.00)	13.45*	5.47
$\overleftrightarrow{Tr}$(298.00, 287.90, 284.40, 272.00)	14.75*	5.63
$\overleftrightarrow{Tr}$(290.20, 295.00, 305.80, 306.80)	13.75*	1.47*
$\overleftrightarrow{Tr}$(304.40, 300.30, 295.60, 292.50)	8.30	1.80*
$\overleftrightarrow{Tr}$(305.50, 303.00, 297.60, 294.20)	8.35	1.48*
$\overleftrightarrow{Tr}$(294.90, 294.60, 293.20, 283.00)	6.65	2.62*
$\overleftrightarrow{Tr}$(259.50, 266.00, 270.00, 284.00)	14.25*	5.13
$\overleftrightarrow{Tr}$(280.00, 275.00, 270.30, 268.00)	8.35*	1.83
$\overleftrightarrow{Tr}$(271.00, 271.00, 251.00, 242.00)	24.50*	2.25*
$\overleftrightarrow{Tr}$(235.00, 228.60, 216.10, 205.20)	20.85*	4.48
$\overleftrightarrow{Tr}$(223.80, 230.00, 243.90, 253.00)	21.55	3.83*
$\overleftrightarrow{Tr}$(222.00, 235.90, 252.00, 272.00)	33.05	8.48*
$\overleftrightarrow{Tr}$(255.30, 262.00, 268.60, 272.80)	12.10*	2.75
$\overleftrightarrow{Tr}$(281.20, 269.50, 258.30, 255.30)	18.55*	3.68*
$\overleftrightarrow{Tr}$(256.50, 258.30, 273.50, 273.50)	16.10	0.45*
$\overleftrightarrow{Tr}$(283.30, 281.00, 266.50, 256.40)	20.70	3.10*
$\overleftrightarrow{Tr}$(250.00, 257.10, 265.20, 272.00)	15.05	3.48*
$\overleftrightarrow{Tr}$(280.60, 277.10, 276.60, 268.30)	6.40*	2.95
$\overleftrightarrow{Tr}$(287.70, 284.00, 277.60, 272.00)	11.05*	2.33
$\overleftrightarrow{Tr}$(273.50, 275.90, 281.10, 281.80)	6.75	0.77*
$\overleftrightarrow{Tr}$(283.80, 282.50, 279.00, 274.60)	6.35	1.43*
$\overleftrightarrow{Tr}$(273.20, 280.00, 280.90, 280.90)	4.30	1.70*
$\overleftrightarrow{Tr}$(281.20, 283.30, 289.10, 289.90)	7.25*	0.72*

Table 3. Representation of Heikin-Ashi Candles

Representation of Heikin-Ashi Candles	Energy measure	Entropy measure
$\overleftrightarrow{Tr}$(279.50, 280.60, 286.78, 295.50)	10.09*	1.96
$\overleftrightarrow{Tr}$(283.69, 283.69, 298.45, 305.00)	18.04	1.64
$\overleftrightarrow{Tr}$(291.07, 291.07, 308.48, 320.60)	23.47	3.03*
$\overleftrightarrow{Tr}$(299.77, 299.77, 310.85, 314.50)	12.90	0.91*
$\overleftrightarrow{Tr}$(305.31, 305.31, 311.58, 316.80)	8.88	1.31
$\overleftrightarrow{Tr}$(308.44, 308.44,311.25, 313.40)	3.88	0.54*
$\overleftrightarrow{Tr}$(309.85, 309.85, 312.75, 317.00)	5.03	1.06
$\overleftrightarrow{Tr}$(308.40, 311.30,312.10, 315.80)	4.10*	1.65
$\overleftrightarrow{Tr}$(311.70, 311.70, 315.68, 318.70)	5.49*	0.76
$\overleftrightarrow{Tr}$(313.69, 313.69, 322.40, 326.80)	10.91	1.10
$\overleftrightarrow{Tr}$(318.04, 318.04,328.30, 336.00)	14.11	1.93*
$\overleftrightarrow{Tr}$(323.00, 323.17, 327.05, 330.80)	5.84	0.98*
$\overleftrightarrow{Tr}$(325.11, 325.11, 329.60, 331.90)	5.64	0.57*
$\overleftrightarrow{Tr}$(326.60, 327.36, 329.40, 331.90)	3.67*	0.81
$\overleftrightarrow{Tr}$(326.60, 328.00, 331.00, 334.00)	4.46	1.47*
$\overleftrightarrow{Tr}$(330.00, 329.14, 324.20, 319.00)	7.97*	1.52
$\overleftrightarrow{Tr}$(326.67, 326.67, 312.63, 305.30)	17.71	1.83
$\overleftrightarrow{Tr}$(319.65, 319.65, 305.75, 297.00)	18.27	2.19*
$\overleftrightarrow{Tr}$(312.70, 312.70, 300.55, 290.70)	17.07*	2.46
$\overleftrightarrow{Tr}$(306.62, 306.62, 275.53, 261.60)	38.06	3.48*
$\overleftrightarrow{Tr}$(298.00, 291.07, 285.58, 272.00)	15.75	5.13*
$\overleftrightarrow{Tr}$(288.32, 288.32, 299.48, 306.80)	14.86	1.86
$\overleftrightarrow{Tr}$(292.50, 293.90, 298.20, 304.40)	8.10*	1.90
$\overleftrightarrow{Tr}$(294.20, 296.05, 300.08, 305.50)	7.66*	1.82
$\overleftrightarrow{Tr}$(298.06, 298.06, 291.43, 283.00)	10.85*	2.11
$\overleftrightarrow{Tr}$(294.74, 294.74, 269.88, 259.50)	30.06	2.59*
$\overleftrightarrow{Tr}$(280.00, 282.31, 273.33, 268.00)	11.65	1.33*
$\overleftrightarrow{Tr}$(277.82, 277.82, 258.75, 242.00)	27.44	4.19
$\overleftrightarrow{Tr}$(268.28, 268.28, 221.08, 205.20)	55.15	3.97*
$\overleftrightarrow{Tr}$(253.00, 244.68, 237.68,223.80)	18.10*	5.55
$\overleftrightarrow{Tr}$(222.00, 241.18, 245.48, 272.00)	27.15*	11.43
$\overleftrightarrow{Tr}$(243.33, 243.33, 264.65, 272.80)	25.40	2.04*
$\overleftrightarrow{Tr}$(253.99, 253.99, 266.08, 281.20)	19.65	3.78
$\overleftrightarrow{Tr}$(256.50, 260.03, 265.45, 273.50)	11.21*	2.90
$\overleftrightarrow{Tr}$(256.40, 262.74, 271.80, 283.30)	17.98*	4.46
$\overleftrightarrow{Tr}$(272.00, 267.27, 261.08,250.00)	14.10*	3.95
$\overleftrightarrow{Tr}$(264.17, 264.17, 275.65, 280.60)	13.95	1.24*
$\overleftrightarrow{Tr}$(269.33, 269.92, 280.33, 287.70)	14.10	1.84*
$\overleftrightarrow{Tr}$(273.50, 275.12, 278.08, 281.80)	5.63*	1.34
$\overleftrightarrow{Tr}$(274.60, 276.60, 279.98, 283.80)	6.29*	1.46
$\overleftrightarrow{Tr}$(273.20, 278.29, 278.75, 208.90)	4.08*	1.81
$\overleftrightarrow{Tr}$(278.52, 278.52, 285.88, 289.90)	9.37	1.01

The results obtained in this example enable to draw two conclusions of a more general natura:

- the implementation of Heikin-Ashi technique has an impact on the structure of imprecision of candlesticks obtained in this way,
- the implementation of Heikin-Ashi technique can cause the increase in imprecision of candlesticks obtained in this way.

5 Summary

The proposed HC numerical model distinguishes all HC types. In addition, the use of this model does not cause a loss of information about the represented HC. The proposed HC numerical model highlights the imprecision of capital market assessments. In the case of economics and finance this imprecision is a natural state of affairs. Therefore. we can conclude that the proposed model reflects the essence of information about the observed capital market well. Therefore, the HC representation described here can be recommended as a numerical tool for analyzing financial data.

Further research on the usefulness of numerical representation of candlesticks, presented in this paper and in [19], seems purposeful. The efficiency of investment decisions taken basing on each type of candles. The research can be conducted with the help of algorithmic trade systems. We recommend this research field as a very promising area of scientific considerations and discussion.

References

1. Detollenaere, B., Mazza, P.: Do Japanese candlesticks help solve the trader's dilemma? J. Bank. Finance **48**, 386–395 (2014). https://doi.org/10.1016/j.jbankfin.2013.03.013
2. Didier, D., Prade, H.: Fuzzy real algebra: some results. Fuzzy Sets Syst. **2**, 327–348 (1979). https://doi.org/10.1016/0165-0114(79)90005-8
3. Fock, H., Klein, C., Zwergel, B.: Performance of candlestick analysis on intraday futures data. J. Future Markets **13**, 28–40 (2005). https://doi.org/10.3905/jod.2005.580514
4. Jasemi, M., Kimigiari, A.M., Memariani, A.: A modern neural network model to do stock market timing on the basis of the ancient investment technique of Japanese Candlestick. Expert Syst. Appl. **38**, 3884–3890 (2011). https://doi.org/10.1016/j.eswa.2010.09.049
5. Kacprzak, D., Kosiński, W., Kosiński, W.K.: Financial stock data and ordered fuzzy numbers. In: Rutkowski, L., Korytkowski, M., Scherer, R., Tadeusiewicz, R., Zadeh, L.A., Zurada, J.M. (eds.) ICAISC 2013. LNCS (LNAI), vol. 7894, pp. 259–270. Springer, Heidelberg (2013). https://doi.org/10.1007/978-3-642-38658-9_24
6. Kamo, T., Dagli, C.: Hybrid approach to the Japanese candlestick method for financial forecasting. Expert Syst. Appl. **36**, 5023–5030 (2009)
7. Klir, G.J.: Developments in uncertainty-based information. Adv. Comput. **36**, 255–332 (1993). https://doi.org/10.1016/s0065-2458(08)60273-9
8. Kosiński, W.: On fuzzy number calculus. Int. J. Appl. Math. Comput. Sci. **16**, 51–57 (2006)
9. Kosiński, W., Słysz, P.: Fuzzy numbers and their quotient space with algebraic operations. Bull. Acad. Pol. Sci. **41**, 285–295 (1993)

10. Kosiński, W., Prokopowicz, P., Ślęzak, D.: Fuzzy numbers with algebraic operations: algorithmic approach. In: Kłopotek, M., Wierzchoń, S.T., Michalewicz, M. (eds.) Proceedings of the IIS' 2002 Sopot., Poland, pp. 311–320. Physica Verlag, Heidelberg (2002)
11. Lee, C.-H.L., Liu, A., Chen, W.-S.: Pattern discovery of fuzzy time series for financial prediction. IEEE Trans. Knowl. Data Eng. **18**, 613–625 (2006)
12. Marshall, B.R., Young, M.R., Rose, L.C.: Candlestick technical trading strategies: can they create value for investors? J. Bank. Finance **30**, 2303–2323 (2006). https://doi.org/10.1016/j.jbankfin.2005.08.001
13. Marszałek, A., Burczyński, T.: Modelling and forecasting financial time series with ordered fuzzy candlesticks. Inf. Sci. **273**, 144–155 (2014). https://doi.org/10.1016/j.ins.2014.03.026
14. Morris, G.L.: Candlestick Charting Explained: Timeless Techniques for Trading Stocks and Futures. McGraw-Hill, New York (2006)
15. Naranjo, R., Santos, M.: A fuzzy decision system for money investment in stock markets based on fuzzy candlesticks pattern recognition. Expert Syst. Appl. **133**, 34–48 (2019)
16. Nison, S.: Japanese Candlestick Charting Techniques. New York Institute of Finance, New York (1991)
17. Piasecki, K.: Revision of the Kosiński's theory of ordered fuzzy numbers. J. Axioms **7**, 1–16 (2018)
18. Piasecki, K.: Relation "greater than or equal to" between ordered fuzzy numbers. Appl. Syst. Innov. **2**, 1–26 (2019)
19. Piasecki, K., Łyczkowska-Hanćkowiak, A.: Representation of Japanese candlesticks by oriented fuzzy numbers. Econometrics **8**(1), 1 (2020). https://doi.org/10.3390/econometrics8010001
20. Piasecki, K., Łyczkowska-Hanćkowiak, A.: Imprecision indexes of oriented fuzzy numbers. In: Uncertainty and Imprecision in Decision Making and Decision Support: New Challenges, Solutions and Perspectives. Advances in Intelligent Systems and Computing. Springer (2021)
21. Prokopowicz, P.: The directed inference for the Kosinski's fuzzy number model. In: Abraham, A., Wegrzyn-Wolska, K., Hassanien, A.E., Snasel, V., Alimi, A.M. (eds.) Proceedings of the Second International Afro-European Conference for Industrial Advancement AECIA 2015. AISC, vol. 427, pp. 493–503. Springer, Cham (2016). https://doi.org/10.1007/978-3-319-29504-6_46
22. Prokopowicz, P., Pedrycz, W.: The directed compatibility between ordered fuzzy numbers - a base tool for a direction sensitive fuzzy information processing. In: Rutkowski, L., Korytkowski, M., Scherer, R., Tadeusiewicz, R., Zadeh, L.A., Zurada, J.M. (eds.) ICAISC 2015. LNCS (LNAI), vol. 9119, pp. 249–259. Springer, Cham (2015). https://doi.org/10.1007/978-3-319-19324-3_23
23. Tsung-Hsun, L., Yung-Ming, S., Tsung-Chi, L.: Profitable candlestick trading strategies. The evidence from a new perspective. Rev. Financ. Econ. **21**, 63–68 (2012). https://doi.org/10.1016/j.rfe.2012.02.001
24. Valcu, D.: Using the Heikin-Ashi technique. Tech. Anal. Stocks Commod. **22**(2), 16–29 (2004)

Three Intuitionistic Fuzzy Estimations of Uncertainty in Service Compositions

Stoyan Poryazov, Velin Andonov(✉), and Emiliya Saranova

Institute of Mathematics and Informatics, Bulgarian Academy of Sciences, Acad. G. Bonchev Str., Block 8, 1113 Sofia, Bulgaria
{stoyan,velin_andonov,emiliya}@math.bas.bg

Abstract. Three estimations of uncertainty of the service of requests by service devices are proposed: intuitionistic fuzzy traffic characterization, intuitionistic fuzzy flow characterization and intuitionistic fuzzy time characterization. They are defined in the form of intuitionistic fuzzy pairs. The proposed characterizations are applied to a service device consisting of two consecutively connected service devices. One of the advantages of the proposed approach is that an estimation of the uncertainty in comprise virtual devices consisting of subdevices on several levels can be obtained using the operations defined over intuitionistic fuzzy pairs.

Keywords: Conceptual modelling · Service characterization · Intuitionistic fuzzy pairs

1 Introduction

In [10], an Intuitionistic Fuzzy (IF) representation of uncertainty in biomedical operations is proposed. The biomedical operations are considered as elements of service systems such as hospital units, telecare/telehealth centers, etc. [14,15]. The focus of the paper is on the representation of uncertainty in the conceptual and analytical models of the operations. The conceptual modelling approach is based on the Service Systems Theory and uses the definitions of the International Telecommunication Union (ITU) related to traffic characterization. The operations with requests are performed by virtual service devices which allows easier conceptual and analytical modelling of compositions of services. Using the methods for conceptual modelling of service systems described in [11,12] and the causal conceptual representation, the authors propose an IF characterization of virtual service devices in the form of Intuitionistic Fuzzy Pair (IFP, see [2]). The degrees of membership, non-membership and uncertainty are expressed as ratios of some of the parameters of the virtual devices. One of the advantages of the proposed approach is that an estimation of the uncertainty in comprise virtual devices consisting of subdevices on several levels can be obtained using the operations over IFPs.

The intuitionistic fuzzy sets are used in the modelling of QoS in service systems, for example of manufacturing grid system (see [16]) where the non-functionality QoS evaluation of resource services is based on users' feelings and

K. T. Atanassov et al. (Eds.): IWIFSGN 2020/BOS/SOR 2020, LNNS 338, pp. 72–84, 2022.
https://doi.org/10.1007/978-3-030-95929-6_6

transaction experiences using Intuitionistic Fuzzy Sets (IFSs). An approach to QoS-aware selection and composition of services, in the presence of both vague user preferences and vague service descriptions is proposed in [13]. Specification of QoS properties, service selection, composition and ranking are fundamental problems of the service oriented paradigm, invariant to technological aspects of the different service technologies, and remain valid also in the context of cloud services [4]. In [17], an approach to the selection of the most suitable service composition in a way that maximizes solutions expressed as functions over fuzzy/crisp QoS attributes, while satisfying user's QoS requirements. Based on the fuzzy sets theory and a genetic algorithm, a triangular fuzzy genetic algorithm is proposed for solving the service composition problem. A feasible method for normalizing the QoS matrix, aggregating formulas of each control structure for eight properties is presented.

In [3], fuzzy theory is used to capture the vagueness, subjectivity, and ambiguity that characteristics of human expressions in eliciting the QoS requirements in a cloud e-market place environment. A review of related works using fuzzy QoS Preference and fuzzy QoS Aspiration is presented.

In [8], strategies and trends of fuzzy QoS-aware cloud service composition in Industry 4.0 and related technologies are reviewed. Especially in [1] fuzzy methods in Resource Allocation Techniques in Next-Generation Broadband Wireless Access Networks are considered. The work [6] introduces fuzzy Analytical hierarchy process using geometric mean method in cloud service selection based-problem.

The aim of the present paper is to summarize and extend some of the proposed concepts in [10]. More precisely, three IF estimations of uncertainty for a given service device are proposed: IF traffic estimation, IF flow estimation and IF time estimation. The proposed estimations can be used to define new QoS indicators. The use of IFPs to quantify the uncertainty in the service allows to obtain estimations of the uncertainty in complex composition of services consisting of embedded service devices on several levels through aggregation or using the operations defined over IFPs (see [2]).

2 Base Virtual Devices

The basic elements of the conceptual models of service systems are the base virtual devices [12]. A general representation of a base virtual device such as server, buffer, switch, etc., is shown in Fig. 1.

Every such device named x has the following parameters:

- Fx – Intensity or incoming rate (frequency) of the flow of requests (i.e., the number of requests per time unit) to device x;
- Px – Probability of directing the requests towards device x;
- Tx – Service time (duration of servicing of a request) in device x;
- Yx – Traffic intensity [Erlang];
- Vx – Traffic volume [Erlang - time unit];
- Nx – Number of lines (service resources, positions, capacity) of device x.

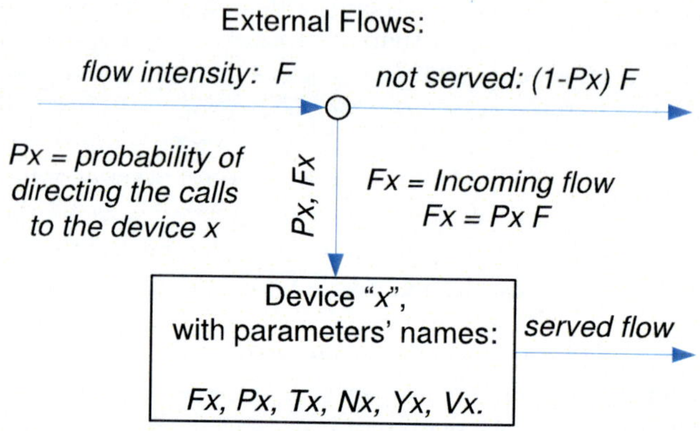

Fig. 1. Graphical representation of a base virtual device.

Base virtual devices of different types are used in the conceptual models (see [12]). The ones used in the present paper together with their graphical representations are shown in Fig. 2.

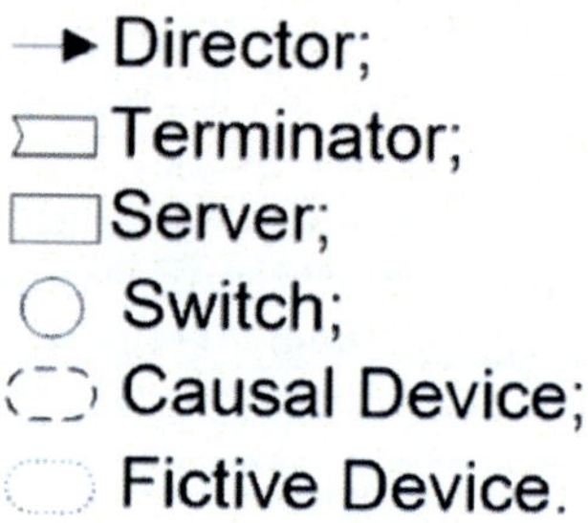

Fig. 2. Base virtual device types and their graphical representations.

Each type of base virtual device has a specific function (see [12]):

- Director – points unconditionally to the next device which the request shall enter but without transferring, changing or delaying it.
- Terminator – eliminates every request that enters it (so it leaves the model without any traces).
- Server – models the delay (service time, holding time) of requests in the corresponding device without their generation or elimination. It models also traffic and time characteristics of the requests processing.
- Switch (Transition) – selects one of its possible exits for each request that enters, thus determining the next device where this request shall go to.
- Causal device – virtual device defined for presentation of causes of service ending, e.g., successful (carried) or not (interrupted, abandoned, etc.).

– Fictive device – presents fictive traffic which is necessary for engineering. For example, not carried traffic is fictive, but it is used for calculating of the offered traffic, which is necessary for real device dimensioning.

The base virtual devices do not contain other devices. Comprise virtual devices are also used in the conceptual models of service systems. The base virtual devices can be embedded on several levels in the comprise devices.

3 Causal Decomposition of the Traffic in Virtual Devices

Following the papers [10,11], the causes of service ending and the corresponding causal devices, are grouped in generalized comprising causal devices. Three causal traffic qualificators are enough for representation of the service in a device: "parasitic", "carried" and "served" (see [5]). Their meaning is justified below (see [10]).

The *parasitic traffic* in a pool of resources is the traffic which was unsuccessfully served in the pool. The parasitic traffic occupies real resources but not for a useful service execution. In this paper, we use the qualifier "uncertain" as more adequate to the user's point of view.

The *carried traffic* in a pool of resources is the traffic which was successfully served in the pool (and carried to the next service device). We distinguish between two types of carried traffic:

– 'zero service', e.g., zero waiting in a buffer if the buffer is empty and there is free requested place for the service in the following device. The requests are receiving zero service in the causal 'zero service device' and may be served without delay.
– 'genuine service' – successful and real service of requests in the pool. The service time is noticeable.

The *served traffic* in a pool of resources is any traffic occupying (using) resources in the pool. The served traffic is a sum of the carried and the parasitic traffic.

Apart from that, the notion *not served traffic* and *offered traffic* should be specified.

The *not served traffic* is the traffic not occupying the recourses in the pool, due to any reason, e.g., 'blocking' [5], in the case of lack of service places in the pool. The intensity of the not served traffic is a multiplication of the not served requests' flow intensity and mean service time of the served requests.

The *offered traffic* is a sum of the served and not served traffic intensities in the pool.

A naming system for the parameters of the devices is also proposed (see [10]). Every parameter's name is a concatenation of one or two qualifiers, parameter's symbol and the device name:

Causal name=<qualifier><qualifier>.<Parameter's Symbol>.<Device Name>.

Some of the qualifiers used for characterization of the traffic are (see [11]):

- crr. = carried;
- nsr. = not served;
- ofr. = offered;
- prs. = parasitic;
- srv. = served.

'Parameter's symbol' is one of the letters P, F, T, Y, V, N (see Sect. 2). Qualifiers are used to characterize the parameters of the devices [12]. The qualifiers may be two, one or none. In case that the parameter's symbol is omitted, the causal name is a name of a device (see Fig. 3). The names of the devices are in small or subscript letters. For example, $crr.Fx$ is the intensity of the carried flow of requests of the device x, $prs.Fx$ is the intensity of the parasitic flow of requests of the device x (see Fig. 3). In the figures, only the names of causal devices may be present. The names of the device parameters are implicit.

A general presentation of a service network portion 'x' is shown in Fig. 3. The virtual device $srv.x$ embedded in it has a limited capacity. Because of the limited capacity, requests rejection due to lack of service places (call attempts blocking) may occur.

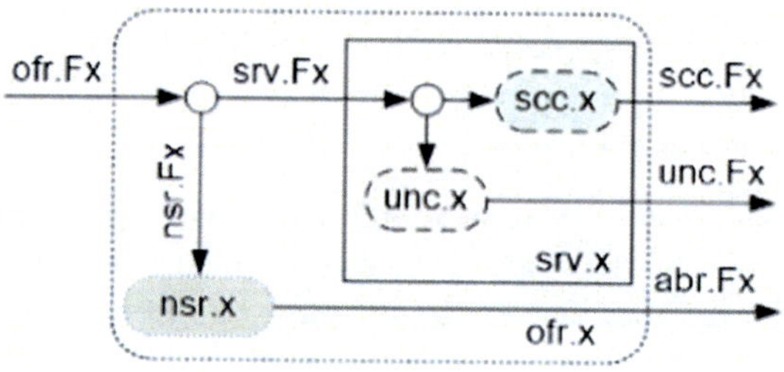

Fig. 3. Causal decomposition of the traffic inside a virtual service device x (see [10]).

The device $nsr.x$ is not a part of the service device.It is a part of the comprise fictive device $ofr.x$ which is necessary for dimensioning and traffic estimation.

The service of requests in a system can be classified into:

- successful – the service belongs to a fully completed target service (qualifier *scc.*);
- unsuccessful (abortive) – the service has either received a random denial or has not finished (without final result) all stages of service (qualifiers *nsr.*, *abr.*).;
- uncertain – all required stages of service are passed but the result of the service is unclear, partial, etc., or shortly – non-determined (qualifier *unc.*).

The qualifier *unc.* is used to characterize the traffic for which the reasons for the parasitic load cannot be determined explicitly. There might be several reasons as to why it cannot be determined explicitly. Some of them are:

- Abandoning/Failure by the users.
- Unsatisfying terms of service according to the users' opinion.
- Technical (objective) reasons – electricity failure, equipment failure, etc.
- Preempted – overtaken by higher priority request.

An important assumption must be stated about the device *unc.x*. It is assumed that in the case of uncertain service the service of the requests is terminated.

4 Three Intuitionistic Fuzzy Service Characterizations

In order to quantify the uncertainty in the service of requests, the notion of Intuitionistic Fuzzy Pair (IFP) is used (see [2]). An IFP is an ordered couple of real numbers $\langle a, b \rangle$ such that $a, b \in [0, 1]$ and $0 \leq a + b \leq 1$ which is used as an evaluation of some object or process. The real numbers a and b are referred to as degrees of validity and non-validity or degrees of membership and non-membership, respectively. The real number $c = 1 - a - b$ is the degree of uncertainty.

Let x be a service device with causal decomposition of the traffic as shown in Fig. 3. Let us define an IFP $\langle \mu_x^y, \nu_x^y \rangle$, where

$$\mu_x^y = \frac{scc.Yx}{ofr.Yx} \tag{1}$$

$$\nu_x^y = \frac{abr.Yx}{ofr.Yx} \,. \tag{2}$$

From Fig. 3 it is clear that the so defined $\langle \mu_x^y, \nu_x^y \rangle$ is indeed an IFP. The degree of uncertainty is defined as

$$\pi_x^y = \frac{unc.Yx}{ofr.Yx} \,. \tag{3}$$

We shall refer to the above IFP as *IF traffic characterization* of the service.

Let us define an IFP $\langle \mu_x^f, \nu_x^f \rangle$, where

$$\mu_x^f = \frac{scc.Fx}{ofr.Fx} \tag{4}$$

$$\nu_x^f = \frac{abr.Fx}{ofr.Fx} \,. \tag{5}$$

From Fig. 3 it is clear that the so defined pair is indeed an IFP. The degree of uncertainty of this pair is given by

$$\pi_x^f = \frac{unc.Fx}{ofr.Fx} \,. \tag{6}$$

We shall refer to the so defined IFP as *IF flow characterization* of the service.

Before we define the third IF characterization, the notion of *partial time* should be introduced. Let us consider k subdevices of a service device x which are not exclusively encompassing the service device g. Let the parameters of each such subdevice be Y_i, F_i, T_i. We have

$$Fg = \sum_{i=1}^{k} F_i \,. \tag{7}$$

Obviously, $Fg \leq Fx$ and $Fg = Fx$ only when the subdevices encompass the service device x.

$$Yg = \sum_{i=1}^{k} F_i T_i \,. \tag{8}$$

Now, since $Yg = Fg\,Tg$, we have

$$Tg = \frac{Yg}{Fg} = \frac{\sum_{i=1}^{k} F_i T_i}{\sum_{i=1}^{k} F_i} \,. \tag{9}$$

The partial service time Tg is defined by

$$prt.Tg = \frac{Yg}{Fx} = \frac{\sum_{i=1}^{k} F_i T_i}{Fx} \,. \tag{10}$$

$prt.Tg$ shows what part of the mean occupation time of the device x belongs the device g. In the special case when the two device x and g are the same, we have

$$prt.Tg = \frac{Yx}{Fx} = Tx \,. \tag{11}$$

The notion of partial time is suitable for the definition of IF time characterization of the real time occupation of service systems. The third (IF time) characterization is given by the pair $\langle \mu_x^t, \nu_x^t \rangle$, where

$$\mu_x^t = \frac{prt.scc.Tx}{srv.Tx} \,, \tag{12}$$

$$\nu_x^t = \frac{prt.abr.Tx}{srv.Tx} \tag{13}$$

and the respective degree of uncertainty is given by

$$\pi_x^t = \frac{prt.unc.Tx}{srv.Tx} \,. \tag{14}$$

We shall refer to the pair $\langle \mu_x^t, \nu_x^t \rangle$ as *IF time characterization* of the service.

5 Intuitionistic Fuzzy Service Characterizations of Sonsecutive Composition of Services

The conceptual and analytical modelling of composition of services is an important topic of research and especially the Quality of Service (QoS) composition. Some traffic quality aggregations of compositions of services are proposed in [9,11] and an IF characterization of consecutive composition of services is presented in [10]. Here, we shall derive analytical expressions for the degrees of membership, non-membership and uncertainty for the IFPs defined in the previous section in the case of consecutive composition of services. The conceptual model of consecutive composition of two services is shown in Fig. 4 (see [10]).

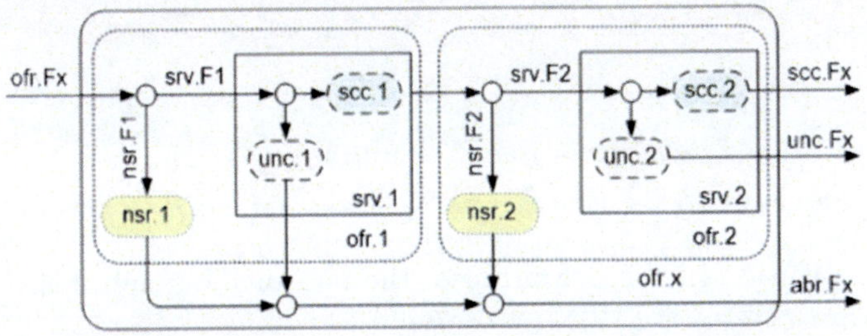

Fig. 4. Consecutive composition of two services inside a virtual service device x (see [10]).

Using the graphical representation of the consecutive composition (see Fig. 4), first we derive expressions for the IF traffic characterization. In [9], traffic quality aggregations for some service compositions are proposed. Here, we obtain intuitionistic fuzzy characterization of uncertainty in a consecutive composition of two operations. The graphical representation of the composition is shown in 4.

5.1 If Traffic Characterization

Using the graphical representation of the composition of services (see 4) and the Little's formula, for the traffic and flow intensities of the successfully served requests we have respectively:

$$scc.Yx = scc.Fx(scc.T_1 + scc.T_2)\,. \tag{15}$$

$$scc.Fx = ofr.Fx(1 - nsr.P_1)(1 - unc.P_1)(1 - nsr.P_2)(1 - unc.P_2)\,. \tag{16}$$

Similarly, for the intensity of the offered traffic we obtain

$$ofr.Yx = ofr.Fx(unc.P_1\,unc.T_1 + (1 - unc.P_1)[scc.T_1 + unc.P_2\,unc.T_2 \\ + (1 - unc.P_2)scc.T_2]) \quad . \tag{17}$$

After substitution of the last two equations above into the definition of the membership degree of the IF traffic characterization (see (1)), we obtain

$$\mu_x^y = \frac{(1 - nsr.P_1)(1 - unc.P_1)(1 - nsr.P_2)(1 - unc.P_2)}{unc.P_1\,unc.T_1 + (1 - unc.P_1)[scc.T_1 + unc.P_2\,unc.T_2 + (1 - unc.P_2)scc.T_2]} . \tag{18}$$

In order to determine the degree of non-membership of the IF traffic characterization, first we obtain an expression for the traffic intensity of the abortive requests:

$$abr.Yx = nsr.Y_1 + unc.Y_1 + nsr.Y_2 \\ = nsr.F_1\,srv.T_1 + unc.F_1\,unc.T_1 + nsr.F_2(scc.T_1 + srv.T_2) . \tag{19}$$

After substitution in the definition of the non-membership degree of the IF traffic characterization (see (2)), we obtain

$$\nu_x^y = \frac{nsr.F_1\,srv.T_1 + unc.F_1\,unc.T_1 + nsr.F_2(scc.T_1 + srv.T_2)}{A} , \tag{20}$$

where

$$A = ofr.Fx[unc.P_1\,unc.T_1 + (1 - unc.P_1)[scc.T_1 + unc.P_2\,unc.T_2 \\ + (1 - unc.P_2)scc.T_2]] \tag{21}$$

To determine the degree of uncertainty π_x^y, we take into account the fact that the uncertain traffic and flow intensities of the comprise device are equal to those of the second embedded device, i.e.,

$$unc.Yx = unc.Y_2 = unc.Fx\,unc.T_2 . \tag{22}$$

$$unc.Fx = unc.F_2 = ofr.Fx(1 - nsr.P_1)(1 - unc.P_1)(1 - nsr.P_2)unc.P_2 . \tag{23}$$

Therefore, the degree of uncertainty is given by

$$\pi_x^y = \frac{(1 - nsr.P_1)(1 - unc.P_1)(1 - nsr.P_2)unc.P_2\,unc.T_2}{unc.P_1\,unc.T_1 + (1 - unc.P_1)[scc.T_1 + unc.P_2\,unc.T_2 + (1 - unc.P_2)scc.T_2]} \tag{24}$$

5.2 If Flow Characterization

To derive an expression for the membership degree of the IF flow characterization first we obtain an expression for the flow intensity of the successfully serviced requests by the comprise device x.

$$scc.Fx = ofr.Fx(1 - nsr.P_1)(1 - unc.P_1)(1 - nsr.P_2)(1 - unc.P_2)\,. \tag{25}$$

After substitution in the expression from the definition of the non membership degree (see (4)), we obtain

$$\mu_x^f = (1 - nsr.P_1)(1 - unc.P_1)(1 - nsr.P_2)(1 - unc.P_2)\,. \tag{26}$$

Similarly, for the non-membership degree of the IF flow characterization (see (5)) we have

$$\nu_x^f = nsr.P_1 + (1 - nsr.P_1)[unc.P_1 + (1 - unc.P_1)nsr.P_2]\,. \tag{27}$$

The degree of uncertainty (see (6)) in this case is given by

$$\pi_x^f = (1 - nsr.P_1)(1 - unc.P_1)(1 - nsr.P_2)unc.P_2 \tag{28}$$

5.3 If Time Characterization

In order to derive expressions for the IF time characterization (see Eqs. (12)–(14)) for consecutive composition of two services, first we derive expression for the mean service time for the successfully served requests by the comprise service device:

$$\begin{aligned} srv.Tx = {} & (1 - nsr.P_2)[unc.P_1 unc.T_1 + (1 - unc.P_1)[scc.T_1 + (1 - nsr.P_2) \\ & [unc.P_2\, unc.T_2 + (1 - unc.P_2)scc.T_2]]]\,. \end{aligned} \tag{29}$$

From the definition of partial time and Fig. 4, we have

$$\begin{aligned} prt.scc.Tx &= \frac{scc.Fx\; scc.Tx}{srv.Fx} \\ &= \frac{(1 - unc.P_1)(1 - nsr.P_2)(1 - unc.P_2)(scc.T_1 + scc.T_2)}{unc.P_1 + (1 - unc.P_1)}\,. \end{aligned} \tag{30}$$

$$\mu_x^t = \frac{(1 - unc.P_1)(1 - unc.P_2)(scc.T_1 + scc.T_2)}{unc.P_1 + (1 - unc.P_1)} \cdot \frac{1}{B}\,, \tag{31}$$

where

$$\begin{aligned} B = {} & unc.P_1\, unc.T_1 + (1 - unc.P_1)[scc.T_1 + (1 - nsr.P_2)[unc.P_2\, unc.T_2 \\ & + (1 - unc.P_2)scc.T_2]]\,. \end{aligned} \tag{32}$$

To determine the degree of non-membership of the IF time characterization (see (13)) first we express the *prt.abr.Tx*.

$$prt.abr.Tx = unc.P_1\, unc.T_1 + (1 - unc.P_1)nsr.P_2\, scc.T_1\,. \tag{33}$$

$$\nu_x^t = \frac{unc.P_1\, unc.T_1 + (1 - unc.P_1)nsr.P_2\, scc.T_1}{C}, \tag{34}$$

where

$$C = (1 - nsr.P_1)[unc.P_1\, unc.T_1 + (1 - unc.P_1)[scc.T_1 + (1 - nsr.P_2)[unc.P_2\, unc.T_2 + (1 - unc.P_2)scc.T_2]]]\,. \tag{35}$$

To determine the degree of uncertainty of the IF time characterization (see (14)) first we derive expression for *prt.unc.Tx*:

$$prt.unc.Tx = (1 - unc.P_1)(1 - nsr.P_2)unc.P_2\, unc.T_2\,. \tag{36}$$

$$\pi_x^t = \frac{(1 - unc.P_1)(1 - nsr.P_2)unc.P_2\, unc.T_2}{D}, \tag{37}$$

where

$$D = (1 - nsr.P_1)[unc.P_1\, unc.T_1 + (1 - unc.P_1)[scc.T_1 + (1 - nsr.P_2)[unc.P_2\, unc.T_2 + (1 - unc.P_2)scc.T_2]]]\,. \tag{38}$$

6 Conclusion

The proposed method and the obtained results allow for the exact evaluation of IF characterization of composition of services if the IF evaluations of the components are known.

In future, the research will continue in two directions:

1. Determining of the applicability and comparison of the IF flow, IF traffic and IF time characterizations in view of choosing the most suitable of them for a specific purpose.
2. Study of the possibilities for expressing of one estimation as a function of other estimations in view of minimizing the numerical complexity.

Acknowledgements. The work of E. Saranova was partially supported by the Task 1.2.5 of the Contract MES DOI-205/23.11.2018.

The work of S. Poryazov was partially supported by the Joint Research Project (No. 18/02.04.2019) of the IMI-BAS and Joint Institute for Nuclear Research, Dubna, Russia.

The work of V. Andonov was partially supported by the Scientific Infrastructure Project, funded by the Bulgarian Ministry of Education and Science (Agreement DO1-367/18.12.2020).

References

1. Al-Surmi, I., Mansoor, A., Ahmed, A.: Toward 5G high utilizations: a survey on OFDMA-based resource allocation techniques in next-generation broadband wireless access networks. EAI Endorsed Trans. Mob. Commun. Appl. **6**(18), e3 (2021). https://doi.org/10.4108/eai.10-2-2021.168713
2. Atanassov, K., Szmidt, E., Kacprzyk, J.: On Intuitionistic Fuzzy Pairs. Notes Intuitionistic Fuzzy Sets **19**(3), 1–13 (2013)
3. Ezenwoke, A., Adigun, M.: Integrating fuzzy theory and visualization for QoS-aware selection of SaaS in cloud e-Marketplaces. Cogent Eng. **8**(1), 1911592 (2021). https://doi.org/10.1080/23311916.2021.1911592
4. Garg, S.K., Versteeg, S., Buyya, R.: A framework for ranking of cloud computing services. Future Gener. Comput. Syst. **29**(4), 1012–1023 (2013). Special Section: Utility and Cloud Computing
5. ITU E.600, ITU-T Recommendation E.600: Terms and Definitions of Traffic Engineering, Melbourne (1988), revised at Helsinki (1993)
6. Helmy, S., Eladl, G., Eisa, M.: Fuzzy Analytical Hierarchy Process (FAHP) using geometric mean method to select best processing framework adequate to big data. J. Theor. Appl. Inf. Technol. 99(1) (2021)
7. Haverkort, B.R.: Performance of Computer Communication Systems: A Model-Based Approach. Wiley, New York (1998)
8. Oztemel, E., Gursev, S.: Literature review of Industry 4.0 and related technologies. J. Intell. Manuf. **31**, 127–182 (2020). https://doi.org/10.1007/s10845-018-1433-8
9. Poryazov, S., Andonov, V., Saranova, E.: Different Traffic quality aggregations for a service composition. In: 2019 Big Data, Knowledge and Control Systems Engineering (BdKCSE), Sofia, Bulgaria, pp. 1–5 (2019)
10. Poryazov, S., Andonov, V., Saranova, E.: Intuitionistic Fuzzy Representation of Uncertainty in Biomedical Operations. In: International Symposium on Bioinformatics and Biomedicine, 8–10 October 2020, Burgas, Bulgaria (in press)
11. Poryazov, S., Andonov, V., Saranova, E.: Traffic quality aggregations of a queuing system. In: 2020 IEEE 10th International Conference on Intelligent Systems (IS), Varna, Bulgaria, pp. 102–110 (2020)
12. Poryazov, S., Saranova, E.: Models of Telecommunication Networks with Virtual Channel Switching and Applications. Prof. Marin Drinov Academic Publishing House, Sofia (2012).In Bulgarian
13. Sora, I., Todinca, D.: Dealing with Fuzzy QoS Properties in Service Composition. http://staff.cs.upt.ro/~ioana/papers/saci2015_fuzzy.pdf
14. Stefanova-Pavlova, M., Andonov, V., Tasseva, V., Gateva, A., Stefanova, E.: Generalized nets in medicine: an example of telemedicine for people with diabetes. In: Angelov, P., Sotirov, S. (eds.) Imprecision and Uncertainty in Information Representation and Processing. Studies in Fuzziness and Soft Computing, vol. 332, pp. 327–357. Springer, Cham (2016). https://doi.org/10.1007/978-3-319-26302-1_20
15. Stefanova-Pavlova, M., et al.: Modeling telehealth services with generalized nets. In: Sgurev, V., Yager, R., Kacprzyk, J., Atanassov, K. (eds.) Recent Contributions in Intelligent Systems. Studies in Computational Intelligence, vol. 657, pp. 279–290. Springer, Cham (2017). https://doi.org/10.1007/978-3-319-41438-6_16

16. Tao, F., Zhao, D., Zhang, L.: Resource service optimal-selection based on intuitionistic fuzzy set and non-functionality QoS in manufacturing grid system. Knowl. Inf. Syst. **25**, 185–208 (2010). https://doi.org/10.1007/s10115-009-0263-6
17. Xu, J., Guo, L., Zhang, R.: QoS-aware service composition using fuzzy set theory and genetic algorithm. Wireless Pers. Commun. **102**, 1009–1028 (2018). https://doi.org/10.1007/s11277-017-5129-8

An Interval-Valued Intuitionistic Fuzzy Estimation of the Area of 2D-Figures Based on Pick's Formula

Krassimir Atanassov(✉), Evgeniy Marinov, and Peter Vassilev

Department of Bioinformatics and Mathematical Modelling,
Institute of Biophysics and Biomedical Engineering, Bulgarian Academy of Sciences,
Acad. G. Bonchev Str., Block 105, 1113 Sofia, Bulgaria
krat@bas.bg

Abstract. We propose an interval-valued intuitionistic fuzzy estimation of the interior of a simple closed curve. This work is based on a previous research focused on an iterative procedure for estimation of the area surrounded by a simple closed curve in the real 2D space. Pick's Formula is employed for calculating the area surrounded by a special types of polygons. The result is a numerical method permitting the implementation of the algorithm in any procedural programming language.

Keywords: Interval-valued intuitionistic fuzzy estimation · Iterative process · Inner and outer polygon

1 Introduction

We propose in this paper an interval-valued intuitionistic fuzzy estimation of the area surrounded by a *continuous simple closed* curve in the real 2D space, i.e. the area of its interior. By a *simple* curve we mean that it has no self intersections and if, moreover, the curve is continuous, this implies that its interior is a *simply connected* domain (cf. Munkres [10, Ch. 9]). The idea is similar to the approximation approach of Pawlak's rough sets (cf. [13]). This work actually can be considered as an extension of our research stated in a series of the papers [5,8,9]. Let us focus on the arbitrarily chosen curve on Fig. 1.

In [9] we introduced a formula of the intuitionistic fuzzy estimation of a simple closed curve surrounded by two polygons (an inner and an outer one) with sides parallel to the coordinate axes.

K. T. Atanassov et al. (Eds.): IWIFSGN 2020/BOS/SOR 2020, LNNS 338, pp. 85–91, 2022.
https://doi.org/10.1007/978-3-030-95929-6_7

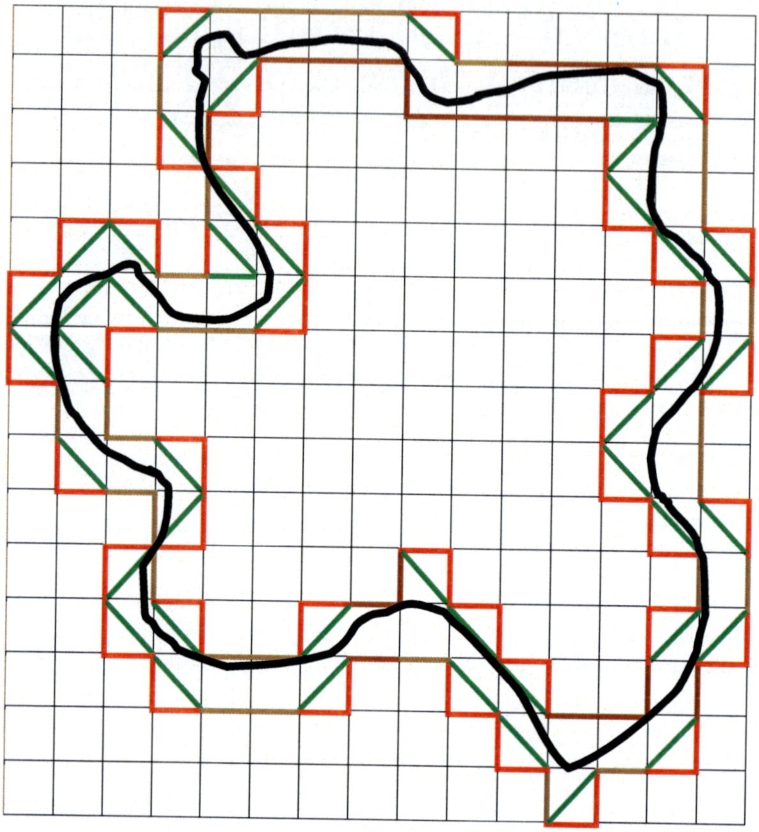

Fig. 1. Simple curve in a grid. We have four polygons surrounding that curve: two polygons (in red) that are parallel to the axes and two polygons (in green) that can pass through diagonals of the grid as well. The corresponding overlap of both types of polygons is also depicted (in brown).

2 Introduction of the Mathematical Notions

A fuzzy set in X (cf. Zadeh [14]) is given by

$$A^{'} = \{\langle x, \mu_{A'}(x)\rangle | x \in X\} \tag{1}$$

where $\mu_{A'}(x) \in [0, 1]$ is the *membership function* of the fuzzy set $A^{'}$. As opposed to the Zadeh's fuzzy set (abbreviated FS), Atanassov extended its definition to an intuitionistic fuzzy set (IFS) (cf. [2] and [3]) A, given by

$$A = \{\langle x, \mu_A(x), \nu_A(x)\rangle | x \in X\} \tag{2}$$

where: $\mu_A : X \to [0, 1]$ and $\nu_A : X \to [0, 1]$ such that

$$0 \leq \mu_A(x) + \nu_A(x) \leq 1 \tag{3}$$

and $\mu_A(x)$, $\nu_A(x) \in [0,1]$ denote a *degree of membership* and a *degree of non-membership* of x to A, respectively. An additional concept for each IFS in X, that is an obvious result of (2) and (3), is called

$$\pi_A(x) = 1 - \mu_A(x) - \nu_A(x) \tag{4}$$

the *degree of uncertainty*, expressing a lack of knowledge of whether x belongs to A or not (cf. [2]). It is obvious that $0 \leq \pi_A(x) \leq 1$, for each $x \in X$. Hesitation margins turn out to be relevant for both applications and the development of theory of IFSs.

Let us denote by $D[0,1]$ the collection of all closed subintervals in $[0,1]$, Interval-valued intuitionistic fuzzy sets (IVIFS), see [4], in X are given by

$$\tilde{A} = \{\langle x, M_A(x), N_A(x)\rangle | x \in X\} \tag{5}$$

where: $M_A : X \to D[0,1]$ and $N_A : X \to D[0,1]$ such that for all $x \in X$,

$$0 \leq \sup M_A(x) + \sup N_A(x) \leq 1. \tag{6}$$

If for any $M \in D[0,1]$ we denote

$$\underset{\sim}{M} := \inf M \text{ and } \tilde{M} := \sup M,$$

then M can be rewritten as

$$M = [\underset{\sim}{M}, \tilde{M}].$$

Hence (6) can be stated in an equivalent way as

$$0 \leq \tilde{M}_A(x) + \tilde{N}_A(x) \leq 1, \tag{7}$$

for any $x \in X$.

In [5,8] we extended the IFS formula from [9] for the area surrounding the closed curve through the introduction of new types of polygons that can better approximate the curve in the same grid step size. In that second type of polygons, the Pick's formula for the computation of surrounded by the polygons's area has been employed in the computation of an estimation of the closed simple curve's interior.

Pick's formula [11] is widely used in practical applications and as showed in [7] it can be used to define polygons approximating a simply connected domain in 2D. In [12, Ch 2.2.1] a similar construction of inner and outer approximation of a curve with polygons with sides parallel to the axes in 2D is shown. As described in [6] there is a method of counting the points inside and on the frontier of two polygons (the inner and outer in our case).

Pick's theorem provides a simple formula for calculating the area S surrounded by a polygon in terms of the number I of grid-points in the interior of the polygon, i.e. not touching any of the sides, and the number B of grid-points on the boundary, i.e. placed on the polygon's perimeter. Assuming that we have a grid with grid-step equal to one, Pick's formula provides the number of unit squares through the following expression:

$$S = I + \frac{B}{2} - 1 \tag{8}$$

Therefore, assuming a grid-step equal to l_0, the formula provides

$$S(l_0) = l_0\left(I + \frac{B}{2} - 1\right) \tag{9}$$

for the area surrounded by the given polygon.

As in [5,8], let us take a 2D Cartesian coordinate system Oxy and an arbitrary simple curve (see Fig. 1) parametrized by

$$\mathbf{r}(t) = (r_1(t), r_2(t)) : [0, 1] \to \mathbb{R} \times \mathbb{R} \tag{10}$$

and grid step size l_0. For that fixed step size in [8] we defined

$$\mathcal{S}_{j_0} = \mathcal{S}(\mathbf{r}, l_0), \tag{11}$$

to be the area of the big meshed rectangular figure surrounding the curve on Fig. 1 and it turns out an extension of (9).

The areas enclosed by the inner and outer polygons (the ones allowed to pass through diagonals) for the given grid-step will be denoted by $\mathcal{A}^i(\mathbf{r}, \frac{l_0}{2^{j_0}})$ and $\mathcal{A}^o(\mathbf{r}, \frac{l_0}{2^{j_0}})$, respectively. $\mathbf{r}(t)$ for $t \in [0, 1]$ is a fixed positively oriented parametrization of the curves and the areas $\mathcal{A}^i$ and $\mathcal{A}^o$ can be calculated by Pick's formula (9). Moreover, the numbers I and B of the interior and boundary vertices for any of the polygons can be determined, for instance, with the algorithm given by D. Alciatore and R. Miranda in [1].

Definition 1 [5]
In the above notations let us define

$$\mu_{(\mathbf{r},l_0)}(1) = \frac{\mathcal{A}^i(\mathbf{r}, \frac{l_0}{2^{j_0+1}}))}{\mathcal{S}_{j_0}} \text{ and } \nu_{(\mathbf{r},l_0)}(1) = \frac{\mathcal{S}_{j_0} - \mathcal{A}^o(\mathbf{r}, \frac{l_0}{2^{j_0+1}})}{\mathcal{S}_{j_0}}.$$

More generally, let us inductively define for any positive integer k,

$$\mu_{(\mathbf{r},l_0)}(k) = \frac{\mathcal{A}^i(\mathbf{r}, \frac{l_0}{2^{j_0+k}}))}{\mathcal{S}_{j_0}} \text{ and } \nu_{(\mathbf{r},l_0)}(k) = \frac{\mathcal{S}_{j_0} - \mathcal{A}^o(\mathbf{r}, \frac{l_0}{2^{j_0+k}})}{\mathcal{S}_{j_0}}.$$

Taking in consideration the last definition we may write down the degree of uncertainty for the kth step as $\pi_{(\mathbf{r},l_0)}(k) = 1 - \mu_{(\mathbf{r},l_0)}(k) - \nu_{(\mathbf{r},l_0)}(k)$. Therefore, we have that

$$\pi_{(\mathbf{r},l_0)}(k) = \frac{\mathcal{A}^o(\mathbf{r}, \frac{l_0}{2^{j_0+k}}) - \mathcal{A}^i(\mathbf{r}, \frac{l_0}{2^{j_0+k}})}{\mathcal{S}_{j_0}},$$

which is exactly the intuitionistic fuzzy estimation of the degree of uncertainty for the corresponding grid-step $\frac{l_0}{2^{j_0+k}}$. It has been proved that for $k_1 < k_2$ we have that $\pi_{(\mathbf{r},l_0)}(k_1) > \pi_{(\mathbf{r},l_0)}(k_2)$, which means that $\pi_{(\mathbf{r},l_0)}$ is a decreasing function on the set of positive integer numbers $\mathbb{N}$. Therefore, we may suppose that we are given a small enough positive real number ϵ_0 based on the curve $\mathbf{r}$, whose interior is supposed to be estimated. Through the described algorithm, we iteratively computed upper and lower estimations (through outer and inner polygons) until a positive integer k has been reached for which $0 \leq \pi_{(\mathbf{r},l_0)}(k) \leq \epsilon_0$. The k-th iteration provides then a satisfactory intuitionistic fuzzy estimation of the curve.

For example, for the curve from Fig. 1, the values are as follows:

$$\mu_{(\mathbf{r},l_0)}(0) = \frac{99}{225} = 0.44 \text{ and } \nu_{(\mathbf{r},l_0)}(0) = \frac{77}{225} = 0.34,$$

3 IVIF Estimation

Let us give an interval-valued intuitionistic fuzzy (IVIF) formula for the estimation of the area surrounded by the curve on Fig. 1. For the already stated,

$$\mu_{(\mathbf{r},l_0)}(0) = \frac{\mathcal{A}^i(\mathbf{r}, \frac{l_0}{2^{j_0}})}{\mathcal{S}_{j_0}} \text{ and } \nu_{(\mathbf{r},l_0)}(0) = \frac{\mathcal{S}_{j_0} - \mathcal{A}^o(\mathbf{r}, \frac{l_0}{2^{j_0}})}{\mathcal{S}_{j_0}},$$

in the previous section, we specified that $\mathcal{A}^i(\mathbf{r}, \frac{l_0}{2^{j_0}})$ and $\mathcal{A}^o(\mathbf{r}, \frac{l_0}{2^{j_0}})$ stand for the inner and respectively, outer polygons' area (the ones that can pass by diagonals of the grid).

Definition 2. Let us, for the pair of polygons (inner and outer) that can only have parallel to the axes sides, introduce the areas they bound by $\mathcal{B}^i$ and $\mathcal{B}^o$, respectively. Therefore, we can denote

$$\underset{\sim(\mathbf{r},l_0)}{\mu}(0) = \frac{\mathcal{B}^i(\mathbf{r}, \frac{l_0}{2^{j_0}})}{\mathcal{S}_{j_0}} \text{ and } \underset{\sim(\mathbf{r},l_0)}{\nu}(0) = \frac{\mathcal{S}_{j_0} - \mathcal{B}^o(\mathbf{r}, \frac{l_0}{2^{j_0}})}{\mathcal{S}_{j_0}},$$

and let us furthermore define,

$$\tilde{\mu}_{(\mathbf{r},l_0)}(0) := \mu_{(\mathbf{r},l_0)}(0) \text{ and } \tilde{\nu}_{(\mathbf{r},l_0)}(0) := \nu_{(\mathbf{r},l_0)}(0).$$

Taking initially the grid with a step of length l_0 (supposed sufficiently small), we pass along the curve with $\mathbf{r}(t)$ letting the parameter t vary from 0 to 1 and, as assumed in the previous section, the curve is positively oriented. Along its path, $\mathbf{r}(t)$ intersects the lines of the grid in different points and may pass through various squares of the grid leaving/entering them through their vertices or edges. As showed in [5,8] this procedure creates in a unique way the four polygons surrounding a simple closed curve as the one on Fig. 1. We are ready to state now the following,

Theorem 1. *For the above defined notions if we denote,*

$$M(k) := [\underset{\sim(\mathbf{r},l_0)}{\mu}, \tilde{\mu}_{(\mathbf{r},l_0)}](k) \text{ and } N(k) := [\underset{\sim(\mathbf{r},l_0)}{\nu}, \tilde{\nu}_{(\mathbf{r},l_0)}](k),$$

then $\langle k, M(k), V(k)\rangle$ *for each* $k \in \mathbb{N}$, *is a well defined IVIF estimation. Furthermore, the limit of* $\tilde{\pi}_{(\mathbf{r},l_0)}(k) := 1 - \tilde{\mu}_{(\mathbf{r},l_0)}(k) - \tilde{\nu}_{(\mathbf{r},l_0)}(k) \leq 1 - \underset{\sim(\mathbf{r},l_0)}{\mu}(k) - \underset{\sim(\mathbf{r},l_0)}{\nu}(k)$ *reaches* 0 *when* k *gets arbitrarily large.*

And therefore this construction provides an arbitrarily close approximation of the area surrounded by the closed curve.

Taking into account the positioning of the four polygons and for fixed j_0 and $\mathbf{r}$ we have that,

$$\mathcal{A}^i(\mathbf{r}, \frac{l_0}{2^{j_0}}) \leq \mathcal{B}^i(\mathbf{r}, \frac{l_0}{2^{j_0}}) \leq \mathcal{A}^o(\mathbf{r}, \frac{l_0}{2^{j_0}}) \leq \mathcal{B}^o(\mathbf{r}, \frac{l_0}{2^{j_0}}) \leq S_{j_0} \tag{12}$$

And therefore from

$$\frac{\mathcal{A}^i(\mathbf{r}, \frac{l_0}{2^{j_0}})}{S_{j_0}} \leq \frac{\mathcal{B}^i(\mathbf{r}, \frac{l_0}{2^{j_0}})}{S_{j_0}} \leq \frac{\mathcal{A}^o(\mathbf{r}, \frac{l_0}{2^{j_0}})}{S_{j_0}} \leq \frac{\mathcal{B}^o(\mathbf{r}, \frac{l_0}{2^{j_0}})}{S_{j_0}} \leq 1$$

and from Definition 1 and Definition 2 first the statement of the theorem is proved. The proof of the second statement follows the same reasoning and steps as it has been done in [5].

For example, for the curve from Fig. 1, the values are as follows:

$$M = [0.39, 0.44] \text{ and } N = [0.29, 0.34]$$

4 Conclusion

An interval-valued intuitionistic fuzzy estimation of the area surrounded by a *continuous simple closed* curve in the real 2D space has been proposed. We also proved the validity of the stated definitions, i.e. that

$$\langle k, M(k), N(k)\rangle$$

for each $k \in \mathbb{N}$, is a well defined IVIF estimation. When k iterates over the family of natural numbers $\mathbb{N}$ analogically to the IF case in [5], we get a numerical procedure that estimates the initial closed curve's area with an arbitrarily small error in the IVIF context. The introduced formulas and estimations of the closed curve can be used for modelling of the area affected by wildfire.

This work actually can be considered as an extension of our research stated in a series of the papers [5,8,9]. And in a next research, we will discuss a modification of the Pick's formula for 3D-figures.

Acknowledgements. This work is supported by the Bulgarian National Science Fund Project DN16/6/2017 "Integrated approach for modeling of the forest fires spread."

References

1. Alciatore, D., Miranda, R.: A winding number and point-in-polygon algorithm. Glaxo Virtual Anatomy Project research report, Department of Mechanical Engineering, Colorado State University (1995)
2. Atanassov, K.: Intuitionistic Fuzzy Sets: Theory and Applications. Springer, Heidelberg (1999). https://doi.org/10.1007/978-3-7908-1870-3
3. Atanassov, K.: On Intuitionistic Fuzzy Sets Theory. Springer, Berlin (2012). https://doi.org/10.1007/978-3-642-29127-2
4. Atanassov, K.: Interval-Valued Intuitionistic Fuzzy Sets. Springer, Cham (2020). https://doi.org/10.1007/978-3-030-32090-4
5. Atanassov, K., Marinov, E., Vassilev, P., Alexandrov, A.: An improved intuitionistic fuzzy estimation of the aread of 2D-figures based on the Pick's formula. In: Proceedings of the 4th International Conference on Numerical and Symbolic Computation - Developments and Applications, Porto, Portugal, 11–12 April 2019, pp. 363–374. ECCOMAS (2019). ISBN: 978-989-99410-5-2
6. Belyaev, A., Fayolle, P.-A.: Counting parallel segments: new variants of Pick's area theorem. Math. Intell. **41**(4), 1–7 (2019). https://doi.org/10.1007/s00283-019-09921-8
7. Abdala Abdala Cousin, A.O., Andrade, D., Zenon, W.T.: Pick's theorem in the classroom. Boletim da Sociedade Paranaense de Matemática **28**(1), 79-83 (2010). https://doi.org/10.5269/bspm.v28i1.10852
8. Marinov E., Velizarova, E., Atanassov, K.: An intuitionistic fuzzy estimation of the area of 2D-figures based on the Pick's formula. In: Atanassov, K., et al. (eds.) Modern Approaches in Fuzzy Sets, Intuitionistic Fuzzy Sets, Generalized Nets and Related Topics, Volume 1: Foundations, SRI-PAS, Warsaw, pp. 145–157 (2014). ISBN: 83-894-7553-7
9. Marinov, E., Velizarova, E., Atanassov, K.: An intuitionistic fuzzy estimation of the area of 2D-figures. Notes Intuitionistic Fuzzy Sets **19**(2), 57–70 (2013)
10. Munkres, J.: Topology, 2nd edn. Prentice Hall, Hoboken (2000)
11. zur Zahlenlehre, G.: Sitzungsberichte Lotos (Prag)/Pick G. Naturwissenschaftlich-Medizinschen Vereines fúr Böhmen **19**, 311–319 (1899)
12. Thomson, B.S., Bruckner, A.M., Bruckner, J.B.: Mathematical Discovery. ClassicalRealAnalysis.com (2011)
13. Zhang, Q., Xie, Q., Wang, G.: A survey on rough set theory and its applications. CAAI Trans. Intell. Technol. **1**(4), 323–333 (2016). https://doi.org/10.1016/j.trit.2016.11.001
14. Zadeh, L.A.: Fuzzy sets. Inf. Control **8**, 338–353 (1965)

The Golden Ratio of Area Method Based on Fuzzy Number Area as a Defuzzyfier

Wojciech T. Dobrosielski(✉)

Kazimierz Wielki University, J.K. Chodkiewicza 30 Street, 85-064 Bydgoszcz, Kujawsko-Pomorskie, Poland
wdobrosielski@ukw.edu.pl

Abstract. This article presents a new method of focusing, which may be implemented in fuzzy controllers. The peculiarity of the proposed method is the use of an universal solution in the form of Golden Ratio and the widely applied method of the centre of gravity. The traditional methods of defuzzying, such as: MOM, FOM, LOM, COG, BADD and many others are used in many fields, however they are not universal solutions. This conclusion enables the author of this paper to present a new solution that we refer to as the Golden Ratio of Area, in abbreviation: GRoA. To verify the efficiency of the GRoA method, the results are compared to existing defuzzying methods. An example of use is the controller process solution of the combustion process, for which only the resulting membership function is presented. To determine the fitness of this method, the measure of dispersion was used as standard deviation. The analysis of methods leads to the conclusion that there are cases that significantly deviate from central values and should be eliminated in the control. Another conclusion is that the application of a pattern other than central value will enable a surer selection of defuzzying methods. The analysis shows that the GRoA method works properly and may be used in fuzzy controller. In the last section, the author presents further propositions of research, for example ordered fuzzy numbers (OFN).

Keywords: Fuzzy logic · Defuzzification · Golden ratio · GRoA · COG

1 Introduction

The main idea of fuzzy sets, fuzzy logic [1,9,14,39] is the possibility of introducing new solutions to facilitate modelling naturally non-exact, ambiguous phenomena. The character of these solutions is found in areas [12,24,41], in which mathematical description is unavailable or too expensive for reasonable execution in exact models. The options to process this category of problems with fuzzy logic instruments are widely used in process control [15], decision-making [16] and expert systems [2,11], robotics [4,17], biological and medical sciences. As it is known, information processing based on fuzzy logic assumptions is achieved by a

K. T. Atanassov et al. (Eds.): IWIFSGN 2020/BOS/SOR 2020, LNNS 338, pp. 92–108, 2022.
https://doi.org/10.1007/978-3-030-95929-6_8

dispersed system with structure based on three main components [5]. The information exchange scenario assumes that the input numerical value that describes the given phenomenon is converted to fuzzy form, which is processed and rebuilt into a non-fuzzy value in the final phase. The transition from the resulting fuzzy set to the real numbers form is referred to as defuzzying in literature [38]. The increasingly wide use of fuzzy logic utilities in applications results in the development of fuzzy logic in itself and its components, including the defuzzying block. In most cases, the selection of the defuzzying method for the given solution depends on the expected sets of function results of the given application behaviour [20,29,30]. Classical defuzzying methods include primarily maxima methods, such as First of Maxima (FOM), Last of Maxima (LOM) or Mean of Maxima (MOM). Another interesting proposition is Centre Of Gravity (COG) and other methods, further described in papers [6,26,33].

Due to the wide range of implications/applications of fuzzy logic in different domains of human life, the creation of new solutions both widens and strengthens the domain itself. This article presents two defuzzying methods, based on the Golden Ratio formula. The last method is a new solution that includes the Golden Ratio and uses the shape of the fuzzy number.

The content of this paper consists of sections, such as: presentation of essential components of fuzzy logic, focus on the defuzzying process, presentation of the new method and analysis of presented examples. The final conclusions and remarks constitute the final section of this paper.

2 Preliminaries

2.1 Fuzzy Sets and Membership Functions

The fuzzy set concept A proposed in 1965 by Zadeh [39], in certain space X, is referred to as a pair set:

$$A = \{(x, \mu(x))\} \; \forall x \in X \tag{1.1}$$

where: μ_A is a membership function that assigns a degree of membership in set A to every component $x \in X$ (assumed space of consideration X), whereas:

$$\mu_A : X \rightarrow [0, 1] \; thus \; \mu_A(x) \in [0, 1] \tag{1.2}$$

The membership function μ_A according to the above definition replaces the conventional set theory function, enabling using less precise notions. To discuss the properties characteristic of the membership function, the following features are considered The support of the fuzzy set is a non-fuzzy set of the universum components, for which the membership function is positive. At the same time, it may be said that the set is empty, when its support is empty.

$$supp(A) = \{x : \mu_A(x) > 0, \, x \in X\} \tag{1.3}$$

Whereas the components of the fuzzy set that are full members of the set will be referred to as fuzzy set core components.

$$core(A) = \{x : \mu_A(x) = 1,\ x \in X\} \tag{1.4}$$

Another component that describes the fuzzy set is the height of the fuzzy set, referred to as the supremum of the membership function.

$$h(A) = height(A) = hgt(A) = sup_{x \in X}(\mu(x)) \tag{1.5}$$

If the height of the fuzzy set is 1, then it is a normal set. There are subnormal sets that are converted to normal forms in the normalisation process.

The discriminant of the fuzzy set may be interpretation based on the so-called $\alpha - cuts$, which defines the degree, in which the given component belongs to the membership function.

2.2 Defuzzification Operator

The real value of the resulting fuzzy set, as conclusion of the fuzzy set, enables controlling the given process. This activity is performed based on a defuzzying component with an appropriate method available. Thus, the defuzzying apparatus may be defined as:

$$W = \{f : X \rightarrow [0, 1]\} \rightarrow X \tag{1.6}$$

where: The defuzzying operator is W, the membership function is f and the universum set, in which membership function are determined, is X.

2.3 Classification of Defuzzying Operators

It is difficult to say, which method of defuzzying is better and which is worse. Thus, the selection of the most appropriate defuzzying method depends on the expected behaviour of the given object and a single universal approach does not exist. The restrictions that may apply to this problem are extensively described in paper [18]. The authors of the study present 13-point restrictions and their propositions apply to the main system types, namely the fuzzy reasoning (or knowledge) systems and fuzzy control systems. Those versions, even though they use fuzzy logic, differ in purposes and requirements. On the other hand, paper [28] shows defuzzying operator properties z, summed to four groups. The first group indicates that the defuzzying operation calculates a single value. In brief, it is an injective function and two fuzzy sets may have the same value. It should be also said that the non-fuzzy value fits in the fuzzy set. Another group of defuzzying operators is presented by the authors as the fact that the membership function determines the defuzzying value. Thus, both concentration and dilation lead to a normal fuzzy set. On the other hand, scaling and translation have no impact on the membership function. In another group, the defuzzying value of two triangular fuzzy sets fits in the range of these sets. The last property of the process described are such defuzzying methods that should assume values from the permissible range in case of barrier information. Another papers that investigated defuzzying operator properties was [29–31], which illustrated the selection of the defuzzying method of the given solution. The presented methods are useful in the design process of fuzzy rules-based systems.

2.4 Discussion of Classical Methods

Due to the fact that fuzzy logic is used in many areas of activity, different defuzzying methods are available. This part of the study briefly presents the essential methods that are discussed in detail in papers, such as [27,35,36].

The diversity of the forms to obtain a non-fuzzy value is focused in three main categories or classes of maxima methods: Random Choice Of Maxima (RCOM), First Of Maxima (FOM), Last Of Maxima (LOM), Middle Of Maxima (MOM) the selection of a non-fuzzy value is made based on the fuzzy set core; class of distribution methods/Centre Of Gravity (COG), Mean Of Maxima (MOM), Basic Defuzzification Distribution (BADD), Generalized Level Set Defuzzification (GLSD), Indexed Centre of Gravity (ICOG), Fuzzy Mean (FM) - transformation of the membership function to probability distribution and area methods/Centre of Area (COA) - division of the area under curve into parts.

In the group of available methods, classical solutions are distinguished:

FOM - First of Maxima
The FOM method is a method of selection of the smallest component from the core of set A, where the defuzzying value is presented by relationship (1.7).

$$FOM(A) = min\ core(A) \tag{1.7}$$

LOM - Last of Maxima
Proper selection of the greatest component value of the core of set A, the LOM method is used, of which the formula is as follows:

$$LOM(A) = max\ core(A) \tag{1.8}$$

MOM - Mean of Maxima
The Eq. (1.9) presents the use of the FOM and LOM method as the method, in which defuzzying values consider the minimum and maximum component of the core of fuzzy set A. The obtained value is an average of these two methods.

$$MOM(A) = \frac{min\ core(A) + max\ core(A)}{2} \tag{1.9}$$

RCOM - Random Choice of Maxima

This method is also referred to as defuzzying from the core, because the defuzzying value always fits in the fuzzy set core. The defuzzying value of this method is a random component of $x \in core(A)$, calculated as probability:

$$RCOM(A) = P(x) = \frac{\lambda(x)}{\lambda(core(A))} \tag{1.10}$$

where λ is the Lebesgue measure in universum X.
MOS - Mean of Support
The **MOS** defuzzying method, in which the defuzzying value is the average of the support of number A.

$$MOM(A) = \frac{supp(A)}{2} \tag{1.11}$$

COG - Centre of Gravity
This method is the most popular at the basis, which determines the centre of gravity of the considered system. In the defuzzying process of fuzzy number A, the **COG** method is expressed by Eq. (1.12).

$$COG(A) = \frac{\int_a^b x\mu_A(x)\mathrm{d}x}{\int_a^b \mu_A(x)\mathrm{d}x} \tag{1.12}$$

BADD - Basic Defuzzification Distribution
Defuzzying method proposed in [13], as extension of the COG and MOM methods. The value of defuzzying from the fuzzy set A is obtained as:

$$BADD(A) = \frac{\int_a^b x\mu_A^\gamma(x)\mathrm{d}x}{\int_a^b \mu_A^\gamma(x)\mathrm{d}x} \tag{1.13}$$

Depending on the parameter $\gamma \in [0, \infty]$, the BADD method may assume the following cases. When $\gamma = 0$, $BADD(A) = MOS(A)$; $= 1$, $BADD(A) = COG(A)$; $\gamma\, then \infty$, $BADD(A) = MOM(A)$.

Every year, the literature on defuzzying methods is supplemented with new items. At this point, it is worth presenting new propositions, such as Optimal Representation - ORD [22] or Optimal Representation Method [21] or RMS in [6]. It also seems advisable to present the papers from the field of fuzzy logic, in particular applications [8,19,40].

The next section of this study presents the new defuzzying method. The method that integrates well-known properties from physics, mechanics.

3 New Defuzzying Methods (Golden Ratio of Area)

The popularity of classical defuzzying methods, encountered in the reduction process of the resulting membership function, is due mainly to the domain of application in the given environment. From the available mechanisms, essentially two approaches are distinguished: the first consists of the maxima methods. Solutions characterised by simplicity both on the interpretation level and on the implementation level. The application of these methods involves fuzzy inference systems. Another group of widely used algorithms are the so-called centre of gravity methods with derivatives. These methods are derived from geometry and are particularly important for solving problems in physics and mechanics. The group of these methods is more appropriate in fuzzy controllers. Ten, it may be said that new defuzzying methods should be based either on simplicity or on a generally known property derived from nature. In a sense, these two main approaches constitute a guideline for the development of the following methods.

The genesis of the development of the Golden Ratio (GR) method is the observation of numerous cases of use of Golden Ratio in fields, such as biology, architecture, medicine or painting. A particular instance of relations is the field of genetics, in which the Golden Ratio is noticed in the DNA strand structure [3] (a deoxyribonucleic acid molecule contains 21 Angstroms of width and 34 Angstroms of length for every full cycle of the DNA double helix).

3.1 Golden Ratio - Introduction to the Solution

In most cases, the natural way of representing information is the determination of the given order in the given set. The solution to this problem is commonly presented in mathematics as number series or sequences. Focusing on a number sequence, one may think of transformation, a function reflecting a set of natural numbers in a set of real numbers. Analogically, the defocussing process in fuzzy controller systems may be thought of in the same way. Another representative of this operation, for the purposes of this section, is the Fibonacci sequence. The essence of the sequence assumes that the sequence begins from two ones and every next sequence constitutes the sum of the two previous sequences. On the other hand, the result of division of every number of the Fibonacci sequence by its precedent, the quotient of approx. $1,618033998875\ldots$ is obtained, which is the number of the Golden Ratio. In turn, a precise limit value is the golden number.

$$\lim_{n \to 0} \frac{k_{n+1}}{k_n} = 1,618033998875 \cdots = \Phi \tag{1.14}$$

Due to the fact that both Fibonacci sequence and Golden Ratio properties are closely interrelated, which is presented in particular in paper [10]. By analogy, it is concluded that the obtained value of the defuzzying stage, controlling the given process using the Golden Ratio method, will be a derivative of a certain general order, as in case of the Fibonacci sequence.

The calculation of the classical formula of the Golden Ratio shown in Fig. 1 assumes that the relation between any whole and the greater part, in this case a segment and the greater part a, is the same as between the greater part and the smaller part b. Assuming that the segment length is $a + b$, where a and b are parts of this ratio, then the following equality:

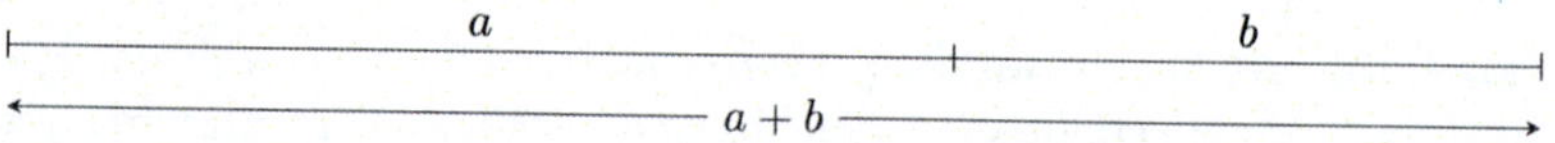

Fig. 1. Interpretation of the golden ratio for one dimension as a point between two points

$$\frac{a+b}{a} = \frac{a}{b} = \Phi \tag{1.15}$$

In this case, one method of finding the value of Φ is to transform the left-hand fraction of Eq. (1.15) into:

$$\frac{a+b}{a} = 1 + \frac{b}{a} = 1 + \frac{1}{\Phi},\ where\ \frac{b}{a} = \frac{1}{\Phi} \tag{1.16}$$

The assumed representation of the Golden Ratio according to equality (1.16), after application of proportion properties and arrangement, the following equation is obtained:

$$\Phi^2 - \Phi - 1 = 0 \tag{1.17}$$

where the only positive solution is number:

$$\Phi = \Phi_1 = \frac{1+\sqrt{5}}{2} = 1,618033998875\ldots \tag{1.18}$$

Thus, the classical Golden Ratio (GR) Method for fuzzy number A is defined as:

Definition 1

$$GR = min(supp(A)) + \frac{|supp(A)|}{\Phi} \\ where\ \Phi = 1,618033998875\ldots \tag{1.19}$$

where: GR is a defuzzying operator, $supp(A)$ is the support of fuzzy set A in universum X and $min(supp(A))$ is the minimum component of set A of the components x for which the membership function value is greater than zero.

At this point, it is worth presenting a graphic interpretation of the proposed defuzzying method, which is shown in the following Fig. 2. As one may see, the method uses the support of the given fuzzy number, on which the Golden Ratio is determined through division. Since the support of the considered fuzzy number is a "relative" number, the resulting defuzzying numbers are inappropriate in itself. Thus, to obtain the correct result, a minimum support component should be added, resulting in an absolute form of the method result GR. In other words, the ratio value is offset by a minimum component, which is $min(supp(A))$. As shown, the proposed Golden Ratio (GR) method is a simple method. This feature results from the use of a widely applied Golden Ratio solution by humans and basic arithmetic operations. Thus, Fig. 2 shows that this apparatus is only connected with the support of the given fuzzy number, not to the shape of the membership function. Thus, the proposed solution is a new group of methods based on the support of a fuzzy number.

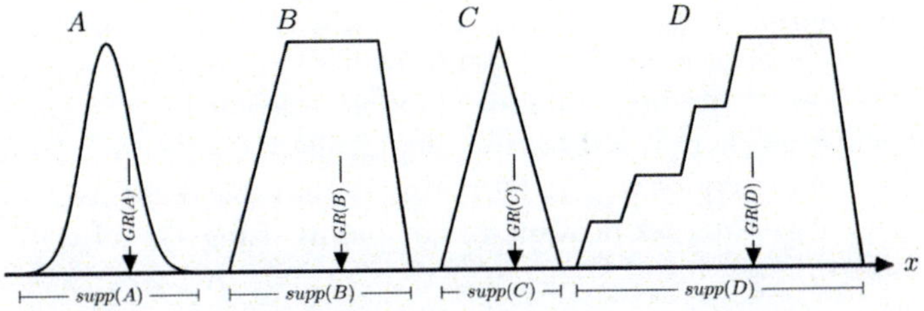

Fig. 2. The example of convex fuzzy numbers A, B, C and the resulting membership function D with an indicated defuzzification using the GR method

3.2 Golden Ratio of Area - Proposition of a New Solution

Literature shows many methods of determining a non-fuzzy value in the defuzzying process. The best known solutions include area methods: *COG (ang. Central of Gravity), COA (ang. Central of Area).* At this point, it is worth noting that methods are not restricted to one property, but applicable to the whole fuzzy number. Thus, the proposed solution of the Golden Ratio of Area (GROA) integrates two main components, such as: area and Golden Ratio methods.

The essence of the solution based on the above premises assumes that for any object (fuzzy number), a shape area is determined in relation to the Golden Ratio. In the previous section, method *GR* only applies to one dimension, in which the base of this ratio is the support. In propositions in this section, the domain considered is the area of the given number. Thus, the Golden Ration is applicable to area division in two parts, so that the ratio of the greater area to the smaller area is the same as the whole to the greater part of the shape area. Thus, it is important that the following relationship is met (1.20). The interpretation of this assumption is presented in Fig. 3, in which any shape is present. The total area P is divided by Φ, resulting in two areas: P_1 and P_2.

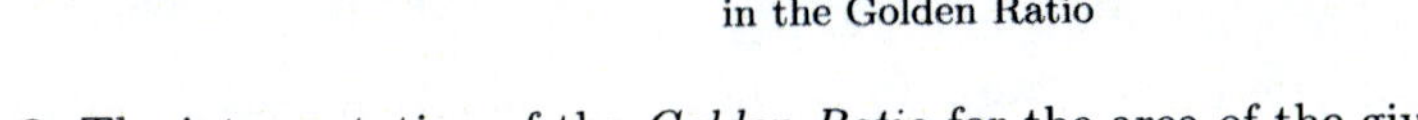

(a) Total area of the given shape

(b) Division of the total area of the shape in the Golden Ratio

Fig. 3. The interpretation of the *Golden Ratio* for the area of the given shape

$$\frac{P_1 + P_2}{P_1} = \frac{P_1}{P_2} = \Phi \tag{1.20}$$

Considering fuzzy numbers, a non-fuzzy value is obtained using the *GROA* method by calculating the area of the given number, which is then divided by value Φ. As a result of the division operation, area $P_1 = \frac{P}{\Phi}$ is obtained and thus area $P_2 = P - P_1$. Subsequently, knowing value P_1, this area is set on a fuzzy number in relation to the beginning of number (a lower limit). The determination of defuzzification is reduced to the calculation of the end of (b upper limit), for which the area is equal to P_1. Formally, it may be presented according to Eq. (1.21), where the defuzzification value sought using the GROA method is the upper limit of integration b.

$$\int_a^b \mu(x)dx = P_1 = \frac{P}{\Phi} \tag{1.21}$$

Thus, the Golden Ratio of Area (GROA) method for the convex fuzzy number A is defined as:

Definition 2. The *Golden Ratio of Area* defuzzying functional for convex fuzzy number A is expressed as:

$$GROA(A) = b \tag{1.22}$$

where: b is the upper limit of integration, which meets the following equality

$$\int_a^b \mu_A(x) = \frac{P_A}{\Phi}, \tag{1.23}$$

P_A is the area of fuzzy number $\Phi = 1,618033998875\ldots$ golden number and $a = min(supp(A))$ as the lower limit of integration.

Appropriately, Eq. (1.23) may be presented as

$$GROA = F(b) = A - F(b) \tag{1.24}$$

where coefficient A is

$$A = \frac{P}{\Phi} = \frac{\int_a^b \mu(x)dx}{\Phi} \tag{1.25}$$

and thus the proposed Golden Ratio of Area method defuzzying formula is stably presented by Eq. (1.26)

$$GROA = F(b) = \frac{P}{\Phi} = \frac{\int_a^b \mu(x)dx}{\Phi} - F(b) \tag{1.26}$$

In the graphic interpretation, the following Fig. 4 may be used. Although the form of the resulting membership function is a relatively idealised construct, it fully reflects the concept. An essential component is the determination of the limits of a fuzzy number for individual components of the resulting membership function, so that the entire area meets the Golden Ratio. More complex shapes require focus on the parts of the given shape.

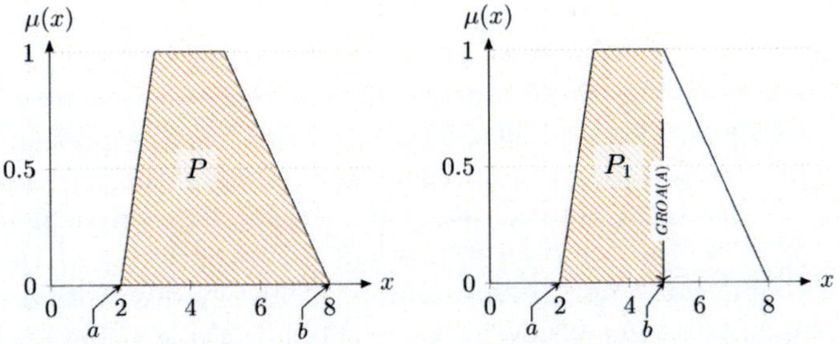

(a) Fuzzy number and its area before division.

(b) Golden ration of area defuzzifcaton value

Fig. 4. Graphic interpretation of the Golden Ratio of Area method

In conclusion of this section, it should be said that the advantage of the solution is the elimination of the imperfection of the GR method described in paper [8] This method shows the lack of impact of the shape of fuzzy number on the defuzzification result. The GR method is primarily applicable to the support of the fuzzy number. Thus, it is known to have advantages, however in a way it simplifies the preceding activities in the aggregation of the individual rules of the fuzzy controller. It may be concluded that the GRoA method is the solution that resolves this problem. In another section of the paper, calculation examples will be presented for the described method and a comparison to other instruments will be made.

4 Case Study: Fuzzy Controller of the Combustion Process

The application of fuzzy logic in the control has been widely discussed for many years. An example of application of the fuzzy controller may be found in papers, such as: [7,23,25,34,37]. However, a completely different case will be discussed, connected with the solid fuel combustion process in the form of coal, also described [32]. The optimum course of the combustion process is obtained when the process is stoichiometric. In this form, the mixture is subject to complete combustion, exhaust fumes contain no fuel or oxygen. This case is also applicable when coefficient $\lambda = 1$. Otherwise, when $\lambda < 1$, rich mixture is present and a shortage of oxygen occurs, when $\lambda > 1$, poor mixture is present and a surplus of oxygen occurs.

Table 1. Main conditions in the combustion process

Case	Type of mixture	Air-fuel equivalence ratio	Amount of oxygen
1	Rich	$\lambda < 1$	Not enough oxygen
2	Optimum	$\lambda = 1$	Optimum oxygen level
3	Poor	$\lambda > 1$	Too much oxygen

The measurement process was performed using a lambda sensor and the controlled component is the fan, which adjusts the amount of oxygen supplied to the combustion chamber. Based on this information, fuzzy controller was proposed. The solution presented here is limited only to the presentation of the resulting membership function and results obtained for the individual defuzzying methods. It has been noticed that with the increase of coefficient λ in the combustion process, the temperature lowers, so the priority of the solution is to maintain coefficient λ on the optimum level. In the description of controller operation, three cases are considered, as indicted in Table 1. In case of rich mixture, the controller enables the fan and starts supplying the oxidiser. In turn, when the mixture is poor, fan activity is reduced. Upon finding that the process is optimum, the fan maintains a constant amount of supply air (Fig. 5).

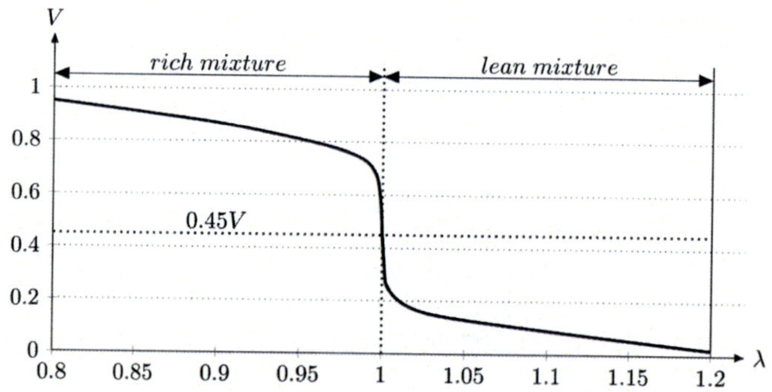

Fig. 5. Output signal of λ sensor

The cases shown in Table 1 correspond to Figs. 6, 7 and 8, while the results are presented in Table 2. The resulting membership function μout is responsible for controlling the blower and consists of three fuzzy sets. μ_A - slow in rectangular shape; μ_B - medium and μ_C - fast in trapezoid shape. The first case applies to a situation, in which low amount of oxygen was found in the combustion process. The resulting membership function presented in Fig. 6 completely includes set μ_C, while the subsequent components have areas smaller by half. The percent ratio of the share of the given set is shown in Table 4, example 1. It may be noticed that most methods are stronger classified in the area of fuzzy set μ_C. This proves proper function of the controller.

Table 2. Results of defuzzification for individual methods

	GROA	GR	COG	FOM	MOM	LOM
Example 1	6.77	6.56	7.37	8.50	9.25	10.00
Example 2	5.21	5.35	5.00	4.70	5.00	5.30
Example 3	3.65	6.56	3.92	2.30	2.50	2.70

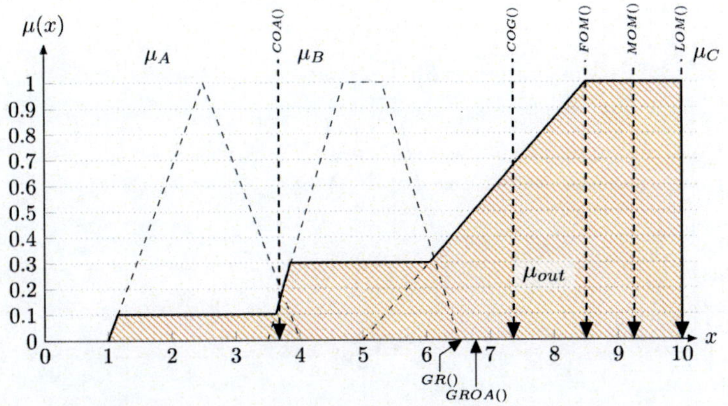

Fig. 6. Result of aggregation for case number 1

By analysing the standard deviation and deviation from the mean, it may be noticed that the GR and LOM methods are beyond this range. This assumes that the mean is the point of reference and the deviation is the permissible range. It may be said that all methods beyond this range should be eliminated from this solution (Table 3).

Table 3. Statistical results for individual methods

	Standard deviation	Mean value	Range, left side	Range, right side
Example 1	1.40	8.07	5.58	9.47
Example 2	0.24	5.09	4.85	5.34
Example 3	1.59	3.60	2.02	5.19

Stable operation of the controller is achieved in case no. 2. The resulting membership function is identified with the fuzzy set μ_B. Other sets are not involved in the control. In this case, the standard deviation is presented in Table 3 is the smallest. Defuzzification values are distributed in the smallest range. For the presented shape of the resulting membership function from Fig. 7, the FOM method in this study is classified as beyond the range of consideration.

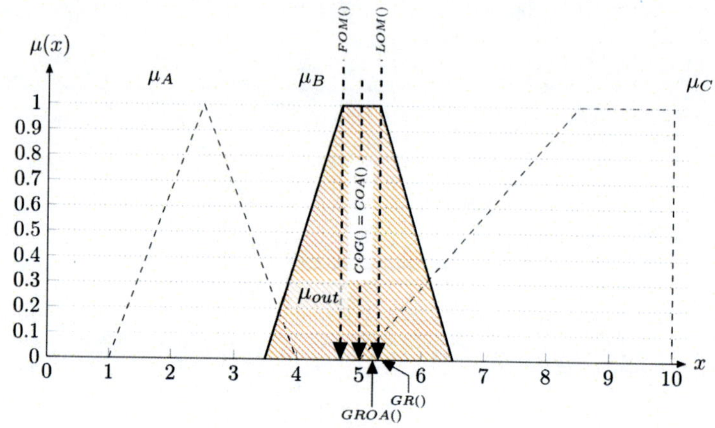

Fig. 7. Result of aggregation for case number 2

The replacement of the trapezoid with a triangular form of the membership function will enable an even higher concentration of defuzzification results, regardless of whether the FOM, LOM, MOM or COG functional is used. The GROA and GR methods constitute a certain deviation, as they aim at maintaining the Golden Ratio by nature.

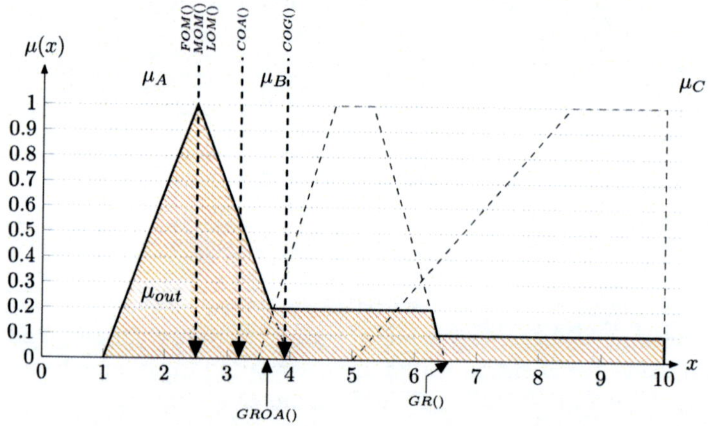

Fig. 8. Result of aggregation for case number 3

Table 4. Percent share of the set area in the given case

Fuzzy number area	Fuzzy set A	Fuzzy set B	Fuzzy set C
	1.5	1.86	3.25
% share for example 1	19.33%	44.09%	100%
% share for example 2	0%	100%	0%
% share for example 3	100%	32.6%	14.77%

The final boundary case to consider is the situation, in which there is too much oxygen. As shown in Fig. 8, the predominant fuzzy set for this case is set μ_A. The dispersion specific to this case is the largest out of the measurements presented. However, the differences between observations do not discriminate against any defuzzying method for this solution.

5 Conclusions

A fuzzy controller cannot exist without the final block, in which defuzzification is completed. For this reason, the development of new solutions in the area of defuzzying operators is perfectly reasonable. The proposed Golden Ratio of Area or GRoA defuzzying method is applicable to two well-known properties, namely Golden Ratio and centre of gravity.

The presented application is connected with fuel combustion and constitutes a background to discuss the operation of the new method. The controlled object will be a blower, which supplies oxygen to the combustion chamber. The method was compared to classical solutions, such as COG, FOM, MOM, LOM, and the previous proposition of the author of GR.

In the evaluation of the results of defuzzying methods, a measure of dispersion was used as standard deviation, which shows the deviation of the given feature from central values. In this case, it is the mean value. The results obtained show that most methods fit the set of dispersion measurements. Cases may be noticed, in which the defuzzification falls out of this range. Such behaviour, in which the value does not fit the statistical range, should be eliminated from the area of interest of the given application. Based on the analysis of the above cases, the application of another central value will be the next step in the consideration of the advisability of application of the given method. An example may be a solution, in which the expert defines the expected value. The author intends to take steps to conduct research on ordered fuzzy numbers (OFN).

References

1. Atanassov, K.T.: Intuitionistic Fuzzy Sets. Springer, Heidelberg (1999). https://doi.org/10.1007/978-3-7908-1870-3_1
2. Baumgart, J.: Implementation of a test environment in R. Stud. Mater. Appl. Comput. Sci. (ISSN 1689-6300) **11**(1), 19–24 (2019)

3. Buldyrev, S.V., Goldberger, A.L., Havlin, S., Peng, C.K., Stanley, H.E.: Fractals in biology and medicine: from DNA to the heartbeat. In: Bunde, A., Havlin, S. (eds.) Fractals in Science, pp. 49–88. Springer, Heidelberg (1994). https://doi.org/10.1007/978-3-642-77953-4_3
4. Campanile, L., Iacono, M., Marulli, F., Mastroianni, M., Mazzocca, N.: Toward a fuzzy-based approach for computational load offloading of IoT devices. J. Univ. Comput. Sci. **26**(11), 1455–1474 (2020)
5. Castro, J.L.: Fuzzy logic controllers are universal approximators. IEEE Trans. Syst. Man Cybern. **25**(4), 629–635 (1995)
6. Chandramohan, A., Rao, M., Arumugam, M.S.: Two new and useful defuzzification methods based on root mean square value. Soft. Comput. **10**(11), 1047–1059 (2006)
7. Czerniak, J.M., Dobrosielski, W.T., Apiecionek, Ł., Ewald, D.: Representation of a trend in OFN during fuzzy observance of the water level from the crisis control center. In: 2015 Federated Conference on Computer Science and Information Systems (FedCSIS), pp. 443–447. IEEE (2015)
8. Dobrosielski, W.T., Szczepański, J., Zarzycki, H.: A proposal for a method of defuzzification based on the golden ratio-GR. In: Atanassov, K., et al. (eds.) Novel Developments in Uncertainty Representation and Processing, vol. 401, pp. 75–84. Springer, Cham (2016). https://doi.org/10.1007/978-3-319-26211-6_7
9. Dubois, D., Prade, H.: Operations on fuzzy numbers. Int. J. Syst. Sci. **9**(6), 613–626 (1978)
10. Dunlap, R.A.: The Golden Ratio and Fibonacci Numbers. World Scientific, New Jersey (1997)
11. Dyczkowski, K.: Cardinalities of interval-valued fuzzy sets and their applications in decision making with imperfect information. In: Intelligent Medical Decision Support System Based on Imperfect Information. SCI, vol. 735, pp. 55–93. Springer, Cham (2018). https://doi.org/10.1007/978-3-319-67005-8_4
12. Ewald, D., Zarzycki, H., Apiecionek, Ł., Czerniak, J.M.: Ordered fuzzy numbers applied in bee swarm optimization systems **26**(11), 1475–1494 (2020). http://www.jucs.org/jucs_26_11/ordered_fuzzy_numbers_applied
13. Filev, D.P., Yager, R.R.: A generalized defuzzification method via bad distributions. Int. J. Intell. Syst. **6**(7), 687–697 (1991)
14. Kacprzyk, J., Yager, R.R., Merigo, J.M.: Towards human-centric aggregation via ordered weighted aggregation operators and linguistic data summaries: a new perspective on zadeh's inspirations. IEEE Comput. Intell. Mag. **14**(1), 16–30 (2019)
15. Kacprzyk, J.: Dynamic programming with imprecise and uncertain information. In: Kulczycki, P., Korbicz, J., Kacprzyk, J. (eds.) Automatic Control, Robotics, and Information Processing. Studies in Systems, Decision and Control, pp. 387–422, vol. 296. Springer, Cham (2020). https://doi.org/10.1007/978-3-030-48587-0_13
16. Kacprzyk, J., Zadrożny, S.: Supporting consensus reaching processes under fuzzy preferences and a fuzzy majority via linguistic summaries. In: Greco, S., Marques Pereira, R.A., Squillante, M., Yager, R.R., Kacprzyk, J. (eds.) Preferences and Decisions. Studies in Fuzziness and Soft Computing, vol. 257, pp. 261–279. Springer, Heidelberg (2010). https://doi.org/10.1007/978-3-642-15976-3_15
17. Kopowski, J., Rojek, I., Mikolajewski, D., Macko, M.: Bydgostian hand exoskeleton - concept and preliminary findings. Stud. Mater. Appl. Comput. Sci. **12**(1), 22–28 (2020). ISSN 1689-6300
18. Leekwijck, W.V., Kerre, E.E.: Defuzzification: criteria and classification. Fuzzy Sets Syst. **108**(2), 159–178 (1999). https://www.sciencedirect.com/science/article/pii/S0165011497003370

19. Molnár, J., Gans, Slavko, O.: Design and implementation self-balancing robot. In: 2020 IEEE Problems of Automated Electrodrive. Theory and Practice (PAEP), pp. 1–4 (2020)
20. Pfluger, N., Yen, J., Langari, R.: A defuzzification strategy for a fuzzy logic controller employing prohibitive information in command formulation. In: [1992 Proceedings] IEEE International Conference on Fuzzy Systems, pp. 717–723. IEEE (1992)
21. Piegat, A., Tomaszewska, M.: Defuzzification with optimal representation method. Przegląd Elektrotechniczny **1**, 110–113 (2017)
22. Piegat, A., Tomaszewska, K.: Optimal representation (ORD) method of intuitionistic fuzzy defuzzification. In: Atanassov, K.T., et al. (eds.) Uncertainty and Imprecision in Decision Making and Decision Support: Cross-Fertilization, New Models and Applications. AISC, pp. 71–82. Springer, Cham (2018). https://doi.org/10.1007/978-3-319-65545-1_8
23. Prokopowicz, P.: The use of ordered fuzzy numbers for modelling changes in dynamic processes. Inf. Sci. **470**, 1–14 (2019)
24. Prokopowicz, P., Mikołajewski, D., Tyburek, K., Kotlarz, P.: Fuzzy-based description of computational complexity of central nervous systems. J. Telecommun. Inf. Technol. **3**, 57–66 (2020)
25. Reznik, L., Ghanayem, O., Bourmistrov, A.: PID plus fuzzy controller structures as a design base for industrial applications. Eng. Appl. Artif. Intell. **13**(4), 419–430 (2000)
26. Rondeau, L., Ruelas, R., Levrat, L., Lamotte, M.: A defuzzification method respecting the fuzzification. Fuzzy Sets Syst. **86**(3), 311–320 (1997). http://www.sciencedirect.com/science/article/pii/S0165011495003991
27. Ross, T.J., et al.: Fuzzy Logic with Engineering Applications, vol. 2. Wiley, New York (2004)
28. Roychowdhury, S., Pedrycz, W.: A survey of defuzzification strategies. Int. J. Intell. Syst. **16**(6), 679–695 (2001)
29. Runkler, T.A.: Extended defuzzification methods and their properties. In: Proceedings of IEEE 5th International Fuzzy Systems, vol. 1, pp. 694–700 (1996)
30. Runkler, T.A.: Selection of appropriate defuzzification methods using application specific properties. IEEE Trans. Fuzzy Syst. **5**(1), 72–79 (1997)
31. Runkler, T.A., Glesner, M.: A set of axioms for defuzzification strategies towards a theory of rational defuzzification operators. In: [Proceedings 1993] Second IEEE International Conference on Fuzzy Systems, vol. 2, pp. 1161–1166 (1993)
32. Ruusunen, M., Leiviskä, K.: Fuzzy modelling of carbon dioxide in a burning process. Control. Eng. Pract. **12**(5), 607–614 (2004)
33. Saade, J.J., Diab, H.B.: Defuzzification techniques for fuzzy controllers. IEEE Trans. Syst. Man Cybernet. B (Cybernetics) **30**(1), 223–229 (2000)
34. Sánchez-Solano, S., Barriga, A., Jiménez, C., Huertas, J.: Design and application of digital fuzzy controllers. In: Proceedings of 6th International Fuzzy Systems Conference, vol. 2, pp. 869–874. IEEE (1997)
35. Talon, A., Curt, C.: Selection of appropriate defuzzification methods: application to the assessment of dam performance. Expert Syst. Appl. **70**, 160–174 (2017)
36. Wang, Y.M.: Centroid defuzzification and the maximizing set and minimizing set ranking based on alpha level sets. Comput. Ind. Eng. **57**(1), 228–236 (2009)
37. Wang, Y., Shi, P., Yan, H.: Reliable control of fuzzy singularly perturbed systems and its application to electronic circuits. IEEE Trans. Circuits Syst. I Regul. Pap. **65**(10), 3519–3528 (2018)

38. Yager, R.R., Filev, D.: On the issue of defuzzification and selection based on a fuzzy set. Fuzzy Sets Syst. **55**(3), 255–271 (1993). http://www.sciencedirect.com/science/article/pii/016501149390252D
39. Zadeh, L.: Fuzzy sets. Inf. Control **8**(3), 338–353 (1965)
40. Zarzycki, H., Dobrosielski, W., Apiecionek, Ł., Vince, T.: Center of circles intersection, a new defuzzification method for fuzzy numbers. Bull. Pol. Acad. Sci. Tech. Sci. **68**(2) (2020)
41. Zarzycki, H., Apiecionek, Ł., Czerniak, J.M., Ewald, D.: The proposal of fuzzy observation and detection of massive data DDOS attack threat. In: Atanassov, K.T., et al. (eds.) Uncertainty and Imprecision in Decision Making and Decision Support: New Challenges, Solutions and Perspectives. AISC, vol. 1081, pp. 363–378. Springer, Cham (2021). https://doi.org/10.1007/978-3-030-47024-1_34

Mathematical Modeling of Processes and Systems

Generalized Net Model of Implementation of Port Knocking on RouterOS

Krasiyan Andreev[1] (✉) and Ivelina Vardeva[2]

[1] National Customs Agency of Bulgaria, Sofia, Bulgaria
[2] Faculty of Technical Sciences, University "Asen Zlatarov", Burgas, Bulgaria

Abstract. The main goal of the development is to study the information processes taking place in computer networks and network security by modelling them with Generalized networks. This will allow to monitor and analyse the ongoing processes in computer networks the possibilities for building network connections the processes of attack when exchanging information, as well as increasing security in the network. The results obtained are the basis of scientifically based recommendations related to the implementation of computer networks, increasing network security and protection of confidential information transmitted through the social network. The need for network security in today's increasingly digital and corporate world is now a must. In practice, all companies or organizations are forced to provide Internet services to their users. In the present research we develop Generalized Net [1, 3] models of the network security feature of port knocking, as a security builder for MikroTik communication devices via the RouterOS operating system. The practical application of generalized networks is a direction of research in the field of generalized networks, together with the theoretical studies of the connections between generalized networks and other concepts. Indexed matrices are used to describe the states of the token [2].

Keywords: Generalized net · Port knocking · Security of network devices · Implementation of port knocking

1 Introduction

Network-based (network-based languages) these languages represent processes as networks. The process flow is easy to understand, but can be problems with abstraction. Most such languages are based on Petri nets, which are widely used for modeling. The theory of generalized networks was introduced by K. Atanassov as an extension of Petri nets. They are a tool for modeling flexible structures in which processes run in parallel over time. The generalized network is built of many transitions, containing conditions under which the nuclei pass from the input to its output places [1, 3].

For the IMs, operations that are analogues of the usual matrix operations of addition and multiplication are definite as well as other specific ones.

We obtain the definitions for standard matrix operation. In the IM case we can use different symbols as indices of the rows and columns and they give us additional information and possibilities for description [2].

K. T. Atanassov et al. (Eds.): IWIFSGN 2020/BOS/SOR 2020, LNNS 338, pp. 111–119, 2022.
https://doi.org/10.1007/978-3-030-95929-6_9

From our experience often have been asked how to protect access to communication devices when necessary in the absence of a secure connection or unauthorized Access point. One option is to use so-called port tapping port knock. As a major difference from the standard port knock, which includes a series of interdependent rules, here we add execution time for each countdown rule, which allows specific actions to be performed for a specific set time.

Computer network security is a must-have component in today's world. The widespread use of the Internet and the steadily increasing speed of access are changing telecommunications in the exchange of information between people, organizations and countries, making security threats a significant risk.

Threats are constantly changing with the development of information technology and security. Changing them requires appropriate response from manufacturers of remedies. All organizations strive for the secure storage and transmission of information and the provision of relevant services to their users, at the same time as zeros and ones.

Sources of threat may be potential anthropogenic, man-made or natural security threats. The threat may be potentially or actually existing to take any action against the object of protection that harms the owner or user, such as loss of integrity or confidentiality [4, 8, 9].

To provide information security, ISO has developed a supplement to the OSI model called Security Architecture. ISO 7498-2:1989 addresses the security of computer networks based on the Open Network Interaction Model.

In the standard:

- The results of the analysis of existing security threats are outlined;
- Specific terminology is introduced to describe security services and mechanisms;
- Recommendations on control and evaluation of the proposed means for ensuring the security of computer networks are presented;
- A network security management concept is proposed.

Various ways of protecting the storage and transmission of information are being developed. Defining a security policy in an organization is the first step in implementing security. The purpose of information and network security is to ensure reliability, integrity and confidentiality.

2 Methodology

The purpose of this article is to present a comprehensive concept for protecting a router at remote access from different points, achieving a high level of security. This article describes options for configuring a firewall using port knocking. The advantages of this type of remote administration are presented and in which cases it is recommended to use it in practice. In order to be able to properly select the appropriate means of protection for communication devices, we first need to have the fullest possible knowledge of a variety of network attacks. Switches and routers work on the second and third layers of the OSI model, respectively, and we will pay the most attention to attacks against them:

- Inability to provide service (DoS attacks);
- Distributed DoS(DDoS attacks);
- Session hijacking attacks.

Anyone can perform the attacks listed above for anyone, and it is imperative that the network security professional understands fully how the attack is conducted in order to counter it [4, 6, 8, 9].

There are many security tools available to help reduce network vulnerability. For example, a firewall can be implemented on the network to prevent multiple attacks. But it is only a small part of the network security infrastructure. Good end-to-end security, regular assessments of the overall level of network vulnerability, good practices for authentication, authorization and accountability, network intrusion detection systems, network behavior monitoring systems - all of which have become an almost mandatory part of building network security and combat active and passive network attacks.

The criteria defined: security, reliability and integrity are mandatory, but different systems and business processes pay different attention to each of these characteristics, depending on which one is more prioritized for their business. Information security is the protection of information, regardless of its form - electronic, printed or otherwise. Computer security focuses on the correct operation of computer systems and networks and the information they process. Information security is the risk management practices associated with the use, operation, storage and transmission of information, as well as the systems and processes used for these purposes. The article fully covers the protection of Mikrotik network devices and the implementation of remote secure access to a network device from different access points [4–10].

RouterOS is a standalone operating system based on the Linux v3.3.5 kernel, designed primarily for MikroTik RouterBOARD.

What Is Port Knocking and What Is It Used for?

Port Knocking is a method that allows remote access to the router only after certain successive attempts have been made to connect to predefined ports. Once we have the correct sequence of connection attempts, RouterOS dynamically adds IP to the allowed address list and gives access to the router. Knocking on a specific port may also contain a text string sent to the device we are tapping on (usually in order to add additional complexity and security). Typical application of the mechanism is that the ports for the management of the default router be closed until the specific procedure described above (known only to the administrator) has been completed. In RouterOS, the firewall is managed by Firewall, which includes sections in the following order: Firewall Rules, NAT, Mangle, Raw, Service Ports, Connections, Address Lists, Layer7 Protocols. Each section controls specific firewall features. However, the result of certain functions in one section can be used to extend functions in another.

The firewall works on the basis of consistent implementation of rules, which are divided into two parts:

- Matcher - something that is checked for compliance;
- Action - what to do if a condition is found.

Conditions and actions may vary depending on what function is performed, in which section of the firewall is running. The conditions are checked for parameters such as:

- sender's MAC address;
- IP address (network or address list);
- type of communication (broadcast, local, multicast, unicast);
- port or range of ports;
- protocol;
- protocol operations (ICMP types and codes, TCP flags, IP options);
- the interface to which the package arrives or through which others are sent.

The actions that take place determine the life of the packages, their subsequent processing, make changes to certain fields in the header and more. The rules are grouped and organized into chains. This allows the packages to be grouped according to one common criterion and to be processed. They can then be transmitted if necessary to other circuits for processing according to other criteria. Pre-defined circuits exist, but administrators can create their own circuits where necessary to optimize traffic processing.

Results. Firewall Lists - The firewall address lists provide the ability to group IPs and networks under a common name. These lists can be used as a condition to look for compliance when creating rules in a filter, broadcasting network connections, tagging, or early packet processing. Address lists can be created manually or dynamically (DHCP). IP addresses (administrators; hackers; country networks; wireless clients; social network addresses, etc.) can be identified and identified through address lists. Address lists help to optimally resolve situations that, for example, require many rules to be created in order to block many IPs. Instead, it is possible to create a single list of all the IP addresses that we want to disable and add one rule to the firewall - blocking them. We can add a list of addresses using the add-dst-to-address-list or add-src-to address-list commands.

Implementation of Port Knocking

In MikroTik RouterOS, port knocking is implemented through specific firewall rules (IP/Firewall/Filter). Performing a configuration that works requires knowing both the address-list and knowing the add-src-to-address-list action in filter. For a higher level of protection, the application of the L7 Firewall is possible. As we explained above, each rule in the RouterOS firewall consists of two parts: matcher and action. The variety of conditions allows for "knocking on ports" at different levels according to the OSI model. Before describing options for Port Knocking, there are also some good practices and recommendations to consider:

- Never start your rules with ICMP a number of port scanning applications first send an ICMP host check message;
- Be sure to accept established and related type connections before the prohibition rule. Otherwise you will have a connection, but only for a certain time - this is set at the last address list;
- Do not directly implement ready-made examples from the Internet - change at least port numbers;

- Make complex rules (combine TCP and UDP) and automate "tapping" in the most convenient way for you: through graphical applications, write scripts, etc.;
- If you like Port Knocking - do not delay its implementation - the bots will not delay their attacks;
- Do not use the L7 Firewall in your first rule or in all rules - you do not need to analyze all packages, but only those that already meet the criteria.

The Netfilter infrastructure included in Linux kernels after version 2.3.15 is the common base in the Linux kernel to which other components (such as the iptables module) are included. Although RouterOS is Linux-based - Netfilter, respectively, the Linux kernel does not offer direct support for port knocking either, but they do contain all of the building blocks needed to implement it. The most important of these is the Recent module, which has the ability to:

- Holds a list of recent packages that meet our requirements
- Decides on the number of recorded incoming packets from a particular source based on whether it is above or below a certain threshold (does it meet a given condition)
- Any number of individually named lists can be saved. Assuming, for example, we have three things that would look like this:
- Knocking 1 - Identifies the addresses of the source that made the first tap, but not the second and third.
- Knocking 2 - Identifies the addresses of the source that performed the first and second knock but not the third.
- Knocking 3 - Identifies the addresses of the source that performed the first, second, and third tap, but has not yet established a connection.

Therefore, the source address cannot exist in more than one sheet, so there are four possible states in which a source can be (Table 1):

Table 1. Possible source states.

State	Name list	Description
0	/	Waiting for the first knock
1	KNOCK1	Waiting for the second knock
2	KNOCK2	Waiting for the third knock
3	KNOCK3	Waiting to creation of connection

The whole six-part process are:

1. Accept traffic that should not be subject to Port Knocking
2. Servicing status "0"
3. Servicing status "1"
4. Service status "2"

5. Servicing status "3"
6. Choosing the right chain

As illustrated in Fig. 1, we will consider a GNs model containing four transitions, which correspond to the following aspects of the above described port knocking.

K represents the knockers' actions;
R represents the router actions;
L represents the lists of check actions;

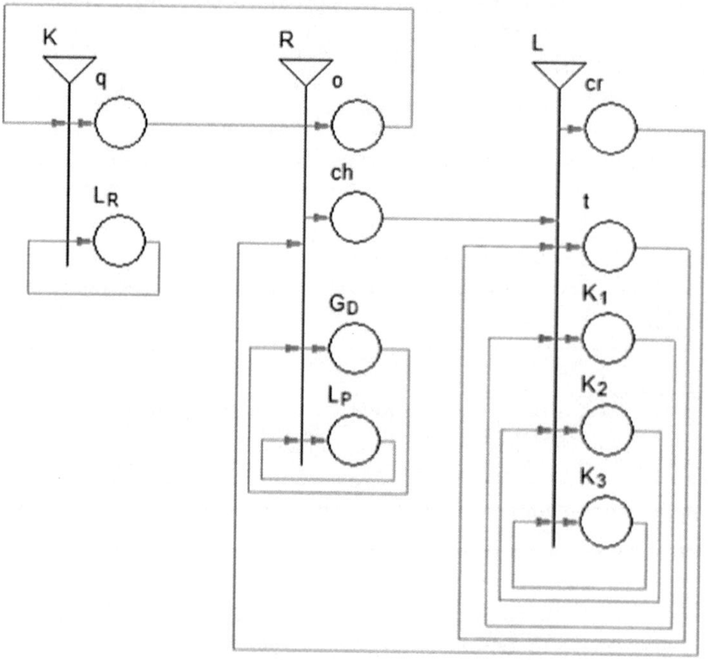

Fig. 1. GN model of implementation of port knocking

Transition K has the following form:

$$R_1 = \langle \{q, L_R\}, \{q, L_R\}, R_1 \vee (q, L_R) \rangle,$$

$$R_1 = \begin{array}{c|cc} & q & L_r \\ \hline q & false & true \\ L_R & W_{L_R,q} & true \end{array}$$

where:

$W_{L_R,q}$ = "there is a request from a knocker to router".

The tokens that enter via place L_R, which is one of the global entrances to the network has the characteristic "*request to knock on a specific port*". The token enter in place q has initial characteristic "*sending request*".

Transition R has the following form:

$$R_2 = \langle\{q, cr, G_D, L_P\}, \{o, ch, G_D, L_P\}, R_2 \vee (q, cr, G_D, L_P)\rangle,$$

R_2 =

	o	ch	G_D	L_P
q	*false*	*false*	*true*	*true*
cr	*false*	*false*	*true*	*true*
G_D	*false*	$W_{G_D,ch}$	*true*	*false*
L_P	$W_{L_P,o}$	$W_{L_P,ch}$	W_{L_P,G_D}	*true*

where:

$W_{G_D,ch}$ = "there is a request to check the received package";
$W_{L_P,o}$ = "the resulting sequential packages are correct for all checks";
$W_{L_P,ch}$ = "current package is sent for check";
W_{L_P,G_D} = "the decisions are meet the requirements".

After transition K the tokens enter via place G_D and L_P have the following characteristics: "decides on recorded incoming packets (does it meet a given condition)" and "list of recent packages that meet our requirements".

Transition L has the following form:

$$R_3 = \langle\{ch, t, K_1, K_2, K_3\}, \{cr, t, K_1, K_2, K_3\}, R_3 \vee (ch, t, K_1, K_2, K_3)\rangle$$

R_3 =

	cr	t	K_1	K_2	K_3
ch	*true*	$W_{ch,t}$	W_{ch,K_1}	W_{ch,K_2}	W_{ch,K_3}
t	*false*	*false*	W_{t,K_1}	W_{t,K_2}	W_{t,K_3}
K_1	*true*	*false*	*false*	*false*	*false*
K_2	*true*	*false*	*false*	*false*	*false*
K_3	*true*	*false*	*false*	*false*	*false*

where:

$W_{ch,t}$ = "there is valid countdown";

W_{ch,K_1} = "there is first check of the package";
W_{ch,K_2} = "there is second check of the package";
W_{ch,K_3} = "there is third check of the package";
W_{t,K_1} = "there is countdown for first check";
W_{t,K_2} = "there is countdown for second check";
W_{t,K_3} = "there is countdown for third check".

After the transition L the tokens that enter places cr, K_1, K_2, K_3 obtain characteristic respectively: "received packet for port verification via knock 1", "received packet for port verification for knock 2" and "received packet via port verification via knock 3".

In Part One, using the filters in the firewall, we make the necessary settings to skip the subsequent packets after the decision to skip the first packet is made - they are part of the same connection and must pass. In parts two to five, if the received packet is on the correct port and the other conditions are fulfilled, we go to the next state - if the conditions are not met, it returns to zero or one (depends on the set time for authentication and waiting time in states from 1 to 3 - usually set to one minute) Fig. 1.

What is common between the different Port Knocking implementation options is that the host-to-host communication in which information flows through virtually closed ports does not return a confirmation to the sender after it has been received. Any combination of ports and transport protocols can be used for the proposed graduations. The total number of possible sequences made for each router that the attacker must "guess" is high enough. Even if the hacker has information that only two ports are used (the first, the easiest method), this reduces things to 64,000 possible TCP, UDP and ICMP ports to choose from. The result of possible combinations for the attacker becomes millions. Another benefit is that even using Port Scanners programs, checking open and closed ports would also not work, since the Port Knocking service uses closed ports while listening. Port Knocking could serve as an extra layer of security, protecting high-risk remote services such as SSH and RDP. Of course, very often the Port Knock service is criticized for hacking a hacker to capture and repeat a successful sequence of port taps or a series of bits security types such as tunnels or allowed IPs. Assuming that a hacker manages to break through Port Knocking protection the worst option would be that in fact the intruder would be able to fight pre-set security measures, knocking on ports comes as an add-on (facilitating the work of the system administrator) to them.

Any operating system with the right clients and server software can benefit from this protection, regardless of whether a similar task is at the planning stage of a network security policy [4–10].

The complexity and capabilities of tools and methods for attacking network devices are steadily increasing, which in turn leads to the continued development of security technologies against them. Specialists in the field need to be aware of the most current threats and ways of eliminating them, which requires continuous updating of their knowledge and skills. Along with this, the upgrading of network devices on a hardware and software level goes hand in hand. It is imperative that any opportunity to add or upgrade a service be considered and, if necessary, implemented in the company's network security policy. Remote access (not just for network devices) is no longer a luxury or a convenience but a necessity.

3 Conclusion

The presented GN model describes the port knocking represents a stealthy authentication system that employs closed ports to carry out identification of trusted users. This novel method provides the means of establishing a connection to an application running on a completely isolated system on which no ports initially are open. In this article, we develop a GN model describing the performance of knocking on RouterOS ports. The presented model can be used to study port knocking processes in the future.

References

1. Atanassov, K.: Generalized Nets. World Scientific, Singapore (1991). [1]
2. Atanassov, K.: Index Matrices: Towards an Augmented Matrix Calculus. Springer, Berlin (2014). https://doi.org/10.1007/978-3-319-10945-9
3. Atanassov, K.: On Generalized Nets Theory. "Prof. M. Drinov" Academic Publishing House, Sofia (2007)
4. Boyadzhiev, D.: MikroTik RouterOS Basics (2016). For The Letters, Sofia, ISSN: 978-619-185-252-9
5. Budiman, A.: Meningkatkan Keamanan VPN Server dengan Menggunakan Metode Port Knocking berbasis MikroTik RouterOS, August 2014. https://repository.unikom.ac.id/28652
6. Hart, T.: Networking with MikroTik: MTCNA Study Guide. Independently Published (2017)
7. Petrov, Y.: Determining the area of the automobile tire contact footprint using generalized nets. Annual of Assen Zlatarov University, Burgas, Bulgaria, v. XLIX(1) (2020)
8. Popeea, T., Olteanu, V., Gheorghe, L., Rughinis, R.: Extension of a port knocking client-server architecture with NTP synchronization. In: 10th RoEeduNet International Conference (RoEduNet), pp. 1–5 (2011)
9. Mursyidah, Husaini, Atthariq, Arhami, M., Hidayat, H.T., Anita, Ramadhona: Analysis and implementation of the Port Knocking method using Firewall-based Mikrotik RouterOS. In: International Conference on Science and Innovated Engineering (I-COSINE), pp. 1–12 (2019)
10. Staneva, L.A.: Minimising using the Method of Quine-McCluskey with Generalised Nets. In: 29th Annual Conference of the European Association for Education in Electrical and Information Engineering, pp. 1–3. IEEE (2019)

Check for updates

Generalized Nets as a Tool for Modelling of Railway Networks. Part 2

Toncho Boyukov[1(✉)] and Krassimir T. Atanassov[2,3]

[1] Faculty of Technical Science, Prof. Asen Zlatarov University, 1 Prof. Yakimov Street, 8010 Burgas, Bulgaria
toncho_b@abv.bg

[2] Department of Bioinformatics and Mathematical Modelling, Institute of Biophysics and Biomedical Engineering, Bulgarian Academy of Sciences, Acad. G. Bonchev Street, Bl. 105, 1113 Sofia, Bulgaria
krat@bas.bg

[3] Intelligent Systems Laboratory, Prof. Asen Zlatarov University, 1 Prof. Yakimov Street, 8010 Burgas, Bulgaria

Abstract. In this paper a new (second) example is given that shows that generalized nets can be used as a tool for modelling of railway networks. In the first part of the research, a generalized net model of the railway network in Southern Bulgaria was given, while here, a generalized net model of the railway network in Northern Bulgaria is described.

Keywords: Generalized nets · Railway

1 Introduction

The railway traffic control is an important activity for each country. There are a lot of models by different tools of the national railway networks (see, e.g., [1,6–8,10–20]). Petri nets are very suitable tool for these models.

As it is mentioned in [9], Generalized Nets (GNs, [2–4]) are a tool for modelling of parallel processes, including as partial cases the standard Petri nets and all their modifications and extensions.

In [9], for the first time, GNs are used as a tool for modelling of processes in the railway transport. Here, we continue this research. The final aim of ours is to construct a detailed GN-model of whole railway network of Bulgaria.

In Sect. 2, we give description of some standard components of the railway network. In Sect. 3, we describe a concrete example with a part of the railway network in Northern Bulgaria.

2 Generalized Net Models of Some Standard Components of the Railway Network

It is easy to see that the railway network of any country has the form of a non-oriented graph. Therefore, we can represent it by a GN. Each train station is represented by a transition with the form of Fig. 1.

K. T. Atanassov et al. (Eds.): IWIFSGN 2020/BOS/SOR 2020, LNNS 338, pp. 120–128, 2022.
https://doi.org/10.1007/978-3-030-95929-6_10

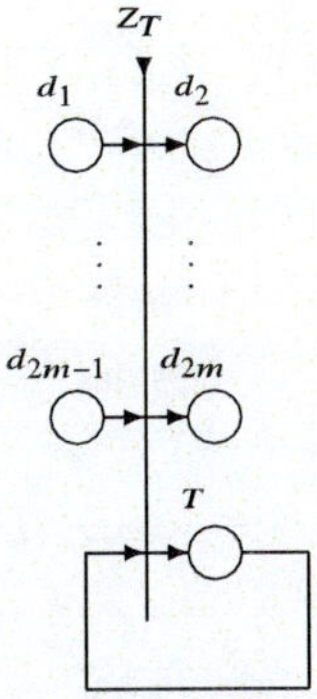

Fig. 1. First transition interpretation of a train station

In it, the train station T has direct connections with m other stations denoted by $S_1, ..., S_m$. Here, places d_{2i-1} and d_{2i} correspond to railway lines from and to station S_i for $i = 1, 2, ..., m$. The index matrix (see [5]) of the transition conditions has the form

$$\begin{array}{c|ccccc} & d_2 & d_4 & \dots & d_{2m} & X \\ \hline d_1 & false & false & \dots & false & true \\ d_3 & false & false & \dots & false & true \\ \vdots & \vdots & \vdots & \vdots & \vdots & \vdots \\ d_{2m-1} & false & false & \dots & false & true \\ T & W_{T,2} & W_{T,4} & \dots & W_{T,2m} & W_{T,T} \end{array},$$

where

$W_{T,2}$ = "the train is directed to train station S_1",
$W_{T,4}$ = "the train is directed to train station S_2", ...,
$W_{T,2m}$ = "the train is directed to train station S_m",
$W_{T,T}$ = "the train must shunt in train station T".

The present (very simple) he GN-model can be made more detailed by the hierarchical operators, defined over the GNs (see [2,3]). In this case, it might obtain the form from Fig. 2, in which all above notations are preserved.

Now, the train station has n parallel interior lines and let place l_k correspond to the k-th of these lines. By this way we can model the separate lines of the trains in the region of station T. In the present case, the index matrix has the form

	d_2	d_4	...	d_{2m}	l_1	...	l_n
d_1	$false$	$false$	...	$false$	$V_{1,1}$	...	$V_{1,n}$
d_3	$false$	$false$	...	$false$	$V_{3,1}$	...	$V_{3,n}$
$\vdots$	$\vdots$	$\vdots$	$\vdots$	$\vdots$	$\vdots$		
d_{2m-1}	$false$	$false$	...	$false$	$V_{2m-1,1}$	...	$V_{2m-1,n}$,
l_1	$W_{1,2}$	$W_{1,4}$	...	$W_{1,2m}$	$U_{1,1}$	...	$U_{1,n}$
$\vdots$	$\vdots$	$\vdots$	$\vdots$	$\vdots$	$\vdots$		
l_n	$W_{n,2}$	$W_{n,4}$	...	$W_{n,2m}$	$U_{n,1}$	...	$U_{n,n}$

where

$V_{1,1} = \cdots = V_{2m-1,1}$ = "the train is directed to the first interior line l_1", ...
$V_{1,n} = \cdots = V_{2m-1,n}$ = "the train is directed to the n-th interior line l_n",
$W_{1,2} = \cdots = W_{n,2}$ = "the train is directed to train station S_1", ...
$W_{T,2m} = \cdots = W_{n,2m}$ = "the train is directed to train station S_m",
$U_{1,1} = \cdots = U_{n,1}$ = "the train is directed to the first interior line l_1", ...
$U_{1,n} = \cdots = U_{n,n}$ = "the train is directed to the n-th interior line l_n",

where the sense of predicates $U_{1,1}, U_{2,2}, \ldots, U_{n,n}$ is that the train does not change its interior line.

So, we can describe the train moves to and from a given train station, and these in it.

In the next Section, we illustrate the possibility to represent the railway network of a given country by GNs, using as an example, the railway network in Northern Bulgaria between Sofia and Varna. In the model that follows, we describe only the most important train connections.

3 Generalized Net Model of a Part of the Railway Network in Northern Bulgaria

The GN-model contains 11 transitions and 45 places. Its tokens represent separate trains that will move from a train station to the next one or with shunt in a train station.

The places denoted by alphabetic indices represent the following train stations in Northern Bulgaria: *So* - Sofia, *Vi* - Vidin, *Me* - Mezdra, *Ru* - Ruse, *GO* - Gorna Oryahovica, *Sh* - Shumen, *Ka* - Kaspichan, *Om* - Omurtag, *Do* - Dobrich, *Su* - Suvorovo, *Va* - Varna. These places are simultaneously inputs and outputs of the transitions. The places, denoted by numerical indices represent the activities of the trains related to entering or leaving the respective train station. Only the following numbers places correspond to town outside Northern Bulgaria: l_2 - Kalafat (Romania), l_8 - Giurgiu (Romania), l_9 - Tulovo, l_{20} - Karnobat, l_{27} - Negru Voda (Romania).

Each token, entering a place, obtains as a current characteristic information about the time for realization of the previous activity.

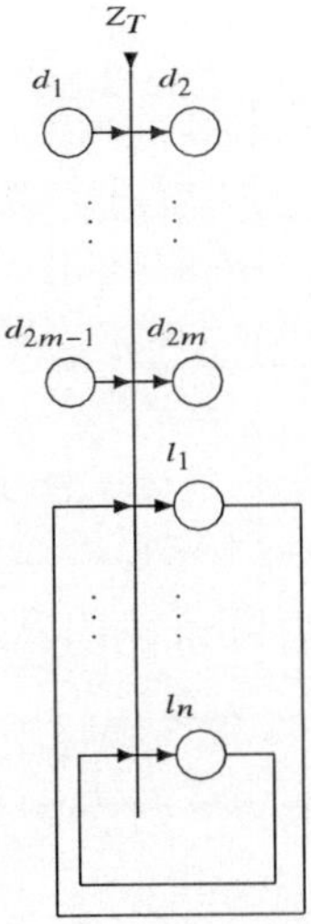

Fig. 2. First transition interpretation of a train station

In places, denoted by alphabetic indices, the token will obtain as additional information the train station from where it arrives and the number of passengers that leave the respective train or catch it.

The GN-transitions are the following.

$$Z_{So} = \langle \{l_5, l_{So}\}, \{l_1, l_{So}\}, \begin{array}{c|cc} & l_1 & l_{So} \\ \hline l_5 & false & true \\ l_{So} & W_{So,1} & W_{So,So} \end{array} \rangle,$$

where

$W_{So,1}$ ="the train is directed to train station of Mezdra",
$W_{So,So}$ ="the train must shunt in train station of Sofia".

$$Z_{Vi} = \langle \{l_2, l_6, l_{Vi}\}, \{l_3, l_4, l_{Vi}\}, \begin{array}{c|ccc} & l_3 & l_4 & l_{Vi} \\ \hline l_2 & false & false & true \\ l_6 & false & false & true \\ l_{Vi} & W_{Vi,3} & W_{Vi,4} & W_{Vi,Vi} \end{array} \rangle,$$

where

$W_{Vi,3}$ ="the train is directed to Romania (Kalafat)",
$W_{Vi,4}$ ="the train is directed to train station Mezdra",
$W_{Vi,Vi}$ ="the train must shunt in train station of Vidin".

$$Z_{Me} = \langle \{l_1, l_4, l_{16}, l_{Me}\}, \{l_5, l_6, l_7, l_{Me}\},$$

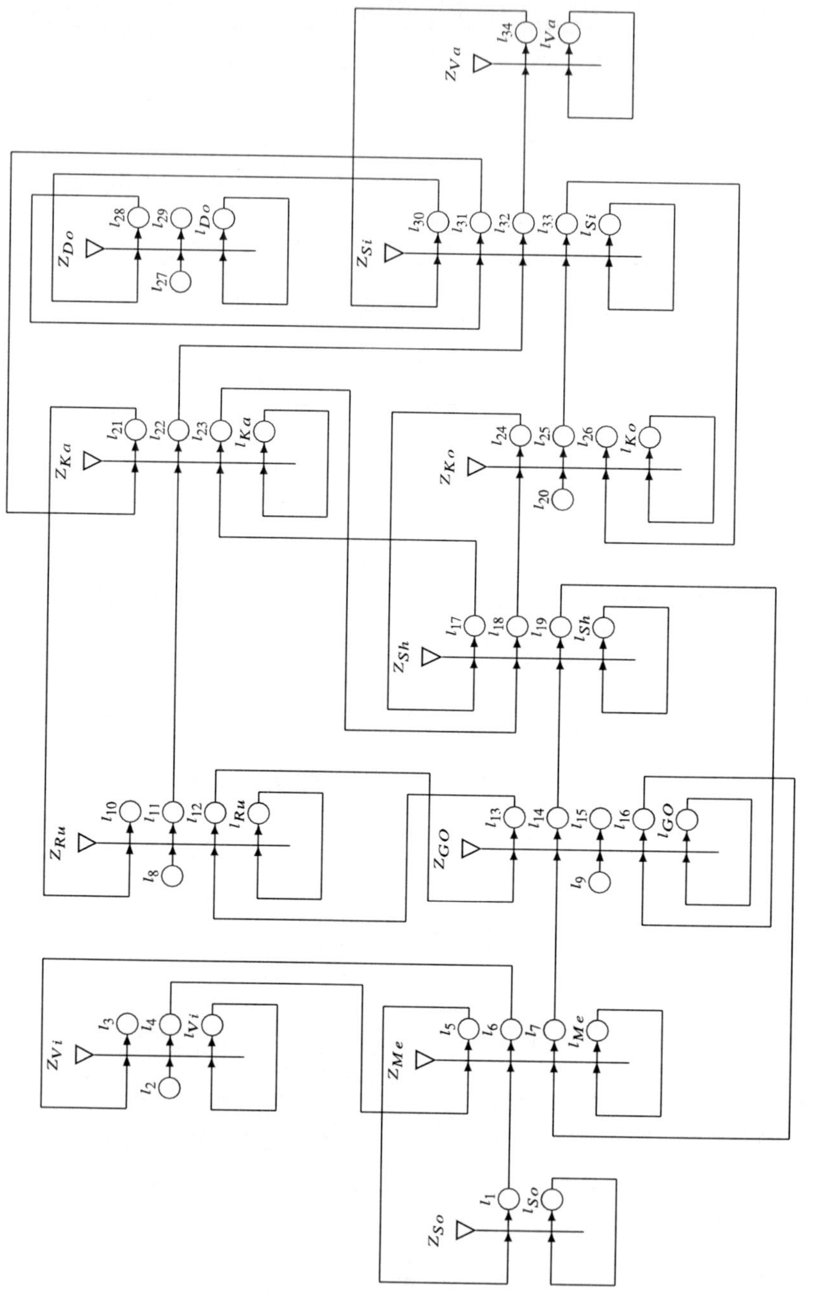

Fig. 3. Generalized net model

$$\begin{array}{c|cccc} & l_5 & l_6 & l_7 & l_{Me} \\ \hline l_1 & false & false & false & true \\ l_4 & false & false & false & true \\ l_{16} & false & false & false & true \\ l_{Me} & W_{Me,5} & W_{Me,6} & W_{Me,7} & W_{Me,Me} \end{array}\rangle,$$

where

$W_{Me,5}$ = "the train is directed to train station Sofia",
$W_{Me,6}$ = "the train is directed to train station Vidin",
$W_{Me,7}$ = "the train is directed to train station Gorna Orjahovica,
$W_{Me,Me}$ = "the train must shunt in train station of Mezdra".

$$Z_{Ru} = \langle\{l_8, l_{13}, l_{21}, l_{Me}\}, \{l_{10}, l_{11}, l_{12}, l_{Ru}\}, \begin{array}{c|cccc} & l_{10} & l_{11} & l_{12} & l_{Ru} \\ \hline l_8 & false & false & false & true \\ l_{13} & false & false & false & true \\ l_{21} & false & false & false & true \\ l_{Ru} & W_{Ru,10} & W_{Ru,11} & W_{Ru,12} & W_{Ru,Ru} \end{array}\rangle,$$

where

$W_{Ru,10}$ = "the train is directed to Romania (Giurgiu)",
$W_{Ru,11}$ = "the train is directed to train station Kaspichan",
$W_{Ru,12}$ = "the train is directed to train station Gorna Orjahovica,
$W_{Ru,Ru}$ = "the train must shunt in train station of Ruse".

$$Z_{GO} = \langle\{l_7, l_9, l_{12}, l_{19}, l_{GO}\}, \{l_{13}, l_{14}, l_{15}, l_{16}, l_{GO}\},$$
$$\begin{array}{c|ccccc} & l_{13} & l_{14} & l_{15} & l_{16} & l_{GO} \\ \hline l_7 & false & false & false & false & true \\ l_9 & false & false & false & false & true \\ l_{12} & false & false & false & false & true \\ l_{19} & false & false & false & false & true \\ l_{GO} & W_{GO,13} & W_{GO,14} & W_{GO,15} & W_{GO,16} & W_{GO,GO} \end{array}\rangle,$$

where

$W_{GO,13}$ = "the train is directed to train station Ruse,
$W_{GO,14}$ = "the train is directed to train station Shumen",
$W_{GO,15}$ = "the train is directed to Southern Bulgaria",
$W_{GO,16}$ = "the train is directed to train station Mezdra",
$W_{GO,GO}$ = "the train must shunt in train station of Gorna Orjahovica".

$$Z_{Sh} = \langle\{l_{14}, l_{23}, l_{24}, l_{Sh}\}, \{l_{17}, l_{18}, l_{19}, l_{Sh}\}, \begin{array}{c|cccc} & l_{17} & l_{18} & l_{19} & l_{Sh} \\ \hline l_{14} & false & false & false & true \\ l_{23} & false & false & false & true \\ l_{24} & false & false & false & true \\ l_{Sh} & W_{Sh,17} & W_{Sh,18} & W_{Sh,19} & W_{Sh,Sh} \end{array}\rangle,$$

where

$W_{Sh,17}$ = "the train is directed to train station Kaspichan",
$W_{Sh,18}$ = "the train is directed to train station Komunari",
$W_{Sh,19}$ = "the train is directed to train station Gorna Orjahovica,
$W_{Sh,Sh}$ = "the train must shunt in train station of Shumen".

$$Z_{Ka} = \langle \{l_{11}, l_{17}, l_{41}, l_{Ka}\}, \{l_{21}, l_{22}, l_{23}, l_{Ka}\}, \begin{array}{c|cccc} & l_{21} & l_{22} & l_{23} & l_{Ka} \\ \hline l_{11} & false & false & false & true \\ l_{17} & false & false & false & true \\ l_{41} & false & false & false & true \\ l_{Ka} & W_{Sh,17} & W_{Sh,18} & W_{Sh,19} & W_{Sh,Sh} \end{array} \rangle,$$

where

$W_{Ka,21}$ = "the train is directed to train station Ruse",
$W_{Ka,22}$ = "the train is directed to train station Sindel",
$W_{KA,23}$ = "the train is directed to train station Shumen,
$W_{KA,Ka}$ = "the train must shunt in train station of Kaspichan".

$$Z_{Ko} = \langle \{l_{18}, l_{20}, l_{33}, l_{Ko}\}, \{l_{24}, l_{25}, l_{26}, l_{o}\}, \begin{array}{c|cccc} & l_{24} & l_{25} & l_{26} & l_{Ko} \\ \hline l_{18} & false & false & false & true \\ l_{20} & false & false & false & true \\ l_{33} & false & false & false & true \\ l_{Ko} & W_{Ko,24} & W_{Ko,25} & W_{Ko,26} & W_{Ko,Ko} \end{array} \rangle,$$

where

$W_{Ko,24}$ = "the train is directed to train station Shumen",
$W_{Ko,25}$ = "the train is directed to train station Sindel",
$W_{Ko,26}$ = "the train is directed to Southern Bulgaria,
$W_{Ko,Ko}$ = "the train must shunt in train station of Komunari".

$$Z_{Do} = \langle \{l_{27}, l_{30}, l_{Do}\}, \{l_{28}, l_{29}, l_{Do}\}, \begin{array}{c|ccc} & l_{28} & l_{29} & l_{Do} \\ \hline l_{27} & false & false & true \\ l_{30} & false & false & true \\ l_{Do} & W_{Do,28} & W_{Do,29} & W_{Do,Do} \end{array} \rangle,$$

where

$W_{Do,28}$ = "the train is directed to train station Sindel",
$W_{Do,29}$ = "the train is directed to Romania (Negru Voda)",
$W_{Do,Do}$ = "the train must shunt in train station of Dobrich".

$$Z_{Si} = \langle\{l_{22}, l_{25}, l_{28}, l_{34}, l_{Si}\}, \{l_{30}, l_{31}, l_{32}, l_{33}, l_{Si}\}, \begin{array}{c|ccccc} & l_{30} & l_{31} & l_{32} & l_{33} & l_{Si} \\ \hline l_{22} & false & false & false & false & true \\ l_{25} & false & false & false & false & true \\ l_{28} & false & false & false & false & true \\ l_{34} & false & false & false & false & true \\ l_{Si} & W_{Si,30} & W_{Si,31} & W_{Si,32} & W_{Si,33} & W_{Si,Si} \end{array}\rangle,$$

where

$W_{Si,30}$ =" the train is directed to train station Dobrich,
$W_{Si,31}$ ="the train is directed to train station Kaspichan",
$W_{Si,32}$ ="the train is directed to train station Varna",
$W_{Si,33}$ ="the train is directed to train station Komunari",
$W_{Si,Si}$ ="the train must shunt in train station of Singel".

$$Z_{Va} = \langle\{l_{32}, l_{Va}\}, \{l_{34}, l_{Va}\}, \begin{array}{c|cc} & l_{34} & l_{Va} \\ \hline l_{32} & false & true \\ l_{Va} & W_{Va,34} & W_{Va,Va} \end{array}\rangle,$$

where

$W_{Va,1}$ ="the train is directed to train station of Sindel",
$W_{Va,Va}$ ="the train must shunt in train station of Varna".

4 Conclusion

The so constructed GN-model can be used for simulation of different situations that can occur between the trains in real-time. In a result, we can obtain ideas e.g., for keeping or changing of the schedule of concrete train(s) in the current moment.

Acknowledgements. This paper has been partially supported by the Bulgarian National Science Fund under the Grant Ref. No. Ref. No. DN 02/10 "New Instruments for Knowledge Discovery from Data and their Modelling".

References

1. van der Aalst, W.M.P., Odijk, M.A.: Analysis of railway stations by means of interval timed coloured Petri nets. Real-Time Syst. **9**, 241–263 (1995)
2. Alexieva, J., Choy, E., Koycheva, E.: Review and bibloigraphy on generalized nets theory and applications. In:-Choy, E., Krawczak, M., Shannon, A., Szmidt, E. (eds.) A Survey of Generalized Nets Raffles KvB Monograph No. 10, pp. 207–301 (2007)
3. Atanassov, K.: General. Nets. World Scientific., London (1991)

4. Atanassov, K.: On Generalized Nets Theory. Prof. M. Drinov Academic Publ. House, Sofia (2007)
5. Atanassov, K.: Index Matrices: Towards an Augmented Matrix Calculus. Springer, Cham (2014)
6. Cheng, Y.A., Yang, L.-A.: Fuzzy petri nets approach for railway traffic control in case of abnormality: evidence from Taiwan railway system. Exp. Syst. Appl. **36**(4), 8040–8048(2009). https://doi.org/10.1016/j.eswa.2008.10.070
7. Fang, W.: A survey on problem models and solution approaches to rescheduling in railway networks. IEEE Trans. Intell. Transp. Syst. **16**(6), 2997–3016 (2015). https://doi.org/10.1109/TITS.2015.2446985
8. Fay, A.: A fuzzy knowledge-based system for railway traffic control. Eng. Appl. Artif. Intell. **13**, 719–729 (2000)
9. Fidanova, S., K. Atanassov, I. Dimov, Generalized nets as a tool for modelling of railway networks. In:-Georgiev, K., Todorov, M., Georgiev, I. (eds.) Advanced Computing in Industrial Mathematics, pp. 23–35. Springer, Cham (2017). https://doi.org/10.1007/978-3-319-65530-7
10. Gutierrez, J., Monzon, A., Pinero, J.: Accessibility, network efficiency, and transport infrastructure planning. Environ. Plann. A **30**(8), 1337–1350 (1998)
11. Islam, D., Zunder, T., Grashoff, P.: Knowledge-base development process for pan-European rail freight corridors- a case study. In: Proceedings of the 13th World Conference on Transport Research WCTR2013Rio, 1–13 January 2014, Rio de Janeiro (2014)
12. Jeremić, D., Milinković, S., Kasalica, S.: Simulating train dispatching logic with high-level petri nets. Tehnički vjesnik **28**(2), 639–648 (2021)
13. Jeremić, D., Milosavljević, M., Milinković, S., Vesković, S., Bundalo, S.: Simulation of a railway mainline junction using High level Petri nets. In: 6th International Conference on Information Society and Technology ICIST 2016, pp. 235–240 (2016)
14. Liu, K., Wang, M., Cao, Y., Zhu, W., Wu, J., Yan, X.: A comprehensive risk analysis of transportation networks affected by rainfall-induced multihazards. Risk Anal. **38**(8), 1618–1633 (2018)
15. Mattsson, L.-G.: Railway capacity and train delay relationships. In: Murray, A.T., Grubesic, T.H. (eds.) Critical Infrastructure, pp.129–150. Springer, Berlin (2007). https://doi.org/10.1007/978-3-540-68056-7
16. Milinković, S., Marković, M., Vesković, S., Ivić, M., Pavlović, N.: A Fuzzy Petri net model for estimation of train delays. Simul. Model. Pract. Theory **33**, 144–157 (2013)
17. Milinković, S., Vesković, S., Marković, M., Ivić, M., Pavlović, N.: Simulation model of a railway junction based on petri nets and fuzzy logic. In: Selected Proceedings of the 12th World Conference on Transport Research Society, Lisbon, Portugal (2010)
18. Pachl, J.: Railway Operation and Control. VTD Rail Pub, Mountlake Terrace (2009)
19. Virtanen, H.: A study in fuzzy petri nets and the relationship to fuzzy logic programming, Reports on Computer Science and Mathematics Åbo Akademi A, 162 (1995)
20. Zhang, Z., Jia, L.: RAMS analysis of railway network: model development and a case study in China. Smart Resil. Transp. **3**(1), 2–11 (2021). https://doi.org/10.1108/SRT-10-2020-0013

Acceleration of the Calculation of Boundary Surface Cross-Sections

Milan Guzan[1](✉), Tibor Vince[1], Ján Molnár[1], Matej Bereš[1], and Branislav Sobota[2]

[1] Department of Theoretical and Industrial Electrical Engineering, Faculty of Electrical Engineering and Informatics, Technical University of Košice, Park Komenského 3, 04200 Košice, Slovak Republic
{milan.guzan,tibor.vince,jan.molnar,matej.beres}@tuke.sk

[2] Department of Computers and Informatics, Faculty of Electrical Engineering and Informatics, Technical University of Košice, Letná 9, 04200 Košice, Slovak Republic
branislav.sobota@tuke.sk

Abstract. The paper presents a new possibility of accelerating the calculation of the cross-section through the boundary surface of the Chua's circuit, generating chaos. Compared to previous research (recommended: 64 bit version of simulation software, parallelization of calculations and software optimization when creating *.exe file) an optimal test cuboid lying on a chaotic attractor is designed, then a second one is added, symmetrically placed also on a chaotic attractor, and finally another, large cuboid is added. The big cuboid is located inside chaotic attractor i.e. within the regions of attractivity corresponding to the chaotic attractor in the individual cross-sections of the boundary surface. The result of using test cuboids is almost double the acceleration of the calculations. This will save money on electricity consumption as well as the purchase of new computers with more powerful but expensive processors. The number, size and suitability of placement of test cuboids in 3D state space allow faster calculation of hundreds of partial cross-sections of the boundary surface used for reconstruction and subsequent visualization of the entire boundary surface in 3D not only on PC screen, but also in virtual reality (virtual cave), for 3D printing with new parameters, or new mathematical models of systems of differential equations generating chaos.

Keywords: Chua's circuit · Boundary surface · Acceleration the calculation

1 Introduction

The human effort to improve things in daily life is, among other things, reflected in the acceleration or optimization of work processes. For example, shortening the operation time on automated lines by even a few tenths of a second is a great benefit, truckers try to transport goods as quickly as possible, etc. The need to accelerate calculations can also be encountered in mathematical calculations of electromagnetic fields, in weather forecasting, or in the calculation of a series of cross-sections of boundary surface (C-SBS) in 3D state space. Why are we interested in calculations of C-SBS series in this article?

K. T. Atanassov et al. (Eds.): IWIFSGN 2020/BOS/SOR 2020, LNNS 338, pp. 129–141, 2022.
https://doi.org/10.1007/978-3-030-95929-6_11

2 Boundary Surface

The boundary surface (BS) separates the individual attractors, occurring in the autonomous electrical circuit, from each other. It provides an exact explanation of the control and failure of sequential circuits [1–5] and clarifies the regularities of the distribution regions of attractivity (RA) of individual attractors in sequential [1–13] and chaos generating circuits [14–19], or in other mathematical models generating chaos [20–28].

The research [15] was the first that brought the answer for a question put in [29] concerning the existence of an object separating CHA from SLC in Chua's circuit (Fig. 1). This object was BS, mentioned in [15] and first documented in the Chua's circuit using C-SBS.

Visualization of RA in 3D actually means calculation of a large number of 2D C-SBS and their subsequent displaying using a large number of images [11–20]. However, by reconstructing the BS in 3D from the calculated 2D C-SBS, it is possible to visualize the whole BS in one (Fig. 2c) or two images [30–33]. Due to the great time-consuming calculation of not only one, but also a large set of C-SBS containing from 100 to 300 cross-sections, it is necessary to constantly consider other possibilities of accelerating the calculations of BS cross-sections. For example, Fig. 2c was reconstructed from 300 C-SBS. The calculation of one C-SBS took up to 5 h using a computer with one CPU (Sempron). This article will therefore point to other, new possibilities for speeding up calculations related to Chua's circuit (Fig. 1), which was the first physical, chaos generating circuit [29].

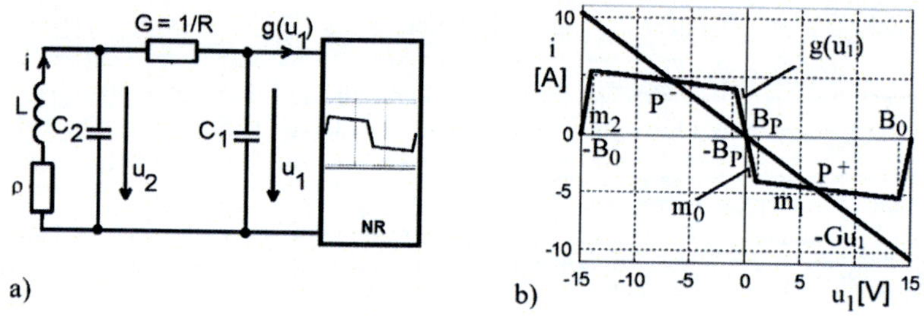

Fig. 1. a) Chua's circuit, b) I–V characteritics $g(u_1)$ with load G in projection to i, u_1 plane.

In general, the calculation of one C-SBS consists in solving a system of differential Eqs. (1) (e.g. using the fourth order Runge-Kutta method) for a grid of M × N initial conditions (IC) shown by dots in the illustrative Fig. 2a. The result of the calculation are represented by colored areas - RA for individual attractors, which are shown in Fig. 2b. The red area represents the region of attractivity for the chaotic attractor (CHA) and the yellow area corresponds to the region of attractivity for infinity or for the stable limit cycle (SLC).

$$C_1 \frac{du_1}{dt} = G(u_2 - u_1) - g(u_1)$$
$$C_2 \frac{du_2}{dt} = G(u_1 - u_2) + i$$
$$L \frac{di}{dt} = -u_2 - \rho i \tag{1}$$

The nonlinear resistor (NR) in (1) is expressed by relation (2) and the parameters that were used to solve the system (1) come from [29] and [16] in (3):

$$g(u_1) = m_2 u_1 + \frac{1}{2}(m_1 - m_0)(|u_1 - B_P| - |u_1 + B_P|) + \frac{1}{2}(m_2 - m_1)(|u_1 - B_O| - |u_1 + B_O|) \tag{2}$$

$$\begin{aligned} &C_1 = 0,11111\text{F},\ C_2 = 1\text{F},\ R = 1/G = 1.42857\,\Omega,\ \rho = 0\,\Omega, \\ &L = 0.142857\text{H},\ B_P = 1\ \text{V},\ B_O = 900\,\text{V},\ m_0 = -0{,}8\text{S},\ m_1 = -0{,}5\text{S},\ m_2 = 5\text{S} \end{aligned} \tag{3}$$

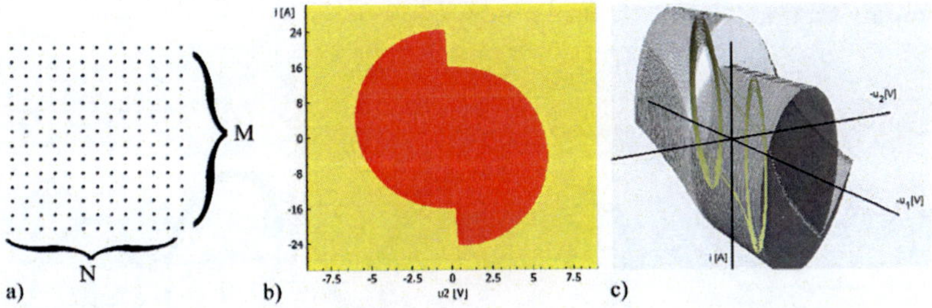

Fig. 2. a) Grid M × N of IC for calculation of one C-SBS, b) calculated C-SBS for $u_1 = 0$ and grid 400 × 400 IC, c) reconstructed BS from 300 calculated C-SBS.

The calculation of one 2D C-SBS is equal to the calculation of several hundred thousand trajectories. Calculating the BS as a whole in 3D therefore places great demands on PC computing power. This implies the need to optimize the calculation of C-SBS so that the results of the calculations are accessible for further processing (e.g. reconstruction of BS in 3D [30–33]) in a relatively short time.

3 Acceleration of C-SBS Calculations Implemented so Far

The calculation of one C-SBS is not possible without detection of particular attractor by the cuboid and also without setting boundaries for infinity, or SLC. At the same time, calculating the system (1) for the M × N IC also means testing the distance of the next solution from the previous one. These are the parameters that significantly affect the speed of the C-SBS calculation in the developed *ChuaViz* software described in [34]. The testing of the distance of individual solutions is provided by the parameters *uhMax* and *ihMax*. Therefore, when calculating one C-SBS, two types of tests for infinity (or SLC) and CHA are proposed in the *ChuaViz* program.

The first test takes place from the very beginning of the trajectory calculation. It is tested whether the representative point (RP) exceeds the minimum and maximum limit corresponding to the values of the state variables i, u_2, u_1. If the RP is outside the selected limit, the trajectory calculation stops and the calculation of a new trajectory for the next, new IC, begins.

The second test does not take place until the *tTest* time has elapsed. During the specified time, it is assumed that the RP is already attracted by the attractor (CHA). If the RP enters the selected cuboid after the time $t > tTest$ has elapsed, it means that it has been attracted by the given attractor, and at the next solution of the system (1) it will no longer move away from the attractor. As in the first case, the trajectory calculation after the entry of the RP into the test cuboid is completed and the system (1) starts to be solved for the next, new IC.

An example of a window of still developed *ChuaViz* software from [34] is shown in Fig. 3. In the left part of the window the parameters (3) are defined, using only 3-segment I–V characteristic (green color in i, u_1 plane) and one test cuboid. The cuboid is indicated by red rectangles. In order to be able to use the 3-segment function $g(u_1)$ instead of 5-segment function, the value of B_0 was set to 900 V in (3). In *ChuaViz*, it is possible to use both, the 5-segment function $g(u_1)$ (for $B_0 = 13$ V) and the function expressed by a cubic polynomial according to [21] by changing the option in the *I–V characteristics* section.

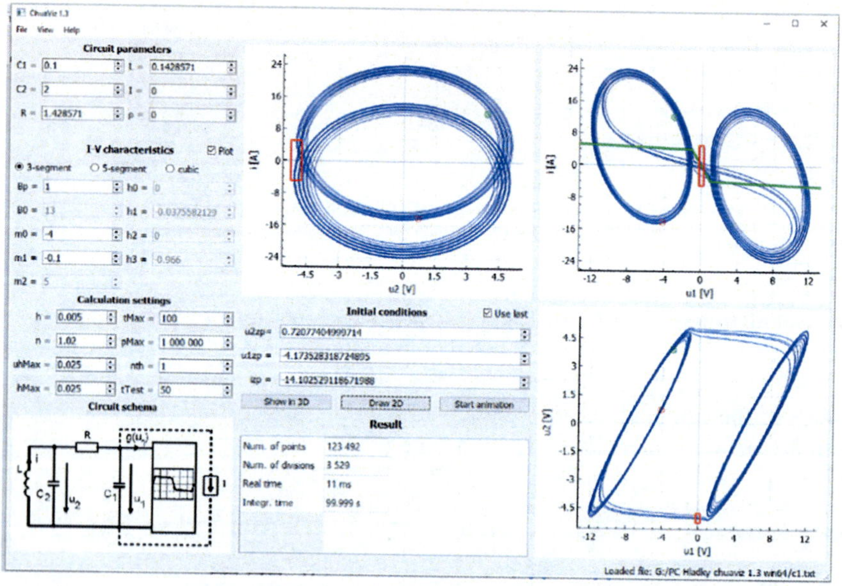

Fig. 3. *ChuaViz* simulation software window.

Given the speed of calculation of one C-SBS, the work [34] shows that:

- The suitable multiplier for the adaptive integration step in solving the system (1) is 1.02,

- faster calculation using 64-bit OS Windows allows to use 64-bit *ChuaViz* software compared to 32-bit version,
- O2 optimization proved to be the fastest software optimization when translating into an executable *.exe file (Visual Studio 2013, C++ language),
- naturally, the calculation of one C-SBS is faster with parallel calculations than with just one CPU.

The above findings helped to speed up the calculation of one C-SBS. However, these are not all ways to speed up the calculation. These will be sketched out in the following section.

4 New Options to Speed up the Calculation of One C-SBS

In this part, other, new possibilities of accelerating the calculations of one C-SBS, which were not realized in [34], will be tested. It was not considered to buy a new computer with a powerful CPU, which would speed up calculations many times over, as the price of the computer would be more than 1700 €, as the CPU itself - Intel Core i9-10980XE, 3.0 GHz, LGA 2066, Boost 4.6 GHz, with 18 cores and 36 threads (18/36) cost 1069 € [35], Intel Xeon Gold 6252, 2.1 GHz, Boost 3.7 GHz (24/48) - 4099 € [36], or AMD EPYC 7742, 2.25 GHz, Boost 3.4 GHz (64/128) costs more than 7500 € [37]. The C-SBS calculations in the following two parts are performed on a laptop with an Intel Core i5-5200U 2.2 GHz (2/4) CPU, Turbo Speed 2.5 GHz, 8 GB RAM, Windows 8.1 64 bit.

Therefore, several of the following untapped possibilities for accelerating C-SBS calculations are considered.

4.1 Appropriate Position of Test Cuboid on CHA

The first step is to find the appropriate position and size of the CHA-detecting cuboid. This consists in determining such a place in 3D on the CHA, where the RP will move most often after the time *tTest* has elapsed (in Fig. 3 *tTest* was set to 50 s). It is therefore necessary to know the dynamics of CHA. This is illustrated in Fig. 3 and Fig. 4.

In 2D projections, the view of the test cuboid in position is often not very clear, so a 3D view of CHA (Fig. 3 – see button *Show in 3D*) and simultaneous verification of the correctness of the test cuboid position was chosen. Figure 4a illustrates their location in 3D with a description of the axes. The integration time of the CHA calculation was $t =$ 100 s. Figure 4b illustrates CHA and cuboid at $t = 1000$ s without axes. Such views will be used in next pictures. Figure 3 and Fig. 4a,b show that the RP will go around a few turns (it's hard to say how many circles, because it's chaos) along a path in the form of large disks, followed by a jump from one disk to another with subsequent cycles along the current disk. A red cuboid is located at the jump point (between the discs). It may seem that the cuboid is large and that a solution from a neighboring IC, which could be attracted to infinity, may get stuck in it. However, this is not true. Figure 4c shows the CHA and also the solution for the new IC - this is the first point (IC), the solution of which by the system (1) is attracted to the infinite. The stop condition is $t = 45$ s and it is obvious that before the expiration of $tTest = 50$ s the RP is very far from the cuboid and CHA.

The black arrows indicate the direction of RP movement from the beginning and the long black arrow even after $t > 45$ s. The condition that the RP motion test does not start until $tTest = 50$ s ensures correct detection of RP motion on the CHA.

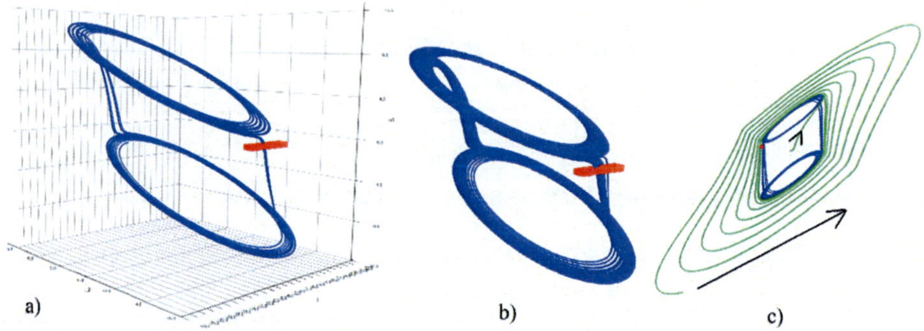

Fig. 4. Trajectories shown in 3D a) CHA with axes, b) CHA without axes, c) CHA and movement of RP to infinity for the integration time of trajectory calculation $t = 45$ s.

Several test cuboids were placed on the CHA, see Fig. 5a. The calculation speed of only one C-SBS corresponding to only one located cuboid is summarized in Table 1, in column: 1 cuboid. It is clear from the table, that when testing cuboids from *n.1* to *n.5*, the fastest calculation of one S-CBS for $u_1 = 0$ is found for the cuboid *n.3* (green) 2:43 min. Each C-SBS will be calculated with a grid of 400 × 400 IC, which represents a calculation of a total of 160 000 trajectories. Will the calculation be faster if all cuboids *n.1*–*n.5* are tested at the same time? Time 2:49 min is a little longer as for *n.3*, where only 1 cuboid is tested during the calculation, but in the latter case up to 5 cuboids are tested simultaneously. The test cuboid *n.3* has a very advantageous position on almost the entire width of the lower disk of CHA. If we combined all the cuboids *n.1*–*n.5* into one, the calculation could be even faster. Therefore, one big (red) test cuboid n.6 was proposed - Fig. 5b. Calculation time is: 2:32 min. It is the shortest calculation time.

Due to the symmetry of CHA, it would be possible to place cuboids *n.6* and *n.1* to *n.5* symmetrically on CHA, as indicated in Fig. 5b and Fig. 5c. The results are listed in Table 1, column: 2 sym.cub. Obviously, the fastest calculation is for test cuboids *n.3* or *n.6* (2:31 or 2:32 min, respectively). A difference of one second represents a measurement error. The resulting calculation times in Table 1 represent the average of the three calculations, differing from each other by 1 s (sometimes by 2 s). The coordinates of cuboids *n.1* to *n.6* are given in Table 2, while the cuboid marked on the last line as: big cuboid will be described later, in Sect. 4.2. From the user's point of view, it is more advantageous and faster to design a test cuboid *n.3*, compared to *n.6*. On the other hand, two test cuboids *n.6* do not affect the computation speed of one C-SBS and therefore it may be more advantageous to test only one cuboid *n.6* instead of two *n.3*. However, as can be seen from the third column in Table 1, in most cases the acceleration of the calculation is not greater than 1.12 times, which is negligible.

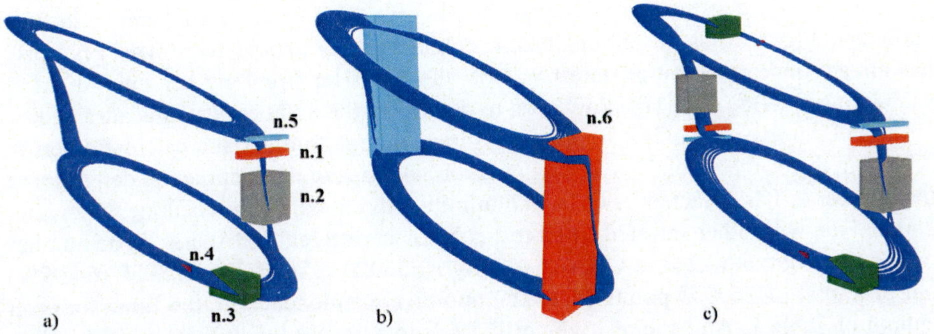

Fig. 5. Placement of test cuboids on CHA a) 5 cuboids on one side *n.1* to *n.5*, b) two large cuboids combining the advantages of cuboids *n.1*–*n.5* with the designation *n.6*, c) symmetrical placement of another five cuboids on the other side of CHA.

Table 1. Speed calculation with different cuboids.

Number of cuboid	1 cuboid (mm:ss)	2 sym.cub. (mm:ss)	Acceleration of calculation 1cub./2 sym.cub.	Acceleration of calculation n.1 1cub./2 sym.cub.
n.1	3:22	3:06	1.09	1.09
n.2	3:01	2:53	1.05	1.17
n.3	2:43	2:31	1.08	1.34
n.4	3:10	2:49	1.12	1.2
n.5	2:54	2:48	1.04	1.2
n.1–n.5	2:49	2:44	1.03	1.23
n.6	2:32	2:32	1.00	1.33

Table 2. Dimensions of test cuboids.

Number of cuboid	u_1 low (V)	u_1 high (V)	u_2 low (V)	u_2 high (V)	i low (A)	i high (A)
n.1	−0.25	0.25	−5.25	5.25	−5	5
n.2	−5.5	−1.75	−5	−4.04	0.618	7.32
n.3	−11.3	−9.48	−4.85	−3.39	−3.37	0.43
n.4	−9	−8.6	−3.4	−3	−8	−7.6
n.5	−1.25	−1	4.75	5.25	−5	5
n.6	−11.7	2.2	−4.75	−4	−5	5
big cuboid	−15	15	−2,5	2,5	−8	8

However, the question may arise: are the calculated C-SBS for the differently located cuboids in Fig. 5 the same, although some cuboids appear to be in huge size? Figure 4c has already indicated that at time $t = 45$ s, the RP is far from any test cuboid. After calculating C-SBS, *ChuaViz* allows you to determine the number of points attracted by CHA and infinity (or SLC). Cuboid *n.4* is small enough to rule out a calculation error. Although this is the second slowest calculation, in terms of the accuracy of determining the number of ICs attracted by CHA and infinite, this calculation is accurate (even after comparison with other selected small test cuboids at different CHA sites). The number of ICs that were attracted to CHA or infinity is: 56020 (35.01%) or 103980 (64.99%) out of a total of 160000 points (IC). The numbers mentioned were the same for each cuboid of Table 1. An example, which IC leads to entry into individual cuboids when tested with two symmetrically placed cuboids (red/aqua as in Fig. 5b) is shown in Fig. 6. The different cases are *n.1–n.5_1side* or *n.1–n.5_2side*, Fig. 6. The cuboids were placed according to Fig. 5a or Fig. 5c. Interestingly, when placing the cuboids according to Fig. 5a, very small red or gray colored regions corresponding to cuboids *n.1* or *n.2* are detected. Purple color for cuboid *n.4* is not visible at all. Green (cuboid *n.3*) and aqua (cuboid *n.5*) color predominate for cuboids located on the edge of both disks, Fig. 5a. If the cuboids are placed according to Fig. 5c (*n.1–n.5_2side*), then the color aqua (for *n.5*) is significantly less present compared to the previous image *n.1–n.5_1side*. Purple already appears (for *n.4*) and again small color areas are visible for cuboid *n.1* and *n.2* (red and gray). The green color, which corresponds to test cuboid *n.3*, clearly predominates. This cuboid captures the most RP. That's why the calculation of one S-CBS is actually the fastest.

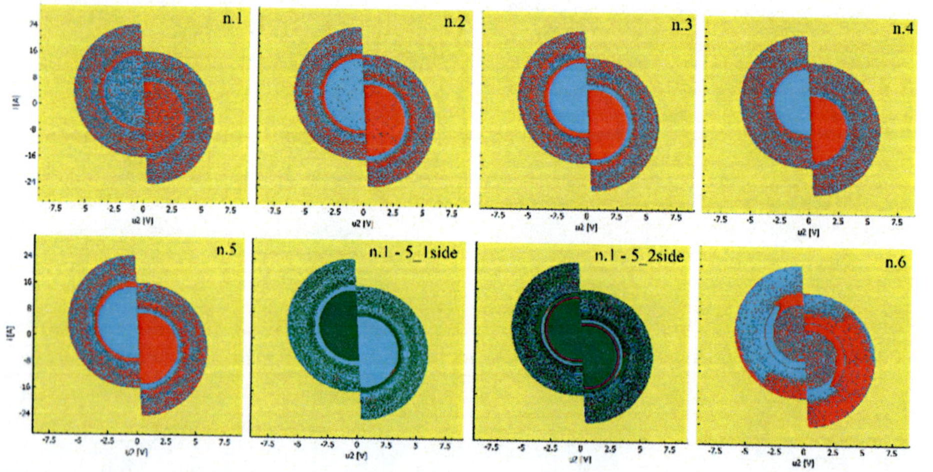

Fig. 6. Example of C-SBS for $u_1 = 0$ as a color map representing the detection of the presence of RP in any of the test cuboids *n.1* to *n.6* (red/aqua) symmetrically placed on the CHA as illustrated in Fig. 5b. The group of five cuboids *n.1–n.5* for figure *n.1–n.5_1side* was placed only on 1 side (Fig. 5a). Figure *n.1–n.5_2side* corresponds to the location of the test cuboids on both sides of the CHA according to Fig. 5c.

4.2 Cuboid Inside the CHA

Another way, how to speed up the C-SBS calculation is to place a big test cuboid so that the condition that every single IC that is in the cuboid will be definitely attracted to the CHA, will be met. This cuboid will not wait to test the RP entry into the cuboid after the *tTest* time has elapsed, but the testing will start as soon as the calculation from the IC starts. If the RP enters this cuboid later, the calculation stops and the calculation from another IC starts. Where and how to place a new test cuboid? The calculation of the C-SBS series will help a lot. Figure 7 shows only a few C-SBS for u_1 = const. listed in the upper right corner. The range $u_1 = \pm 15$ V covers the entire 3D space that we are interested in during the reconstruction of the BS.

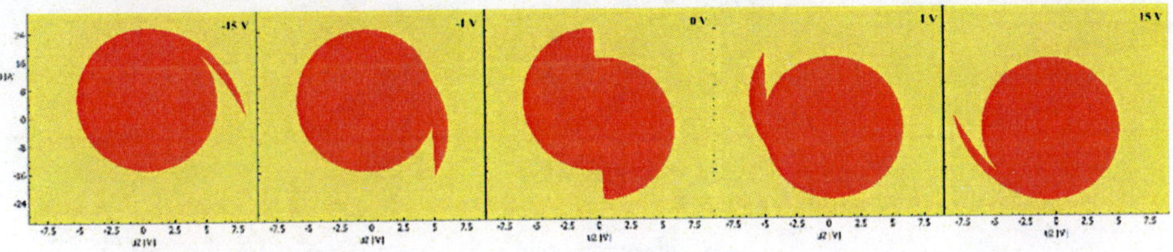

Fig. 7. Several calculated C-SBS in the range from $u_1 = -15$ V to + 15 V.

C-SBS (Fig. 7) are calculated with a grid of 400 × 400 ICs using one cubid n.1. The calculation of indicative C-SBS can be accelerated using a coarser grid, e.g. 100 × 100 or 50 × 50. It is clear that by choosing a coarse grid the computation time will be shortened from 16 to 64 times, but the color interface will not be as smooth as in Fig. 7. For example, for a 100 × 100 grid, the C-SBS calculation time for $u_1 = 0$ is only 12s and for a 50 × 50 grid it is only 3s compared to a time of 3:22 min for *n.6* in Table 1. From Fig. 7 it is clear that somewhere in the middle of the individual pictures, in the red RA for the CHA, it would be possible to place the test cuboid so as not to interfere with the yellow area.

The projection into the plane *i*, u_2 of the proposed cuboid for the range $u_1 = \pm 15$ V is illustrated in Fig. 8. An overall view of all projection planes, including two symmetrically placed cuboids on CHA labeled *n.1*, is shown in Fig. 9a. Figure 9b shows the location of the cuboids in 3D.

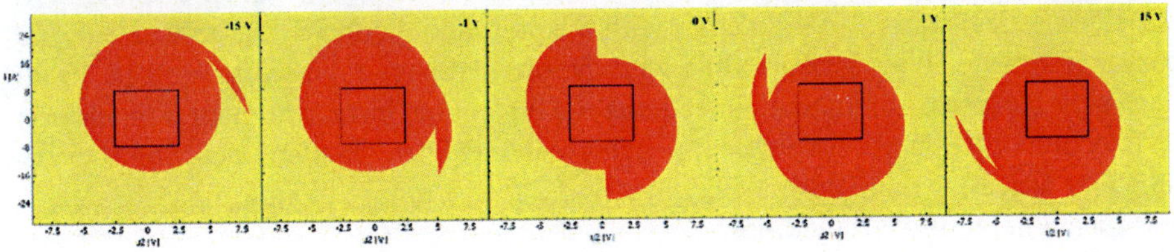

Fig. 8. Indicated projection of a new test cuboid in plane *i*, u_2 located only in red RA for CHA.

From Fig. 8 and Fig. 9, it is thus clear that in the C-SBS calculation, it is possible to test the position of the RP right from the beginning of the calculation, because the

IC may lie inside a big test cuboid. With the new, big cuboid designed in this way, a faster calculation of one C-SBS is assumed. So, the question is: How many times will the calculation of one C-SBS speed up compared to the results in Table 1?

Again, the speeds of the calculations were tested for the same cuboids as in Table 1 (without all *n1*–*n5* at the same time) with the difference that a big cuboid was added to them. Its dimensions are given in Table 2. The test results are recorded in Table 3 together with a multiple of the acceleration when tested with one and two symmetrically placed cuboids, on one or both sides of the CHA, as illustrated in Fig. 5a and Fig. 5c.

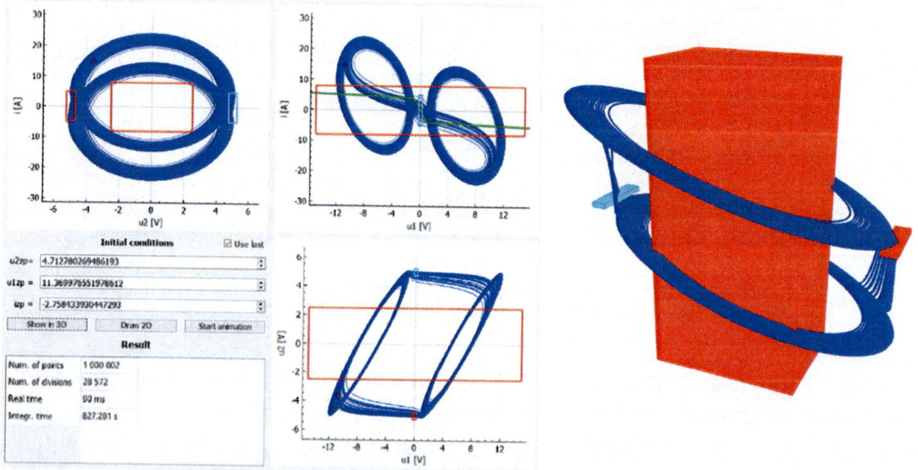

Fig. 9. Design of a test cuboid, which should lie entirely inside the CHA a) individual 2D sections in two Monge projections, b) 3D view of two symmetrical cuboids n.1 (tested after the time tTest) and one big cuboid, tested right from the beginning of the calculation.

In this context, it will be interesting to find out how large an area of red RA is covered by a big cuboid and what part of the red area will be covered by the detection of two small cuboids on CHA. Again, illustrative control C-SBSs were calculated for $u_1 = 0$ and the results are apparent from Fig. 10. To distinguish the small red cuboid on CHA from the big red cuboid, the color of the small cuboid was changed to green. Because C-SBS are very similar, for Fig. 10 was chosen C-SBS only for cuboids *n.1*, *n.3* and *n.6*. On C-SBS for 2 symmetrical cuboids *n.6* together with the present big cuboid, the influence of big cuboids from Fig. 5b is clearly visible with the predominant color aqua or green. From Fig. 10 it is clear from them that, indeed, the red big cuboid has a dominant influence on the detection of RP in 3D state space, which accelerated the calculation of almost every C-SBS by at least 1 min. With a calculation of about 3 min without a big cuboid, this means a not negligible acceleration.

It is clear from Table 3 that after the addition of the big cuboid, the C-SBS calculations at $u_1 = 0$ accelerated again. The ratio of the speed of C-SBS calculations shows that the acceleration is:

- 1.48 to 1.82 times, at the ratio of computation times for 1 cuboid from Table 1 (times can also be found in the second column of Table 3) and 1 cuboid + big cuboid,
- 1.45 to 1.72 times, at the ratio of computation times for 2 symmetrical cuboids with 2 symmetrical cuboids + big cuboid,
- 1.48 to 1.87 times, at the ratio of computation times of the slowest computation with 1 cuboid n.1 (3:22 min) and different (from n.1 to n.6) 2 symmetrical cuboids + big cuboid, and, finally
- 1.41 to 1.48 times, at the ratio of computation times of the fastest calculation for 1 cuboid n.6 (2:32 min) and different (from n.1 to n.6) 2 symmetrical cuboids + big cuboid.

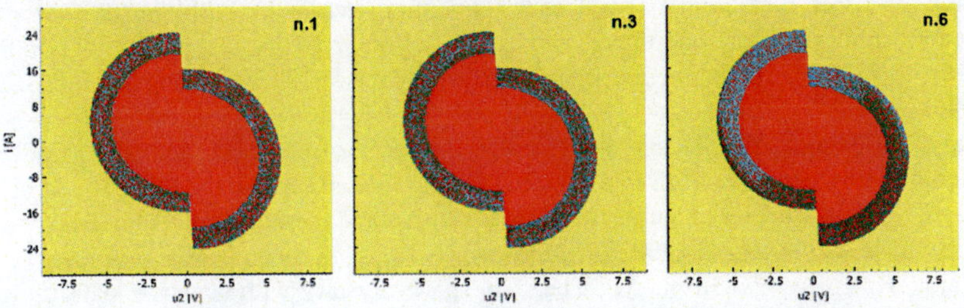

Fig. 10. Example of C-SBS for $u_1 = 0$ as a color map representing the detection of the presence of RP in any of the test cuboids *n.1*, *n.3* and *n.6* (green/aqua) symmetrically placed on the CHA as illustrated in Fig. 5a,b and the big cuboid located according to Fig. 8 and Fig. 9.

Table 3. Speed up the calculation after adding a big cuboid (Fig. 8, Fig. 9).

Number of cuboid	From Table 1 1 cub. & 2 sym. Cub. (mm:ss)	1 cuboid + big cub. (mm:ss)	Acceleration with big cuboid	2 sym.cub. + big cuboid (mm:ss)	Acceleration with cuboid	Total acceleration compared to *n.1* & *n.6*
n.1	3:22 & 3:06	1:51	1.82	1:48	1.72	1.87 & 1.41
n.2	3:01 & 2:53	1:47	1.69	1:45	1.65	1.84 & 1.45
n.3	2:43 & 2:31	1:45	1.55	1:44	1.45	1.57 & 1.46
n.4	3:10 & 2:49	1:51	1.71	1:47	1.58	1.76 & 1.42
n.5	2:54 & 2:48	1:45	1.66	1:44	1.62	1.67 & 1.46
n.6	2:32 & 2:32	1:43	1.48	1:43	1.48	1.48 & 1.48

5 Conclusion

Properly placing cuboids on the CHA and inside the CHA, it is possible to achieve up to 2 times faster calculation of C-SBS with unchanged computer hardware. Although the user would not choose the optimal placement of test cuboids on the CHA, it is certainly necessary to consider placing another, large test cuboid inside the CHA, if the morphology of the CHA and BS allows it. The presence of a large cuboid will speed up the calculation of several tens to hundreds of C-SBS 1.45–1.72 times, which is not a negligible acceleration of calculations. The quickly calculated set of C-SBS is then ready for reconstruction of BS in 3D, subsequent visualization in virtual reality, or for 3D printing.

References

1. Špány, V.: Special surfaces and trajectories of the multidimensional state space. In: Proceedings of VŠT in Košice, Košice, no. 1, pp. 123–152 (1978). (in Slovak)
2. Špány, V., Pivka, L.: Boundary surfaces in sequential circuits. In: Beitrage zur Theoretischen Elektrotechnik, Das Internationale Symposium in Ilmenau, Ilmenau, pp. 90–99 (1988)
3. Špány, V., Pivka, L.: Dynamic properties of flip-flop sensors. J. Electr. Eng. **47**(7–8), 169–178 (1996)
4. Špány, V., Galajda, P., Guzan, M.: Boundary surfaces of one-port memories. In: Tesla III Milenium: Proceedings of 5th International Conference, Beograd, pp. IV-131–137 (1996)
5. Guzan, M.: Limit cycles and trajectories in a multivalued memory cell. Acta Electrotechnica et Informatica **2**(1), 36–41 (2002). (in Slovak)
6. Galajda, P., Špány, V., Guzan, M.: The state space mystery with negative load in multiple-valued logic. Radioengineering **8**(2), 2–7 (1999)
7. Galajda, P., Špány, V., Guzan, M.: The state space mystery with virtual saddle point in memory cell. In: Proceedings of the DSP - MCOM 2005, Košice, pp. 147–150 (2005)
8. Galajda, P., Guzan, M., Špány, V.: The state space description of the MVL memory circuits. In: Education, Science and Economics at Universities: Integration to International Educational AREA, International Conference, Wydawnictvo Naukove NOVUM, Plock, Poland, pp. 351–359 (2010)
9. Guzan, M.: Boundary surface and stable manifold in sequential circuits. In: Radioelektronika 2011, Proceedings of 21-th International Conference, Brno, Czech Republic, pp. 219–222 (2011)
10. Galajda, P., Guzan, M., Špány, V.: The control of a memory cell with the multiple stable states. In: Radioelektronika 2011, Proceedings of 21-th International Conference, Brno, Czech Republic, pp. 211–214 (2011)
11. Guzan, M.: Boundary surface of a ternary memory in the absence of limit cycles. In: Radioelektronika 2012, Proceedings of 22-nd International Conference, Brno, Czech Republic, pp. 1–4 (2012)
12. Guzan M.: Boundary surface of 5-valued memory. J. Eng. **2013** (2013). https://www.hindawi.com/journals/je/2013/626824/. Accessed 12 Nov 2020. Article Number 626824
13. Guzan, M.: Analysis of 6(4) - valued memory. Elektronika ir Elektrotechnika **20**(6), 89–92 (2014)
14. Petrzela, J.: Multi-valued static memory with resonant tunneling diodes as natural source of chaos. Nonlinear Dyn. **94**(3), 1867–1887 (2018). https://doi.org/10.1007/s11071-018-4462-0
15. Špány, V., Pivka, L.: Boundary surfaces in sequential circuits. Int. J. Circuit Theory Appl. **18**, 349–360 (1990)
16. Špány, V., et al.: Chua's singularities: great miracle in circuit theory. Int. J. Bifurc. Chaos **20**(10), 2993–3006 (2010)
17. Guzan, M.: Variations of boundary surface in Chua's circuit. Radioengineering **2**(3), 814–823 (2015)

18. Hlavačka, M., Guzan, M.: Hidden attractor and regions of attraction. In: 27-th International Conference on Radioelektronika, Brno, Czech Republic, pp. 82–85 (2017)
19. Guzan, M., Hlavačka, M., Maliakova, M.: Regions of attraction for two hidden attractors. In: Proceedings of the International Conference on Modern Electrical and Energy Systems, MEES 2019, Kremenchuk, Ukraine, pp. 346–349 (2019)
20. Petržela, J.: On the canonical circuit realizations of fourth-order and fifth-order hyperjerk chaotic function. In: 29-th International Conference on Radioelektronika, Pardubice, Czech Republic, pp. 39–44 (2019)
21. Bilotta, E., Pantano, P.: A Gallery of Chua attractors, World Scientific Series on Nonlinear Science, Singapore, vol. 61 (2008)
22. Ontanon-Garcia, L.J., Campos-Canton, E.: Widening of the basins of attraction of a multistable switching dynamical system with the location of symmetric equilibria. Nonlinear Anal.-Hybrid Syst. **26**(nov2017), 38–47 (2017)
23. Petržela, J., Polak, L.: Minimal realizations of autonomous chaotic oscillators based on transimmittance filters. IEEE Access **7**, 17561–17577 (2019)
24. Lai, Q., Nestor, T., Kengne, J., Zhao, X.W.: Coexisting attractors and circuit implementation of a new 4D chaotic system with two equilibria. Chaos Solitons Fract. **107**, 92–102 (2018)
25. Campos-Canton, E.: Chaotic attractors based on unstable dissipative systems via third-order differential equation. Int. J. Modern Phys. C **27**(1), 1–11 (2016)
26. Li, C., Sprott, J.C.: How to bridge attractors and repellors. Int. J. Bifurc. Chaos **27**(10), 1–11 (2017)
27. Yadava, K., Prasad, A., Dev Shrimali, M.: Control of coexisting attractors via temporal feedback. Phys. Lett. A **382**, 2127–2132 (2018)
28. Petržela, J., Gotthans, T.: New chaotic dynamical system with a conic-shaped equilibrium located on the Plane Structure. Appl. Sci **7**(10), 976 (2017)
29. Matsumoto, T., Chua, L.O., Kumoro, M.: The double scroll. IEEE Trans. Circ. Syst. CAS-**32**(8), 798–818 (1985)
30. Sobota, B., Guzan, M.: Virtualization of Chua's circuit state space. In: Recent Advances in Chaotic Systems and Synchronization - From Theory to Real World Applications, pp. 127–163. Elsevier, United Kingdom (2019)
31. Guzan, M., et al.: Boundary Surface of Chua's circuit in 3D state space. In: International Conference on Modern Electrical and Energy System, Kremenchuk, Ukraine, pp. 399–391 (2017)
32. Guzan, M., Astalos, J., Dvorcak, P.: Reconstructing boundary surface of Chua's circuit by Paraview. Electromech. Saving Syst. **15**(3), 163–167 (2016)
33. Sobota, B.: 3D modelling of Chua's circuit boundary surface. Acta Electrotechnica et Informatica **11**(1), 44–47 (2011)
34. Racz, Zs., Sobota, B., Guzan, M.: Parallelizing boundary surface computation of Chua's circuit. In: 27th International Conference on Radioelektronika, Brno, Czech Republic, pp. 1–4 (2017)
35. Intel Core i9-10980XE. https://www.alza.sk/intel-core-i9-10980xe-d5708794.htm. Accessed 12 Nov 2020
36. Intel Xeon Gold 6252. https://www.alza.sk/intel-xeon-gold-6252-d5630396.htm?o=1. Accessed 12 Nov 2020
37. AMD EPYC 7742. https://www.alza.sk/amd-epyc-7742-d5686762.htm?o=1. Accessed 12 Nov 2020

Decision Making and Decision Support Systems

On Problems and Methods of Coordinated Scheduling and Location

Jerzy Józefczyk(✉), Mirosław Ławrynowicz, and Grzegorz Filcek

Department of Computer Science and Systems Engineering, Wroclaw University of Science and Technology, Wybrzeże Wyspiańskiego 27, 50-370 Wroclaw, Poland
{jerzy.jozefczyk,miroslaw.lawrynowicz,grzegorz.filcek}@pwr.edu.pl

Abstract. We investigate a combined approach for job scheduling on parallel machines and the location of machines referred to in the literature as ScheLoc. We sum up the up-to-date achievements and propose a general classification of such a two-part problem involving different solving methods. A sequential approach requires both parts' successive consideration, unlike the joint proposal with their simultaneous investigation. The uncertain version is also proposed, aiming to replace the two-part deterministic problem with a single counterpart with interval release dates of jobs. The mentioned solution approaches are illustrated by specific job scheduling and machine location sub-problems.

1 Introduction

Systems (systemic) thinking has been present in science and research activities for many decades, at least since the beginning of the last age, when Jan Smuts has pioneered the concept of holism [1]. It mainly refers to complex research objects composed of parts (elements) mutually connected, i.e., having the inner structure. Then, systems thinking assumes the examination of an object as a whole, taking into account its structure. We regard systems thinking as the opposite of reductionism that works via the decomposition of a complex object. Consequently, knowledge of an object is the sum of knowledge of elements.

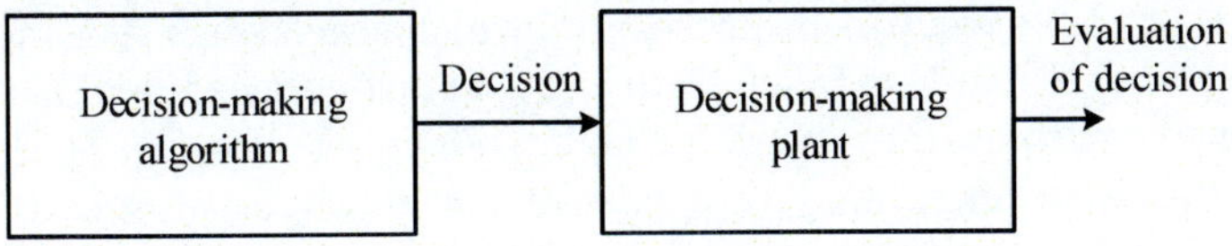

Fig. 1. Decision-making system

Systems thinking can also be applied to operations research objects (plants) where decision-making problems are examined. This paper uses the general scheme for the decision-making system, representing a decision-making problem and composed of a decision-making plant (an object of examination) and

K. T. Atanassov et al. (Eds.): IWIFSGN 2020/BOS/SOR 2020, LNNS 338, pp. 145–163, 2022.
https://doi.org/10.1007/978-3-030-95929-6_12

decision-making algorithm Fig. 1. In the other part of this paper, the optimal decision-making problems are only considered, where the evaluation of decisions is understood as a criterion. We focus the considerations on such optimization problems when more than a single decision should be taken. There are many such problems in operations research and real-world applications. Let us only list some examples: location-routing, inventory-location, inventory-routing, routing-scheduling, production-inventory, production-transportation [2].

Coordinated scheduling and location, referred to as ScheLoc, is another example discussed in this paper [3]. The necessity to make two interconnected decisions causes the complexity that can be managed by systems thinking. A decision on machines' locations as sites where jobs can be executed accompanies the fundamental decision on jobs' schedules. Such a composition of decisions is motivated by some real-world applications. The unloading of ships or evacuation of people from hazardous areas can serve as examples [4,5]. The unloading of ships by cranes is an example of a real-world application for the former mentioned case. Ships waiting for the unloading at sea can be treated as jobs. Locations at an embankment for cranes as executors and the order of ship's service after reaching the embankment need the determination.

The remainder of the work is organized as follows. A general ScheLoc problem followed by a literature review is formulated in Subsect. 2.1. Both different cases of the problem and solution algorithms are there presented. Overall remarks on methods employed for uncertain optimization problems are the content of Subsect. 2.2. The proposition for a unified classification of different ScheLoc's versions, together with some approaches for their solving, is given in Subsect. 3.1. The considerations apply systems thinking. The procedures, called optimization schemes, are then illustrated in Subsect. 3.2. Section 4 concludes the deliberations and provides the outlines of other possible methods.

2 Preliminaries

2.1 Scheduling and Location Problems

2.1.1 General Description and Classification

Considerations refer to a generalized version of job scheduling on parallel machines, also called executors or processors, which crucial case deals with determining an optimal schedule σ taken from a set of feasible schedules D_σ; see [6] for a comprehensive presentation. Now, let us introduce sets $\mathrm{J} = \{J_1, J_2, ..., J_j, ..., J_n\}$ and $\mathrm{M} = \{M_1, M_2, ..., M_i, ..., M_m\}$ of n jobs and m machines, respectively. Then the schedule $\sigma \in \mathrm{D}_\sigma$ determines for every job J_j a starting time t_j and a machine assigned to its processing. First of all, jobs are characterized by processing times $p = [p_{ij}]_{i=\overline{1,m},j=\overline{1,n}}$, where p_{ij} stands for the processing time of the job J_j by the machine M_i. Additionally, due dates $d = [d_j]_{j=\overline{1,n}}$, weights $w = [w_j]_{j=\overline{1,n}}$, and release dates $r = [r_j]_{j=\overline{1,n}}$ can feature the jobs. The sequence of parameters $\mathrm{A_S} = (p, d, w, r)$, together with the schedule σ, enables us to calculate the completion times of jobs $[C_j(\sigma; \mathrm{A_S})]_{j=\overline{1,n}}$. Job scheduling criteria

$\bar{Q}_S([C_j(\sigma; A_S)]_{j=\overline{1,n}}) \triangleq Q_s(\sigma, A_S)$, which base on such times, make it possible to select the best schedules from a set of feasible ones D_σ. The form of D_σ guarantees the fulfillment of requirements imposed on jobs and concerning preemptions, precedence constraints, single assignments of machines to a job, the possibility of simultaneous processing by a machine at most an only job, and others. For simplicity, the apparent dependence of D_σ and A_S is omitted in the notation. Then, the overall job scheduling sub-problem referred to as PS deals with searching for the optimal schedule, where $\sigma \in D_\sigma$ and the sequence of parameters A_S is given, i.e.,

$$\min_{\sigma \in D_\sigma} Q_S(\sigma, A_S). \tag{1}$$

The mentioned generalization of the considered sub-problem assumes that elements of r, i.e., the release dates, are not given a priori, and their values are connected with a priori unknown locations of machines where jobs should be performed. It is motivated by such real-world applications where each job J_j is deployed at a given site (location) a_j in a plane or space. To be executed by M_i, the job has to move from a_j with speed v_j to M_i's location x_i. The movement between a_j and x_i is incurred by cost c_{ij} as an element of $c = [c_{ij}]_{i=\overline{1,m}, j=\overline{1,n}}$. Launching the location for the machine M_i costs g_i, and $g = [g_i]_{i=\overline{1,m}}$. Sites a_j, x_i, respectively, for all jobs, all machines, and job velocities v_j are elements of vectors $a = [a_j]_{j=\overline{1,n}}$, $x = [x_i]_{i=\overline{1,m}}$ and $v = [v_j]_{j=\overline{1,n}}$. Vectors a and v together with costs c, g form a sequence of parameters $A_L = (a, v, c, g)$. Unlike sites a the machines' locations x are unknown and should be selected from a given set of available locations. Location criteria $Q_s(x, A_L)$ usually comprise transportation costs and (or) location (investment) costs of launching particular sites.

A location sub-problem PL

$$\min_{x \in D_x} Q_L(x, A_L), \tag{2}$$

is solved for a given sequence of parameters A_L and the domain D_x representing available machines' positions taken from the discrete or continuous area and possibly dependent on A_L. The particular type of PL depends on Q_L and D_x.

Generalized release dates for jobs, which rely on machines' locations, connect both sub-problems. Namely, $r_{ij}(x_i) = \rho_j + v_j^{-1} d(a_j, x_i)$ is the release date for J_j scheduled on M_i, where d is the distance between locations of J_j and M_i, and ρ_j stands for a given ready time for J_j. Consequently, the set of feasible schedules D_σ and the criterion (1) depend on x that is denoted as $D_{\sigma x}$ and $Q_{SL}(\sigma, x; A_S)$, respectively. Please note that if locations of jobs and machines are not considered, then $r_j = \rho_j$.

As mentioned in Sect. 1, just characterized coordinated problem, being the composition of the sub-problems PL and PS, is called in the literature as ScheLoc. We propose the extended Graham's notation $\alpha|\beta|\gamma|\delta$ [7] to represent different versions of ScheLoc. The first three fields manifold Graham's notation and describe machines, jobs, and job scheduling criteria. The second field includes information on release dates r_j by adding an adequate notation based on r_j. For example, r_j-V and $r_j^- \le r_j \le r_j^+$ denote respectively variable (non-fixed) release dates and their unknown values belonging to intervals of known bounds. The last field represents PL. Discrete and continuous location areas are generally distinguished in the literature. The first case assumes the finite number of possible locations, unlike the second case when any point from a continuous area can be chosen for a facility (machine, executor). A network is a sub-case of the former possibility.

Specific instances of PL with discrete location areas are based on uncapacitated facility location problems (UFLP) and are differentiated by a form of criterion. A full version of UFLP employs both a transportation cost and an investment cost. The former cost can be understood widely, including distance and time-related measures. A single cost taken from both just mentioned ones can serve as a criterion as well. Abandon of investment cost with a fixed number of deployed machines leads to the so-called p-median problem. Facility location problems for a case with continuous location areas use similar criteria. However, it is worth noting that limitation to transportation cost only gives, in consequence, distinct counterparts with a single facility. Therefore, the evaluation usually contains a part assessing the relationship among all deployed facilities. The following instances δ can be distinguished while taking into account the specificity of the considered coordinated problem investigated till now in the literature or worth considering in the future:

- UFLP (full UFLP),
- UFLP_T (UFLP with the restriction to a transportation cost),
- UFLP_I (UFLP with the limitation to an investment cost),
- P_MEDIAN (p-median problem),
- CFLP (facility location with continuous location area).

Table 1. The track record on coordinated scheduling-location

$1|r_j$-V$|C_{max}|$UFLP_T	[3]
$1|r_j$-V$|C_{max}|$CFLP	[4,8]
$1|p_j = p, r_j$-V$|C_{max}|$CFLP	[8]
$P|r_j$-V$|C_{max}$	[14]
$P|r_j$-V$|C_{max}|$UFLP_T	[10]
$P|r_j$-V$|\sum C_j|$CFLP	[12,13]
$R|r_j$-V$|C_{max}|$P_MEDIAN	[11]
$R|r_j$-V$|C_{max}$	[11]
$P|r_j$-V$|\sum C_j|$UFLP_I	[15]

2.1.2 Literature Review

Considered Problems

The scarce collection of ScheLoc problems have been investigated so far and is represented in the literature. Table 1 provides the essential works grouped according to the introduced notation. Investigations have started with a single machine case [3,4,8]. The authors of all the works have carried out a locational geometrical analysis for a single machine deployment. It has resulted in the calculation of release dates for all jobs based on distances among locations of jobs and the machine. Then, they have solved a classical job scheduling that has been a polynomial-time optimization problem. Research in [10] have extended such an approach for identical machines. Unlike the previous works, they have obtained clusters of jobs assigned to individual machines resulting from an iterative procedure with distance analysis. It is worth noting that authors of all these considerations solved individual cases of discrete or continuous facility location problems with the cost criteria limited to distances between jobs and a machine. Problems proposed in [12] and [13] have continued similar investigations for more machines and the criterion in the form of the sum of completion times. Another approach can be found in [14], where a criterion leading, in consequence, to the calculation of release dates for jobs has not been used at all. Values of release dates necessary for solving corresponding job scheduling have been determined from the quality of the job scheduling point of view. The author has considered both discrete and continuous location areas. Both approaches encountered in previous investigations have been applied and compared by [11] for unrelated machines and the makespan as the criterion. The idea of the first approach has been dealt with the solution for the p-median problem enabling the calculation of release dates and searching for the solution of $R||C_{\max}$. The authors have compared results with the approach enabling the joint determination of machines' locations and a schedule of jobs to minimize the makespan without employing a distinct criterion for the deployment. As the latter approach could suffer from the deficiency of the location evaluation, a bi-criteria approach has been proposed by [15], where apart from the sum of completion times, the total investment cost has also been taken into account.

Used Algorithms
Locational sub-problems PL for a single machine have been solved mainly by geometrical analysis evaluating distances among positions for discrete and continuous location areas [3,4,8]. The values of release dates, calculated based on regions in the available area where the release dates do not change, are data for the ERD (Earliest Release Dates) rule used for solving job scheduling. For example, [8] propose the dedicated 'big triangle small triangle method' [9].

The version with many machines has been introduced in [14], where both discrete and continuous planar areas are considered. For the criterion $C_{\max}$, the mathematical models are presented, and the solver GAMS is employed as a solution tool. The authors give examples of results limited to ten jobs and three machines. The job scheduling on many parallel identical machines, the makespan $C_{\max}$ as the criterion, and discrete locations for machines are investigated in [5,10]. The constructive algorithm has been proposed that is based on the iterations between location and assignment sub-problems. The Master Thesis Piasecki [12] and the following conference paper [13] extend the investigations on the minimization of $P|r(x)|\sum C_j$ with the development of many parallel identical machines on the continuous area. The evolutionary algorithm solves the resultant strongly NP-hard problem. The results for small instances are compared with optimal outcomes generated by the solver CPLEX CP Optimizer. A more advanced complex issue is discussed in [11]. Namely, many unrelated parallel machines should be located in a discrete area, and apart from C_{max}, the additional criterion has been introduced in the form of the sum of release dates for jobs. Two approaches are proposed, solved, and compared. The first one, referred to as the sequential approach, comprises two steps: solving PL sub-problem to minimize the sum of release dates and PS to minimize C_{max} with the release dates calculated based on the previously addressed sub-problem PL. The known algorithms have been employed to solve both sub-problems. In the second approach, called the systemic approach, both sub-problems are solved jointly to minimize C_{max}. Machine locations are determined to minimize C_{max}, not the sum of release dates, like in the sequential approach. The memetic algorithm based on Tabu Search metaheuristics has been proposed as the solution tool. The conducted experimental comparison confirmed the substantial advantage of the systemic approach in comparison with the sequential approach.

A new bi-criteria joint location-scheduling problem is discussed in [15]. Apart from the sum of completion times, the problem includes the additional criterion that evaluates machine locations' cost. A NSGA II-based [16] evolutionary algorithm is proposed to solve the problem.

Four criteria are taken into consideration in [17] while assessing locational and assignment decisions for military applications. The solution algorithm is based on the NSGA II approach for solving the resulting multi-objective optimization problem. In comparison with previous works, the contribution of this paper can be explained as follows. First of all, a new optimization problem is solved. A single criterion has evaluated all variables (decisions) in previous works. The distances and release dates, which have been used in some works to determine machine locations, were additional criteria. The makespan C_{max}, or the sum of completion times $\sum C_j$, was the main criterion, and it assessed the schedule of jobs. The quality of the deployment of machines was only indirectly evaluated by C_{max} or $\sum C_j$, as they were the basis for calculating release dates.

2.2 Solution Methods for Parametric Uncertainty

Uncertainty is a fundamental paradigm constituting analysis and decision-making problems more challenging on the one hand but closer to real-world systems on the other hand. The uncertainty can be interpreted, described, and handled differently and investigated differently for particular applications. A representation of uncertainty and an evaluation of an optimization variable, i.e., a criterion, have to be chosen from various possibilities. They constitute a specific uncertain optimization problem or, in general, an uncertain decision-making problem.

The following brief presentation is limited to parametric uncertainty. It refers to the minimization of function $f(u, \beta)$ for the optimization variable u, which belongs to the set of feasible variables U, where β is the value of a parameter (vector parameter, matrix parameter, in general).

2.2.1 Selected Representations of Uncertainty

Various representations can describe such uncertainty. The probabilistic version is the most popular. It is then assumed that β is the realization of a random variable $\tilde{\beta}$ characterized by the probability distribution function $F(\beta) = \Pr\{\tilde{\beta} \leq \beta\}$ or the density function $\phi(\beta) = dF(\beta)/d\beta$ if it exists, where Pr stands for the probability and $F(\beta)$ is defined over the set B of possible values β. So, $\langle \mathrm{B}, F(\beta)\rangle$ is the probabilistic representation of uncertain parameter β. Such a possibility makes this representation very reliable, but it is connected with its weakness when real-world data concerning β do not exist or are not available. The lack of real-world data β can be replaced by an expert who can be assumed as the source of knowledge on uncertain parameters β. Such an approach is proposed in fuzzy-based representations and their derivative versions. As a number from the interval $[0, 1]$ expressed by an expert, the degree of truth means that the parameter β takes a given value. An expert's opinion for all possible values β is called the membership function $\mu : \mathrm{B} \to [0, 1]$. This representation in the form of $\langle \mathrm{B}, \mu(\beta)\rangle$ is the subjective one as it reflects an individual opinion on β, and it has not to be connected with real-world values β.

Consequently, the final result of the optimization strongly depends on an expert's credibility (quality). The case considered by many researchers omitting the drawbacks of both previously mentioned representations assumes that a set of feasible values B is the only information on uncertain parameters β. The interval representation is an actual example of this case. It is assumed that every element in β belongs to the interval in which bounds are given. Every specific realization of β belonging to the Cartesian product of intervals for β's individual elements is called a scenario.

2.2.2 Criteria Under Parametric Uncertainty

The value of $f(u, \beta)$ cannot serve as the evaluation of u due to the non-fixed value of β. The selection of adequate criterion (evaluation) for such a non-deterministic version is crucial. Many propositions can be found in the literature. The majority of them are based on the substantiation of deterministic counterparts. Let us mention the most critical cases. The substantiation consists of transforming uncertain criterion to the deterministic value with different aggregation operators, like minimum, maximum or average, on uncertain parameters β. For example, the substantiation imposed on $f(u, \beta)$ can result in the following criteria for the non-deterministic case: $\max_{\beta \in \mathrm{B}} f(u, \beta)$, $\min_{\beta \in \mathrm{B}} f(u, \beta)$, $\int_{\mathrm{B}} (u, \beta) d\beta$. It is important to note that listed expressions do not depend on uncertain parameters β but only on optimization variables u, being their univocal assessment. The version with the operator maximum gives an important class of problems called minmax optimization problems where the maximum on parameters β needs the minimization on u.

The substantiation can refer not only to deterministic criteria but also to some expressions based on them. The regret (opportunity loss) is an example. This notion introduced by Savage is the difference between the value of $f(u, \beta)$ and its optimal value, i.e.:

$$f(u,\beta) - \min_{v \in U} f(v,\beta). \tag{3}$$

It is calculated for every $u \in U$ and every scenario $\beta \in B$. The optimal solution for a single β would equal (3) to zero. On the other hand, optimal solutions can be different for different scenarios β. The aggregation operators can be used to make (3) independent of scenarios β. The most popular and sensible is the operator of maximum, which leads to robust (worst-case) solutions, i.e.:

$$\max_{\beta \in B} \left[f(u,\beta) - \min_{v \in U} f(v,\beta) \right]. \tag{4}$$

The expression (4) takes the value of (3) maximum for scenarios β as assessing the current u. Consequently, the optimization variable u minimizing (4) is sought. The operators of minimization and averaging can be adequately used. It is easy to see that taking (4) as the uncertain optimization criterion implies challenging optimization problems comprising three nested optimization subproblems. Time-reasonable solution algorithms may be obtained only for particular instances when deterministic counterparts are easy, i.e., polynomially solvable optimization problems. Unfortunately, easy deterministic optimization problems are infrequent and pretty far from real-world applications. Therefore, heuristic algorithms can only be used.

Let us mention other possibilities to solve the uncertain optimization problems (see [18] as an example). As far as applying heuristic solution algorithms can relax the minmax approaches with very rigorous problem formulation, relaxations can refer directly to criteria that can cause having easier optimization problems. Another method for evaluating decisions with application to uncertain parameter optimization introduced by Roy is called (b, w)–robustness. However, it can be applied only to a finite set of scenarios, i.e., where $B = \{\beta_1, \beta_i, ..., \beta_k, ..., \beta_K\}$, and k is the number of all scenarios. The (b, w) robust schedule maximizes the number of scenarios for which the value of criterion does not exceed b, and the value of criterion for every scenario is not greater than w. The determination of a robust solution can be expressed in the form of the following optimization task

$$\max_{u \in U} \sum_{l=1}^{k} \mathbf{1}(b - f(u,\beta_l)) \tag{5}$$

subject to $f(u,\beta_l) \le w$, $l = 1, 2, ..., k$ where $\mathbf{1}(\cdot)$ is the unit function, i.e., $\mathbf{1}(x) = 1(0)$ for $x \ge 0$ $(x < 0)$.

Two consecutive approaches, namely p-robustness, where p is the numerical parameter of this approach, and lexicographic α-robustness, enable us to obtain a set of robust solutions, not a single solution as in the previous approach. According to the former one, the collection of p-robust decisions is as follows:

$$D_p(u) = \left\{ u \in U : \forall_{l=1,2,...,k}\ f(u,\beta_j) - \min_{v \in U} f(v,\beta_l) \le p \right\}. \tag{6}$$

It is easy to see that the p-robustness approach is a notably more straightforward case of the minmax regret as the latter seeks a single schedule minimizing

p. The lexicographic α-robustness requires sorting $f(u, \beta_l)$ in the non-increasing order that gives the sequence of values $f(u, \beta_l) \geq f(u, \beta_{l+1})$, $l = 1, 2, ..., k$. Then, the set of lexicographic α-robust solutions can be written as

$$\overline{\mathrm{D}}_\alpha(u) = \left\{u \in \mathrm{U} : \forall_{j \leq k}\ f(u, \beta_j) - \min_{v \in \mathrm{U}} f(v, \beta_j) \leq \alpha\right\}. \quad (7)$$

Consequently, the optimization variable u is lexicographic α-robust if the jth greatest ordered value of criterion is greater than the best jth greatest ordered value of criterion no more than α. This property has to be true for all $j \leq k$. The lexicographic α-robust measure is less rigorous than the p-robust measure, which means a robust schedule can exist for the lexicographic, unlike the p-robust solution.

3 Coordinated Scheduling and Location

3.1 Proposed Solution Approaches

Up-to-date considerations and results enable us to distinguish two fundamental approaches for solving job scheduling on parallel machines with machines' unknown locations.

The first one, called the sequential approach, assumes PL and PS's consecutive solving (Fig. 2). It means that the positions x for machines are firstly determined. Consequently, it enables us to calculate all machine-driven release dates r_{ij} followed by the substantiation of r_{ij} to r_j and a classic job scheduling solution with different release dates for jobs. Briefly speaking, this approach, also called in the literature as an integrated approach, separates PL and PS in time, preserving their interconnection. This approach's general scheme can be more formally expressed in the form of an optimization scheme S_Seq.

Sequential optimization scheme S_Seq

Require: J, M, $\mathrm{A_S}$, $\mathrm{A_L}$, D_x, $\mathrm{D}_{\sigma x}$
Ensure: $\tilde{\sigma}$, $\tilde{x}$, $\tilde{Q}_\mathrm{L} \triangleq Q_\mathrm{L}(\tilde{x}; \mathrm{A_L})$, $\tilde{Q}_\mathrm{SL} \triangleq Q_\mathrm{SL}(\tilde{\sigma}, \tilde{x}; \mathrm{A_S})$
1: Solve the locational problem $\min_{x \in \mathrm{D}_x} Q_\mathrm{L}(x, \mathrm{A_L})$ to obtain $\tilde{x}$ and $\tilde{Q}_\mathrm{L}$.
2: Calculate $r_{ij}(\tilde{x}_i) = \rho_j + v_j^{-1} d(a_j, \tilde{x}_i)$, $i = \overline{1, m}$, $j = \overline{1, n}$.
3: Set the release dates of jobs using the selected operator Φ to have $r_j = \Phi(r_{1j}, ..., r_{mj})$.
4: Solve the job scheduling problem $\min_{\sigma \in \mathrm{D}_{\sigma x}} Q_\mathrm{SL}(\sigma, \tilde{x}; \mathrm{A_S})$ with different release dates r_j to determine $\tilde{\sigma}$ and $\tilde{Q}_\mathrm{SL}$.

The operator 'max' can serve as an example of Φ. Then, $r_j = \max(r_{1j}, ..., r_{mj})$.

Figures 3 and 4 depict the idea of the second coordinated approach when both decisions σ and x are determined simultaneously as a pair $(\sigma, x) \in \mathrm{D}$, i.e., without the separation in time. The set $\mathrm{D} = \mathrm{D}_{\sigma x} \times \mathrm{D}_x$ of decision pairs $(\sigma, x) \in \mathrm{D}$ contains all constraints enabling the determination of feasible schedules and machines' locations. Its particular form depends on the cases of PS and

PL considered as well as the method of the complex problem's modeling. Different criteria can evaluate both decisions. Two cases may be distinguished: the evaluation by only the essential criterion Q_{SL} (Fig. 3) or by both criteria Q_{SL} and Q_{L} (Fig. 4). The latter possibility leads to the bi-criteria optimization problem when decisions evaluated from both sub-problems points of view are sought via minimization of vector criteria $[Q_{\mathrm{SL}}(\sigma, x; \mathrm{A_S}), Q_{\mathrm{L}}(x; \mathrm{A_L})]^{\mathrm{T}}$. Analogously to the first approach, optimization schemes S_Jnt (the joint single-criterion approach) and S_Jnt_B (the joint bi-criteria approach) more precisely characterize the first and the second case, respectively.

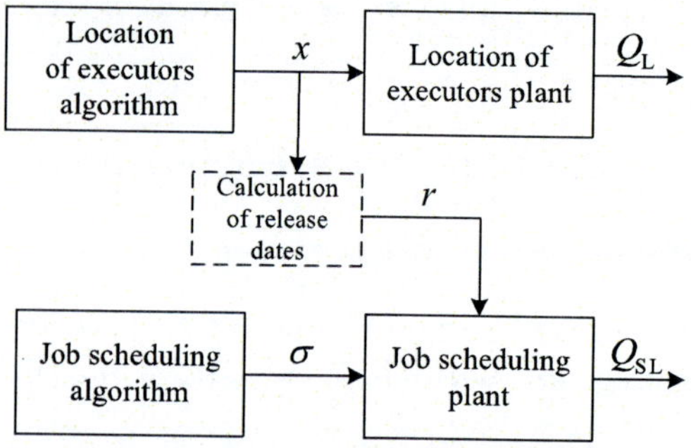

Fig. 2. Sequential approach for solving ScheLoc

Joint optimization scheme S_Jnt

Require: J, M, $\mathrm{A_S}$, $\mathrm{A_L}$, D

Ensure: $\hat{\sigma}$, $\hat{x}$, $\hat{Q}_{\mathrm{SL}} \triangleq Q_{\mathrm{SL}}(\hat{\sigma}, \hat{x}; \mathrm{A_S})$

1: Solve the joint scheduling-location problem $\min_{(\sigma,x)\in \mathrm{D}} Q_{\mathrm{SL}}(\sigma, x; \mathrm{A_S})$ to obtain $\hat{\sigma}$, $\hat{x}$ and $\hat{Q}_{\mathrm{SL}}$.

Joint bi-criteria optimization scheme S_Jnt_B

Require: J, M, $\mathrm{A_S}$, $\mathrm{A_L}$, D

Ensure: Set of λ pairs of decisions $\mathrm{ARG_{PF}} = \{(\sigma_l^*, x_l^*) : l = \overline{1, \lambda}\}$ and corresponding Pareto front $\mathrm{PF} = (Q_l^* = [Q_{\mathrm{SL},l}^*, Q_{\mathrm{L},l}^*]^{\mathrm{T}} : Q_{\mathrm{SL},l-1}^* < Q_{\mathrm{SL},l}^*, l = \overline{2, \lambda})$.

1: Solve the joint scheduling-location problem $\min_{(\sigma,x)\in \mathrm{D}}[Q_{\mathrm{SL}}(\sigma, x; \mathrm{A_S}), Q_{\mathrm{L}}(x; \mathrm{A_L})]^{\mathrm{T}}$ to get $\mathrm{ARG_{PF}}$ and PF.

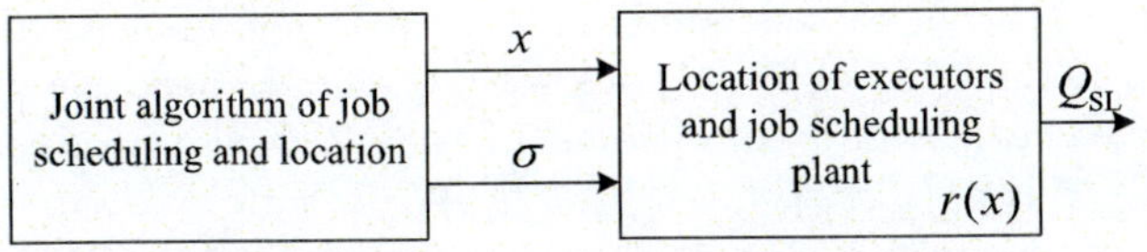

Fig. 3. Joint approach for solving ScheLoc with an only criterion Q_{SL}

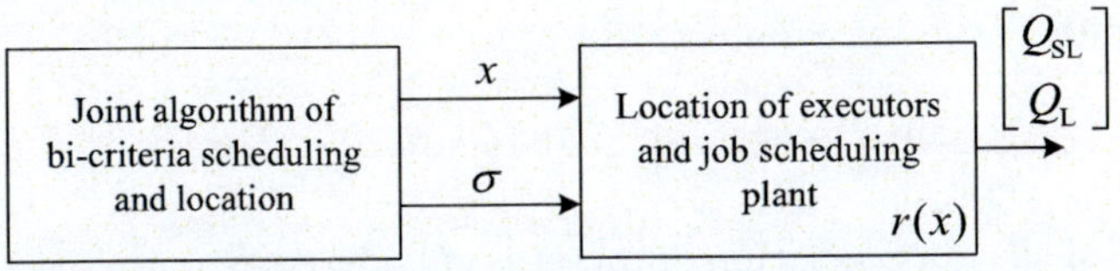

Fig. 4. Bi-criteria joint approach for solving ScheLoc

The obvious comparison of the optimization schemes S_Seq and S_Jnt can be given.

Property. The optimization scheme S_Seq does not outperform the optimization scheme S_Jnt, i.e., $Q_{\mathrm{SL}}(\tilde{\sigma}, \tilde{x}; \mathrm{A_S}) \geq Q_{\mathrm{SL}}(\hat{\sigma}, \hat{x}; \mathrm{A_S})$.
For the justification, please note that

$$Q_{\mathrm{SL}}(\hat{\sigma}, \hat{x}; \mathrm{A_S}) = \min_{(\sigma,x)\in \mathrm{D}} Q_{\mathrm{SL}}(\sigma, x; \mathrm{A_S}) \leq \min_{(\sigma,x)\in \mathrm{D}'} Q_{\mathrm{SL}}(\sigma, x; \mathrm{A_S})$$
$$= \min_{\sigma\in \mathrm{D}_{\sigma\tilde{x}}} Q_{\mathrm{SL}}(\sigma, \tilde{x}; \mathrm{A_S}) = Q_{\mathrm{SL}}(\tilde{\sigma}, \tilde{x}; \mathrm{A_S}), \quad (8)$$

where $\mathrm{D}' = \{(\sigma, x) : \sigma \in \mathrm{D}_{\sigma\tilde{x}} \wedge \tilde{x} = \arg\min_{x\in \mathrm{D}_x} Q_{\mathrm{L}}(x; \mathrm{A_L})\}$ and $\mathrm{D}' \subseteq \mathrm{D}$. Such evaluation immediately confirms the property.

To sum up, the proposed sequential (integrated) and joint approaches are examples of the coordinated scheduling and location announced in this work's title.

Finally, we propose an entirely different approach to cope with ScheLoc, which eliminates the necessity to determine locations for machines and, in consequence, to solve the sub-problem PL. Instead, it is assumed that values of release dates belong to known and given intervals. Consequently, a non-deterministic version of the sub-problem PS should only be solved (Fig. 5). We employ the interval min-max regret approach to realize this idea, which enables us to solve optimization problems with parametric uncertainty [23].

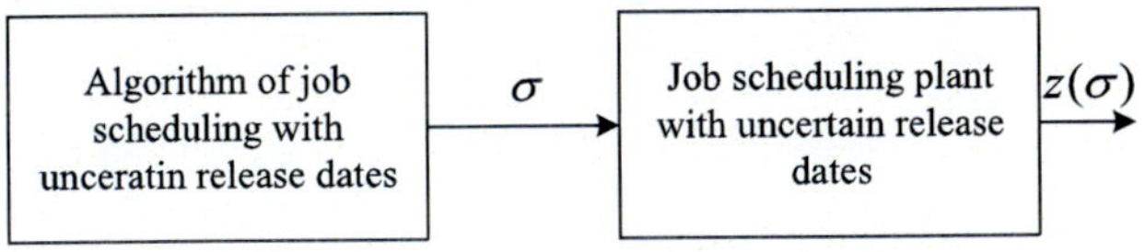

Fig. 5. Uncertain approach for solving ScheLoc that is addressing an uncertain version of PS

Using this approach, we assume that an a priori unknown value of release date r_j belongs to the interval $[r_j^-, r_j^+]$ of known bounds. Then, any value of vector r r is called a scenario and is an element of the Cartesian product of intervals for all r_j, i.e., $r \in \mathrm{R} = [r_1^-, r_1^+] \times ... \times [r_n^-, r_n^+]$. We calculate a regret for every σ and r. It is the difference between the criterion $\hat{Q}_\mathrm{S}(\sigma, r) \triangleq Q_\mathrm{S}(\sigma, \mathrm{A_S})$ and the optimal value of this criterion $Q'_\mathrm{S}(r) = \min_{\pi \in \mathrm{D}_\sigma} \hat{Q}_\mathrm{S}(\pi, r)$. It is crucial noting that now criterion Q_S does not depend on x. The optimal schedule σ'' is the result of minimization of the maximum regret $z(\sigma)$, i.e.

$$\min_{\sigma \in \mathrm{D}_\sigma} z(\sigma) = \min_{\sigma \in \mathrm{D}_\sigma} \max_{r \in \mathrm{R}} \left[\hat{Q}_\mathrm{S}(\sigma, r) - Q'_\mathrm{S}(r)\right]. \tag{9}$$

The outcome of the maximization of regret is called a worst-case scenario referred to as r^σ. The following optimization scheme fulfills the presented approach.

Uncertain optimization scheme S_Unc

Require: J, M, $\mathrm{A_S}$, D_σ, R
Ensure: σ'', r^σ, $z(\sigma'')$, $Q^n_\mathrm{S} \triangleq Q_\mathrm{S}(\sigma'', r^\sigma)$.
1: Solve $\min_{\pi \in \mathrm{D}_\sigma} \hat{Q}_\mathrm{S}(\pi, r)$ to get $Q'_\mathrm{S}(r)$.
2: Obtain the worst-case scenario $r^\sigma = \arg\max_{r \in R}[\hat{Q}_\mathrm{S}(\sigma, r) - Q'_\mathrm{S}(r)]$ and calculate $z(\sigma) = \hat{Q}_\mathrm{S}(\sigma, r^\sigma) - Q'_\mathrm{S}(r^\sigma)$.
3: Solve $\min_{\sigma \in \mathrm{D}_\sigma} z(\sigma)$ to obtain σ'', $z(\sigma'')$, and Q''_S.

It is necessary to note that the value $z(\sigma'')$ evaluates this optimization scheme. However, it cannot be directly compared with the scheduling criterion Q_S. Therefore, Q''_S allow us to compare this optimization scheme with other ones.

We assume the knowledge of the set of all feasible scenarios R. The way of its determination is not fixed, but it is a challenge. Undoubtedly, the bounds of intervals can be calculated, taking into account the vectors a and v.

The more in-depth formulations of both sub-problems PS and PL enable us to consider ScheLoc problems that have been solved recently and are planned for future research. Some examples, recently discussed in the Division of Intelligent Decision Support Systems at the Wroclaw University of Science and Technology, are presented in the next section.

3.2 Selected Results

This sub-section discusses three particular ScheLoc problems (cases) and different solution approaches using adequate optimization schemes. Mainly, they are taken from other cited works, where more details can be found. Of course, more than one scheme can be proposed for a specific problem.

3.2.1 Case $P|r_j\_V|\sum C_j|UFLP\_I$

Such a case with discrete location areas is considered by [15]. The joint bi-criteria optimization scheme S_Jn_B is used. A binary matrix $s = [s_{jik}]_{j,k=\overline{1,n},i=\overline{1,\mu}}$ represents the decision variable σ. The entry x_i of matrix x responsible for machines' locations is equal to 1(0) if the ith location is selected from a set of μ possible locations (otherwise). Analogously, the entry s_{jik} is equal to 1(0) if the jth job is scheduled on the ith machine as the kth (otherwise). The sequences of parameters take the forms $A_S = (p, r)$, $A_L = (a, v, g)$ and the Euclidean distances between locations are figured.

The location criterion $Q_L(x; A_L) = \sum_{i=1}^{\mu} c_i x_i$ only contains the investment cost. The job scheduling criterion is calculated according to the formula

$$Q_{SL}(s, x; A_S) = \sum_{i=1}^{\mu}\sum_{k=1}^{n} x_i C_{i,k}(s, x). \tag{10}$$

The auxiliary variable $C_{i,k}(s, x)$ stands for the completion time of a job performed by the ith machine as the kth. The conditions (11)–(15) define the set of feasible schedules D

$$1 \le \sum_{i=1}^{\mu} x_i \le \mu, \tag{11}$$

$$\sum_{i=1}^{\mu}\sum_{k=1}^{n} s_{jik} = 1, \quad j = 1, 2, \dots, n, \tag{12}$$

$$\sum_{j=1}^{n} s_{jik} \le x_i, \quad i = 1, 2, \dots, \mu, \; k = 1, 2, \dots, n, \tag{13}$$

$$C_{i,k}(s, x) \ge \sum_{j=1}^{n} (r_j + p_j) s_{jik}, \quad i = 1, 2, \dots, \mu, \; k = 1, 2, \dots, n, \tag{14}$$

$$C_{i,k}(s, x) \ge C_{i,k-1}(s, x) + \sum_{j=1}^{n} p_j s_{jik}, \quad i = 1, 2, \dots, \mu, \; k = 2, 3, \dots, n. \tag{15}$$

The number $m = \sum_{i=1}^{\mu} x_i$ of launched locations for machines is not fixed. Sites are chosen from the set D_x, satisfy the constraint (11), and affect (13)–(15).

Constraints (12) and (13) ensure that each job is carried out on one position of a single machine, and no more than one job can be performed on the launched location of each machine, respectively. Moreover, in a feasible solution, the kth in order job on the ith machine cannot start before its release date or the completion time of a job, scheduled on the $(k-1)$th position of the same machine. This condition is met by constraints (14) and (15).

The reduction of the discussed problem to strongly NP-hard $1|r_j|\sum C_j$ [19] is shown in [15]. The NSGA II-based heuristic algorithm has been elaborated for larger instances. For minor cases, one can use the solver MATLAB, which could return exact solutions and Pareto fronts in a reasonable time. Table 2 presents a comparison of both solution tools. PF's point $(Q_{SL,l}, Q_{L,l})$ closest to (0, 0) and the distance *Dist* between them serve as performance indices for the evaluation. The upper index NS indicates the outcomes of the heuristic algorithm. It is worth noting that the performance indices evaluate a trade-off point, not a whole PF. The conducted computational experiments confirmed the usefulness of the proposed heuristic algorithm. Such a corollary also implies from Table 2, where results generated by the evaluated algorithm is worse no more than 1.2 times comparing to the outcomes generated by the solver (the last column in Table 2). Some results in Table 2 line up. They are marked there in bold types.

Table 2. Dependence of $Q_{\mathrm{SL},l}$, $Q_{\mathrm{L},l}$ and *Dist* on n and μ for both algorithms

(n, μ)	$Q^*_{\mathrm{SL},l}$	$Q^*_{\mathrm{L},l}$	$Q^{\mathrm{NS}}_{\mathrm{SL},l}$	$Q^{\mathrm{NS}}_{\mathrm{L},l}$	$Dist^*$	$Dist^{\mathrm{NS}}$	$Dist^{\mathrm{NS}}/Dist^*$
(10, 4)	266.776	182	**266.776**	**182**	322.945	**322.945**	1.000
(15, 4)	432.052	285	**432.052**	**285**	517.584	**517.584**	1.000
(20, 4)	596.663	385	624.574	**385**	710.093	733.701	1.033
(25, 4)	866.800	385	956.164	**385**	948.455	1030.765	1.087
(10, 6)	265.559	177	**265.559**	**177**	319.140	**319.140**	1.000
(15, 6)	406.760	257	448.850	**257**	481.147	517.218	1.075
(20, 6)	458.741	406	477.497	452	612.600	725.684	1.185
(25, 6)	611.681	452	716.951	536	760.564	911.585	1.199

3.2.2 Cases $R|r_j_V|C_{max}|$P_MEDIAN and $R|r_j\_V|C_{max}$

The conference paper [11] illustrates other ScheLoc problems, i.e., involving a locational criterion $R|r_j_V|C_{max}|$P_MEDIAN and a single criterion problem $R|r_j\_V|C_{max}$. The optimization schemes S_Seq and S_Jnt are applied, respectively. A discrete location area with the Euclidean distances among locations is taken into account. The set D_x offers μ feasible locations ready for deploying m machines accurately. Moreover, the parameters' sequences are limited to $A_S = (p, r)$, $A_L = (a, v)$, which means that no direct location costs are taken into account. Instead, the criterion for PL

$$Q_L(x; A_L) = \sum_{i=1}^{\mu} \min_{x \in D_x} r_{ij} \tag{16}$$

directly uses machine-based release dates $r_{ij}(x_i) = \rho_j + v_j^{-1} d(a_j, x_i)$. The criterion for PS takes the recursive form

$$Q_{SL}(\sigma, x; A_s) = \max_{1 \le i \le m} C_{\sigma_i^{n_i}}(x_i), \tag{17}$$

where

$$C_{\sigma_i^l}(x_i) = p_{\sigma_i^l} + \max\left\{r_{ij}, C_{\sigma_i^{l-1}}(x_i)\right\}, \quad l = \overline{1, n_i}, \ C_{\sigma_i^0}(x_i) = 0, \tag{18}$$

and n_i is the number of jobs assigned to M_i. The schedule $\sigma = \{\sigma_1, \sigma_2, ..., \sigma_i, ..., \sigma_m\}$ takes now into account m individual schedules for machines

$\sigma_i = (\sigma_i^1, \sigma_i^2, ..., \sigma_i^l, ..., \sigma_i^{n_i})$ of the lengths n_i. The set of feasible schedules $D_{\sigma x} = \{\sigma : \sigma_{i_1}^{l_1} \neq \sigma_{i_2}^{l_2}, \sum_{i=1}^{m} n_i = n\}$ comprises m permutations without repetitions of all jobs.

According to the current notation, the form of criterion Q_L and the demand for launching m executors define the so-called p-median problem (m-median, according to the contemporary notation), polynomially solvable for fixed m [20,21]. Unfortunately, the exact algorithm for the p-median problem requires unacceptable computation time. On the contrary, the considered $R|r_j\text{-}V|C_{max}$ is NP-hard, which straightforwardly implies from the NP-hardness of $R|r_j|C_{max}$ even for two machines. Therefore, the optimization scheme S_Seq employs two heuristic algorithms, i.e., the Gamma heuristic (GH) and the Particle Swarm Optimization (PSO), respectively, for m-median and $R|r_j\text{-}V|C_{max}$ [24,25]. A Tabu Search based Memetic Algorithm (TSMA) has been developed and applied for the optimization scheme S_Jnt. Examples of comparing both schemes are given in Table 3, where the values of criteria $\hat{Q}_{SL}$ (the joint approach) and $\tilde{Q}_{SL}$ (the sequential approach) together with corresponding computational times in seconds $\hat{t}_{SL}$, $\tilde{t}_{SL}$ and performance indices $\delta_Q = [(\tilde{Q}_{SL} - \hat{Q}_{SL})/\hat{Q}_{SL}]100\%$, $\delta_t = [(\tilde{t}_{SL} - \hat{t}_{SL})/\hat{t}_{SL}]100\%$ are presented. The results illustrate the joint approach's substantial advantage, which turned out to be at least 22% better than the sequential approach in terms of the makespan. Except for the more minor instances, S_Jnt also outperforms S_Seq taking into account the computational time.

Table 3. Dependence of Q_{SL} and t on n for joint and sequential approaches

n	$\hat{Q}_{SL}$	$\tilde{Q}_{SL}$	δ_Q	$\hat{t}_{SL}$ [s]	$\tilde{t}_{SL}$ [s]	δ_t
100	39.764	52.780	32.733	9.221	7.035	–23.707
200	71.551	100.954	41.094	16.934	14.697	–13.210
300	146.625	188.781	28.751	25.755	26.968	4.710
400	235.912	297.873	26.264	34.777	36.089	3.773
500	300.651	369.762	22.987	41.048	47.114	14.778
600	348.684	462.543	32.654	49.912	58.281	16.768
700	405.646	527.986	30.159	58.995	72.092	22.200
800	489.871	659.451	34.617	67.665	83.668	23.650
900	579.543	778.992	34.415	80.019	99.769	24.682
1000	645.671	868.621	34.530	88.894	117.249	31.898

3.2.3 Uncertain Case $1|r_j^- \leq r_j \leq r_j^+|C_{max}$

The last case illustrates the optimization scheme S_Unc compared to S_Jnt for a single machine and the criterion C_{max}. Again, a discrete location area with the Euclidean distances among locations and a fixed number μ of possible sites for a single machine is assumed. The problem formulation adopts the notation from the previous sub-section to a single machine case. In consequence, matrices with data like p are now corresponding vectors. The recursive formula

$$\hat{Q}_S(\sigma, r) \triangleq C_n(\sigma, r), \tag{19}$$

defines the makespan $\hat{Q}_S(\sigma, r)$, where

$$C_k(\sigma, r) = p_{\sigma(k)} + \max\left\{C_{k-1}(\sigma, r), r_{\sigma(k)}\right\}, \quad k = 2, 3, \ldots, n, \ C_0(\sigma, r) = 0. \tag{20}$$

According to the information provided in Subsect. 2.2, the regret for σ under the scenario r refers to the difference $\hat{Q}_S(\sigma, r) - Q_S^{'}(\sigma_r^{'})$, where $\sigma_r^{'}$ is the optimal permutation under the scenario r, $Q_S^{'} \triangleq \min_{\pi \in D_\sigma} Q(\pi, r)$ stands for the optimal makespan for the deterministic counterpart, and D_σ is the set of permutations without repetitions of all the jobs. Consequently, the worst-case regret of makespan for scenarios R takes the form:

$$z(\sigma) = \max_{r \in R}\left[\hat{Q}_S(\sigma, r) - Q_S^{'}(\sigma_r^{'})\right]. \tag{21}$$

Additionally, $r^\sigma = \arg\max_{r \in R}[\hat{Q}_S(\sigma, r) - Q_S^{'}(\sigma_r^{'})]$ is the so-called worst-case scenario for the permutation σ. The application of optimization scheme S_Unc requires solving $\min_{\sigma \in D_\sigma} z(\sigma)$ to obtain $\sigma^{''}$. Table 4 compares the results of the S_Unc with the joint optimization scheme S_Jnt applied for $1|r_j\_V|C_{max}$. In [26], it is proved that $\min_{\sigma \in D_\sigma}(\sigma, r)$ for a single machine can be solved in polynomial time with the substantial restriction of the number of scenarios that have to be taken into account during the optimization procedure. Moreover, the polynomial algorithm $O(n \log n)$ realizing the S_Unc scheme is proposed by [26]. The S_Jnt scheme for the considered specific example is easy to solve by ERD (Earliest Release Dates) rule launching μ times, i.e., for every possible location of a single machine. Consequently, the resultant time complexity is $O(\mu n \log n)$. According to the performance index $[(\hat{Q}_{SL} - Q_S(\sigma^{''}))/Q_S(\sigma^{''})]100\%$, the results generated by S_Unc and presented in Table 4 are slightly worse, i.e., no more than 1.3%. More information for S_Jnt than for S_Unc justify the benefit. Moreover, the S_Unc algorithm seeks a permutation minimizing the maximum regret z, not the makespan Q_{SL}. Please note that outcomes get closer when the number of jobs increases. It is well-seen for the 'any intervals' case. By contrast, for the 'disjoint intervals' example, one can indicate instances when both optimization schemes 'algorithms generate the same jobs' permutations. Then, the makespan values are identical, as shown at the right-side columns of Table 4 (figures marked in bold types).

From the computational time point of view, the comparison is not evident. S_Unc is faster for the particular case of disjoint intervals, unlike the first part of the experiments reported on the left-hand side of Table 4. All the times are given in milliseconds.

Initial investigations show that S_Unc is not competitive to S_Jnt for a single machine. The existence of a fast polynomial algorithm for S_Jnt is the justification. Such comparison for many machines is not evident due to at least the NP-hard time complexity of Scheloc. Adequate considerations are in progress.

Table 4. Dependence of makespan and computational time on n for S_Jnt and S_Unc, $\mu = 50$

n	$\hat{Q}_{SL}$	$Q_S(\sigma'')$	$\hat{t}_{SL}$ [ms]	t_{SL} [ms]	$\hat{Q}_{SL}$	$Q_S(\sigma'')$	$\hat{t}_{SL}$ [ms]	t_{SL} [ms]
40	2741.530	2776.632	8	10	**556.55**	**556.55**	8	9
80	5719.114	5747.075	11	13	**709.33**	**709.33**	12	13
120	8737.921	8785.887	26	27	**811.91**	**811.91**	27	28
160	11505.807	11533.889	30	32	**987.38**	**987.38**	31	33
200	14482.500	14503.128	36	37	**1108.99**	**1108.99**	37	38
240	17534.556	17557.584	44	46	**1293.77**	**1293.77**	42	44
280	20638.523	20680.748	53	56	**1671.41**	**1671.41**	54	55
320	23102.561	23138.121	64	66	**1822.76**	**1822.76**	63	65
360	**26457.697**	**26457.697**	79	81	**2098.41**	**2098.41**	79	80
400	**29299.100**	**29299.100**	93	95	**2223.91**	**2223.91**	94	96

Results for 'disjoint intervals' are presented in columns 6–9.

4 Other Cases and Final Remarks

The presented approaches do not exploit all possibilities to tackle ScheLoc problems. Let us mention some identified challenges and open issues.

A. The deployment of machines can be connected with the assignment of jobs to individual machines. Such a generalized assignment sub-problem should be followed by m separate job scheduling sub-problems on single machines. The apparent advantage deals with the substantial simplification of PS. The detailed analysis is an open issue, in particular, on the connection between location and assignment. This idea's first attempt can be found in [10] where clusters of jobs are determined around launched sites for machines.

B. Exploiting the sequential approach arises the necessity to figure the values of release-dates. Such a calculation can be carried out based on distances among all locations of jobs and machines. Machine-dependent release dates r_{ij} can be straightforwardly figured for every pair (J_j, M_i). Unfortunately, a single release date r_j is needed for every job to have classical job scheduling problems. The substantiation via selecting maximum, minimum, or mean values seems to be a fundamental proposition. Other cases need more profound research. Otherwise, to our best knowledge, we would have a new issue with machine-dependent release dates, which requires the comparison to similar known problems. Contingent originality and relevance of the topic would be the outcome of such a comparison.

C. The intrinsic feature of Scheloc deals with the dependence of release dates r_{ij} on distances among executors and jobs and jobs' speeds v_j, i.e., $r_{ij}(x_i) = \rho_j + v_j^{-1} d(a_j, x_i)$. More general calculation formulae seem to be worth considering in future work, e.g., the ones independent on v_j.

D. The elimination of sub-problem PL from considerations of ScheLoc is called the S_Unc optimization scheme. The characterized in the previous section minmax method with interval uncertainty is one of the more possible approaches. The probabilistic (stochastic) method seems to be promising. Then, release dates for jobs would be values of random variables of a given probability distribution. The justification of such an approach and first results are discussed by [22]. The authors consider stochastic job execution times and the expected value of $\sum C_j$ together with machines' total location costs as two criteria. They apply a simple scalarization, and, in consequence, a single criterion optimization problem is only solved. A genetic algorithm and the sample average approximation algorithm were elaborated for the large and small instances.

E. The way of release date values acquisition is substantial for uncertain approaches and cases, nevertheless the used representation method. The probabilistic approach's application has to be supported by real-world data on release dates that could be regarded as objective information. The use of intervals as available information of release dates needs not to be confirmed by real-world data. Bounds of intervals can be interpreted as subjective information given, for example, by an expert. The expert can analyze the mutual relationship among given locations of jobs and expected positions of machines. To sum up this thread, additional research on the acquisition of release dates for uncertain counterparts of ScheLoc problems is indispensable.

Current research, conducted in the Division of Intelligent Decision Support Systems at the Wroclaw University of Science and Technology, deals with deterministic versions of $\mathrm{R}|r_j\text{-V}|\mathrm{C}_{\max}|\delta$. Different forms of the evaluation of locational sub-problem PL are used, and the Simulated Annealing-based and Tabu Search-based heuristic algorithms are developed for the optimization scheme S_Jnt. Moreover, the results obtained for the uncertain version of $1|r_j^- \le r_j \le r_j^+|\mathrm{C}_{\max}$ (the optimization scheme S_Unc) is extended for $\mathrm{R}|r_j^- \le r_j \le r_j^+|\mathrm{C}_{\max}$.

The majority of research issues for different job scheduling problems, using mentioned in this paper optimization schemes, are open and ready to be undertaken.

References

1. Józefczyk, J., Hojda, M.: Systems approach in complex problems of decision-making and decision-support. In: Kulczycki, P., Korbicz, J., Kacprzyk, J. (eds.) Automatic Control, Robotics, and Information Processing. SSDC, vol. 296, pp. 589–615. Springer, Cham (2021). https://doi.org/10.1007/978-3-030-48587-0_19
2. Nagy, G., Salhi, S.: Location-routing: issues, models and methods. Eur. J. Oper. Res. **177**, 649–672 (2007)
3. Hennes H., Hamacher H.W.: Integrated Scheduling and Location Models: Single Machine Makespan Problems. Report on Wirtschaftsmathematik, p. 82, University of Kaiserslautern (2002)
4. Elvikis, D., Hamacher, H.W., Kalsch M.T.: Simultaneous scheduling and location (ScheLoc): the planar ScheLoc makespan problem. J. Sched. **12**(4), 361–374 (2009)
5. Hessler, C.J.: Scheduling-Location Algorithms with Applications in Evacuation Planning. Verlag Dr, Hut (2017)
6. Pinedo, M.L.: Scheduling: Theory. Algorithms. and Systems. Springer-Verlag, New York (2012). https://doi.org/10.1007/978-1-4614-2361-4

7. Graham, R.L., Lawler, E.L., Lenstra, J.K., Rinnooy, Kan A., H.G.: Optimization and approximation in deterministic sequencing and scheduling: a survey. Ann. Discrete Math. **5**, 287–326 (1979)
8. Kalsch, M.T., Drezner, Z.: Solving scheduling and location problems in the plane simultaneously. Comput. Oper. Res. **37**(2), 256–264 (2010)
9. Drezner, Z., Suzuki, A.: The big triangle small triangle method for the solution of nonconvex facility location problems. Oper. Res. **52**(1), 128–135 (2004)
10. Hessler, C.J., Deghdak, K.: Discrete parallel machine makespan ScheLoc problem. J. Comb. Optim. **34**(4), 1159–1186 (2017)
11. Ławrynowicz, M., Józefczyk J.: A memetic algorithm for the discrete scheduling-location problem with unrelated executors. In: Proceedings of the 24th International Conference on Methods and Models in Automation and Robotics MMAR. Międzyzdroje. Poland, 26–29 August 2019
12. Piasecki, B.: Application of AI-based algorithms for joint problem of task scheduling and deployment of executors Master Thesis. Wroclaw University of Science and Technology. Poland (2018). (in Polish)
13. Piasecki, B., Józefczyk, J.: Evolutionary algorithm for joint task scheduling and deployment of executors. In: Automation of Discrete Processes. Theory and Applications Silesian University of Technology, vol. 1, pp. 169–178 (2018). (in Polish)
14. Rajabzadeh, M., Ziaee, M., Bozorgi-Amiri, A.: Integrated approach in solving parallel machine scheduling and location (ScheLoc) problem. Int. J. Ind. Eng. Comput. **7**(4), 573–584 (2016)
15. Filcek, G., Józefczyk, J., Ławrynowicz, M.: An evolutionary algorithm for joint bi-criteria location-scheduling problem. Int. J. Ind. Eng. Comput. **12**, 159–176 (2021)
16. Deb, K.: Multi-Objective Optimization Using Evolutionary Algorithms. John Wiley & Sons, Chichester (2001)
17. Wesolkowski, S., Francetic, N., Grant, S.C.: Trade training device selection via multi-objective optimization. In: IEEE Congress on Evolutionary Computation (2014)
18. Kalai, R., Lamboray, C., Vanderpooten, D.: Lexicographic a-robustness: An alternative to min-max criteria. Euro. J. Oper. Res. **220**(3), 722–728 (2020)
19. Lenstra, J.K., Rinnooy Kan, A.H.G., Brucker, P.: Complexity of machine scheduling problems. Ann. Discrete Math. **1**, 343–362 (1977)
20. Garey, M.R., Johnson, D.S.: Computers and Intractability A Guide to the Theory of NP-Completeness. , W. H. Freeman and Company, New York (1979)
21. Downey, R.G., Fellows, M.R.: Fundamentals of Parameterized Complexity. Springer, London (2013)
22. Liu, M., Liu, X., Zhang, E., Chu, F., Chu, Ch.: Scenario-based heuristic to two-stage stochastic program for the parallel machine ScheLoc problem. Int. J. Prod. Res. **57**(6), 1706–1723 (2019)
23. Kouvelis, P., Yu, G.: Robust Discrete Optimization and Its Applications. Kluwer Academic Publishers, London .(1997)
24. Rosing, K.E., Revelle, C.S., Schilling, D.A.: A gamma heuristic for the p-median problem. Eur. J. Oper. Res. **117**(3), 522–532 (1999)
25. Lin, Y.K.: Particle swarm optimization algorithm for unrelated parallel machine scheduling with release dates, Math. Probl. Eng, **2013**, 409486 (2013)
26. Bachtler, O., Krumke, S.O., Le, H.M.: Robust single machine makespan scheduling with release date uncertainty. Oper. Res. Lett. **48**, 816–819 (2020)

A Comprehensive Approach to Assessing Higher and Lower Order Thinking Skills

Emil Hadzhikolev[1], Stanka Hadzhikoleva[1](✉), Daniela Orozova[2,3], and Kostadin Yotov[1]

[1] Faculty of Mathematics and Informatics, Plovdiv University Paisii Hilendarski, 236 Bulgaria Bul., Plovdiv, Bulgaria
{hadjikolev,stankah}@uni-plovdiv.bg, kostadin_yotov@abv.bg

[2] Faculty of Computer Science and Engineering, Burgas Free University, 62, San Stefano Street, Burgas, Bulgaria
orozova@bfu.bg

[3] Institute of Information and Communication Technologies, Bulgarian Academy of Sciences, 1, 15 Noemvri Street, Sofia, Bulgaria

Abstract. The article proposes a formal model of a software process for multicomponent assessment. It aims to divide the process into several well-defines steps, defining reusable abstract meta-models and assessment models. Specific tests are determined by setting specific test units that follow the framework of the meta-model and the test model. An important emphasis in the model is the ability to define different ways of forming final assessments based on fuzzy sets.

Keywords: Generalized Net (GN) model · Assessing HOTS · Fuzzy logic assessment

1 Introduction

The mass transition to online learning caused by the COVID'19 pandemic posed a new challenge for teachers –***planning and conducting objective assessment in an electronic environment***. Objective assessment has many characteristics: true score, reliability and validity, honesty, differentiation, versatility of assessment, etc. [1–3]. It provides teachers with a better opportunity both to assess the acquisition of knowledge by the students and to analyze the effectiveness of teaching. On the other hand, it gives the learners a sense of justice and motivates them to gain and demonstrate complex knowledge and skills.

Objective assessment involves ***studying numerous predetermined, specific cognitive skills***. Depending on the subject area and the teacher's preferences, they can be ***determined by means of a taxonomy***, e. g. Bloom's taxonomy [4], Webb's Depth of Knowledge [5], SOLO Taxonomy [6], Finks Taxonomy [7], etc. They may be ***based on more general concepts***, such as Higher Order Thinking Skills (HOTS), including skills for critical, logical, reflective, metacognitive, and creative thinking [8], from the three higher levels in Bloom's taxonomy (Analysis, Synthesis, and Evaluation) and Lower Order Thinking Skills (LOTS) – including the lower levels of Bloom's taxonomy, i. e.

K. T. Atanassov et al. (Eds.): IWIFSGN 2020/BOS/SOR 2020, LNNS 338, pp. 164–177, 2022.
https://doi.org/10.1007/978-3-030-95929-6_13

Knowledge, Comprehension, and Application. In specific domains, it is appropriate for ***the teacher to create his/her own classification*** [9–11].

Different types of cognitive skills are assessed through a wide variety of test units – tests, solving tasks and cases, creating programs, working on projects, answering questions, writing essays and many others. ***The way of forming the final assessment based on the grades from different types of test units with which different knowledge and competencies are assessed***, also determines the success of the assessment process.

The article proposes a ***formal model of a software process for multicomponent assessment***. It aims to divide the process into several well-defines steps, defining reusable abstract meta-models and assessment models. Specific tests are determined by setting specific test units that follow the framework of the meta-model and the test model. An important emphasis in the model is the ability to define different ways of forming final assessments based on fuzzy sets.

2 Multicomponent Models for Assessment in Learning

Plenty of models for assessing students' knowledge and skills have been described in scientific literature. Various techniques for multicriteria assessment have been explored by Mardani and team [12].

One approach to multicomponent assessment based on Bloom's taxonomy, revision by Anderson and Krathwohl [13], is presented in [14], ***in which the components are the individual levels of the taxonomy***. This paper explores different ***techniques for calculating the final assessment***, as a ***linear function on multiple components***. The grade ***E is formed by the formula*** $E = \sum_{i=1}^{n} a_i y_i$, ***where*** y_i ***are the grades by components categorized by Bloom***, $a_i \geq 0$ are their respective weights, $\sum_{i=1}^{n} a_i = 1$, and n is the number of components whose value is 6 when assessed by Bloom. The assessor could choose ***four main assessment models*** or their modifications, each of which has its advantages and disadvantages: ***flat model*** – all weights are equal; ***progressive model*** – the weights increase with each level of Bloom's taxonomy; ***basic model*** – assessments from Bloom's intermediate levels have higher weight, and the high-end and low-end ones – too basic knowledge or too complex skills – have lower weight; ***regressive model*** – the weights decrease with each level of Bloom's taxonomy.

Models for Multicomponent Fuzzy Evaluation are proposed in [15]. The final grade is formed by using linear functions and fuzzy logic to hierarchically, in separate steps, calculate assessments for next-level components. The use of artificial intelligence methods allows for partial automation of the assessment process [16].

Other models are presented in [17–19]. The current model complements and expands them with opportunities to apply new approaches to form final assessments of learners' knowledge and skills based on fuzzy sets.

The article ***proposes a model of software process for multicomponent assessment***, which includes 6 stages:

1. ***Creating a meta-model of a test.*** The test meta-model is defined by types of assessment components and assessment models. The types of assessment components determine what knowledge and skills are assessed, e. g. according to a specific taxonomy or a classification created by the teacher, suitable for a specific domain.

Assessment models can be based on different formulas with different weights for the assessment components, fuzzy logic, etc.

2. ***Creating a test model.*** The test model includes characteristics for the target group, the subject of study, the beginning and duration of the test, etc. At this stage, for each of the types of assessment components of the meta-model, a corresponding number is set of test units which will contain the test instances.
3. ***Test configuration.*** Specific test units are set here. They are entered or selected from a question database. For each specific test unit, the type of assessment elements it belongs to must be indicated. At this stage, the specific students who will take the test are set.
4. ***Conducting the test*** is a process in which specific tests are generated. They must include a specific number of test units of each type of assessment components – as specified in the model and meta-model. Students enter their answers and solutions.
5. ***Assessment process.*** It includes manual assessment of open questions and tasks, if any, and automatic assessment of the test according to the selected assessment model.
6. ***Analysis of the results and assessment of the test.*** The test creator uses the obtained results to assess the test and the effectiveness of the applied assessment approaches.

The model is described with the Generalized Nets notation [20–23]. They are an extension of the Petri Nets and are widely used for modeling processes in various domains.

3 Generalized Net Model for Assessment of Higher and Lower Order Thinking Skills

The Generalized Net A, describing the process of modeling and constructing a test to assess the students' knowledge and skills, contains the following set of transitions:

$A = \{Z_1, Z_2, Z_3, Z_4, Z_5, Z_6\}$, where:

- Z_1 – creating a meta-model of a test;
- Z_2 – creating a model of a test;
- Z_3 – test configuration;
- Z_4 – conducting the test;
- Z_5 – assessment process;
- Z_6 – analysis of the results and assessment of the test.

The following tokens are used to describe the processes:

- α-token – users of the software system – administrators, test creators, authors of questions, students;
- β-token – types of assessment components, for assessing practical or theoretical skills, higher and lower order thinking skills, skills categorized according to a specific training taxonomy or other;
- γ-token – assessment models – weighted average of the assessment components, assessment based on weighting coefficients, fuzzy logic, etc.;

- δ-token – meta-model of a test;
- η-token – model of a test;
- λ-token – database of test questions;
- μ-token – assessment component (for example: test question, task, case, essay, etc.);
- ϕ-token – time parameters for starting the test.

The Generalized Net model is presented in Fig. 1.

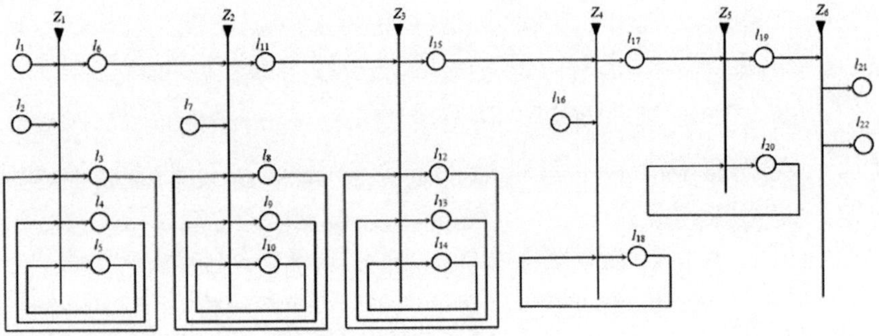

Fig. 1. Generalized net model for assessment of higher and lower order thinking skills

The process is started by a software system administrator presented via α-token, which enters the generalized net through place l_1.

In place l_2 a request is started to create a new meta-model of a test, using a δ-token with initial characteristic:

"unique identifier of a test meta-model"

In place l_3 α-token cycles with characteristic:

"list of available users of the software system"

In place l_4 and in place l_5 β-token and γ-token respectively are cycling, with initial characteristic:

"available types of assessment components" and "available assessment models".

During the activation of the transition Z_1, the α-token from place l_3 is divided into 2 cores. One stays in place l_3, and the other receive a new characteristic:

"test creator",

and merges with the δ-token from place l_2.

Similarly, each of the β-token and γ-token from places l_4 and l_5 is divided into two cores. One core remains to cycle in the corresponding place, and the other merges with δ-token from place l_2 and receive a new characteristic:

"list of assessment component types" and "list of assessment models".

Thus, when the transition Z_1 is activated, in place l_6, a δ-token enters with characteristic:

"administrator, unique identifier of the test meta-model, list of assessment component types, list of assessment models, test creator".

$Z_1 = \langle\{l_1, l_2, l_3, l_4, l_5\}, \{l_3, l_4, l_5, l_6\}, r_1\rangle$,

$r_1 =$

	l_3	l_4	l_5	l_6
l_1	*false*	*false*	*false*	$W_{1,6}$
l_2	*false*	*false*	*false*	$W_{2,6}$
l_3	*true*	*false*	*false*	$W_{3,6}$
l_4	*false*	*true*	*false*	$W_{4,6}$
l_5	*false*	*false*	*true*	$W_{5,6}$

where:
$W_{1,6} =$ "a user is authorized by the system as an administrator";
$W_{2,6} =$ "a new test meta-model is initialized";
$W_{3,6} =$ "a test creator is selected";
$W_{4,6} =$ "the list of assessment component types specific to the test meta-model is set";
$W_{5,6} =$ "assessment models are set for the test meta-model".

Transition Z_2 presents the process of creating a test model.

The δ-token enters through place l_6 and merges with the available δ-tokens in place l_9.

In place l_7 a request for creation of a new test model is started by entering a η-token with initial characteristic:

"unique identifier of a test model".

In place l_8, the η-token receive new characteristics:

"discipline, target group, test duration, test start, list of test question creators".

In place l_{10} a list of meta-models of tests cycles, represented by a δ-token. During the operation of the transition, the δ-token is divided into 2 cores. One stays in l_{10}, and the other goes in l_{11} and receive the characteristic:

"administrator, unique identifier of the test meta-model, list of assessment component types, list of assessment models, test creator".

β-token enters place l_9 and gets a new characteristic:

"list of assessment components number",

which determines an important characteristic for the test model, namely – how many assessment components of each type the test instances will contain.

When activating the transition Z_2, the η-token enters from the initial place l_7 and moves to place l_8 without receiving a new characteristic.

On the next activation of the transition, the η-token from place l_8 merges with the δ-token from place l_9 and enters place l_{11}, with the following characteristic:

"unique identifier of a test model, discipline, target group, test duration, test start, list of test question creators, unique identifier of the test meta-model, list of assessment component types, list of assessment models, list of assessment components number".

$Z_2 = <\{l_6, l_7, l_8, l_9, l_{10}\}, \{l_8, l_9, l_{10}, l_{11}\}, r_2>,$

$r_2 =$

	l_8	l_9	l_{10}	l_{11}
l_6	*false*	*false*	*false*	$W_{6,11}$
l_7	*false*	*false*	*false*	$W_{7,11}$
l_8	*true*	*false*	*false*	$W_{8,11}$
l_9	*false*	*true*	*false*	$W_{9,11}$
l_{10}	*false*	*false*	*true*	$W_{10,11}$

where:

$W_{6,11}$ = "default data for the meta-model is selected (administrator, unique identifier of the test meta-model, list of assessment component types, list of assessment models, test creator)";

$W_{7,11}$ = "a request for creating a test model is initiated and a test identification number is set";

$W_{8,11}$ = "characteristics of the test model are set (discipline, target group, test duration, test start, list of test question creators)";

$W_{9,11}$ = "a number of assessment components is determined for each assessment component type";

$W_{10,11}$ = "a meta-model of the test is selected".

In transition Z_3 the process of the test configuration is presented.

In place l_{12}, λ-token cycles with characteristic:

"database with available test questions".

It is divided into two cores. One stays to cycle in place l_{12}, and the other receive the characteristic:

"test question, classification according to the current type of assessment elements, unique identifier of a test model".

In place l_{13}, the μ-token cycles to create a new test component. It is divided into two cores. One stays to cycle in place l_{13}, and the other receive the characteristic:

"unique identifier of a test question, test question, classification accor ing to the current type of assessment elements, unique identifier of a test model".

In place l_{14} the α-token cycles with a list of the software system users. It is divided into two cores. One stays to cycle in place l_{14}, and the other receive a new characteristic:

"list of students".

When activating transition Z_3, the η-token from place l_{11} merges with the λ-token from place l_{12}, μ-token from place l_{13} and α-token from place l_{14}, and enters place l_{15} with the following characteristic:

"unique identifier of test model, discipline, target group, test duration, test start, list of test question creators, unique identifier of the test meta-model, list of assessment component types, list of assessment models, list of assessment components number, list of questions, list of students".

$Z_3 = \langle\{l_{11}, l_{12}, l_{13}, l_{14}\}, \{l_{12}, l_{13}, l_{14}, l_{15}\}, r_3\rangle,$

$r_3 =$

	l_{12}	l_{13}	l_{14}	l_{15}
l_{11}	*false*	*false*	*false*	$W_{11,15}$
l_{12}	*true*	*false*	*false*	$W_{12,15}$
l_{13}	*false*	*true*	*false*	$W_{13,15}$
l_{14}	*false*	*false*	*true*	$W_{14,15}$

where:
$W_{11,15}$ = "new model configuration data is selected";
$W_{12,15}$ = "a test question from the database has been added to the test";
$W_{13,15}$ = "a new test question has been created and added to the test";
$W_{14,15}$ = "the students who will take the test are selected".
When the transition Z_4 is activated from the initial place l_{16} a ϕ-token enters with characteristic:

"time when the test is activated for execution"

In place l_{18} a μ-token with initial characteristic:

"test question",

on each cycle of activation of the transition receives a new feature:

"completed test question".

η-token from place l_{15} merges with the ϕ-token from place l_{16} and μ-token from place l_{18} and enters place l_{17} with the following characteristic:

"unique identifier of test model, discipline, target group, test duration, test start, list of test question creators, unique identifier of the test meta-model, list of assessment component types, list of assessment models, list of assessment components number, list of test questions, completed test questions, list of students".

$Z_4 = \langle\{l_{15}, l_{16}, l_{18}\}, \{l_{17}, l_{18}\}, r_4\rangle,$

$r_4 =$

	l_{17}	l_{18}
l_{15}	$W_{15,17}$	$W_{15,18}$
l_{16}	$false$	$W_{16,18}$
l_{18}	$W_{18,17}$	$false$

where:
$W_{15,17}$ = "the test was completed successfully at the specified time";
$W_{15,18}$ = "a test question has been answered";
$W_{16,18}$ = "the time has come to start the test";
$W_{18,17}$ = "all test question have been answered".

In transition Z_5, an iterative manual assessment of the open test questions is presented. In place l_{20}, a μ-token with characteristic:

"test question",

receive a new characteristic:

"assessment of test question".

η-token from place l_{17} merges with the μ-token from place l_{20} and enters place l_{19} with the following characteristic:

> *"unique identifier of test model, discipline, target group, test duration, test start, list of test question creators, unique identifier of the test meta-model, list of assessment component types, list of assessment models, list of assessment components number, list of test questions, completed test questions, assessment of test questions, list of students".*

In the previous Sect. 2, different techniques for formation and calculation of the final assessment are considered. An original approach to current assessment of students based on fuzzy assessments is presented in Sect. 4 of the paper.

$Z_5 = <\{l_{17}, l_{20}\}, \{l_{19}, l_{20}\}, r_5>$,

$r_5 =$

	l_{19}	l_{20}
l_{17}	$W_{17,19}$	$W_{17,20}$
l_{20}	$false$	$true$

where:
$W_{17,19} = W_{20,19} =$ "all test questions have been assessed";
$W_{17,20} =$ "a test question has been assessed";
After the η-token enters place l_{19}, the transition Z_6 is activated.
In place l_{21} it receives the following characteristic:

"analysis of test results",

and in place l_{22} it receives the characteristic:

"test assessment".

$Z_6 = \langle\{l_{19}\}, \{l_{21}, l_{22}\}, r_6\rangle,$

$$r_6 = \begin{array}{c|cc} & l_{21} & l_{22} \\ \hline l_{19} & W_{19,21} & W_{19,22} \end{array}$$

where:
$W_{19,21} =$ "analysis of the results has been made";
$W_{19,22} =$ "the test has been assessed".

4 Use Case of Hierarchical Multicomponents Assessment with Fuzzy Logic

In our experiment we use a generally accepted classification, determining the skills of the upper three levels of Bloom's taxonomy (Analysis, Synthesis, and Evaluation) as higher-order thinking skills (HOTS), and those from the lower three levels (Knowledge, Comprehension, and Application) – as lower-order thinking skills (LOTS). The subject of assessment is the theoretical knowledge and practical programming skills. We use four types of assessment components:

- ***tLOTS*** – theoretical LOTS are assessed, the grade is points in the interval [0, 30];
- ***pLOTS*** – practical LOTS are assessed, the grade is an integer in the interval [0, 15];
- ***tHOTS*** – theoretical HOTS are assessed, the grade is an integer in the interval [0, 15];
- ***pHOTS*** – practical HOTS are assessed, the grade is a whole number in the interval [2, 6].

The *pHOTS* assessment is formed by a practical programming task and is in the interval [2, 6]. The points for the other three assessment components are formed through test solving. The final grade *finalN* is an integer on the rating scale [2, 6].

In our experiment, we use a model of ***hierarchical tree organization of the assessment components***. In this model, each major component, as well as the final assessment, depends on a multitude of sub-components at the previous level. In that, each sub-component affects only one higher-level component. An alternative is a model *of hierarchical graph organization of the assessment components* which allows the possibility for one component to influence the assessment of several components from arbitrary higher levels, and to depend on components at different lower levels.

The formalized assessment logic is presented in Fig. 2. If necessary, by defining additional dependencies between components at different levels, the model may be changed.

The main components forming the final assessment ***for the theoretical-practical hierarchical multicomponent assessment*** are ***theory*** and ***practice***. Sub-components of the theory are the ***normalized forms*** of the primary components ***tLOTS*** and ***tHOTS***, and of the practice – the normalized forms of ***pLOTS*** and ***pHOTS***. The purpose of the normalized forms if to reduce the assessment value of the primary assessment components to a common assessment scale (in this case from 2 to 6).

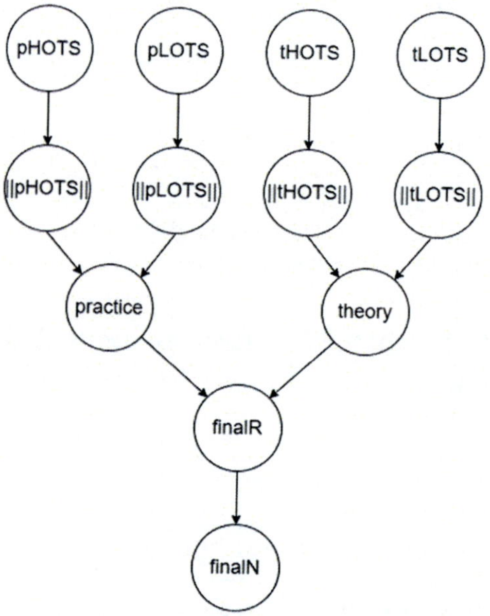

Fig. 2. Model of theoretical and practical hierarchical multicomponent assessment

In the first functional level, forms of the main assessment components are normalized. The specific normalizing functions can implement ***a liberal or strict approach*** for assessment, or their modification, depending on the assessor's judgment. In the ***liberal approach***, the points from the primary assessment components are reduced directly to the interval [2, 6]:

(1.1) $\|tLOTS\| = 2 + 4\,tLOTS/30$
(1.2) $\|tHOTS\| = 2 + 4\,tHOTS/15$
(1.3) $\|pLOTS\| = 2 + 4\,pLOTS/15$
(1.4) $\|pHOTS\| = pHOTS.$

In the ***strict assessment approach***, the points are initially mapped in the interval [0, 6]. Grades below 2,5 are rounded to 2. Accordingly, the membership functions for the second level components are:

$$(2.1)\ \|tLOTS\| = \begin{cases} 2, & 6\frac{tLOTS}{30} < 2.5 \\ 6\frac{tLOTS}{30}, & 6\frac{tLOTS}{30} \geq 2.5 \end{cases}$$

$$(2.2)\ \|tHOTS\| = \begin{cases} 2, & 6\frac{tHOTS}{15} < 2.5 \\ 6\frac{tHOTS}{15}, & 6\frac{tHOTS}{15} \geq 2.5 \end{cases}$$

$$(2.3)\ \|pLOTS\| = \begin{cases} 2, & 6\frac{pLOTS}{15} < 2.5 \\ 6\frac{pLOTS}{15}, & 6\frac{pLOTS}{15} \geq 2.5 \end{cases}$$

(2.4) $\|pHOTS\| = pHOTS.$

Presented with IF-THEN logic, the formulas look like this:

(3.1) $\|\boldsymbol{tLOTS}\|: IF\ 6\,tLOTS/30 < 2.5\ THEN\ \|tLOTS\|\ IS\ 2\ ELSE\ \|tLOTS\|\ IS\ 6\,tLOTS/30$
(3.2) $\|\boldsymbol{tHOTS}\|: IF\ 6\,tHOTS/15 < 2.5\ THEN\ \|tHOTS\|\ IS\ 2\ ELSE\ \|tHOTS\|\ IS\ 6\,tHOTS/15$
(3.3) $\|\boldsymbol{pLOTS}\|: IF\ 6\,pLOTS/15 < 2.5\ THEN\ \|pLOTS\|\ IS\ 2\ ELSE\ \|pLOTS\|\ IS\ 6\,pLOTS/15$

The specific rules we use to assess in practice and theory are:

(4.1) $(\boldsymbol{practice}(\|\boldsymbol{pLOTS}\|, \|\boldsymbol{pHOTS}\|)$:
$IF\ \ (\|pHOTS\| + \|pLOTS\|)/2 < 2.5\ THEN\ practice\ IS\ 2$
$ELSE\ IF\ \|pLOTS\| = 2\ OR\ \|pHOTS\| = 2\ THEN\ practice\ IS\ 3$
$ELSE\ IF\ abs(\|pHOTS\| - \|pLOTS\|) \geq 2\ THEN\ practice\ IS\ ((\|pHOTS\| + \|pLOTS\|)/2 - \alpha)$
$ELSE\ practice\ IS\ ((\|pHOTS\| + \|pLOTS\|)/2)$

(4.2) $\boldsymbol{theory}(\|\boldsymbol{tLOTS}\|, \|\boldsymbol{tHOTS}\|)$:
$IF\ \ (\|tHOTS\| + \|tLOTS\|)/2 < 2.5\ THEN\ theory\ IS\ 2$
$ELSE\ IF\ abs(\|tHOTS\| - \|tLOTS\|) \geq 2\ THEN\ theory\ IS\ ((\|tHOTS\| + \|tLOTS\|)/2 + \beta)$
$ELSE\ theory\ IS\ ((\|tHOTS\| + \|tLOTS\|)/2)$

where $\alpha, \beta \epsilon [0,0.5]$

In our experiments, we chose the following values for the coefficients α and β: $\alpha = 0.5, \beta = 0.3$.

To determine the final grade, we have chosen the theory and practice to be equal and we have formulated the rules as follows:

(5) $\boldsymbol{finalR(theory,\ practice)}$:

$IF\ \ (theory + practice)/2 < 2.5\ THEN\ finalR\ IS\ 2$
$ELSE\ IF\ theory = 2\ OR\ practice = 2\ THEN\ finalR\ IS\ 3$
$ELSE\ finalR\ IS\ ((theory + practice)/2)$

We define the final score $finalN$ by rounding $finalR$ to the closest integer:

(6) $\boldsymbol{finalN(finalR\,)} = ROUND(finalR)$

5 Conclusion

The objective assessment of students' knowledge and skills in a digital environment is a challenge that seeks its solutions. Teachers asks themselves many questions – how to assess specific knowledge and skills, what tools to use, how to form the final grade, etc.

This article proposes a model that aims on one hand to summarize the assessment process, and on the other – to allow for higher personalization of the way assessments are formed. This is achieved by defining meta-models and assessment models that set frameworks for creating specific tests and ways to assess them. A software system implementing the described process will provide teachers with a flexible platform for experimenting with various standard and other author's models for constructing tests and assessing students. The ability to analyze and assess the models used, will give teachers a tool for choosing the optimal and objective assessment model.

Acknowledgments. The work is funded by the MU21-FMI-004 at the Research Fund of the University of Plovdiv "Paisii Hilendarski".

References

1. Schaughency, E., Smith, J.K., van der Meer, J., Berg, D.: Classical test theory and higher education: five questions. In: Handbook on Measurement, Assessment, and Evaluation in Higher Education, Routledge (2017)
2. Suskie, L.: Fair Assessment Practices: Giving Students Equitable Opportunties to Demonstrate Learning, Adventures in Assessment, vol. 14 (2002)
3. Ruskov, S., Ruskova, Y.: The problem of evaluation in higher education. In: XXI International Scientific-Technical Conference Trans and Motauto 2013, vol. 3, pp. 73–79 (2013)
4. Bloom, B., et. al.: Taxonomy of educational objectives: the classification of educational goals. Handbook I: Cognitive domain. New York: David McKay Company (1956)
5. Hess, K.: A Guide for Using Webb's Depth of Knowledge with Common Core State Standards, Common Core Institute (2013)
6. Biggs, J., Collis, K.: Evaluating the Quality of Learning. Academic Press, The SOLO taxonomy. New York (1982)

7. Fink, D.: "A self directed guide to designing course for significant learning". San Francisco: Jossey – Bass (2009)
8. King, F., Goodson, L., Rohani, F.: Higher Order Thinking Skills - Definition, Teaching Strategies, Assessment. Publ, Center for Advancement of Learning and Assessment (1997)
9. Bayoumy, H.M.M., Albeladi, G.A.: Clinical judgment skills among junior-level nursing students enrolled in adult health nursing courses: errors and risk level classification. Nurse Educ. Pract. **48**, 102888 (2020)
10. Kilgus, S., Bonifay, W., Eklund, K., et al.: Development and validation of the intervention skills profile - skills: a brief measure of student social-emotional and academic enabling skills. J. Sch. Psychol. **83**, 66–88 (2020)
11. Pérez-Escamirosa, F., et al.: Objective classification of psychomotor laparoscopic skills of surgeons based on three different approaches. Int. J. Comput. Assist. Radiol. Surg. **15**(1), 27–40 (2019). https://doi.org/10.1007/s11548-019-02073-2
12. Mardani, A., Jusoh, A., Nor, K., Khalifah, Z., Zakwan, N., Valipour, A.: criteria decision-making techniques and their applications – a review of the literature from 2000 to 2014. Econ. Res. Ekonomska Istraživanja **28**(1), 516–571 (2015)
13. Anderson, L., Krathwohl, D. (eds.): A Taxonomy for Learning, Teaching and Assessing: A Revision of Bloom's Taxonomy of Educational Objectives. Addison Wesley Longman, New York (2001)
14. Hadzhikoleva, S., Hadzhikolev, E., Kasakliev, N.: Using Peer Assessment to Enhance Higher Order Thinking Skills. Tem J. **8**(1), 242–247 (2019)
15. Hadzhikolev, E., Hazdhikoleva, S., Yotov, K., Orozova, D.: Models for multicomponent fuzzy evaluation, with a focus on the assessment of higher-order thinking skills. Tem J. **9**(4), 1656–1662 (2020)
16. Hadzhikolev, E., Yotov, K., Trankov, M., Hadzhikoleva, S.: Use of neural networks in assessing knowledge and skills of university students. In: Proceedings of ICERI2019 Conference, 11th-13th November, Seville, Spain, pp. 7474–7484 (2019)
17. Orozova, D.: Appropriate e-test system selection model. Comptes rendus de l'Acade'mie bulgare des Sciences **72**(6), 811–820 (2019)
18. Orozova D., Hristova, N.: Generalized net model for dynamic decision making and prognoses. In: XXI International Symposium on Electrical Apparatus and Technologies SIELA 2020, 3–6 June 2020, Bourgas, Bulgaria, pp. 330–333 (2020)
19. Popchev, I., Orozova, D.: Towards a multistep method for assessment in e-learning of emerging technologies. Cyber. Inf. Technol. **20**(3), 116–129 (2020). https://doi.org/10.2478/cait-2020-0032
20. Atanassov, K.: Generalized Nets. World Scientific, Singapore (1991)
21. Atanassov, K., On Generalized Nets theory, Prof. Marin Drinov Publishing House, Sofia (2007)
22. Atanassov, K., E. Sotirova, Generalized Nets, Prof. Marin Drinov Publishing House, Sofia (2017)
23. Atanassov, K., Hadjiski, M.: Generalised nets and intelligent systems. Int. J. Gen Syst. **39**(5), 457–470 (2010)

Evaluating the General Claim Process Through Temporal Intuitionistic Fuzzy Pairs

Hristo Blidov and Lyubka Doukovska(✉)

Intelligent Systems Department, Institute of Information and Communication Technologies, Bulgarian Academy of Sciences, Acad. Georgi Bonchev Street, Block 2, 1113 Sofia, Bulgaria
{hristo.blidov,lyubka.doukovska}@iict.bas.bg

Abstract. This statement aims to review and evaluate the development of court proceedings in civil disputes or the so-called "general claim process" from the initiation of the case to its conclusion by the court of first instance through the use of temporal intuitionist fuzzy pairs. The obtained results can be defined as expanding the fields of application of the theory of intuitionistic fuzzy sets.

Keywords: Intelligent systems · Intuitionistic fuzzy set · Temporal intuitionistic fuzzy pair · General claim process

1 Introduction

The General claim process is a procedure before a civil court for resolving a legal dispute between equal subjects - individuals or legal entities. The resolution of the dispute is carried out by a third party independent and not related to the disputing parties - most often a judge or an arbitrator [1].

There are three main legal systems in the world - the continental or European (Romano-Germanic) legal family based on the continental legal system, the common law family based on common (Anglo-Saxon) law and precedent, and the religious-traditional legal family. Legal systems, sometimes referred to as legal systems, are the way in which law is organized, structured and enforced through the legal system.

The most common legal system is the so-called the continental or European (Romano-Germanic), abbreviated "continental legal system", and its territorial scope extends to the whole of Europe with the exception of the United Kingdom of Great Britain and Northern Ireland.

Generally speaking, the continental legal system includes material law and procedural law. The material law covers the system of norms, which reflect actually existing public relations and actually existing phenomena (theft, marriage, purchase - sale, etc.). The theory of the legal system has been studied jointly and as part with the general theory of law by Prof. Rossen Tashev. In his works the author traces the formation and development of the basic legal systems; the structure, dynamics, internal conflicts in the legal system and its relations with the surrounding social environment [2].

K. T. Atanassov et al. (Eds.): IWIFSGN 2020/BOS/SOR 2020, LNNS 338, pp. 178–184, 2022.
https://doi.org/10.1007/978-3-030-95929-6_14

Material law provides for the rights and obligations of individual participants in legal relations, as well as the legal consequences that ensure the implementation of these rules. Substantive law is divided into separate branches - civil, criminal, financial, tax law, etc.

Procedural law, in turn, determines the order and forms for realization of substantive law. The procedural order is necessary to ensure the lawful application of the relevant substantive law.

The procedural law regulating the civil law relations between the persons is regulated by a normative act - most often a Code.

Although each country applying a continental legal system has its own rules and procedures, it should be borne in mind that these procedures are essentially very similar. The original source of these rules was the French Civil Code, better known as the Code Napoleon, which was created in 1804 and was later adopted by many other countries in Europe and around the world [3]. Code Napoleon has a significant role in the development of civil proceedings in Europe (Italy, Belgium, the Netherlands, Spain, Portugal, Romania, Germany, etc.).

For example, the Bulgarian model of a continental legal system was introduced by the current Civil Procedure Code of the Republic of Bulgaria (CPC, promulgated SG, issue 12 of 08 February 1952), which is the relevant normative act and appears in sync with the adopted in other European countries models. It has been amended several times and has been repealed by § 3 of the transitional final provisions of the new Civil Procedure Code - SG, no. 59 of 20 July 2007, in force since 01.03.2008. Despite the differences in the procedures themselves in each country, it can be assumed that there is a high degree of similarity, consistency and logic of the process between them (procedures).

The doyen in the field of Bulgarian civil procedural law in the modern Bulgarian history is indisputably Prof. Jhivko Stalev. His life and professional path are dedicated to the study and development of the Bulgarian civil process. Significant of his work is the textbook on Bulgarian civil procedural law [4]. There is no Bulgarian lawyer who has not studied his in-depth works in the field, but also Prof. Jhivko Stalev is well known to professional and scientific circles throughout Europe.

The parties in the general claim process are the persons on whose behalf and against whom the case is conducted - the plaintiff and the defendant. In certain cases, the involvement of a third party, called an "interested party in the proceedings", is also possible, assisting one of the parties in the case. The Civil Court is not a party to the proceedings, but is the body that exercises the function of managing and resolving the legal dispute.

The court proceedings (the process) begin with the referral of the court by the plaintiff with a claim. The plaintiff is the person (natural or legal) who claims that his civil law guaranteed by law has been violated.

In order to be successful, legal proceedings should be brought before the court competent to decide the specific legal dispute. In civil proceedings there are 2 types of jurisdiction – generic jurisdiction and local jurisdiction. Several factors are important for determining the generic jurisdiction, the most common of which is related to the price of the claim. If the claim of the plaintiff is in a larger amount, then the district court will be competent. Otherwise, the district court is competent to hear the case. Of course, there are other hypotheses for determining the generic jurisdiction, but this is the most common.

Regarding local jurisdiction, the general legal hypothesis is that the claim is brought before the court in the area where the defendant's permanent address or seat is located. As with generic jurisdiction, the Civil Procedure Code provides for more than 10 different hypotheses related to the determination of local jurisdiction, but I dwell on this, as the most commonly applicable.

Often, due to ignorance, unintentionally or even intentionally, the plaintiff can refer to a court that is incompetent - generic, local or both. In these cases, the court seized is obliged to forward the claim ex officio to the court competent to hear the legal dispute.

The claim is considered officially filed upon receipt of the claim in court. Once the claim has reached the competent court, the role of the judge in the management and administration of the trial begins.

Here it should be borne in mind that the newly received claim is processed through the so-called "a system for random distribution of cases in court" and thus, on a random basis, the case is assigned to the judge (judge-rapporteur) who will hear the legal dispute.

The first action that the judge should perform upon receipt of the claim is to check its regularity, i.e. whether it meets the requirements for content and appendices to the application set in the Civil Procedure Code.

The irregular claim suffers from the lack of the requisites set in the law. For example, it is not signed by the plaintiff. The absence of any of the annexes to the claim - a document for paid state fee or a power of attorney (in case the claim is signed by a lawyer), is also a reason to assume that the application is irregular.

When the statement of claim does not meet the legal requirements, the judge should give written instructions to the plaintiff to correct the irregularities within one week. If the instructions of the court are not strictly followed or are not fulfilled in time, the judge terminates the initiated case and returns the claim to the plaintiff.

After checking the regularity, the court checks whether the claim is admissible.

The subject against whom the claim is brought is called the "defendant". In his action, the applicant claims that it was the defendant who infringed his rights.

Therefore, once the judge found that the claim was regular and admissible, he sent the claim to the defendant for a response.

Both sides in the process are hierarchically equal. The equality of the parties is one of the main principles in the general claim process. Guided by this principle, the court is obliged to provide the parties with an equal opportunity to exercise the rights granted to them. The court applies the law equally to all. Therefore, once the defendant has been served with the claim, he has the right to defend himself, and the defendant's procedural conduct is to a large extent directly related to the final outcome of the case.

The defendant could be disinterested and not take any protective action in the court proceedings. Failure to reply to the statement of claim confers a significant advantage on the plaintiff in the proceedings. Once missed, the time limit for carrying out a particular procedural action cannot be restored.

The most common hypothesis is that the defendant, within the time allowed by the court, organizes his defense in the process and takes the necessary action to challenge the allegations set out in the statement of claim, taking an opinion (answers to the claim) on its admissibility and validity.

Another possibility for the defendant is also relatively common, as it is the most complicated in terms of procedural actions. This is due to the fact that the parties to the case acquire a double quality at the same time - the original plaintiff subsequently becomes a defendant in the counterclaim, and the original defendant becomes a plaintiff in a counterclaim.

The defendant, within the time limit for replying to the statement of claim, in turn and together with the answer, files in court a counter-claim against the original plaintiff. The counterclaim should meet all the requirements as well as the initial claim, and the court should check the regularity and admissibility of the counterclaim.

If the counterclaim is regular and admissible, the court shall send it to the original claimant for a response. If the claim is irregular or inadmissible, the court returns it to the defendant, sending to the plaintiff only the defendant's initial response.

After the completion of the exchange of papers, the case moves to the stage of consideration of the case in open court.

In paper [5] an application to the apparatus of generalized nets is proposed for modeling the first phase of the general claim process. The advantages of using such model are very different, some of them are expressed in helping to detect and eliminate unnecessary complications in the process itself, leading to its simplification and optimization, presenting a complex process simpler and more understandable, for scientific and teaching needs etc.

2 Temporal Intuitionistic Fuzzy Pair

The concept of a Temporal Intuitionistic Fuzzy Pair (TIFP) is introduced in [6, 7].

Let $T = \{t_1, t_2, \dots\}$ be a fixed time-scale with finite (in the present research) or infinite number of elements. Let E be a fixed universe. The standard Intuitionistic Fuzzy Set [8, 9] is defined as the object with the form:

$$A = \{<x, \mu_A(x), \nu_A(x)> | x \in E\}.$$

In [8, 9], the object:

$$A(T) = \{<<x, t>, \mu_A(x, t), \nu_A(x, t)> | x \in E \,\&\, t \in T\}$$

is defined as a Temporal Intuitionistic Fuzzy Set (TIFS).

In [10], the object $x = <a, b>$, where $a, b \in [0, 1]$ and $a + b \leq 1$ is called an Intuitionistic Fuzzy Pair (IFP).

By analogy with the last two concepts, here, we define the concept of a TIFP by:

$$x(t) = <a(t), b(t)>,$$

where $a(t), b(t) \in [0, 1]$ and $a(t) + b(t) \leq 1$, where $t \in T$ (the defined above time-scale). Therefore, for a fixed IFS $<a, b>$, we can define the set of TIFPs by:

$$B(T) = \{<x(t), a(t), b(t)> | t \in T\}$$

and obviously, this set is an ordinary IFS with universe T.

3 Results of the Application of the Temporal Intuitionistic Fuzzy Pair Apparatus

Let within a certain period of time "*X*", one court "*Y*" is seized with "*N*" in number of claims. The court is obliged to check the regularity of the claim.

There are different hypotheses:

- the claim is initially regular and the court did not find any irregularities in it;
- the claim is irregular. In this case, the court is obliged to instruct the plaintiff to rectify the irregularity within a period specified by law, and the options for development are the following:
- if the plaintiff follows the instructions of the court - then the court officially initiates the court case and gives it a number - for example, "gr.d. № 5/2020."/before that there is no court case number, but only an incoming number and date of the claim - for example "Entr. № 13/10.01.2020"/,
- if the plaintiff does not follow the instructions of the court - the court returns the claim and no court case is initiated at all.

It should be borne in mind that the more common hypothesis is that the court's instructions should be followed and the court should formally initiate a court case (about 80–90% of the filed claims continue to the next phase of the court's investigation).

The temporal intuitionistic fuzzy evaluation is:

$$\left\langle \frac{b(t)}{a(t)}, \frac{c(t)}{a(t)} \right\rangle,$$

where a(t), is the total number of claims submitted so far, b(t) – the number of the regular claims submitted so far, a(t), c(t) – the number of the number of the claims rejected so far. Therefore, the evaluation is a TIFP because:

$$\frac{b(t)}{a(t)} + \frac{c(t)}{a(t)} \leq 1,$$

and $a(t) - b(t) - c(t)$ is the number of claims, returned for correction that have not yet been filed in a corrected form. It is possible that some of them will be re-submitted, as the irregularities on them will be corrected, and another part - will remain irregular.

Once it has been established that the claim is regular, the court should examine whether the claim is admissible. The reason for inadmissibility could be the fact that in the same court or in different courts there are two pending cases between the same parties, on the same grounds and for the same request. In such a case, the filed case is later terminated and the claim is returned to the plaintiff.

The options are as follows:

- if the claim is inadmissible - the court terminates the case;
- if the claim is admissible - the court sends the claim to the defendant for a response.

At this stage of the proceedings, on average, about 5–10% of the claims are dropped.

Only after the judge found that the claim was regular and admissible, he sent the claim to the defendant, giving the latter the opportunity to take a position on the claim by filing a response or counterclaim.

Once there is already pending court proceedings, it is logical for it to develop and, respectively, to end with a court act. There are different options for terminating (prematurely or not) court proceedings as follows:

- waiver of the claim by the plaintiff;
- withdrawal of the claim by the plaintiff;
- suspension of the court proceedings at the request of the parties, which has not been resumed by them after the expiration of 6 months;
- court agreement;
- the claim of the plaintiff is fully respected;
- the applicant's claim is dismissed in its entirety;
- the applicant's claim is upheld in part.

It is important to keep in mind that period "X" due to delay (due to procedural conduct of the parties or the court), the case may not be closed. This means that for the period under review, perhaps about 60% of the initiated cases are expected to be completed at first instance.

The temporal intuitionistic fuzzy evaluation is:

$$\left\langle \frac{e(t)}{d(t)}, \frac{f(t)}{d(t)} \right\rangle,$$

where $d(t)$ is the total number of cases, $e(t)$ – is the number of cases in which a final decision has been reached (positive or negative for the claimant or has ended with a termination agreement, $f(t)$ – the number of cases in which the process was terminated prematurely.

Therefore, the evaluation is a TIFP because:

$$\frac{e(t)}{d(t)} + \frac{f(t)}{d(t)} \leq 1,$$

and $d(t) - e(t) - f(t)$ is the number of cases, the number of cases that have not been closed so far. Some of them may be completed in some of the possible ways, and some may not be completed.

The temporal intuitionistic fuzzy evaluation for the plaintiff's request is:

$$\left\langle \frac{h(t)}{g(t)}, \frac{i(t)}{g(t)} \right\rangle,$$

where $g(t)$ is the total number of claims, $h(t)$ – is the number of claims in which the claim of the plaintiff is fully or partially upheld, $i(t)$ – the number of rejected claims.

Therefore, the evaluation is a TIFP because:

$$\frac{h(t)}{g(t)} + \frac{i(t)}{g(t)} \leq 1,$$

and $g(t) - h(t) - i(t)$ is the number of cases not completed in period "*X*". Some of them may be completed in some of the possible ways, and some may not be completed.

The temporal intuitionistic fuzzy evaluation from the point of view of the plaintiff is:

$$\left\langle \frac{k(t)}{j(t)}, \frac{l(t)}{j(t)} \right\rangle,$$

where $j(t)$ is the total number of claims, i.e., $j(t) = g(t)$, $k(t)$ – is the number of claims in which the claim of the plaintiff is fully respected, $l(t)$ – the number of rejected claims, i.e., $l(t) = i(t)$. Therefore, the evaluation is a TIFP because:

$$\frac{k(t)}{j(t)} + \frac{l(t)}{j(t)} \leq 1,$$

and $j(t) - k(t) - l(t)$ is the number of cases in which the claim of the plaintiff is only partially upheld.

4 Conclusion

This study presents the results of the application of temporal intuitionistic fuzzy pairs to assess the development of civil litigation or the so-called "general claim process" from the initiation of the case to its conclusion by the court of first instance. It is evident that the approach used is universal and allows to achieve an independent assessment of the observed system. The obtained results can be defined as expanding the fields of application of the theory of intuitionistic fuzzy sets.

Acknowledgements. The authors are grateful for the support provided by Grant DN02/10 "New tools for data mining and their modeling", funded by the Bulgarian National Science Fund.

References

1. Civil Procedure Code. Link. https://lex.bg/laws/ldoc/2135558368
2. Tashev R.: Theory of the Legal System, Sibi, p. 316 (2007), ISBN: 954-730-303-1
3. Code Napoleon, 21st of March 1804
4. Stalev J.: Bulgarian Civil Procedural Law, 1st edn., 10th edn. 1448p (2020). ISBN: 9789542831532
5. Blidov, H., Doukovska, L., Atanassov, K.: Generalized net model of the first phase of the general claim process. In: Proceedings of the 10-th International Conference on Intelligent Systems – IS 2020, Varna, Bulgaria, IEEE Xplore, pp. 626–629 (2020). ISBN: 978-1-7281-5456-5, ISSN: 1541-1672, https://doi.org/10.1109/IS48319.2020.9200126
6. Atanassov, K., Atanassova, V.: Temporal intuitionistic fuzzy pairs. Proc. Jangjeon Math. Soc. **24**(1), (2021, in press)
7. Bureva, V., Atanassova, L., Atanassov, K.: Game method for modelling with temporal intuitionistic fuzzy evaluations for locating the wildfire ignition point. Notes Intuit. Fuzzy Sets **26**(4), 90–106 (2020). https://doi.org/10.7546/nifs.2020.26.4.90-106. Print ISSN: 1310-4926, ISSN: 2367-8283
8. Atanassov, K.: Intuitionistic Fuzzy Sets. Springer, Heidelberg (1999)
9. Atanassov, K.: On Intuitionistic Fuzzy Sets Theory. Springer, Berlin (2012)
10. Atanassov, K., Szmidt, E., Kacprzyk, J.: On intuitionistic fuzzy pairs. Notes Intuit. Fuzzy Sets **19**(3), 1–13 (2013)

Investigating the Efficiency of Market Indicators in Trading Systems

Przemysław Juszczuk[1(✉)] and Lech Kruś[2]

[1] Department of Machine Learning, University of Economics in Katowice, 1 Maja 50, 40-287 Katowice, Poland
przemyslaw.juszczuk@ue.edu.pl
[2] Systems Research Institute, Polish Academy of Sciences, Newelska 6, 01-447 Warsaw, Poland
krus@ibspan.waw.pl

Abstract. In this paper, we evaluate the usability of different market indicators. To do so, we propose and analyze three methods that can be used to estimate the efficiency of market indicators for specific trading instruments. In these methods, different measures of efficiency are applied, referring to the correctness of signals generated by the given indicator, their profitability, and the maximal drawdown concept. Formulations of these measures are presented and implemented in an experimental trading system. The system has been used in tests based on real-world data acquired from the forex market.

1 Introduction

Nowadays we observe the still-growing popularity of trading platforms available for decision-makers. A large number of different instruments along with many market indicators leads to the high accessibility of trading tools for everyone. A variety of available instruments covers classical stock indices, bonds, funds, contracts, or even cryptocurrencies. One should know, that risks related to these instruments differ a lot. In some cases, it could easily lead to serious losses in the decision-maker's account.

Despite the large variety of instruments available for the decision-makers, the tools used to predict the prices are the same for different markets and instruments. Among the most popular approaches, we can give technical analysis, fundamental analysis, and sentiment analysis. The purpose of trading systems based on any of these concepts (or their mixture) is to predict the potentially good entry point on the market, which eventually leads to achieving profits.

Technical analysis focuses on various mathematical methods allowing to transform price instruments into values of indicator which is used to generate profitable buy or sell signals. One of the most serious drawbacks of these methods is the delay between the signal occurring on the indicator, and the actual price situation on the chart. Despite that, in paper [28] we observe the still-growing interest in using the technical indicators concepts.

K. T. Atanassov et al. (Eds.): IWIFSGN 2020/BOS/SOR 2020, LNNS 338, pp. 185–203, 2022.
https://doi.org/10.1007/978-3-030-95929-6_15

The fundamental analysis is focused on deriving signals based on some macro and microeconomic events occurring in the country related to the selected instrument like the domestic product or home sales. Many political events like elections or central bank interventions [7] can be easily put in the fundamental analysis as well. Sentiment analysis may be interpreted as a concept close to the fundamental analysis. It is rather a new approach, in which the decision is made on the basis of the group sentiment (often called the wisdom of the crowd).

Papers like [13] discuss the capacity of these methods to derive efficient decisions. The author purported, that markets behave randomly, and any predictions made on the basis of different indicator concepts not necessarily lead to profits on the market. However, as a counter-example we mention the article [27] in which authors suggested, that the forex market as the currency market can actually be effective, thus there is a possibility to build a profitable decision support system.

The concept of fully automatic trading systems (called also algorithmic trading) capable to derive profitable signals is very popular. The algorithmic trading is a relatively new field of science. This concept is based on replacing the role of human in the process of making a decision on the market. A general description of this method may be found in [25]. The algorithmic trading as an approach for optimization of placing orders in the context of minimizing the transaction costs was presented in [8]. The direct impact of the algorithmic trading on the foreign exchange market was presented in [1]. However, one of the crucial elements of such an automatic approach is speed. Thus often used the concept of High-Frequency trading, which can be understood as the classical system with extremely fast execution time. The High-Frequency trading: [3,11,19,31], and algorithmic trading [33] nowadays are crucial in the process of deriving an effective trading system. However, very high demand for the classical trading systems, whose main role is to support the decision process [24] is still observed.

Deriving the accurate signals fast is often the bottleneck of these fully automated trading systems. Thus these approaches often rely on approximate mechanisms like metaheuristics. Examples of such works include genetic programming [18] and evolutionary algorithms [4]. Such approaches are based on the concept of setting the optimal values of technical indicators. Newer works in this field rely on the concept of multi criteria evolutionary algorithms [5] and the particle swarm optimization [2]. Approximate approaches are very popular, and there is still a growing tendency of using metaheuristics. The newest works in this field are presented in [12] and [20]. A different, soft multicriteria computing approach is proposed in [17] to support decisions on the Forex market.

In this paper, we propose a new approach to measuring trading systems efficiency. To do so, we firstly derive a formal description of three different measures focused on different aspects of the trading systems. Next, we discuss the differences in the system evaluation based on these different measures. Additionally, we derive numerical experiments on real data from the market. Further, we analyze obtaining results taking into account that the trading period is precisely selected on the basis of the observed trend direction. Therefore our main goal is to show, that crucial action is to properly indicate a set of instruments with high potential to achieve profits, rather than to derive the "holy grail" capable to always achieve profits.

The outline of the paper is as follows. In Sect. 2 we discuss related works focused on various kinds of trading methods. In Sect. 3 we present basic information about rule-based trading systems and the concept of signals derived for the decision-maker. Section 4 presents the details of three different approaches used to measure system efficiency. Section 5 includes details of numerical experiments and a short discussion. The last section concludes.

2 Related Works

Algorithmic trading is a relatively new field of science. According to this idea, the role of the human in the trading process on the market is replaced by a computer-based system. An example definition of algorithmic trading can be found in [25]. Initially, this concept was used for example to generate liquidity on the markets [16]. More detailed information about the concept of algorithmic trading and its impact on the forex market can be found in [9].

The concept of fully automated trading creates very complex problems analyzed from various perspectives. Since the first papers including the use of neural networks like [10], where the neural network was used as a preliminary signal, this subject gains more and more attention. Newer approaches include the use of the neural network as a prediction tool (see [32]) not directly related to the problem of generating signals. Concept connecting the neural network-based prediction and the technical analysis can be found in [29]. A more complex approach using the neural networks can be found in [12]. Where self-organizing maps were used as a classification tool related to the candlestick chart formations.

Technical analysis indicators still belong to the most popular trading tools, and concepts like the Relative Strength Index (firstly described in [30]) are still often used in the modern trading systems. More complex methods can be found in [6] (grammar evolution) or [4] (evolutionary computation). These methods are often based on deriving a new set of rules, deriving new indicators, or setting the optimal values of existing indicators. The newest works related to this field can be found in [21]. However, works related to new market indicators are presented very rarely. An example of such an article is [14] in which a new indicator correlated with risk factors is presented.

The rule generation process based on the fundamental analysis is rather difficult. However, this analysis is very popular in the case of complex systems related to text mining. Examples of such works can be found in [23], while the more general review is presented in [22].

Available papers are focused on deriving a good entry point for the transactions. The papers dealing with effective system evaluation are relatively rare. In [15] authors discuss the machine learning methods and focus on the concepts like the accuracy of prediction and cumulative return accumulated at the end of the trading window. In [1] the accuracy along with the average profit on the single trade (in percent) and duration of the trade is given. A more systematic review discussing different papers and performance metrics used in these approaches can be found in [26]. However, most of these works are focused on complex systems, rather than evaluating the true efficiency of the single market indicator.

3 Rule-Based Trading Systems

An instrument on the forex market is the currency pair, which represents the ratio of two different currencies to each other. For example, EURUSD with the price 1.4200 means, that for the single Euro we can buy 1.4200 USD. The smallest observed change in price is called "pips" and it is equal to 0.0001 for any currency pair except pairs with Japan Yen, for which pips is equal to 0.01. By the trading system, we mean any approach capable to open transactions on the market on the basis of predefined rules in some time period T. The whole time period T is divided into smaller time frames t called readings which can last anything from 1 min to 1 month. In this article, we investigate two different time frames: H4 which is equal to 4 h, and D1 which is equal to 1 day. Every new action on the market is possible all the time, however, it is assumed, that possible trading signals are derived, when a new t is given.

The most popular trading rules (like these presented in [21]) are based on the behavior of the market indicator, where the relation between two successive values of the indicator can be used as a trigger in the trading system rules. For example, the value of the market indicator RSI (Relative Strength Index) in reading t is calculated as follows:

$$RSI_n(t) = 100 - \frac{100}{1 - \frac{avg_{gain}}{avg_{loss}}}, \tag{1}$$

where avg_{gain} is a sum of gains over the past n readings; avg_{loss} is a sum of losses over the past n periods. The relation between indicator value and currency pair value is given in Fig. 1. Please note, that signals derived by the indicator are marked in the lower part of the figure as grey rectangles.

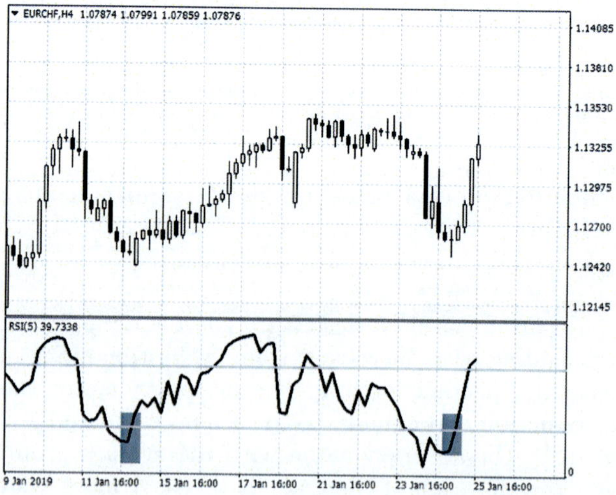

Fig. 1. The example currency pair with the corresponding RSI market indicator (seen in the lower part of the figure)

We focus on market indicators, for which every rule can be defined in the same manner. In general, the trading rule for the indicator can be presented as follows:

$$BUY\ if\ (ind(t-1) < c) \wedge (ind(t) > c), \tag{2}$$

$ind(t-1)$ is the value of indicator ind in the reading $t-1$; c is a constant value for the given indicator. Details about the exact rules related to selected analyzed in this paper market indicator can be found in Appendix 1.

In the below approach we concentrate on a set of technical indicators. However fundamental indicators could be used as well. To simplify the whole evaluation process of the trading system, the following assumptions have been made:

- a single trading indicator is analyzed at a time;
- trading rules are based on the simple condition, that the trading signal is generated when the indicator crosses the defined level (please see Appendix 1 for more details);
- only BUY signals are analyzed.

According to the above assumptions for each selected time interval and selected currency pair, we analyze all signals generated by a single indicator. The procedure is repeated for all analyzed indicators.

4 System Efficiency

By the system efficiency, we understand here capability of the trading system to derive satisfactory signals for the decision-maker. By the satisfactory, we mean the signals, that allow making decisions that are beneficial for the decision-maker. Below we present a few quantities related to the system efficiency:

- number of generated signals;
- profit or loss achieved in some measurable units (in our case in pips) – called further "profit";
- number of profitable signals derived (as a percentage of the overall number of signals) – called further "accuracy";
- the maximal observed loss of profit in successive transactions during the whole trading period – called further "max drawdown".

The first one related to the number of generated signals is not as important as the remaining ones, however, it is used to calculate the system accuracy. The formula for accuracy measure is given as follows:

$$acc(T,p) = \frac{TP}{FP+TP}, \tag{3}$$

where TP is the number of signals, for which the price direction was properly predicted (for which the actual price after p readings is greater than the initial price observed); FP is the number of wrongly predicted signals; p is the waiting time, which is equal to the number of readings, after which the result is measured and T is the analyzed time period. The concept of this measure is presented in Fig. 2. At the vertical axis, we observe the price value for the selected instrument. At the horizontal axis, we can see the time period t, where the price of the instrument changes between $t = 1$ and $t = 17$. At time $t = 17$ the signal is generated. Now, depending on the p-value (waiting time parameter), the signal can be evaluated after 1, 3, or 5 time intervals. Since we do not measure the range of the price change between the time $t = 17$ (when the signal is generated), and the time, when the signal is evaluated, every price value higher than this observed in $t = 17$ will be considered as a profitable one. In such a situation, the shorter waiting time (small p values) could include more trend disturbance and lead to some less-predictable price moves. While the longer waiting time gives a sense of stability in the case of the signal. The second discussed measure is the profit achieved in the time period T with the assumption, that after the waiting time equal to p the currency pair value differs at least by ϵ. In this case, besides a profitable signal and a loss signal, we consider the third type of signal called by us WAIT. The WAIT signal corresponds to not sufficient change in the currency pair value. This profit measure is similar to the accuracy. However, we do not count profitable/non-profitable signals, but rather the sum of pips of all negative and positive signals.

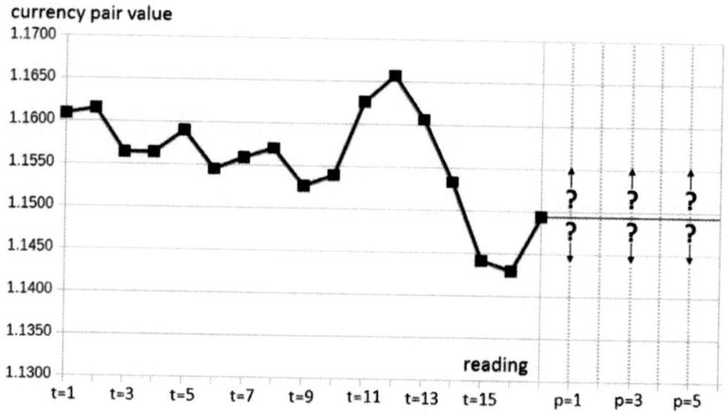

Fig. 2. Estimating the TP/FP signal for different p values (accuracy measure)

The formula is given as follows:

$$profit(T, p, \epsilon) = \sum_{i=1}^{n}(sig_i(t)), \tag{4}$$

for signals i satisfying condition ($sig_i\ (t) > \epsilon$) or ($sig_i\ (t) < -\epsilon$) where n is the number of signals, for which there was a currency pair value difference larger than ϵ; $sig_k(t)$ is the currency pair value observed in reading t after waiting time equal p. Basically, we consider all signals generated by the trading system, for which after waiting time p there was at least ϵ difference in currency pair value (positive or negative). Profit (measured in pips) for all such signals is summed and presented as the measured value for time period T. This concept for different ϵ values is presented in Fig. 3. In this figure, the vertical axis once again represents the price change, while the horizontal axis represents the time period. The signal is generated at $t = 17$. However, the signal evaluation is now extended by the additional parameter ϵ, which means the minimal price difference between the given p and $t = 17$, that must occur, to recognize the signal as a valid. Obviously, this measure greatly reduces the number of valid signals, however allows us to minimize the impact of the trend disturbance on the results. While extending the minimal price range defined by the ϵ, we expect, that the minimal assumed price change will be observed. In the case of the side trend, low market value, or simply, small daily volatility on the market, the number of such random signals will be reduced.

The last proposed measure is related to the ability of the trading system to ensure as small losses as possible during the trading period of T. The main difference between this measure and the two proposed early in this paper is the units used to evaluate the trading system. Previously we rather calculated the number of signals (profitable or not) or profit or loss generated by the signals in the given trading period. In the case of this new measure, we observe changes in the account balance of the decision-maker using the selected trading system. It is made for the assumed signals and an assumed method of closing position related to this signal (like after p time intervals, or after price change higher than ϵ).

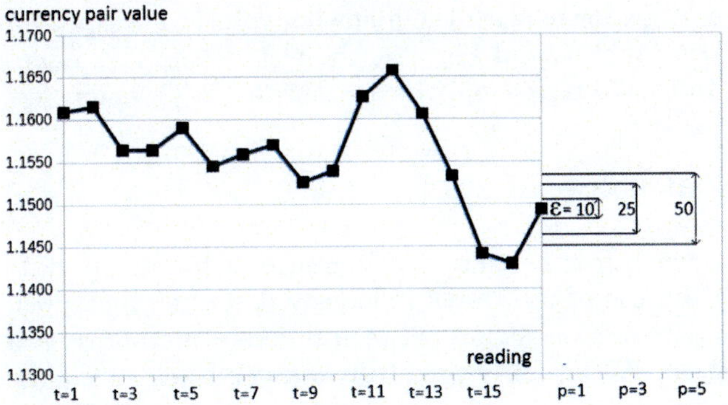

Fig. 3. Three different ϵ ranges, which should be exceeded, to recognize the signal

Please see Fig. 4 presenting this concept. On the left side of the figure with black and grey color we see the profitable signals (black color) and loss signals (grey color) during the trading period T. However, the final measure is not related to the account value observed at the end of the trading period T. To acquire the max drawdown value we calculate and collect all drops in the account balance observed during the whole trading period (see right side of Fig. 4). Finally, the drop with the maximal loss value is selected and presented to the decision-maker as the max drowdown.

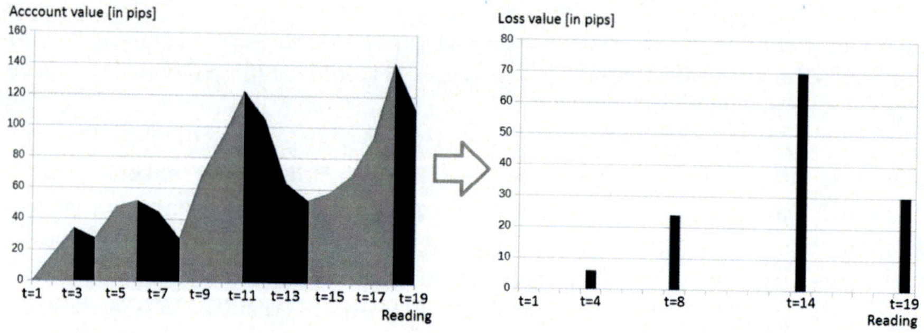

Fig. 4. Max drawdown concept presented as the account value in successive signals

The maximal drawdown is the maximum value observed from all these losses. This measure is described by the following formula:

$$draw(T) = max(v_1, v_2, ..., v_n),\ where\ v_i = h_i - l_i, \tag{5}$$

where h_i is the i-th peak value observed in the account value, while l_i is the lowest value before the new height is derived (the lowest value following directly after the h_i. Thus we select the highest drop observed in the whole trading during time span T. In Fig. 4 on the left side we can see four such peaks: $h_i = 34$, $h_2 = 52$, $h_3 = 123$, and $h_4 = 142$. At the end of each grey fragment there are local lowest values marked as $l_1 = 28$, $l_2 = 28$, $l_3 = 53$, and $l_4 = 112$. Thus we achieve V set of drawdowns with the following values: $V = (6, 24, 70, 30)$ which are marked on the right side of Fig. 4 as the histogram. The highest value is v_3 corresponding to the 70 pips drop which is our $draw(T)$ value.

5 Numerical Experiments

In this section, we experimentally verify results achieved by different market indicators with the use of the three previously described measures. Our goal is to emphasize differences occurring between different measures used to evaluate the trading system. We basically present two different cases explained below:

- Comparison and analysis of three different measures used to estimate the trading system efficiency. Among these measures, the first two (described in the previous section) are related directly to the number of profitable signals acquired by the trading system during some predefined trading period. While the last measure called the max drawdown is used to measure, what is the "worst-case scenario" for the given trading system in the analyzed period. In other words, we estimate, what is the maximum loss on the account balance observed for the whole trading period;
- Analysis of results derived by the trading system in the case, where we strictly focus on the time interval, when the strong bullish trend is observed.

These two cases are used to check our thesis, that there is no single efficiency measure, which can be effectively used to estimate the usability of the trading system. There is a need to switch between these measures, or even focus on multicriteria measure capable to discover additional dependencies between different measures and concentrate on trading systems capable to derive the good signals for different efficiency measures. We also show, that even the general trend identification could lead to an increase in the trading system efficiency. Thus, the proper selection of the trading period should not be marginalized.

In our experiments, we used data of 15 different currency pairs covering the time span of 9 years from 1st January 2010 to 31st December 2018. We analyzed 5 different market indicators (detailed results for 2 of them can be found in Appendix 2) and two different time windows D1 – where each new reading is derived at the end of the day and H4 – with new reading derived after every 4 h. Please note, that for the D1 time window the overall number of available readings was equal to 2400, while for the H4 it was 12000 readings.

Additionally, we analyzed a number of signals generated by all indicators. To emphasize the impact of the trend on the overall results we performed additional tests for the same set of currency pairs for the D1 time window. However, we kept the smaller number of readings covering only the fragments, when the bullish trend was present.

We present results for three different indicators: CCI – Commodity Channel Index, RSI – Relative Strength Index, and DM – DeMarker (results for two additional indicators: Bulls indicator and Moving Average of Oscillator – OsMA can be found in Appendix 2).

Table 1. The accuracy of the prediction for the D1 time window and three selected indicators

D1	CCI			RSI			DM		
	p = 1	p = 3	p = 5	p = 1	p = 3	p = 5	p = 1	p = 3	p = 5
AUDCHF	63%	16%	21%	74%	10%	19%	68%	40%	43%
AUDJPY	61%	18%	24%	69%	6%	17%	69%	40%	47%
AUDUSD	57%	19%	23%	64%	18%	18%	68%	37%	37%
EURAUD	54%	22%	26%	69%	17%	31%	66%	24%	32%
EURCHF	46%	14%	28%	63%	16%	31%	63%	28%	40%
EURGBP	58%	20%	31%	69%	16%	25%	71%	38%	34%
EURUSD	56%	15%	22%	69%	18%	21%	67%	28%	35%
GBPCHF	56%	14%	18%	70%	20%	36%	70%	23%	35%
GBPJPY	53%	13%	19%	60%	14%	31%	73%	32%	32%
GBPNZD	51%	16%	25%	76%	16%	26%	72%	28%	37%
GBPUSD	54%	23%	25%	68%	8%	29%	69%	25%	39%
NZDCAD	55%	20%	25%	75%	13%	25%	70%	49%	60%
NZDJPY	55%	19%	23%	65%	13%	20%	68%	31%	31%
NZDUSD	55%	17%	23%	47%	12%	21%	64%	38%	46%
USDCHF	57%	17%	22%	66%	16%	13%	68%	31%	30%

In the first part of the experiments, we investigated the efficiency of indicators with the use of accuracy measure. We focused on three different waiting times equal to $p = 1$, $p = 3$ and $p = 5$. The results for the D1 can be seen in Table 1. What can be observed, is that the best results were achieved when the waiting time was equal to $p = 1$ (the result is measured after the single reading). In almost all cases the achieved results were above 50%, which means that even the single market indicator was capable to effectively predict the potential price movement.

The accuracy measured after delay p in this experiment is related to the duration of the signal. The position opened on the basis of the derived signal is closed after p readings. Thus for the lower p values, we agree with the higher trend disturbance impact on the results. On the other hand, the lower value of p means, that in the case of a small price change, the probability, that the signal will be profitable (after p, the instrument price will be higher than the price observed for the signal) is higher. On the basis of Table 1 we observe a simple dependency, that the low value of p (and fast close of the trade) leads to a larger number of profitable signals. However, please note, that the size of the profit in this particular experiment is not under consideration. While it is obvious, that faster close of the position will lead to relatively small profits.

A similar accuracy analysis was performed for the H4 time window. The results can be found in Table 2. Please note, that despite a large number of signals (12000 overall number of readings for the H4) the accuracy is similar to this achieved for the D1 time window. Once again we observe relatively high accuracy values for the $p = 1$, which further decreases, while p is increased. Thus the observations can be given: achieved results are similar for different time windows and for different indicators.

Table 2. The accuracy of the prediction for the H4 time window and three selected indicators

H4	CCI			RSI			DM		
	p=1	p=3	p=5	p=1	p=3	p=5	p=1	p=3	p=5
AUDCHF	56%	17%	24%	67%	14%	18%	63%	38%	40%
AUDJPY	55%	17%	25%	71%	17%	24%	64%	34%	37%
AUDUSD	56%	18%	25%	67%	20%	27%	69%	38%	40%
EURAUD	57%	18%	25%	66%	14%	20%	68%	34%	37%
EURCHF	55%	20%	25%	74%	15%	20%	70%	41%	38%
EURGBP	54%	17%	23%	70%	16%	22%	67%	35%	36%
EURUSD	56%	18%	23%	65%	21%	25%	71%	39%	38%
GBPCHF	53%	19%	23%	68%	13%	20%	72%	38%	38%
GBPJPY	54%	15%	21%	69%	17%	26%	69%	34%	34%
GBPNZD	54%	19%	26%	63%	18%	22%	70%	34%	40%
GBPUSD	56%	17%	23%	67%	22%	22%	70%	36%	39%
NZDCAD	55%	19%	27%	72%	19%	29%	67%	39%	41%
NZDJPY	57%	17%	24%	67%	18%	20%	68%	38%	40%
NZDUSD	56%	16%	24%	68%	19%	21%	68%	33%	40%
USDCHF	56%	18%	26%	63%	17%	25%	65%	33%	36%

The second part of the experiments covers the analysis of profit achieved by different indicators. In this case, we measure the overall performance of the indicator rather than a number of good/bad signals. The detailed formulation of this measure is presented in the previous section. In the experiments we analyzed three different values: $\epsilon = 10$, which means at least 10 pips of price change (rise or drop); $\epsilon = 25$ (25 pips), and finally $\epsilon = 50$ (50 pips). For all these experiments we assumed the constant p value equal to 3. We investigated here the whole time span of 9 years for D1 but only selected trending fragments for all currency pairs. Results can be seen in Table 3.

One should know, that the different ϵ values did not have a visible impact on the final result. This observation is in some way counter-intuitive, thus we rather expect, that large ϵ should reduce the impact of the trend disturbance at the results. The overall poor quality of results could be possibly triggered by the fact, that most of the time during the analyzed 9-years time span we observed a side trend. However, this observation should be deeply analyzed in the future for the data, where all kinds of trends could be easily identified and separated.

Crucial is that for the same profit measure we performed the analysis for the limited time span (see the beginning of this section), where currency pairs were in the bullish trend. We once again focused on the D1 time window. The results can be seen in Table 4. For most cases trading systems still did not manage to achieve profits, however, the potential loss is rather low in comparison to the full-time span covering the 9 years (even with the assumption, that the shorter time interval generates a smaller number of signals). In this particular experiment, the range of changes denoted as ϵ had rather a small impact on the results.

Table 3. Profit achieved by different market indicators for the whole analyzed time span (9 years) and the time window equal to D1. (* denotes values in pips below −1000)

D1	CCI			RSI			DM		
ϵ	10	25	50	10	25	50	10	25	50
AUDCHF	−826	−817	−799	−224	−222	−227	−228	−227	−217
AUDJPY	*	*	*	*	*	*	*	*	*
AUDUSD	−749	−740	−727	−211	−213	−197	−116	−118	−124
EURAUD	−753	−747	−762	−237	−242	−249	−186	−188	−191
EURCHF	−434	−419	−380	−52	−49	−49	−351	−349	−328
EURGBP	−359	−366	−362	−71	−65	−39	−94	−90	−80
EURUSD	−669	−670	−636	−166	−167	−159	−152	−151	−139
GBPCHF	*	*	−975	−293	−297	−304	−202	−203	−198
GBPJPY	*	*	*	*	*	*	*	*	*
GBPNZD	*	*	*	−380	−379	−379	−266	−270	−272
GBPUSD	−749	−742	−746	−285	−285	−282	−164	−161	−168
NZDCAD	−478	−477	−458	−99	−97	−94	62	55	58
NZDJPY	*	*	*	*	*	*	*	*	*
NZDUSD	−662	−671	−690	−170	−170	−166	5	4	−27
USDCHF	−710	−705	−698	−96	−94	−82	−130	−125	−123

Achieving the highest profit in the selected time span seems to be crucial for the efficient trading system. However, the second important aspect is the ability of the system to achieve profits in consecutive signals. The last efficiency measure we investigated in the experiments is the maximal drawdown. To do so, we selected four different currency pairs and three market indicators for the bullish trend in D1 time window. In Fig. 5 we can see four selected currency pairs, for which we used the data from Table 4 and $\epsilon = 10$ pips.

In the case of the last efficiency measure it is crucial to evaluate, what was the maximum number of successive signals for which the system recorded losses (and to measure this total loss). In Fig. 5(a) we can see, that for the EURCHF currency pair it was possible to achieve over 300 pips of profit, while the maximal drawdown (shown in light grey) was almost 450 pips. One should be aware, that the overall good result of the system should not be analyzed separately, thus during the whole trading system uptime, the 450 pips loss was observed. This means, that there occurred signal (or signals), for which the loss was greater than the final profit achieved by the system. Over 50% of drawdown is observed in Fig. 5(b) for the EURUSD currency pair. Despite the good final result (a few hundreds of pips in both cases) the maximal drawdown is very high and may not be accepted by the decision-maker. For the third case (Fig. 5(c)) and GBPCHF the final result was small and equal 70 pips, however, the drawdown was relatively small.

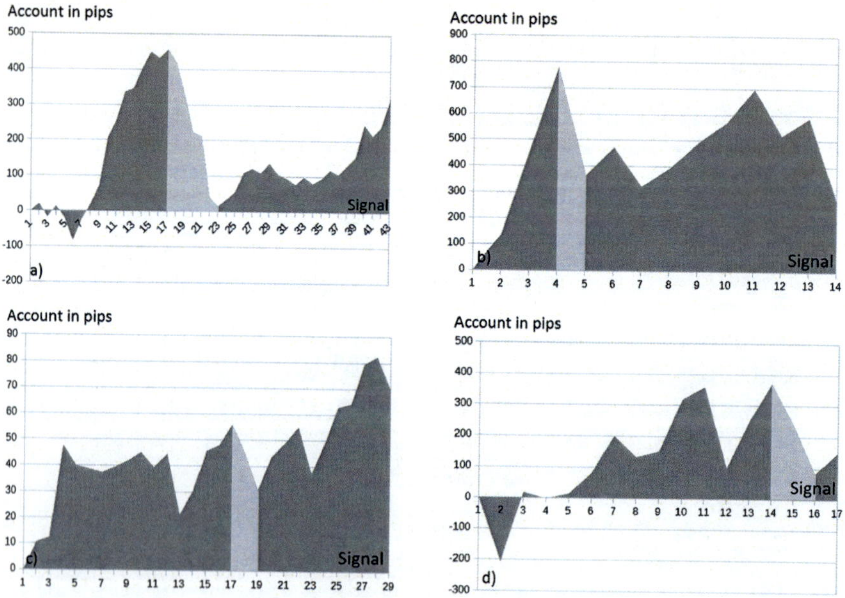

Fig. 5. Profit achieved by different market indicators for the selected bullish trend and the time window equal to D1. Details about these time intervals are given in brackets. * denotes values in pips below -1000

In general, the decision-maker should rather look for the more stable trading systems capable to derive the small profit for successive signals, than focus on systems, for which the potential loss is very high.

Table 4. Number of signals derived by all systems for three different approaches

D1	CCI			RSI			DM		
Time interval/ϵ	10	25	50	10	25	50	10	25	50
AUDCHF (10 2011–11 2011)	−15	−13	−10	−20	−17	−17	13	14	14
AUDJPY (8 2011–2 2015)	*	*	*	*	*	*	*	*	*
AUDUSD (4 2010–9 2011)	−202	−198	−195	−85	−87	−87	−36	−36	−40
EURAUD (4 2012–4 2014)	−9	−7	−17	−11	−12	−12	51	53	53
EURCHF (2 2015–3 2018)	−111	−104	−77	2	2	2	−16	−17	−13
EURGBP (6 2015–3 2018)	−75	−76	−87	−33	−33	−26	−34	−35	−35
EURUSD (4 2010–9 2011)	−101	−103	−104	−39	−41	−41	−29	−29	−29
GBPCHF (3 2013–8 2014)	−86	−86	−81	8	8	8	35	34	40
GBPJPY (5 2012–9 2015)	*	*	*	−243	−243	−243	−196	−171	−255
GBPNZD (5 2013–11 2015)	−236	−236	−228	−51	−51	−44	−108	−110	−108
GBPUSD (5 2013–8 2014)	−31	−29	−29	−15	−15	−10	10	11	11
NZDCAD (4 2010–2 2014)	−142	−140	−161	−3	−3	−6	32	32	37
NZDJPY (7 2012–2 2015)	−772	−732	−704	−201	−215	−179	−603	−580	−457
NZDUSD (4 2010–7 2013)	−312	−315	−335	−54	−54	−54	62	60	46
USDCHF (6 2015–5 2017)	−153	−153	−152	−20	−20	−8	−20	−17	−10

6 Conclusions and Summary

In this paper, we presented one of the crucial aspects related to the trading systems: how to properly evaluate the efficiency of the trading system. To do so, we investigated data from the forex market. The forex market is considered as the most liquid market in the world with very high volatility. Thus it is a perfect benchmark environment to evaluate different efficiency measures. We acquired real quotations for 15 currency pairs from the time span of 9 years.

We focused on the rule-based trading systems and selected market indicators. We presented the most common trading rules related to these systems.

We formulated and analyzed three different efficiency measures: The first one (called the accuracy) is strictly based on the number of signals, for which the price direction was predicted properly. The second measure relates to the possible profit under the assumption that the evaluation is made only in the case, where some visible price movement (rise or drop) will occur. Finally, the third measure was based on the concept of maximal drawdown, which is the measure of the possible maximal successive loss during the whole trading period.

For all the three efficiency measures we performed detailed analysis including the efficiency of selected indicators and final profit(loss) achieved by each market indicator for 15 analyzed currency pairs. On the basis of the achieved results we concluded, that the crucial aspect is not to properly select the set of market indicators, but to trade in selected periods, for which visible trend is observed. Moreover, the final result achieved by the trading system at the end of the trading period is not as important, as to maintain the small successive profits over time. We conclude, that there is no doubt, that the profit achieved by the trading system is the crucial criterion to evaluate the given trading approach, however, the maximal drawdown is a very important aspect of trading and it should be included as the additional criterion. This observation indicates the necessity of multicriteria analysis.

Our future works will include the approach, in which more than one of the efficiency measures described in this paper should be analyzed jointly as the multicriteria analysis problem. In such a case, all instruments analyzed by the trading systems and all indicators used in the trading system will be considered as a set of different variants in the multicriteria space, where different criteria are related to different measures. We purport, that the approach, in which different measures are analyzed jointly should greatly improve the overall quality of signals derived by the trading systems.

Appendix 1

In this Appendix, we present details for rules used in the numerical experiments for all five market indicators.

- rule for the CCI indicator:

$$BUY\ if\ (CCI(t-1) < -100) \wedge (CCI(t) > -100),$$

- rule for the RSI indicator:

$$BUY\ if\ (RSI(t-1) < 0.3) \wedge (RSI(t) > 0.3),$$

- rule for the DeMarker indicator:

$$BUY\ if\ (DM(t-1) < 0.3) \wedge (DM(t) > 0.3),$$

- rule for the Bulls indicator:

$$BUY\ if\ (Bulls(t-1) < 0) \wedge (Bulls(t) > 0),$$

- rule for the OsMA indicator:

$$BUY\ if\ (OsMA(t-1) < 0) \wedge (OsMA(t) > 0),$$

Appendix 2

This Appendix presents detailed results for two remaining market indicators (Table 5, 6, 7, 8 and 9).

Table 5. The overall number of signals generated for the D1 time window (2400 readings) and H4 time window (12000 readings) – two indicators

	D1		H4	
	Bulls	OsMA	Bulls	OsMA
AUDCHF	165	101	742	499
AUDJPY	150	94	754	517
AUDUSD	160	93	811	528
EURAUD	163	95	796	503
EURCHF	180	95	891	522
EURGBP	178	108	772	502
EURUSD	159	106	737	509
GBPCHF	169	97	779	516
GBPJPY	154	89	766	502
GBPNZD	153	101	829	525
GBPUSD	168	101	772	507
NZDCAD	174	101	783	537
NZDJPY	154	89	717	499
NZDUSD	165	103	759	516
USDCHF	156	99	792	500

Table 6. The accuracy of the prediction for the D1 time window and two selected indicators

D1	Bulls			OsMA		
	p = 1	p = 3	p = 5	p = 1	p = 3	p = 5
AUDCHF	93%	45%	40%	68%	30%	33%
AUDJPY	89%	43%	47%	70%	32%	30%
AUDUSD	93%	45%	43%	68%	26%	31%
EURAUD	90%	42%	43%	77%	32%	33%
EURCHF	87%	42%	49%	59%	16%	26%
EURGBP	88%	37%	46%	80%	19%	31%
EURUSD	92%	34%	41%	66%	26%	38%
GBPCHF	89%	42%	45%	74%	30%	29%
GBPJPY	89%	41%	39%	70%	25%	27%
GBPNZD	93%	44%	41%	70%	32%	40%
GBPUSD	86%	43%	43%	63%	32%	37%
NZDCAD	88%	48%	48%	57%	30%	32%
NZDJPY	94%	42%	45%	62%	28%	34%
NZDUSD	90%	42%	50%	67%	29%	39%
USDCHF	88%	30%	30%	74%	22%	29%

Table 7. The accuracy of the prediction for the H4 time window and two selected indicators

H4	Bulls			OsMA		
	p = 1	p = 3	p = 5	p = 1	p = 3	p = 5
AUDCHF	90%	39%	43%	72%	29%	33%
AUDJPY	92%	40%	42%	72%	30%	34%
AUDUSD	90%	40%	44%	66%	32%	35%
EURAUD	88%	40%	43%	76%	30%	30%
EURCHF	86%	44%	44%	67%	31%	33%
EURGBP	88%	40%	42%	70%	32%	34%
EURUSD	86%	40%	43%	69%	29%	35%
GBPCHF	89%	40%	41%	66%	32%	34%
GBPJPY	89%	43%	44%	67%	31%	35%
GBPNZD	86%	42%	43%	69%	28%	29%
GBPUSD	90%	44%	44%	71%	33%	35%
NZDCAD	90%	44%	48%	69%	34%	35%
NZDJPY	89%	42%	44%	67%	32%	36%
NZDUSD	88%	40%	44%	72%	33%	34%
USDCHF	88%	40%	45%	71%	28%	35%

Table 8. Profit achieved by different market indicators for the whole analyzed time span (9 years) and the time window equal to D1 – 2 indicators

D1	Bulls			OsMA		
	E = 10	E = 25	E = 50	E = 10	E = 25	E = 50
AUDCHF	−560	−556	−531	−296	−286	−272
AUDJPY	−3604	−3621	−3725	−3119	−3039	−3046
AUDUSD	−331	−334	−322	−375	−379	−367
EURAUD	−323	−319	−327	−359	−358	−343
EURCHF	−69	−72	−76	−174	−167	−133
EURGBP	−81	−83	−44	−238	−237	−188
EURUSD	−405	−406	−387	−269	−272	−275
GBPCHF	−381	−382	−371	−447	−441	−452
GBPJPY	−7229	−7296	−7354	−4092	−4008	−3906
GBPNZD	−527	−520	−520	−565	−562	−579
GBPUSD	−196	−185	−218	−268	−268	−279
NZDCAD	17	8	32	−101	−101	−78
NZDJPY	−1771	−1774	−1549	−1982	−2009	−2009
NZDUSD	−86	−101	−111	−304	−309	−312
USDCHF	−473	−455	−436	−327	−322	−302

Table 9. Profit achieved by different market indicators for the selected bullish trend and the time window equal to D1 – 2 indicators

	Bulls			OsMA		
	E = 10	E = 25	E = 50	E = 10	E = 25	E = 50
AUDCHF	−62	−59	−60	43	47	47
AUDJPY	−2043	−2096	−2073	−796	−758	−791
AUDUSD	−14	−12	−14	−124	−123	−119
EURAUD	−9	−4	−8	−33	−35	−39
EURCHF	32	23	18	−29	−30	−14
EURGBP	−11	−11	−9	−47	−49	−51
EURUSD	−170	−172	−168	27	27	27
GBPCHF	70	63	57	−45	−45	−54
GBPJPY	−981	−1012	−1060	−645	−607	−517
GBPNZD	22	24	29	−179	−180	−180
GBPUSD	−21	−19	−25	−60	−60	−57
NZDCAD	−15	−17	−15	−24	−25	−14
NZDJPY	−199	−215	−179	−380	−382	−561
NZDUSD	−2	−8	−22	−104	−106	−109
USDCHF	−130	−126	−118	−128	−125	−128

References

1. Arevalo, R., Garcia, J., Guijarro, F., Peris, A.: A dynamic trading rule based on filtered flag pattern recognition for stock market price forecasting. Expert Syst. Appl. **81**, 177–192 (2017)
2. Bagheria, A., Peyhanib, H.M., Akbaric, M.: Financial forecasting using ANFIS networks with quantum-behaved particle swarm optimization. Expert Syst. Appl. **41**(14), 6235–6250 (2014)
3. Bernales, A.: Make-take decisions under high-frequency trading competition. J. Finan. Mark. **45**, 1–18 (2019)
4. Bodas-Sagi, D.J., Fernández, P., Hidalgo, J.I., Soltero, F.J., Risco-Martín, J.L.: Multiobjective optimization of technical market indicators. In: GECCO 2009 Proceedings of the 11th Annual Conference Companion on Genetic and Evolutionary Computation Conference, 1999–2004 (2013). Brooks, R., Financial Management, London: Pearson Education, 2nd Ed
5. Bodas-Sagi, D.J., Fernández-Blanco, P., Hidalgo, J.I., Soltero-Domingo, F.J.: A parallel evolutionary algorithm for technical market indicators optimization. Nat. Comput. **12**(2), 195–207 (2013)
6. Brabazon, A., O'Neill, M.: Evolving technical trading rules for spot foreign-exchange markets using grammatical evolution. CMS **1**(3), 311–327 (2004)
7. Broto, C.: The effectiveness of forex interventions in four Latin American countries. Emerg. Mark. Rev. **17**, 224–240 (2013)
8. Brownlees, C.T., Cipollini, F., Gallo, G.M.: Intra-daily volume modeling and prediction for algorithmic trading. J. Financ. Economet. **9**(3), 489–518 (2010)

9. Chaboud, A., Chiquoine, B., Hjalmarsson, E., Vega, C.: Rise of the machines: algorithmic trading in the foreign exchange market. J. Am. Finance Assoc. **69**(5), 2045–2084 (2014)
10. Chan, K.C.C., Teong, F.K.: Enhancing technical analysis in the forex market using neural networks. In: IEEE International Conference on Neural Networks, vol. 2, pp. 1023–1027 (1995)
11. Chu, J., Chan, S., Zhang, Y.: High frequency momentum trading with cryptocurrencies. Res. Int. Bus. Financ. **52**, 101176 (2020)
12. Ciskowski, P., Zaton, M.: Neural pattern recognition with self organizing maps for efficient processing of forex market data streams. In: Rutkowski, L., Scherer, R., Tadeusiewicz, R., Zadeh, L.A., Zurada, J.M. (eds.) Artificial Intelligence and Soft Computing, ICAISC 2010. LNCS, vol. 6113, pp. 307–314. Springer, Heidelberg (2010). https://doi.org/10.1007/978-3-642-13208-7_39
13. Fama, E.: The behavior of stock-market prices. J. Bus. **38**, 34–105 (1965)
14. Fliess, M., Join, C.: Towards new technical indicators for trading systems and risk management. In: 15th IFAC Symposium on System Identification (2009)
15. Gerlein, E.A., McGinnity, M., Belatreche, A., Coleman, S.: Evaluating machine learning classification for financial trading: an empirical approach. Expert Syst. Appl. **54**, 193–207 (2016)
16. Hendershotta, T., Riordan, R.: Algorithmic trading and the market for liquidity. J. Financial Quant. Anal. **48**(4), 1001–1024 (2013)
17. Juszczuk, P., Kruś, L.: Soft multicriteria computing supporting decisions on the Forex market. Appl. Soft Comput. **96**, 106654 (2020)
18. Lee, Ch.S., Loh, K.Y.: GP-based optimisation of technical trading indicators and profitability in FX market. Neural Inf. Process. **3**, 1159–1163 (2002)
19. Malceniece, L., Malcenieks, K., Putnins, T.J.: High frequency trading and comovement in financial markets. J. Financ. Econ. **134**(2), 381–399 (2019)
20. Moscinski, R., Zakrzewska, D.: Building an efficient evolutionary algorithm for forex market predictions. In: Jackowski, K., Burduk, R., Walkowiak, K., Wozniak, M., Yin, H. (eds.) IDEAL 2015. LNCS, vol. 9375, pp. 352–360. Springer, Cham (2015). https://doi.org/10.1007/978-3-319-24834-9_41
21. Ozturk, M., Toroslu, I.H., Fidan, G.: Heuristic based trading system on Forex data using technical indicator rules. Appl. Soft Comput. **43**, 170–186 (2016)
22. Nassirtoussi, A.K., Aghabozorgi, S., Wah, T.Y., Ling, N.D.Ch.: Text mining for market prediction: a systematic review. Expert Syst. Appl. **41**, 7653–7670 (2014)
23. Nassirtoussi, A.K., Aghabozorgi, S., Wah, T.Y., Ling, N.D.Ch.: Text mining of news-headlines for FOREX market prediction: a Multi-layer Dimension Reduction Algorithm with semantics and sentiment. Expert Syst. Appl. **42**, 306–324 (2015)
24. Ng, W.W.Y., Liang, X.-L., Li, J., Yeoung, D.S., Chan, P.P.K.: LG-Trader: stock trading decision support based on feature selection by weighted localized generalization error model. Neurocomputing **146**, 104–112 (2014)
25. Nuti, G., Mirghaemi, M., Treleaven, P., Yingsaeree, C.: Algorithmic trading. Computer **44**(11), 61–69 (2011)
26. Pradeepkumar, D., Ravi, V.: Soft computing hybrids for FOREX rate prediction: a comprehensive review. Comput. Oper. Res. **99**, 262–284 (2018)
27. Shmilovici, A., Kahiri, Y., Ben-Gal, I., Hauser, S.: Measuring the efficiency of the intraday forex market with a universal data compression algorithm. Comput. Econ. **33**(2), 131–154 (2009)
28. Taylor, M.P., Allen, H.: The use of technical analysis in the foreign exchange market. Int. J. Financ. Econ. **11**, 327–338 (2006)

29. Thawornwonga, S., Enkea, D., Daglia, C.: Neural networks as a decision maker for stock trading: a technical analysis Approach. Int. J. Smart Eng. Syst. Des. **5**(4), 313–325 (2003)
30. Wilder, J.W.: New Concepts in Technical Trading Systems. Trend Research (1978)
31. Yang, H., Ge, H., Luo, Y.: The optimal bid-ask price strategies of high-frequency trading and the effect on market liquidity. Res. Int. Bus. Financ. **53**, 101194 (2020)
32. Yao, J., Tan, Ch.L.: A case study on using neural networks to perform technical forecasting of forex. Neurocomputing **32**(1–4), 79–98 (2000)
33. Zhou, H., Kalev, P.S.: Algorithmic and high frequency trading in Asia-Pacific, now and the future. Pac. Basin Financ. J. **53**, 186–207 (2019)

Supporting the Process of Negotiations with the Use of Multi-criteria Analysis

Andrzej Łodziński(✉)

Warsaw University of Life Sciences - SGGW, Warsaw, Poland

Abstract. The paper presents the method for supporting the process of negotiations with the use of multi-criteria analysis. The negotiation process is modeled using multi-criteria optimization. The method used to find the solutions is constituted by a procedure involving the interactive selection of certain proposals. The parties submit their proposals concerning the subjects of the negotiations. These proposals comprise the values of parameters of the multi-criteria optimization problem. The selection of solutions is accomplished by solving the optimization problem with parameters defining the aspirations of each party involved in the negotiations. Finally, the solutions obtained by the parties are evaluated.

Keywords: Negotiation process · Multi-criteria optimization · Equitably efficient decision · Scalarizing function · Reference point method

1 Introduction

The paper presents the application of multi-criteria optimization analysis to decision selection in the process of negotiations. Negotiations enable a decision to be reached in cases where the interests of the participants differ. Negotiations are carried out in order to reach a result that is more favorable than that which might have been achieved without negotiations. Negotiating parties can benefit by reaching a mutual agreement, as compared to a situation in which they act separately. A suitably arranged agreement is better for the parties than no agreement at all; moreover, some agreements are more favorable for both parties than other ones. In general, through the negotiations, the parties wish not only to reach an agreement, but to reach an optimal agreement, i.e., one that would be best for both parties.

Negotiations are characterized by a lack of clear solutions and by the necessity of taking the preferences of all parties into account. The process of negotiations can be modeled using game theory, in which case the solution is the Nash cooperative or Raiffa-Kalai-Smorodinsky solution (Luce and Raiffa 1966; Malawski et al. 1997; Raiffa 1998; Straffin 2004).

The paper is devoted to the application of multi-criteria optimization in supporting the decision making in negotiation processes. The process of negotiations

K. T. Atanassov et al. (Eds.): IWIFSGN 2020/BOS/SOR 2020, LNNS 338, pp. 204–214, 2022.
https://doi.org/10.1007/978-3-030-95929-6_16

is modeled using the multi-criteria optimization problem. The method of decision selection is based on interactive selection of certain proposals for solutions, i.e. the algorithm requires the reaction of parties during this process. The parties submit their proposals concerning the subjects of the negotiations. These proposals comprise the values of parameters of the multi-criteria optimization problem, and thus the problem is solved. Then, the parties evaluate the solution, accepting or rejecting it. In the latter case, the parties submit new proposals, with the new values of parameters and the problem is solved again for these new parameters. The process of selection of decision is not a one-time but an iterative process involving the parties involved, one which incorporates learning about the negotiated problem.

2 Modeling of the Negotiations Process

The negotiation process is modeled as an interactive decision-making process. Each party presents its proposals for solutions. The negotiation process then consists of seeking a common decision, which reconciles the interests of both parties. The parties attempt to find a common compromise solution. The decision requires the voluntary consent of both parties and is arrived at together, not unilaterally. Both parties have to accept this decision.

During the negotiation process many different purposes are implemented using the same set of feasible solutions. The negotiation process is modeled by introducing a decision variable that describes the solution, as well as two evaluation functions evaluating the solution from the point of view of each party. During the negotiations, each proposal is evaluated by each party by means of its evaluation function. This function is a measure of the satisfaction of a party with a given solution, evaluating the degree of realization of each subject of negotiations by each party. Higher values of this function indicates greater satisfaction from the side of the relevant party; thus each function is maximized. The basis for evaluation and solution selection are two functions of evaluation: the criteria for both parties.

The following are given:

party 1 and party 2 are the parties in the negotiations;
n is the number of subjects for the negotiations;
$x \in X_0$ is a solution: a decision to which the parties are to agree, belonging to the set of feasible decisions $X_0 \subset R^n$, $x = (x_1, x_2, \ldots, x_n)$; each coordinate x_i, $i = 1, \ldots, n$ defines the $i - th$ subject of negotiations;
$f1 : X_0 \to R^1$ is the evaluation function of a decision x by party 1;
$f2 : X_0 \to R^1$ is the evaluation function of a decision x by party 2;

The problem of decision selection has a multi-criteria character. The decision is characterized by a composite evaluation function, wherein the first component is a function of evaluation of the decision by one party, while the second a function of evaluation of the decision by the other party. Each party wants to maximize its evaluation function, but must take into account the existence of the other party. The selection of the solution is accomplished using both evaluation functions.

The negotiation process is considered a problem of multi-criteria optimization with the function of purpose $f = (f1, f2)$:

$$\max_x\{(f1(x), f2(x)) : \quad x \in X_0\} \tag{1}$$

where:

X_0 is the set of feasible decisions;
$x \in X_0$ is a vector of decision variables;
$f = (f1, f2)$ is the vector function which maps the decision space X_0 into the evaluation space $Y_0 \subseteq R^2$.

The function $f = (f1, f2)$ assigns to each decision variable vector $x \in X_0$ an evaluation vector $y = f(x)$, this vector being composed of two elements, y_1 and y_2, which measures the quality of decision x. The formulation of the multi-criteria optimization problem is expressed in the decision space. This is a natural representation of the decision problem, where the goal is to choose the right decision.

The expression $f = (f_1, f_2)$, equivalent, in effect, to the specification of y_1 and y_2, indicates the transformation of the set of feasible decisions X_0 into the set of achievable assessment vectors Y_0.

The multi-criteria optimization model (1) can be rewritten in an equivalent form in the space of achievement vectors. Consider the following problem:

$$\max_x\{(y_1, y_2) : \quad y \in Y_0\} \tag{2}$$

where:

$x \in X_0$ denotes the vector of decision variables;
$y = (y_1, y_2)$ is the achievement vector;
$Y_0 = f(X_0)$ is the set of achievement vectors.

The set of results achieved, Y_0, is given in implicit form, i.e., through the set of feasible decisions X_0 and the mapping of the model $f = (f_1, f_2)$.

The purpose of problem (1) is to help in the selection of such a decision that takes into account the best interests of both parties (Lewandowski and Wierzbicki 1989; Ogryczak 2002; Wierzbicki 1984).

3 An Equitably Effcient Decision

The solution in the negotiation process should satisfy certain properties that the parties accept as reasonable. The solution should be:

- An optimal solution in the sense of Pareto, i.e., such that it would be impossible to improve the solution for one party without making the solution worse for the other party;

- An anonymity solution, i.e., one that should not depend on the way the parties are numbered, so that neither is more important; the parties are treated in the same way in the sense that the solution does not depend on the name of, or other factors specific to, a given party;
- An equalizing solution, i.e., an evaluation vector characterized by lesser variation in terms of coordinates of evaluation is preferred in comparison to a vector with the same number of, but with a greater level of diversity of, coordinates;
- The solution should take the strength of the parties to the negotiations into account.

A decision, which satisfies these conditions is an equitably efficient decision. This is Pareto-optimal decision which satisfies additional conditions – the anonymity property and the transfer principle.

The non-dominated solutions (optimum Pareto) are defined as follows:

$$\hat{Y}_0 = \{\hat{y} \in Y_0 : (\hat{y} + \tilde{D}) \cap Y_0 = \varnothing\} \tag{3}$$

where:

$Y_0 = f(X_0)$ is the set of achievement vectors,
$\tilde{D} = D \setminus \{0\}$ is a positive cone without the top; as a positive cone, one can adopt $\tilde{D} = R^2_+$ (Górecki 2000; Lewandowski and Wierzbicki 1989).

In the multi-criteria problem (1), which is used to select a decision in the negotiation process, the relation of preferences should satisfy additional properties: the anonymity property and the transfer principle.

This preference relation is called an anonymous relation if, for every assessment $y = (y_1, y_2) \in R^2$ and for any permutation P of the set $\{1, 2\}$, the following property holds:

$$(y_{P(1)}, y_{P(2)}) \approx (y_1, y_2). \tag{4}$$

No distinction is made between the results that differ in their arrangement. Evaluation vectors with the same coordinates, but stated in a different manner, are indistinguishable.

The relationship of preferences should satisfy the principle of transfer, meaning that the following condition is satisfied:

for the evaluation vector $y = (y_1, y_2 \in R^2)$:

$$y_{i'} > y_{i''} \Rightarrow y - \varepsilon \cdot e_{i'} + \varepsilon \cdot e_{i''} \succ y \text{ for } 0 < y_{i'} - y_{i''} < \varepsilon. \tag{5}$$

Equalizing transfer is a slight deterioration of a superior coordinate of the evaluation vector and the simultaneous improvement of an inferior coordinate, yielding an evaluation vector which is strictly preferred over the initial evaluation vector. This is the structure of equalizing: the evaluation vector with a lesser degree of diversity of coordinates is preferred in relation to the vector with the same sum, but with a greater degree of diversity of coordinates.

A non-dominated vector, satisfying the anonymity property and the axiom of equalizing transfer is called equitably non-dominated vector. The set of equitably non-dominated vectors is denoted by $\hat{Y}_{0e}$. In a decision space, equitably efficient decisions are specified. The decision $\hat{x} \in X_0$ is called an equitably efficient decision if the corresponding evaluation vector $\hat{y} = f(\hat{x})$ is an equitably non-dominated vector. The set of equitably efficient decisions is denoted by $\hat{X}_{0e}$ (Ogryczak 2002).

The relation of equitable domination can be expressed as the relation of inequality for cumulative, ordered evaluation vectors. This relation can be determined with the use of transformation $\overline{T} : R^2 \rightarrow R^2$, which accumulates the coordinates in decreasing order in the evaluation vector.

The transformation $\overline{T} : R^2 \rightarrow R^2$ is defined as follows:

$$\overline{T}_i(y) = \sum_{l=1}^{2} T_l(y) \text{ for } i = 1, 2 \tag{6}$$

where:

$T(y)$ is the vector with nonicreasing ordered coordinates of the vector y, i.e. $T(y) = (T_1(y), T_2(y))$, where $T_1(y) \leq T_2(y)$ and there is a permutation P of the set $\{1, 2\}$, such that $T_i(y) = y_{P(i)}$ for $i = 1, 2$.

The relation of equitable domination $\succ_e$ is a simple vector domination for the evaluation vectors with coordinates, which are the accumulated values of the ordered evaluation vector (Ogryczak 2002).

The evaluation vector y^1 dominates the vector y^2 in an equitable manner if the following condition is satisfied:

$$y^1 \succ_e y^2 \Leftrightarrow \overline{T}(y^1) \geq \overline{T}(y^2) \tag{7}$$

Solving the problem of decision selection in the negotiation process consists in the determination of an equitably efficient decision that satisfies the preferences of both parties.

4 Scalarization of the Problem

For the determination of equitably efficient solutions of multi-criteria problem (1), a specific multi-criteria problem is solved. This is a problem with the vector function of the cumulative, ordered evaluation vectors, i.e. the following problem:

$$\max_x \{(\overline{T}_1(y), \overline{T}_2(y)) : \quad y \in Y_0\} \tag{8}$$

where:

$y = (y_1, y_2)$ is an evaluation vector;
$\overline{T}(y) = (\overline{T}_1(y), \overline{T}_2(y))$ is a cumulative, ordered evaluation vector,
Y_0 is the set of achievable evaluation vectors.

An efficient solution of multi-criteria optimization problem (8) is an equitably efficient solution of multi-criteria problem (1).

To determine the solution of multi-criteria problem (8), the scalarization of this problem with the scalarizing function $s : Y \times \Omega \rightarrow R^1$ is introduced, namely:

$$\max_x\{(s(y,\overline{y}) : \quad x \in X_0\} \tag{9}$$

where:

$y = (y_1, y_2)$ is an evaluation vector;
$\overline{y} = (\overline{y}_1, \overline{y}_2)$ are the control parameters for individual evaluations.

This is a problem of single objective optimization with a specially introduced scalarizing function of two variables: the evaluation vector $y \in Y_0$ and a control parameter $\overline{y} \in \Omega \subset R^2$. This function is $s : Y \times \Omega \rightarrow R^1$. The parameter $\overline{y} = (\overline{y}_1, \overline{y}_2)$ is available to the parties, enabling them to review the set of equitably efficient solutions.

The optimal solution of problem (9) should be a solution to multiple criteria problem (8). A scalarizing function should satisfy the properties of completeness and sufficiency. The property of sufficiency means that, for each control parameter $\overline{y}$, the solution of the scalarization problem is an equitably efficient solution, i.e., $\hat{y} \in \hat{Y}_{0e}$. The property of completeness means that, through appropriate changes of the parameter $\overline{y}$, any solution $\hat{y} \in \hat{Y}_{0e}$ can be achieved. Such a function completely characterizes the equitably efficient solutions. Inversely, each maximum of such a function is an equitably efficient solution. Each equitably efficient solution can be achieved with certain appropriate values of control parameters $\overline{y}$.

The complete and sufficient parameterization of the set of equitably efficient solutions $\hat{Y}_{0w}$ can be achieved, using the method of the reference point for problem (8). This method makes use of aspiration levels as control parameters. Aspiration levels are values of evaluation function that satisfy the decision maker.

The scalarizing function in the method of reference point is as follows:

$$s(y,\overline{y}) = \min_{1 \leq i \leq 2}(\overline{T}_i(y) - \overline{T}_i(\overline{y})) + \varepsilon \cdot \sum_{i=1}^{2}(\overline{T}_i(y) - \overline{T}_i(\overline{y})) \tag{10}$$

where:

$y = (y_1, y_2)$ is an evaluation vector;
$\overline{T}(y) = (\overline{T}_1(y), \overline{T}_2(y))$ is a cumulative, ordered evaluation vector;
$\overline{y} = (\overline{y}_1, \overline{y}_2)$ is a vector of aspiration levels;
$T(\overline{y}) = T_1(\overline{y}, T_2(\overline{y}))$ is cumulative, ordered vector of aspiration levels;
ε is an arbitrary small, positive adjustment parameter.

Such a scalarizing function is called the function of achievement. The aim is to find a solution that approaches the specific requirements, i.e., the aspiration levels, as closely as possible.

Maximizing this function with respect to y determines the equitably efficient solutions $\hat{y}$ and $\hat{x}$. Note that the equitably efficient decision $\hat{x}$ depends on the aspiration levels $\overline{y}$ (Lewandowski and Wierzbicki 1989; Ogryczak 2002; Wierzbicki 1984).

5 The Set of Negotiations

The aim of the negotiation process is not only to achieve an agreement, even one that is beneficial for both parties, but to arrive at a solution that meets the expectations of parties to the maximum extent possible and, in particular, that is not worse than a solution attainable without negotiations.

Before starting the negotiations, parties should consider what result they will achieve if negotiations are unsuccessful: the *status quo* point. This point is the result which can be achieved by each party without negotiations with the other party. If the parties can achieve the result $ys = (y1s, y2s)$ without negotiations (i.e. party 1 can achieve the result $y1s$, party 2 the result $y2s$), neither of the parties will agree to an inferior result. During negotiations, the parties want to improve the solution in relation to this point. The *status quo* point determines the strength of the parties in the negotiations and their impact on the result.

The set of negotiations is a collection of equitably dominated evaluation values dominating the *status quo* point.

The set of negotiations is therefore defined as follows:

$$B(\hat{Y}_{0e}, ys) = \{\hat{y} = (\hat{y}1, \hat{y}2) \in \hat{Y}_{0e} \wedge \hat{y}1 \geq y1s \wedge \hat{y}2 \geq y2s\} \tag{11}$$

where:

$\hat{y} = (\hat{y}1, \hat{y}2) \in \hat{Y}_{0e}$ is the equitably non-dominated vector;
$ys = (y1s, y2s)$ is the *status quo* point - the result, which can be achieved by both parties without agreement.

The set of negotiations encompasses the points from the set of equitably non-dominated results, which give each party at least as much as it can achieve individually (without negotiations).

The parties wish to find a decision, $\hat{x} \in X_0$, such that the corresponding evaluation vector $\hat{y} = (\hat{y}1, \hat{y}2) = (f1(\hat{x}), f2(\hat{x}))$ belongs to the set of negotiations $B(\hat{Y}_{0e}, ys)$ (Luce and Raiffa 1966; Raiffa 1998).

6 The Method of Selection of the Solution

The solution to the multi-criteria optimization problem (8) is a set of equitably efficient decisions. In order to resolve the problem, one solution should be selected, one that will be evaluated by both parties. Since the solution is a whole set, the parties select the solution with the help of an interactive computer-based procedure, which affords for a controlled overview of the whole set. Each party, participating in the negotiations, determines its proposed solutions in the form of aspiration levels. These are the values of evaluation of individual negotiation issues that each party would like to achieve. These values are control parameters of the scalarizing function. For these values, the system indicates different equitably efficient solutions for analysis which correspond to current values of the control parameters. The aim is to find the solutions which meet, as closely as possible, the specific requirements, i.e., aspiration levels.

The method of decision selection is as follows:

1. The initial arrangements.
2. Iterative algorithm-proposals for further decisions:
   a. Interaction with the system—parties define their proposals for individual subjects of negotiations as aspirations levels $\overline{y}1$ and $\overline{y}2$.
   b. Calculations—identifying another solution from the set of negotiations, $\hat{y} = (\hat{y}1, \hat{y}2) \in B(\hat{Y}_{0e}, ys)$.
   c. Evaluation of the obtained solutions $\hat{y} = (\hat{y}1, \hat{y}2)$—the parties may or may not accept the solution. If not, the parties submit new proposals, providing new values for their aspiration levels $\overline{y}1$ and $\overline{y}2$, and a new solution is determined (see Sect. 2.2).
3. Determination of the decision that meets the requirements of both parties.

A decision selection is not a single act of optimization, but a dynamic process of searching for solutions. This means that the parties learn, and can change their preferences, during the process. Comparing the results of the evaluation $\hat{y}1$ and $\hat{y}2$ to their aspiration points $\overline{y}1$ and $\overline{y}2$, we see that each party has information about what is and what is not achievable, and how far the parties' proposals $\overline{y}1$ and $\overline{y}2$ are from the possible solutions $\hat{y}1$ and $\hat{y}2$. This enables the parties to make appropriate modifications in their proposals and to arrive at their new aspiration points. These levels of aspiration are determined adaptively during the learning process. The process ends, when the parties find a decision enabling them to achieve results that meet their aspirations or that are, in a sense, sufficiently close to these aspirations.

The method of decision selection is shown in Fig. 1.

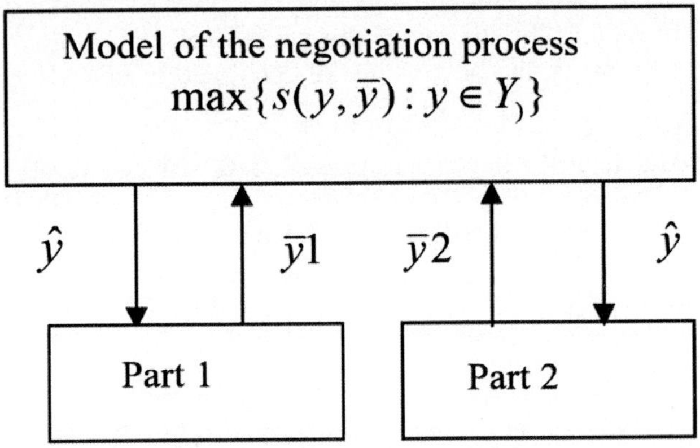

Fig. 1. The method of decision selection

This method imposes no rigid scenarios on the parties and permits them to change their preferences while solving the problem. As we see, parties learn about the problem during the negotiations. The computer does not replace the parties in the selection of a solution. It should be observed that the entire process of solution selection is controlled by both parties. It aims to find a solution that satisfies both parties.

7 An Example of Negotiations

The following example illustrates the method of decision selection in the process of negotiations (Górecki 2000).

The negotiation problem is as follows:

party 1 and party 2 are the parties participating in the negotiations;
$n = 2$ is the number of subjects to be negotiated;
$x = (x_1, x_2) \in X_0$ is a decision, to which the parties are to agree and which belongs to the set of feasible decisions $X_0 \subset R^2$, x_1 is the decision concerning the first subject of negotiations, x_2 is the decision on the second subject of negotiations;
$X_0 = \{(x_1, x_2) \in R^2 : -2 \cdot x_1 + 3 \cdot x_2 \leq 27, 6 \cdot x_1 + 7 \cdot x_2 \leq 175, 0 \leq x_1 \leq 21, 0 \leq x_2 \leq 13\}$ is the set of feasible decisions;
$f1 : X \to R^1 f1(x) = \frac{20}{21}x_1 - \frac{2}{3}x_2$ is an evaluation function of decision x by party 1;
$f2 : X \to R^2 f2(x) = -\frac{4}{21}x_1 + \frac{2}{3}x_2$ is an evaluation function of decision x by party 2;
$ys = (ys1, ys2) = (10, 1)$ is the *status quo* point.

As a first step of the multi-criteria analysis, a single-criterion optimization of the evaluation function of each party is performed. The result is a matrix of implementation goals, incorporating the values of all of the criteria of each party, obtained by means of the solution of two single-criterion problems. This matrix enables the estimation of the range of changes of each evaluation function on the feasible set, as well as provides some information about the conflicting nature of the evaluation functions. The matrix of the implementation of goals generates a utopia vector representing the best values of each of the separate criteria.

Table 1. Matrix of the implementation of goals with the utopia vector

Optimized criterion	Solution	
	$y1$	$y2$
Evaluation by party 1 $y1$	20.00	−4.00
Evaluation by party 2 $y2$	−2.85	7.52
Utopia vector	20.00	7.52

Source: The author's own calculations

Table 2. Interactive analysis of the search for a solution

Iteration	Evaluation of party 1 $y1$	Evaluation of party 2 $y2$
1. Aspiration point $\overline{y}$	20	7.52
Solution $\hat{y}$	15.33	0.66
2. Aspiration point $\overline{y}$	15	7
Solution $\hat{y}$	14.99	0.83
3. Aspiration point $\overline{y}$	14	6.5
Solution $\hat{y}$	13.33	1.33
4. Aspiration point $\overline{y}$	13	6.5
Solution $\hat{y}$	12.99	1.83
5. Aspiration point $\overline{y}$	12	5
Solution $\hat{y}$	11.99	2.33

Source: Own calculations

Table 1 shows the clear advantage of party 1 in the negotiations here considered.

At the beginning of the analysis, the parties specify their preferences in terms of the aspiration points, composing the utopia vector. The resulting solution clearly prefers the first party and is not acceptable for the other party, being less desirable than its *status quo* point. To improve the solution, the first party reduces its requirements in the next iteration. The result is a slight deterioration of the solution for the first party and a slight improvement of the solution for the other party. The solution is still less desirable than the *status quo* point of the other party. In subsequent iterations, the first party again reduces its requirements and the obtained solutions are now more desirable than the *status quo* point of the other party, and thus are more rewarding. The analysis shows that the solution depends significantly on the first party, which has a stronger position in negotiations and can impose a solution. For iteration 5, the relevant decisions are as follows: $\hat{x}^5 = (18.81, 8.88)$ (Table 2).

Final selection of a specific solution depends on the parties' preferences. The example shows that this method enables the parties to explore the capabilities of decision selection during the interactive analysis and to seek a mutually satisfactory solution.

8 Conclusions

The paper presents a method of modeling a process of negotiations in the form of a multi-criteria optimization problem, a process that is then used to support the decision selection. Modeling the negotiation process as a multi-criteria optimization problem enables us to create the variants of a decision and to track their consequences.

The method of interactive analysis, based on the reference point, is applied to a multi-criteria problem with a cumulative, ordered evaluation vector. This enables us to arrive at solutions, which are tailored to the parties' preferences. The numerical example shows that the proper computational problem can be solved efficiently by standard optimization software and illustrates the potentialities of the traction of the parties.

This procedure does not determine the final solution, but supports and teaches the parties about the specific negotiation problem. The final decision is to be made by the parties involved in the negotiations.

References

Fisher, R., Ury, W., Patron, B.: Getting to YES. Negotiating agreement without giving in (in Polish). PWE, Warsaw (2002)

Górecki, H.: Optimization and control of dynamic systems (in Polish). Institutional Educational Science Publishing House of the Academy of Mining and Metallurgy, Kraków (2000)

Lewandowski, A., Wierzbicki, A.: Aspiration Based Decision Support Systems. Lecture Notes in Economics and Mathematical Systems, vol. 331, Springer, Heidelberg (1989). https://doi.org/10.1007/978-3-662-21637-8

Luce, D., Raiffa, H.: Games and decisions (in Polish). PWN, Warsaw (1966)

Malawski, M., Wieczorek, A., Sosnowska, H.: Competition and Cooperation. Game Theory in Economics and the Social Sciences (in Polish). PWN, Warsaw (1997)

Ogryczak, W.: Multicriteria optimization and Decisions under Risk. Control and Cybernetics, vol. 31, no. 4 (2002)

Raiffa, H.: The Art and Science of Negotiations. Harvard University Press, Cambridge Mass (1998)

Straffin, Ph.D.: Game Theory (in Polish). Scholar, Warsaw (2004)

Young, H.P.: Equity In Theory and Practice (in Polish). Scholar, Warsaw (2003)

Wachowicz, T.: E-negotiations. modeling, analysis and support. Karol Adamiecki University of Economics in Katowice, Katowice (2006)

Wierzbicki, A.: Negotiation and Mediation in Conflicts: II. Plural Rationality and Interactive Decision Processes. In: Grauer M., Thompson M., Wierzbicki A.P. (eds) Plural Rationality and Interactive Decision Processes. Lecture Notes in Economics and Mathematical Systems, vol. 248. Springer, Berlin (1985). https://doi.org/10.1007/978-3-662-02432-4_9

A Comparative Study of Two Nature-Inspired Algorithms for Routing Optimization

Hubert Zarzycki[1](✉), Dawid Ewald[2], Oskar Skubisz[3], and Piotr Kardasz[4]

[1] Tadeusz Kosciuszko Military Academy of Land Forces in Wroclaw, Wroclaw, Poland
hubert.zarzycki@awl.edu.pl

[2] Institute of Technology, Department of Computer Science, Kazimierz Wielki University, Bydgoszcz, Poland
dawidewald@ukw.edu.pl

[3] Tadeusz Kosciuszko Military Academy of Land Forces in Wroclaw, Wroclaw, Poland
oskar.skubisz@uwr.edu.pl

[4] Development and Innovation Research Foundation, Wroclaw School of Information Technology, Wroclaw, Poland
p.kardasz@kbri.pl

Abstract. The paper presents and compares two algorithms inspired by nature - genetic and ant. They were applied to the problem of determining the optimal route. After the implementation of the algorithms, tests were carried out and the results were correct from the point of view of the solution sought. Then the results were analyzed and the advantages and disadvantages of both algorithms were indicated. The article helps to better understand which of the analyzed algorithms is better suited to specific tasks.

Keywords: Ant algorithm · ACO · Genetic algorithm · Routing optimization · Traveling salesmen problem

1 Introduction

The beginnings of the evolution of living organisms on earth are dated back to around 3.8 billion years. Countless species of animals and plants began and ended in this period. New organs, behavioral models and adaptive abilities were developed that ensured survival in often extreme climatic conditions. Over time, these mechanisms have been constantly transformed, achieving a higher and higher degree of perfection. Currently, the natural world is full of empirical data, natural solutions, algorithms and techniques, the discovery and application of which in practice can bring many tangible benefits.

Progressive technological development more and more often makes people reflect on the use of phenomena, features and models observable in the natural world in usable areas. Selected features of the functioning of the nervous system, selected models of behavior and some of the features of the functioning of living organisms are currently the basis for creating various types of artificial intelligence algorithms, including those used in the optimization of various types of tasks.

K. T. Atanassov et al. (Eds.): IWIFSGN 2020/BOS/SOR 2020, LNNS 338, pp. 215–228, 2022.
https://doi.org/10.1007/978-3-030-95929-6_17

The article analyzes selected algorithms inspired by nature. The potential of such algorithms is endless and can be used in many areas such as logistics, telecommunications, medicine, finance [33, 35], mechanics, power engineering [7], computer science [23], AI controls [11, 13, 14, 36] etc. It is because each of the elements of the nature that surrounds us can be used in the construction of optimization algorithms [1]. These algorithms adapt to the natural environment to increase the chance of survival. They are looking for the most effective way to optimize all sorts of complex tasks. This type of algorithm includes algorithms such as the genetic algorithm (GA), ant and bee algorithms based on swarm intelligence (SI) [15, 18, 24, 27], including ant colony optimization (ACO) [2, 5, 9, 22], bee colony (ABC) [16–18] and particle swarm optimization (PSO), bat algorithm (BA), pigs [8], firefly algorithm firefly, gravity algorithm and neural networks. The article presents a comparison of the ant algorithm from the area of swarm intelligence and the genetic algorithm.

2 Genetic Algorithm

The basis of the concept of nature-inspired algorithms is an ingeniously simple assumption. By using the processes taking place in biological cells of organisms, it is possible to program using methods similar to computational algorithms used in computer systems. Based on this assumption, scientists managed to create a programming method for logical elements written on nucleic acids. The DNA code is digital, but not binary. Binary code uses the values 0 and 1 (which is why it is called binary), DNA code uses 4 values: T, C, G and A. While a binary byte mainly consists of eight binary digits, a DNA "byte" (called a codon) contains three signs. Since each character can be assigned one of four values, the DNA codon has sixty-four possible values, as opposed to binary code where there are two hundred and fiftysix binary values [32].

The GA genetic algorithm uses the stochastic search method. The principle of operation is based here on the natural competition between individuals of a given group to take control of limited natural resources. The success of the winner is usually determined by his gene pool, and the preservation of species continuity by such individuals results in the spread of their genes. Through the selective selection and reproduction of individuals with the best gene pool, the population will have a chance to obtain more and more natural resources more effectively. The individuals best adapted to the environment gain a relatively greater chance of reproducing their offspring. Individuals that are poorly adapted or have limited access to resources will have a limited population size, thus minimizing the chance of their survival. According to this assumption, the gene pool derived from adapted or highly adapted individuals will spread among an increasing population of individuals with each subsequent generation. The right configuration of good characteristics, passed on by different fathers, can sometimes lead to the appearance of an individual with characteristics far beyond the average achieved by other descendants. This individual will be characterized by a greater or more perfect number of adaptive traits than the set of adaptive traits of one of his parents. It should therefore be recognized that a group of individuals maintaining species continuity will develop better and better and will adapt to their place of existence more and more effectively. AG are intended to use this simple analogy.

These algorithms cover the entire population of individuals, each of which represents a possible decision to solve a given problem. Each individual is evaluated separately on the basis of their adaptive characteristics, according to the degree or parameter determining the most favorable result or the highest utility for the entire process. For example, a measure of adaptation could be the force-to-weight ratio for a "bridge" design. In nature, this is equivalent to judging how efficient an organism is in competing for resources. Individuals with the most effective set of adaptive traits will pass these traits to their offspring by "crossing" with other individuals in the population. This results in the emergence of new individuals that share common features inherited from their parents. Less adapted individuals with a smaller pool of desirable genes will reduce their numbers. This will result in their gradual disappearance from the population over time. In this way, a new group of individuals develop. Successive generations of individuals have more interdependent features than the previous generation. Thus, during a generational change, the good characteristics will be passed on and will develop throughout the population. Crossing the most adapted individuals leads to the fact that the most prospective plots of the exploration area are examined. Anyway, the population will encounter optimal solutions to the tasks. There are many ways to implement the idea of biological evolution in AG.

The standard algorithm is divided into three steps:

1. Crossbreeding.
2. Selection of individuals.
3. Formations of the new generation. Adaptation assessment is conducted. These steps are repeated until the result is satisfactory or until one of the following conditions occurs:
4. The number of generations (cycles) will reach its maximum value.
5. Time is up.

Creating a New Population. An initial population is formed which, quite possibly, will turn out to be in poor condition. However, there is great confidence that the algorithm will correct this problem. They must be consistent with the assumptions and adapted to reproduction.

Multiplication. All parents reproduce. Two parents are required to obtain a descendant. The descendant (child) inherits the characteristics of the parents.

Mutation. Mutations are similar to the reproductive process. A number of individuals are selected and preselected for certain operations.

Choice. Then we start choosing those who will survive. The fate of the survivors after the selection is predetermined, indicating in the form of a parameter. The remaining individuals must be removed.

For the traveling salesman task, a model was chosen in which 30% (thirty) of agents die, and in their place the best ones come with some modifications.

3 The Ant Algorithm

Ants interact with each other to effectively get food and carry out other tasks. There are no dominant individuals in the colony, no overseers or executors, no individuals who give instructions and coordinate actions. The colony is completely self-organized. Each of the ants only knows the local situation. Neither knows about the overall situation in the colony. An ant knows only what it has learned by itself or obtained directly or implicitly from nearby ants.

The implicit interactions of ants, called stigmergies, make it possible to find a short path from an anthill to a food source. Every time ants move from an anthill to their food, they leave a trail of pheromones. Other ants, sensing such marks on the ground, will instinctively follow them. Subsequent ants passing by also leave a trail of pheromones. So the more ants pass a certain path, the more attractive this path will be. The shorter the path to the food source, the less time it takes for an ant to follow this path. For this reason, the marks left on the shortest path are clearer (figure below). Each ant remembers the nodes it has passed through. The nodes visited in the ant's memory constitute a taboo lists. When selecting a node for the next step, the ant ignores all the nodes it has visited. So, after each edge crossing, the prohibition list is updated with a new node. Before the next iteration (generation) of the algorithm, the ant's memory is reset and all nodes are again available for visiting (Fig. 1).

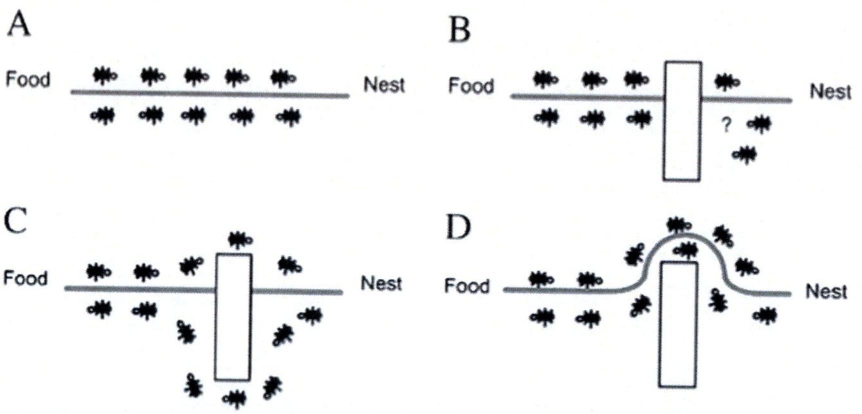

Fig. 1. Visualization of the ant algorithm, source: https://medium.com/@aengel/howto-optimize-flow-shop-problem-with-ant-colony-optimization-fdea9443149c.

In addition to the tabu list, when choosing the next edge to cross, the ant is guided by the attractiveness of the edge. This attractiveness depends on two factors. The first is the distance between the knots (edge weight), the second is the aggregate pheromone marks left on the edge by ants from earlier generations. Of course, unlike the edge weights, which are constant, the pheromone traces are updated with each iteration (generation) of the algorithm. By walking through the path, the ants strengthen the pheromone trail on it. Additionally, just like in nature, it is assumed that pheromones gradually evaporate over time [3, 4, 10, 20, 21].

Formulas describing the classical ant algorithm are given below Edges:
The ant will move from the node ‘i’ to the node ‘j’ with the probability:

$$p_{i,j} = \frac{\left(T_{i,j}^{\alpha}\right)\left(N_{i,j}^{\beta}\right)}{\sum\left(T_{i,j}^{\alpha}\right)\left(N_{i,j}^{\beta}\right)} \tag{1}$$

Where: $T_{i,j}$ - is the number of pheromones at the edge i, j;
α - parameter controlling the influence of $T_{i,j}$;
$N_{i,j}$ – edge i, j attractiveness i, j (equal to $1/d_{i,j}$, where d is the distance);
β - parameter controlling the influence of $N_{i,j}$.
After successfully completing the route, the ant leaves a mark on all edges it has traveled. This footprint is inversely proportional to the distance traveled:

$$T_{i,j} = (1-p)T_{i,j} + \Delta T_{i,j} \tag{2}$$

Where: $T_{i,j}$ - is the number of pheromones on the edge i, j;
p - pheromone evaporation rate;
$\Delta T_{i,j}$ - the amount of deposited pheromones, usually described as in the Eq. (3).

$$\Delta T_{i,j}^{k} = \begin{cases} 1/L_k \\ 0 \end{cases} \tag{3}$$

Where: L_k - value of the k-th ant (usually a distance).
It is worth emphasizing that pheromones evaporate. This means that the intensity of the pheromones at all edges decreases with each iteration of the algorithm. Finally, each iteration must update the pheromone intensity values.

Ant algorithms are used in many areas, including transport, technical and economic optimization. They are widely used in the areas of logistics, where their optimal way of searching for the shortest paths to reach the destination has been appreciated. The ant algorithm uses modeling based on the method of the ant moving along a selected path. This approach may find application in finding rational solutions to the problems of optimization of graph interpretation. Based on the results of a number of conducted experiments, it can be concluded that the effectiveness of using ant algorithms increases with the increase in the dimension of the optimization problems being solved. Good results are now obtained for non-stationary systems characterized by time-varying parameters, for example for the calculation of telecommunication and computer networks. The ant algorithm was also used to develop the optimal structure of the GPS research network as part of creating precise geodetic technologies. Currently, thanks to the use of ant algorithms, good results have been obtained in solving optimization problems, such as network traffic optimization problems, traveling salesman problem, transport problem, scheduling problem, square assignment problem, graph coloring problem and many others.

4 Route Optimization

There are many types of route problems including vehicle routing, logistics, network transmission, telecommunication etc. [25, 28, 29]. They need appropriate computer equipment to carry out simulations and calculations [6, 34]. In order to carry out the research described in this work, specialized software was prepared. The application was solving the traveling salesman problem. This problem is one of the best known combinatorial optimization problems. The main difficulty is finding the optimal route, assuming that the road goes through all the specified cities at least once back to the original starting point. One of the conditions for solving the problem is the profitability criterion of the route (the shortest, cheapest, etc.) The second condition is the appropriate distance and cost matrices, etc. It was assumed that the route must pass through each city only once - in this case, the Hamiltonian cycles were used (the cycle covering all the vertices of the graph, a cycle that passes through each vertex of a given graph exactly once.). The traveling salesman problem is about optimization and is one of the key challenges for the logistics and transport planning industry.

Similarly, using the genetic algorithms and the ant algorithm, it is possible to perform similar calculations. This simulation aims to map natural heuristic processes. The algorithms in this case do not sort all options. Their goal is to improve the existing conditions on the basis of previously gathered experience.

Nine cities acting as checkpoints were introduced into the program. Each city was associated with the remaining nine connections. Along with shortening the distance to the last city, the length of the route increases.

Using this method helps to carry out checks without complicated calculations in order to check the shortest distance that can be traveled. The achieved results allow for comparison with non-optimal algorithm results. The study adopted such a system of cities that the best route was determined by points formed in the order from city1 to citymax. For example, the most optimal route should include cities 1, 2, 3, 4, 5, 6, 7, 8, 9. Based on the following equation, a distance was created for each pair of cities:

$$distance = (cityFrom + cityTo) * range$$

- distance - distance between selected cities.
- cityFrom - the city from which the distance is checked.
- cityTo - the city to which the connection will be made.
- range - by how much the distance between cities will be increased.

Thanks to this method, it is possible to evenly increase the distance between cities, while maintaining control over the data obtained during the tests. It was assumed that cities change distance by twenty points. Where the distance between the first and second city is twenty, and between the first and third - forty, etc. For the genetic algorithm, a criterion was adopted according to which the selected distance between cities determined the attractiveness of the route. And in the ant algorithm, in addition to the distance, an additional criterion was introduced. The pheromones that the ants leave on the path were used as a model here, it is checked whether there are pheromones left by other ants traveling earlier.

In order to simplify the calculations in the program and to limit the amount of stored data, the program does not register connections with all cities. It was only necessary to register new connections for the selected city, because we have access to previously created data for this city. For example, for city number three: city number one and city number two will not be checked because the connection of these cities to city number three has already been investigated.

5 Analysis of the Obtained Results

In the practical part, the C # language and the Microsoft Visual Studio 2019 environment were used. The algorithms were written in a console project.

For each of the two analyzed algorithms, the city of origin (zero) was indicated. The first test was to examine the speed of one iteration to determine which of the algorithms is faster and to show how long it takes to run additional iterations. The obtained result showed that both algorithms successfully solved the problem. The data has been optimized and the best paths in the graph have been identified. However, the results showed that the execution time and the accuracy of the data obtained using these algorithms differ.

At the stage of analyzing the data obtained as a result of comparing both algorithms, it was found that the optimal results were obtained using the ant colony optimization algorithm. This algorithm required fewer iterations. The tables below present the data collected while working with implemented algorithms.

Table 1. Genetic algorithm. Iteration × 100.

Generation no.	Distance before generation	Distance after generation	Difference	Error%	Time
1	*900*	*900*	*0*	*100*	*0.00991*
2	*900*	*800*	*100*	*44,4*	*0.00849*
3	*920*	*920*	*0*	*100*	*0.00849*
4	*920*	*920*	*0*	*100*	*0.00990*
5	*840*	*740*	*100*	*16.6*	*0.01274*
Average:				**72.20**	**0.00991**

Table 2. Genetic algorithm. Iteration × 1000.

Generation no.	Distance before generation	Distance after generation	Difference	Error%	Time
1	*840*	*780*	*60*	*50*	*0.01141*
2	*840*	*800*	*60*	*83,3*	*0.01283*
3	*860*	*860*	*0*	*100*	*0.01137*
4	*860*	*820*	*40*	*71*	*0.01288*
5	*880*	*760*	*120*	*25*	*0.01283*
Average:				**65,9**	**0.01226**

Table 3. Genetic algorithm. Iteration × 100 000.

Generation no.	Distance before generation	Distance after generation	Difference	Error%	Time
1	*900*	*840*	*60*	*66*	*0.34397*
2	*880*	*780*	*100*	*37*	*0.38098*
3	*880*	*840*	*40*	*75*	*0.33612*
4	*860*	*860*	*0*	*100*	*0.34311*
5	*840*	*800*	*40*	*66*	*0.35299*
Average:				**69.1**	**0.35143**

Table 4. Ant algorithm. Iteration × 100.

Generation no.	Distance before generation	Distance after generation	Difference	Error%	Time
1	*860*	*760*	*100*	*28*	*0.03501*
2	*860*	*740*	*120*	*14*	*0.05250*
3	*800*	*740*	*60*	*25*	*0.05520*
4	*800*	*720*	*80*	*0*	*0.06597*
5	*820*	*720*	*100*	*0*	*0.04847*
Average:				**13.5**	**0.05143**

Table 5. Ant algorithm. Iteration × 1000.

Generation no.	Distance before generation	Distance after generation	Difference	Error%	Time
1	*820*	*720*	*100*	*0*	*0.08886*
2	*860*	*860*	*0*	*100*	*0.09559*
3	*760*	*860*	*−100*	*350*	*0.09424*
4	*860*	*880*	*−20*	*114*	*0.09559*
5	*820*	*720*	*100*	*0*	*0.12386*
Average:				**112.8**	**0.09963**

Table 6. Ant algorithm. Number of iterations depends on the speed of finding the result.

Generation no.	Distance before generation	Distance after generation	Difference	Error%	Time
1	*880*	*720*	*160*	*0*	*0.02827*
2	*820*	*720*	*100*	*0*	*0.01481*
3	*840*	*720*	*120*	*0*	*0.02153*
4	*860*	*720*	*140*	*0*	*0.02018*
5	*760*	*720*	*40*	*0*	*0.01076*
Average:				**0**	**0.01911**

The Tables 1, 2, 3, 4, 5 and 6 present the results of genetic and ant algorithms. The data covers five generations. In order to enable the precise analysis of the obtained results, information was collected during one generation for one hundred, one thousand and one hundred thousand iterations. The "Distance before generation" represents the distance obtained at the beginning and how the algorithm optimized this path in the "Distance after generation" column. "Difference" - shows for how many units the road is optimized. "Error" - the percentage value indicates what error occurred after the implementation of the algorithm. 100% - "Distance after generation", 0% - the minimum possible received distance. "Time" is the computation time until the last iteration. In the ant algorithm, the agent has to go through nine cities in one iteration. In the case of a genetic algorithm, it is the life cycle of one generation (birth, death, mutation). The test results showed that the use of the genetic algorithm achieves a better result of the average time needed to complete one iteration than the use of the ant algorithm. For the genetic algorithm, the elapsed time measured in milliseconds, for one generation it is 0.00991 (with 100 iteration times), and for an ant −0.03501, the difference is 0.02509 ms (three and a half times more). With one generation and a small number of iterations, the difference seems small and imperceptible to notice, but if more iterations are performed, the difference will be noticeable. At a thousand repetitions, the genetic algorithm showed an average of 0.01226 ms, while the ant algorithm showed 0.09963 ms, which is nearly nine times slower. This is a significant difference with a relatively small number of iterations. With the increase in the number of iterations, the difference remained in the same proportion (Fig. 2).

For the ant algorithm, no data was given to the table when iterating at ×100000. In the case of ant algorithm, the data is mixed and is degraded from the ×1000 iteration, and the percentage of errors increases with increasing iteration. Table 5 shows this error where for the third and fourth generation an error of 350% and 114% was achieved. In this case, further tests were not justified. Instead of ×100000 for the ant algorithm, the "break" method was used, which turned off the calculations when the algorithm finds the correct answer. In both cases, when calculating the algorithms, the processor load was the same and amounted to about 14%, and the memory load was 12.7 MB. The second test was to demonstrate the accuracy of the data when working with algorithms. At this stage, the results turned out to be inconsistent. The ant algorithm achieved surprisingly

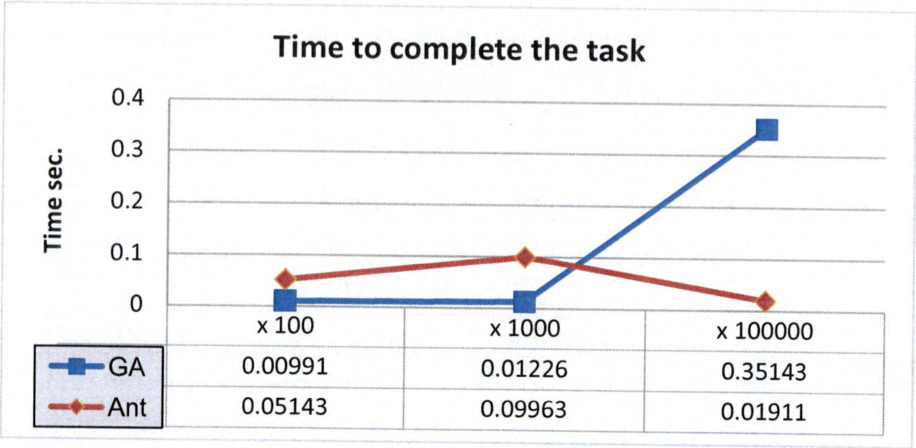

Fig. 2. Time of task execution with both algorithms at different number of iterations. (For the ant algorithm, with ×100000 given, break occurs earlier when the correct answer is found)

good results in this respect. To achieve the result indicated in the task, up to six iterations are required with five generations of the algorithm. Based on the results of the previous analysis, the genetic algorithm turned out to be nine times faster and can calculate more iterations than the ant algorithm. However, the genetic algorithm does not find the optimal result. Instead, it provides the closest answer it can find. For this reason, it is difficult to define specific stop conditions for the GA algorithm and it should be allowed to calculate to the last specified iteration. The conducted research shows that the genetic algorithm has a high percentage of incorrect data, about 60% - the ant algorithm performs much better in this regard.

It is worth noting that during the route discovery process, the ants choose the most attractive nodes (near the optimal node) and discard nodes with low attractiveness (far from the optimal node). As a result of this process, the following effect is achieved. The number of optimal paths from one generation to the next will evolve rapidly [30]. The ant algorithm will get correct results faster than the genetic algorithm. With one hundred iterations, the mean error is 13.5%. Although the ant algorithm requires an average of six iterations, it rarely requires more computation to get the optimal result. In some cases, the error is due to the fact that the answer has been found, for example, in the sixth iteration, and the algorithm continues to compute until the assumed hundred iterations are met and the result appears as for the ×1000 case. As you go through a thousand iterations, you can see that the error percentage is increasing very quickly. For the ×100000 index, the ant algorithm has a break method that turns off the computation because the ants find the shortest path and the error rate is 0%.

6 Comparison of Algorithms and Conclusions

The obtained results clearly show that the genetic algorithm is less computationally intensive than the ant algorithm. This index did not help the genetic algorithm to get the best results faster [26]. The use of the ant algorithm means that more accurate results can be obtained. The ant algorithm was able to get the correct answer it was looking for with relatively few iterations. The use of the genetic algorithm for similar conditions did not lead to the optimal result. The conclusions indicate the necessity to modify the GA algorithm so that it is possible to obtain results closer to or equal to the optimal one.

The results obtained using the ant algorithm are more accurate than the data obtained using the genetic algorithm. The ant algorithm showed the best possible path between cities, while allowing to travel this path with the smallest sum of distance units. The use of the genetic algorithm has produced many good solutions, but they are not better than those obtained from the ant algorithm.

The implementation of the algorithms is not difficult, but their application should take into account various aspects of their operation. The ant algorithm is characterized by the need to store a lot of information about the state of ants and cities that have already been visited. The amount of memory resources needed increases with the complexity of the given conditions. The genetic algorithm is less demanding in this respect. To create the next generation of agents, you need to modify their own genes or remove agents from the generation list. It is also worth emphasizing that the GA algorithm may have a large number of possible modifications. To solve a specific problem more efficiently, you can try to modify the algorithm with different parameters.

Each of the two algorithms under consideration has its advantages and disadvantages. The best and worst sides of the algorithms and their possible uses are discussed below.

The ant algorithm will work well where the goal of the task is to find the most attractive path, for example for the use of artificial intelligence in games or for logistics planning. This algorithm will provide precise data on the optimal routes in a relatively short time. Difficulty may arise during the implementation phase as this algorithm must, by definition, store and update the complete data for each ant. In a similar task, the result obtained using the genetic algorithm has worse data accuracy. The GA algorithm is justified when the goal of the task is not to find the shortest path, but to quickly find many useful paths that can be used alternatively. The ant algorithm will usually show the best route, but if over time for some reason it is not possible to follow this path, the result becomes outdated, then the ant algorithm must be restarted to calculate the shortest path adapted to the changed conditions. It will be different when using the genetic algorithm where this problem will not appear at all. The GA algorithm will offer several ways to solve the problem at once. They will not be the best, but they will be optimized and, if necessary, you can choose from several routes. Such functionality could be useful in real-time [12, 31] and control systems where a set of good results and not necessarily optimal results are expected quickly. So, if the goal of the task is anything other than finding the shortest path, then the ant's algorithm will have to adapt to the new task. In such a situation, the ant will still have to go through all the key points specified in the condition. The connection of the points must guarantee the digital ant the ability to move between them and leave a chemical marker - pheromone for subsequent agents.

The genetic algorithm does not introduce such limitations. Its job is to determine who will have the best number of chromosomes, select the best agents, then cross them and expand the colony further. The creator of such an algorithm should have no problem using the logic of the algorithm to solve various problems. In addition, genetic algorithms have many possibilities for modification in a set of different methods of crossing and mutating factors. For example, when looking for an attractive path, a genetic algorithm is used that develops its colony as microorganisms, that is, they do not intersect, but bud. But there is one caveat: the descendant must have slightly different genes. And the classic implementation means that the agents will cross over as we observe in nature. And for tasks not related to finding the most attractive path, consider using a genetic algorithm. Because this algorithm will be easy to run. While the ant algorithm requires extra effort to perform adaptation to specific conditions.

The speeds of the algorithms vary considerably. Tests have shown that the ant algorithm takes longer to complete one simulation than the genetic algorithm. The change is that the accuracy of the data in the ant algorithm is higher, it requires fewer simulations to get a better path. Since a genetic algorithm will often not provide the best results, the user of the algorithm cannot know how good the data obtained in a hundred iterations is. In two hundred iterations, the data may be much better, but there is a chance that it may not improve or improve slightly.

References

1. Apiecionek, Ł, Zarzycki, H., Czerniak, J.M., Dobrosielski, W.T., Ewald, D.: The cellular automata theory with fuzzy numbers in simulation of real fires in buildings. In: Atanassov, K.T., et al. (eds.) IWIFSGN 2016. AISC, vol. 559, pp. 169–182. Springer, Cham (2018). https://doi.org/10.1007/978-3-319-65545-1_16
2. Bianchi, L., Gambardella, L.M., Dorigo, M.: An ant colony optimization approach to the probabilistic traveling salesman problem. In: Guervós, J.J.M., Adamidis, P., Beyer, H.G., Schwefel, H.P., Fernández-Villacañas, J.L., (eds) Parallel Problem Solving from Nature — PPSN VII. PPSN 2002. LNCS, vol. 2439. Springer, Heidelberg (2002). https://doi.org/10.1007/3-540-45712-7_85
3. Colorni, A., Dorigo, M., Maniezzo, V.: Distributed Optimization by Ant Colonies, Politecnico di Milano (1991)
4. Czerniak, J.M., Dobrosielski, W.T., Zarzycki, H., Apiecionek, Ł.: A proposal of the new owlANT method for determining the distance between terms in ontology. In: Filev, D., et al. (eds) Intelligent Systems 2014. AISC, vol. 323. Springer, Cham (2015). https://doi.org/10.1007/978-3-319-11310-4_21
5. Czerniak, J.M., Apiecionek, Ł., Zarzycki, H.: Application of ordered fuzzy numbers in a new OFNAnt algorithm based on ant colony optimization. In: Kozielski, S., Mrozek, D., Kasprowski, P., Małysiak-Mrozek, B., Kostrzewa, D., (eds) Beyond Databases, Architectures, and Structures. BDAS 2014. CCIS, vol. 424. Springer, Cham (2014). https://doi.org/10.1007/978-3-319-06932-6_25
6. Czerniak, J.M., Zarzycki, H., Apiecionek, Ł., Palczewski, W., Kardasz, P.: A Cellular Automata-Based Simulation Tool for Real Fire Accident Prevention Mathematical Problems in Engineering (2018)
7. Czerniak, J.M., Zarzycki, H., Dobrosielski, W., Szczepański, J.: New fuzzy numbers comparison operators in energy effectiveness simulation and modeling systems. In: ECMS 2018 Proceedings European Council for Modeling and Simulation (2018)
8. Czerniak, J.M., Zarzycki, H., Ewald, D., Augustyn, P.: Application of OFN numbers in the artificial duroc pigs optimization (ADPO) method. In: Atanassov, K., et al. (eds) Uncertainty and Imprecision in Decision Making and Decision Support: New Challenges, Solutions and Perspectives. IWIFSGN 2018. AISC, vol. 1081. Springer, Cham (2021). https://doi.org/10.1007/978-3-030-47024-1_31

9. Dorigo, M., Stutzle, T.: Ant Colony Optimization, p. 69. The MIT Press Cambridge Massachusetts Institute of Technology, London (2004)
10. Dorigo, M., Gambardella, M.L.: Ant colony system: a cooperative learning approach to the traveling salesman problem. IEEE Trans. Evol. Comput. **1**(1), 53–66 (1997)
11. Dobrosielski, W.T., Szczepański, J., Zarzycki, H.: A proposal for a method of defuzzification based on the golden ratio—GR. In: Atanassov, K., et al. (eds) Novel Developments in Uncertainty Representation and Processing. AISC, vol 401. Springer, Cham (2016). https://doi.org/10.1007/978-3-319-26211-6_7
12. Dobrosielski, W.T., Jacek, M., Czerniak J.M., Zarzycki, H., Szczepański, J.: Fuzzy Numbers applied to a heat furnace control, theory and applications of ordered fuzzy numbers - A Tribute to Professor Witold Kosiński. Studies in Fuzziness and Soft Computing, vol. 356. Pp. 269–288, Springer, Cham (2017). https://doi.org/10.1007/978-3-319-59614-3_16
13. Dobrosielski, W.T., Czerniak, J.M., Szczepanski, J., Zarzycki, H.: Triangular Expanding, a new defuzzification method on ordered fuzzy numbers. In: Kacprzyk, J., Szmidt, E., Zadrożny, S., Atanassov, K., Krawczak, M. (eds.) Advances in Fuzzy Logic and Technology 2017. EUSFLAT 2017, IWIFSGN 2017. AISC, vol. 642, pp. 605–619, Springer, Cham (2018). https://doi.org/10.1007/978-3-319-66830-7_54
14. Dobrosielski, W.T., Czerniak, J.M., Szczepański, J., Zarzycki, H.: Triangular expanding, a new defuzzification method on ordered fuzzy numbers. In: Kacprzyk, J., Szmidt, E., Zadrożny, S., Atanassov, K., Krawczak, M. (eds.) Advances in Fuzzy Logic and Technology 2017. EUSFLAT 2017, IWIFSGN 2017. AISC, vol. 641. Springer, Cham (2018). https://doi.org/10.1007/978-3-319-66830-7_54
15. Engelbrecht, A.P.: Fundamentals of Computational Swarm Intelligence. Wiley (2005)
16. Ewald, D., Czerniak, J.M., Zarzycki, H.: Approach to solve a criteria problem of the ABC algorithm used to the WBDP multicriteria optimization. In: Angelov, P., et al. (eds) Intelligent Systems'2014. AISC, vol 322. Springer, Cham (2015). https://doi.org/10.1007/978-3-319-11313-5_12
17. Ewald, D., Czerniak, J.M., Zarzycki, H.: OFNBee method used for solving a set of benchmarks. In: Kacprzyk, J., Szmidt, E., Zadrożny, S., Atanassov, K.T., Krawczak, M. (eds.) IWIFSGN/EUSFLAT -2017. AISC, vol. 642, pp. 24–35. Springer, Cham (2018). https://doi.org/10.1007/978-3-319-66824-6_3
18. Ewald, D., Zarzycki, H., Apiecionek, Ł, Czerniak, J.M.: Ordered fuzzy numbers applied in bee swarm optimization systems. J. Univ. Comput. Sci. **26**(11), 1475–1494 (2020)
19. Fedorenko, M.: Analiza porównawcza wybranych algorytmów inspirowanych naturą, WSIZ, Wroclaw (2020)
20. Hu, X., Zhang, J., Li, Y.: Orthogonal methods based ant colony search for solving continuous optimization problems. J. Comput. Sci. Technol. **23**(1), 2–18 (2008)
21. Iredi, S., Merkle, D., Middendorf, M.: Bi-criterion optimization with multi colony ant algorithms. In: Zitzler, E., Thiele, L., Deb, K., Coello Coello, C.A., Corne, D., (eds) Evolutionary Multi-Criterion Optimization. EMO 2001. LNCS, vol. 1993. Springer, Berlin (2001). https://doi.org/10.1007/3-540-44719-9_25
22. Iżuk, B., Piechowiak, M.: The impact of ant colony optimization parameters on the connections efficiency in networks, studies and materials in applied computer science, vol. 12, no. 2 (2020)
23. Lebiediewa, S., Zarzycki, H., Dobrosielski, W.T.: A new approach to the equivalence of relational and object-oriented databases. In: Atanassov, K., et al. (eds) Novel Developments in Uncertainty Representation and Processing. AISC, vol. 401. Springer, Cham (2016). https://doi.org/10.1007/978-3-319-26211-6_8
24. Merkle, D.: Swarm Intelligence: Introduction and Application, Springer Verlag (2008)
25. Negulescu, S.C., Oprean, C., Kifor, C.V., Carabulea, I.: Elitist Ant System for Route Allocation Problem. p. 63, Lucia Blaga University of Sibiu (2008)

26. Ochelska-Mierzejewska, J.: A comparison of ant colony optimization and genetic algorithm for solving the traveling salesman problem. J. Appl. Comput. Sci. **24**(1), 51–66 (2016)
27. Palmer, J.: Smart future for swarm robots, Technology Reporter, BBC News, August (2008)
28. Piechowiak, M., Zwierzykowski, P.: The evaluation of unconstrained multicast routing algorithms in Ad-Hoc networks. In: Kwiecień, A., Gaj, P., Stera, P., (eds) Computer Networks. CN 2012. CCIS, vol. 291. Springer, Berlin, Heidelberg (2012). https://doi.org/10.1007/978-3-642-31217-5_36
29. Piechowiak, M., Zwierzykowski, P.: The evaluation of multicast routing algorithms with delay constraints in mesh network. In: 8th IEEE, IET International Symposium on Communication Systems, Networks and Digital Signal Processing CSNSDP 2012, Poznań, Poland (2012)
30. Sariff, N.B., Buniyamin, N.: Genetic algorithm versus ant colony optimization algorithm, s. 132, Universiti Teknologi MARA (2010)
31. Śmigielski, G., Dygdała, R., Zarzycki, H., Lewandowski, D.: Real-time system of delivering water-capsule for firefighting. In: Kobayashi, S., Piegat, A., Pejaś, J., El Fray, I., Kacprzyk, J. (eds) Hard and Soft Computing for Artificial Intelligence, Multimedia and Security. ACS 2016. AICS, vol. 534. Springer, Cham (2017). https://doi.org/10.1007/978-3-319-48429-7_10
32. Whitley, L.D., Vose, M. D.: Foundations of Genetic Algorithms, pp. 303 Morgan Kaufmann publishers (1995)
33. Zarzycki, H., Czerniak, J.M., Dobrosielski W.T.: Detecting Nasdaq composite index trends with OFNs. In: Prokopowicz, P., Czerniak, J., Mikołajewski, D., Apiecionek, Ł., Slezak, D. (eds) Theory and Applications of Ordered Fuzzy Numbers. Studies in Fuzziness and Soft Computing, vol. 356. Springer, Cham (2017). https://doi.org/10.1007/978-3-319-59614-3_11
34. Zarzycki, H., Czerniak, J.M., Lakomski, D., Kardasz, P.: Performance comparison of CRM systems dedicated to reporting failures to IT department. In: Madeyski, L., Śmiałek, M., Hnatkowska, B., Huzar, Z. (eds) Software Engineering: Challenges and Solutions. AISC, vol. 504. Springer, Cham (2017). https://doi.org/10.1007/978-3-319-43606-7_10
35. Zarzycki, H., Dobrosielski, W.T.: Use of Ordered Fuzzy Numbers to observe quotations on financial markets. In: AISC. Springer, Cham (2021)
36. Zarzycki, H., Dobrosielski, W.T., Vince, T., Apiecionek, Ł.: Center of Circles Intersection, a new defuzzification method on fuzzy numbers, bulletin of the polish academy of sciences. Technical Sciences (2020)

Selection of T-Norms for Calculating Belief Measure and Their Influence on Support Decision with Uncertainty

Sebastian Porębski(✉)

Department of Cybernetics, Nanotechnology and Data Processing,
Faculty of Automatic Control, Electronics and Computer Science,
Silesian University of Technology, Gliwice, Poland
sebastian.porebski@polsl.pl

Abstract. The work deals with information processing with belief function of Dempster-Shafer theory (DST) generalized for a fuzzy environment. Within the domain of explainable artificial intelligence, different set theories are proposed that try to upgrade the system readability while keeping its reliability: fuzzy, intuitionistic, rough, soft to name a few. Usually they are connected to particular knowledge-based inference mechanisms: probabilistic, evidential, possibilistic, neuro-fuzzy reasoning. This chapter focuses on effective and verified decision support solution which origins from DST and allows processing uncertain knowledge and imprecise information. In this approach, where aforementioned sources of knowledge are described in a fuzzy form, accurate measure of their correspondence needs to be examined. Hence, study validates the method of fuzzy set inclusion measure calculation, particularly exploring different triangle operations. Comparison of experimental differences reveals that minimum operation can be outperformed by other t-norms and their choose is highly problem-depended. The final results let us to increase quality of decision support systems in their further development.

Keywords: Interpretable decision support systems · Fuzzy sets · Approximate reasoning · Fuzzy inclusion measure

1 Introduction

Decision support systems (DSS) serve in many areas of life and are required to be reliable and interpretable [1]. Decision support is a task that can be solved by decision-making process similar to a classification. Nevertheless, majority of classifiers find it difficult to fit into a group of useful DSS if interpretability is the key criterion.

This paper deals with decision support task using fuzzy if-then rules. The fuzzy set theory has been adopted and is often used in the fields of processing imperfect data. Fuzzy set theory is often generalized to the theory of intuitionistic sets [2,11,14,18], rough sets [7,10,12] and all of these ideas are independently

K. T. Atanassov et al. (Eds.): IWIFSGN 2020/BOS/SOR 2020, LNNS 338, pp. 229–240, 2022.
https://doi.org/10.1007/978-3-030-95929-6_18

developed as well as hybridized [5]. In many works on DSS [7,18], in addition to information processing methods, some attention is devoted to inference methods. If the decision support system calculates a supported decision, it should connect the result with some certainty factor. In the Dempster-Shafer theory (DST) [4], that generalize the probability to the notions of its lower and upper types, the indicated decisions are assigned the so-called "degree of belief" [19].

The DST is a proven method of making decisions with the uncertainty of knowledge. When confronted with data imprecision, techniques are still being explored to improve the aforementioned priorities required for the usability of DSS. An attempt to assess the impact of selected models of comparing uncertain knowledge written in the form of fuzzy focal elements and imprecise information modeled by fuzzy sets is a particularly important aspect in the paper. Among the works exploring the topic of comparing formalisms modeled as fuzzy sets, the measures of similarity and distance dominate [11]. In the definition of the belief function from the DST, regardless of the modeling method, the matching of knowledge and data (verified during decision support) is calculated on the basis of the inclusion operation. This inclusion calculation has evolved from classical sets, fuzzy sets in the knowledge base, to fuzzy sets both used in the knowledge base and for modeling imprecision of input information. In this work, one of the proposed methods of calculating the inclusions of fuzzy sets is used and attempts were made to improve it by means of different triangular operations [6]. The presented results show with certainty that the choice of the t-norm in the inclusion operation cannot be left to chance and must be tuned at the stage of building the DSS.

2 Belief Functions for Precise and Imprecise Evidence

At the heart of the DST is the concept of the basic probability assignment (BPA). It allows assigning a probability value to the product of events without taking into account the dependence of individual events. In this work, events are modeled by fuzzy sets and the value of the basic probability expresses their connection with the decision or hypothesis under consideration.

Let us present the set C of the considered hypotheses (decisions) as D_1, $D_2, \cdots, D_c, \cdots, D_C$. For each cth decision ($c = 1, \cdots, C$), we can present a separate BPA, i.e. determine the importance of individual events supporting a specific decision. The work deals with n-dimensional problem solving, hence the information vectors can be defined as $\boldsymbol{x}_i = [x_{i1}, \cdots, x_{ij}, \cdots, x_{in}]$. For each of the C hypotheses n single events can be defined which are modeled by fuzzy sets that can be denoted commonly as:

$$\mathbb{A}_c = \{A_{1c}, A_{2c}, \cdots, A_{jc}, \cdots, A_{nc}\}, \tag{1}$$

where A_{jc} is the fuzzy constraint [23] of the jth feature value x_j of input information $\boldsymbol{x}$ when cth hypothesis is considered. In other words, membership function μ_{jc} of the fuzzy set A_{jc} determine what is a typical value of x_j within the cth hypothesis. With elements listed in (1) we can determine power set:

$$
\begin{gathered}
2^{\mathbb{A}_c} = \{\, A_{1c}, A_{2c}, \cdots, A_{jc}, \cdots, A_{nc}, \\
A_{1c} \wedge A_{2c}, \cdots, A_{1c} \wedge A_{nc}, \cdots, A_{jc} \wedge A_{j+1,c}, \cdots, A_{n-1,c} \wedge A_{nc}, \\
A_{1c} \wedge A_{2c} \wedge A_{3c}, \cdots, A_{j-1,c} \wedge A_{jc} \wedge A_{j+1,c}, \cdots, A_{n-2,c} \wedge A_{n-1,c} \wedge A_{nc}, \\
A_{1c} \wedge A_{2c} \wedge \cdots \wedge A_{jc} \wedge \cdots \wedge A_{nc}\},
\end{gathered} \tag{2}
$$

then the BPA will be a mapping function $m : 2^{\mathbb{A}_c} \rightarrow [0, 1]$. Among the $2^{\mathbb{A}_c}$ there are the so-called focal elements. Focal element is the event (or their combination) for which part of the knowledge (part of the belief) about the hypothesis D_c is focused. Formally, an element from the set $2^{\mathbb{A}_c}$ is a focal element if it is assigned a non-zero basic probability. Thus, by means of the set of focal elements denoted as S_c we can express such a subset of $2^{\mathbb{A}_c}$ that:

$$
S_c = \{s \in 2^{\mathbb{A}_c} | m(s) > 0\} = \{s_{1c}, \cdots, s_{kc}, \cdots, s_{r_c c}\} \tag{3}
$$

where $m(s)$ is the basic probability value of the focal element s. The sum of the BPA equals one for all focal elements. Thus all knowledge about c-th hypothesis is concentrated in individual $s \in S$. The number of focal elements chosen for cth hypothesis is denoted as r_c. The maximal value of $r_c = 2^n - 1$ that is the size of power set 2^{A_c}. However, when building the knowledge base in the DSS, the final number of focal elements for each supported decision can be much smaller than $2^n - 1$.

Modeling knowledge by means of fuzzy focal elements allows understanding of knowledge in the form of interpretable concepts, and these in turn can be used in fuzzy if-then rules. To write one fuzzy if-then rule you only need a fuzzy focal element s_{kc} and its basic probability value m_{kc}. The latter will also show the importance of the rule in the DSS. If-then rule based on the k-th focal element for the c-th hypothesis has a following form:

$$
\text{IF } s_{kc} \text{ THEN } D_c \text{ with probability } m_{kc}, \tag{4}
$$

where s_{kc} is the focal element expressing the fuzzy constraint of the values of the vector $\boldsymbol{x}$. For example, if the fuzzy focal element relates to a complex condition (composed by the terms A_{1c}, A_{2c}, A_{5c}), the premise of the rule created on s_{kc} is:

$$
x_1 \text{ is } A_{1c} \text{ and } x_2 \text{ is } A_{2c} \text{ and } x_5 \text{ is } A_{5c}.
$$

In the latter part of the work the set of features included in s_{kc} is denoted as $\mathbb{J}_{kc}$, which for above example is $\{1, 2, 5\}$.

Classic belief measure [19] is the function that for some input information $\boldsymbol{x}_i$ sums values of the basic probability of those focal elements that are true for the input case. This truth can be calculated as the inclusion of $\boldsymbol{x}_i$ in s_{kc}, hence:

$$
Bel^{(c)}(\boldsymbol{x}_i) = \sum_{k=1}^{r_c} I(\boldsymbol{x}_i \subset s_{kc}) \cdot m_{kc}, \tag{5}
$$

where $\boldsymbol{x}_i \subset s_{kc}$ is the set inclusion, $I(\cdot)$ is indicator function and m_{kc} is the weight of focal elements (its basic probability value). Due to the fact that $I(\cdot) = 0$ or 1 and $0 \leq m_{kc} \leq 1$ thus $0 \leq Bel^{(c)}(\boldsymbol{x}_i) \leq 1$. In the classic approach $\boldsymbol{x}_i$ is the vector of n values (e.g. one-argument sets, singletons), hence the value of the inclusion with complex s_{kc} is aggregated and $I(\boldsymbol{x}_i \subset s_{kc}) = 1$ when $\forall_{j \in \mathbb{J}_{kc}} x_j \subset A_{jc}$.

The first step in capturing fuzzy information shortly after the most important works by Dempster and Shafer [4,19] was Zadeh's [23] description of the concepts of a weighted belief and plausibility measure. When modeling the fuzzy focal element (i.e. the events A_j as fuzzy sets), the value of inclusion has to be generalized. Using the possibility measure [23], the measure of the inclusion of the $\boldsymbol{x}_i$ value in a fuzzy focal element, i.e. $I(\boldsymbol{x}_i \in s_{kc})$, takes the following form:

$$I(\boldsymbol{x}_i \subset s_{kc}) = \min_{j \in \mathbb{J}_{kc}} [x_{ij} \subset A_{jc}] = \min_{j \in \mathbb{J}_{kc}} [\mu_{jc}(x_{ij})] , \tag{6}$$

that is, inclusion is as large as the membership of a single event x_{ij} to the concept of A_{jc}, and when we deal with the event vector $\boldsymbol{x}_i$, we aggregate the membership with the minimum function. For example, if we are dealing with a focal element that describes three logical conditions:

$$s_{kc} : \{A_{1c} \wedge A_{2c} \wedge A_{5c}\} ,$$

then

$$I(\boldsymbol{x}_i \in s_{kc}) = \min [\mu_{1c}(x_{i1}), \mu_{2c}(x_{i2}), \mu_{5c}(x_{i5})] .$$

In some works, the above dependence was reduced to the form of full or zero inclusion conditioned by the fulfillment of a specific matching threshold (uncertainty threshold) [20]. However, the threshold require its tuning during knowledge discovery, hence this work tries to process imprecision as it is without judging whether the matching level is satisfying or not.

The next step necessary from the point of view of imprecise knowledge processing is the assumption that the input data processed by the DSS may also be imprecise. Hence instead of vector $\boldsymbol{x}_i$ we are dealing with collection $\widetilde{\boldsymbol{X}}_i = \left[\widetilde{X}_{i1}, \widetilde{X}_{i2}, \cdots, \widetilde{X}_{ij}, \cdots, \widetilde{X}_{in}\right]$. Generally, let's assume that the input value, even if it seems precise, can come from a measurement series or be an implementation of a random variable with a known distribution, then replacing x_{ij} with a fuzzy set (fuzzy number) $\widetilde{X}_{ij}$ is quite easy. If we have any knowledge about the possible distribution of the variable, we can determine the width of the triangular membership function, i.e. $\widetilde{X}_{ij}$ can be described by the membership function $\widetilde{\mu}_{ij}(x) = \max\left[\min\left[\frac{x-a}{b-a}, \frac{c-x}{c-b}\right], 0\right]$, where a, b, c, determine the vertices of the triangular membership function. This work searches for the best technique of the calculation of the belief measure in which both the focal element is uncertain and the input information is imprecise. So both can be modeled using fuzzy sets. The question that arises is how to model the input information, i.e. with what formalism to describe its imprecision. The more important question that is posed and tried to answer in this paper is how to calculate the matching level of fuzzy information and fuzzy knowledge (or basically the measure of inclusions if the belief measure needs to be calculated). Among the ideas taken, in previous works the generalization of inclusions proposed by [17] is used, having the following character:

$$I(\widetilde{\boldsymbol{X}}_i \subset s_{kc}) = \min_{j \in \mathbb{J}_{kc}} [\widetilde{X}_{ij} \subset A_{jc}] = \min_{j \in \mathbb{J}_{kc}} \left[\frac{\sum\limits_{x_j \in supp(\widetilde{X}_{ij})} \min\left[\mu_{jc}(x), \widetilde{\mu}_{ij}(x)\right]}{\sum\limits_{x_j \in supp(\widetilde{X}_{ij})} \widetilde{\mu}_{ij}(x)} \right]. \tag{7}$$

The main part of the experiments of this work is to check if the calculation of fuzzy information inclusions in the fuzzy focal element using the minimum function in (7) is most effective. Due to the fact that the minimum function is only one of the triangular operations called t-norms that can be calculated for two fuzzy sets, the generalized definition of (7) will look as following:

$$I(\widetilde{\boldsymbol{X}}_i \subset s_{kc}) = \min_{j \in \mathbb{J}_{kc}} \left[\frac{\sum\limits_{x_j \in supp(\widetilde{X}_{ij})} \star_T \left[\mu_{jc}(x), \widetilde{\mu}_{ij}(x)\right]}{\sum\limits_{x_j \in supp(\widetilde{X}_{ij})} \widetilde{\mu}_{ij}(x)} \right]. \tag{8}$$

Among the t-norms to be used in place of $\star_T \left[\mu_{jc}(x), \widetilde{\mu}_{ij}(x)\right]$, the following four were selected:

- Minimum: $\star_{T_M} \left[\mu_{jc}(x), \widetilde{\mu}_{ij}(x)\right] = \min\left[\mu_{jc}(x), \widetilde{\mu}_{ij}(x)\right]$,
- product: $\star_{T_P} \left[\mu_{jc}(x), \widetilde{\mu}_{ij}(x)\right] = \mu_{jc}(x) \cdot \widetilde{\mu}_{ij}(x)$,
- Łukasiewicz: $\star_{T_L} \left[\mu_{jc}(x), \widetilde{\mu}_{ij}(x)\right] = \max\left[\mu_{jc}(x) + \widetilde{\mu}_{ij}(x) - 1, 0\right]$,
- Einstein: $\star_{T_E} \left[\mu_{jc}(x), \widetilde{\mu}_{ij}(x)\right] = \frac{\mu_{jc}(x) \cdot \widetilde{\mu}_{ij}(x)}{2 - (\mu_{jc}(x) + \widetilde{\mu}_{ij}(x) - \mu_{jc}(x) \cdot \widetilde{\mu}_{ij}(x))}$.

In the course of the experiments it will be checked how individual triangular operations influence 1) inclusion of events and focal elements, 2) a belief measure calculated for complex focal elements. In the latter case, examples will be presented for synthetic data (data series) and two different benchmark data sets for classification.

3 Numerical Experiments

The knowledge about the data from three categories was modeled as three focal elements described by the Gaussian membership function complementary covering the x_j domain [10]. The input information modeled with a triangular membership function. In order to check the differences between the individual results, the input information, i.e. the fuzzy number $\widetilde{X}_{ij}$, was shifted across the x_j value domain to calculate the inclusion with each focal element from the three decision categories for each of its positions. The resulting inclusion value charts are shown in Fig. 1. This figure presents two analogous yet independent examples.

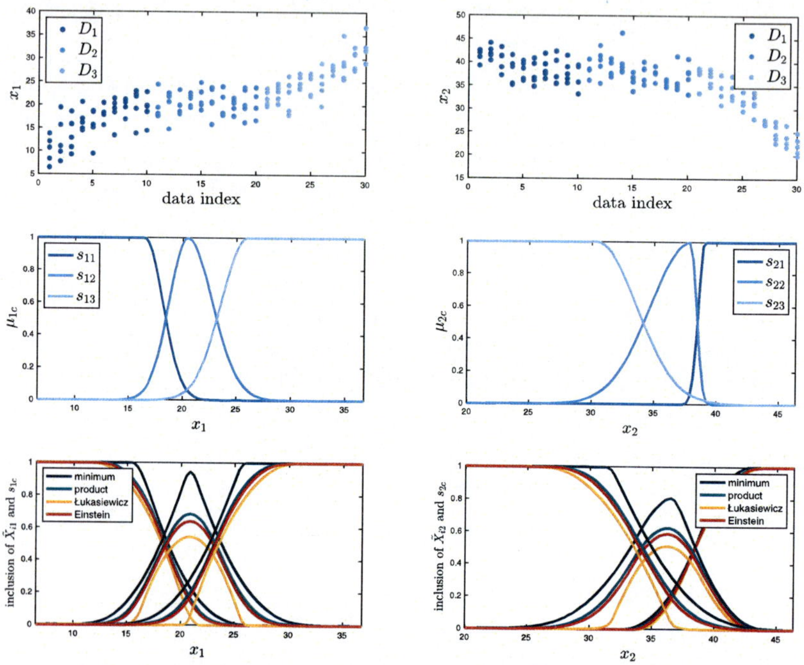

Fig. 1. Comparison of different t-norms calculated for two independent synthetic data series (left and right part of the figure). The information data series are divided into three classes (D_1, D_2 and D_3 on the top diagrams). Fuzzy focal elements s_{kc} are presented with their membership functions in middle diagrams. Results of the calculation of inclusion measure for different t-norms are presented in the bottom diagrams of the figure.

According to the bottom part of Fig. 1. we can compare the changes of inclusion value for four different t-norms. For both examples (left and right) we can see that minimum t-norm allows obtaining the highest inclusion values, especially for the interval where D_2 should be indicated. However bell-shapes for D_2 are also the widest for the minimum norm. According to this, Łukasiewicz norm expresses opposite situation, i.e. low inclusion values allows for thin bell shape of this inclusion result. Two remaining results (for product and Einstein norms) has similar shapes with average peaks.

The influence of the t-norm on the calculation of inclusions promotes the key element of decision support, i.e. the value of the belief measure. The comparison of all belief values related to C decisions affects the ultimately indicated decision in the support system. Figure 2 presents the data from Fig. 1 treated this time as two features of the same input vector, i.e. two-dimensional data. The decision surfaces will depend on which of the belief measures has reached the highest value for a given point (the creation of these decision surfaces was the result of calculating inclusions and belief measures for each point in the whole domain of x_1 and x_2 values.

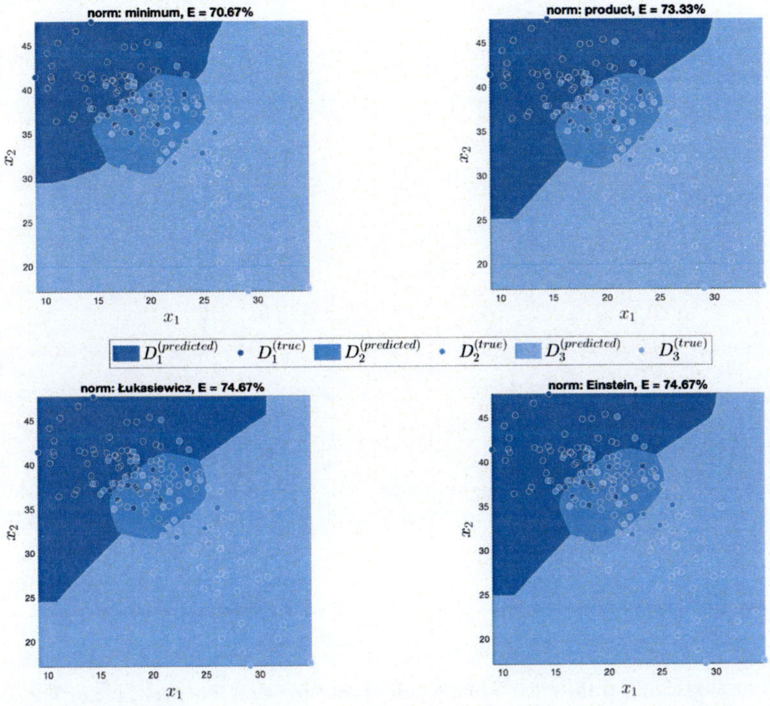

Fig. 2. Decision support according to four different t-norms within inclusion measure calculation for synthetic data series. Three areas indicate final decision which is supported according to the highest belief value. Each result is related to different classification quality E.

The most interesting are the shapes of the decision boundaries that reflect the final decision stated according to comparison of the belief measure values. As we can see on the first subplot in Fig. 2 where minimum norm is utilized in inference process, shapes of decision boundaries are not smooth enough in the area where decision D_2 is chosen as the final. This feature is assured by the rest of norms with smoother decision boundary in this conflicting area. On the other hand, decision boundaries are more sharp for the x_1 and x_2 coordinates that are not crucial since data cases does not occur there. These observation follows the calculated number of correctly classified data. Different t-norms allow obtaining slightly higher classification accuracy (denoted as E and calculated simply as the percentage number of correctly classified data cases).

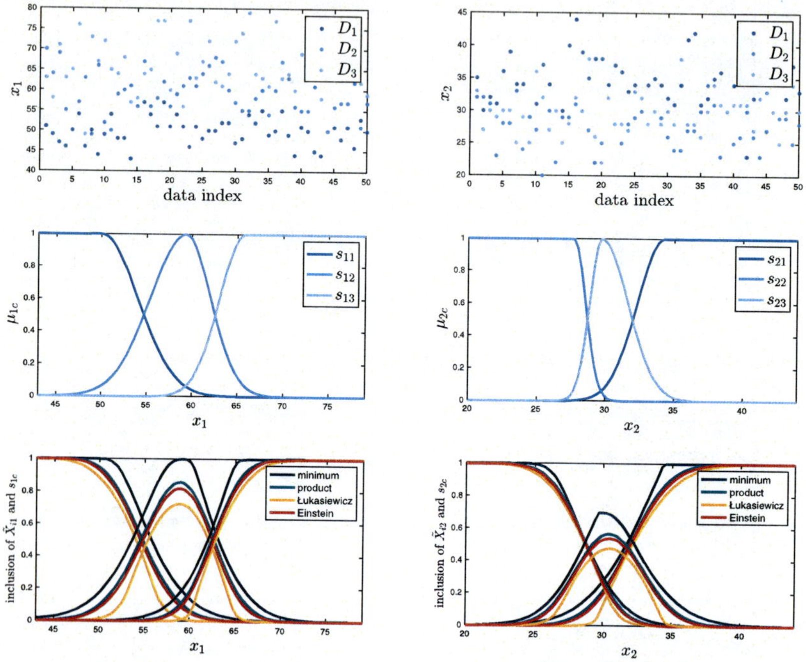

Fig. 3. Comparison of different t-norms calculated for two feature data of Iris data set (left and right part of the figure). The data are divided into three classes (D_1, D_2 and D_3 on the top diagrams). Fuzzy focal elements s_kc are presented with their membership functions in middle diagrams. Results of the calculation of inclusion measure for different t-norms are presented in the bottom diagrams.

A similar procedure was carried out for the classification data (the first two dimensions were selected from the Iris benchmark database, and the entire calculation procedure is the same as in the previous examples).

Results that are presented for synthetic data set are analogous for those obtained with real-data example (Fig. 3 and 4). T-norms influence the inclusion results (Fig. 3) and again, Łukasiewicz norm allows shaping decision boundaries for the best (but not significantly better) supported decision quality (see Fig. 4).

The third group of data is the classification problem for XOR data. These are synthetically generated data and are presented in the left diagram of Fig. 5. Correct class recognition requires non-linear classification (see right part of Fig. 5). The selected inference model allows for perfect decision support, however, the aim of this step is to investigate the difference in the shapes of belief measures for these data.

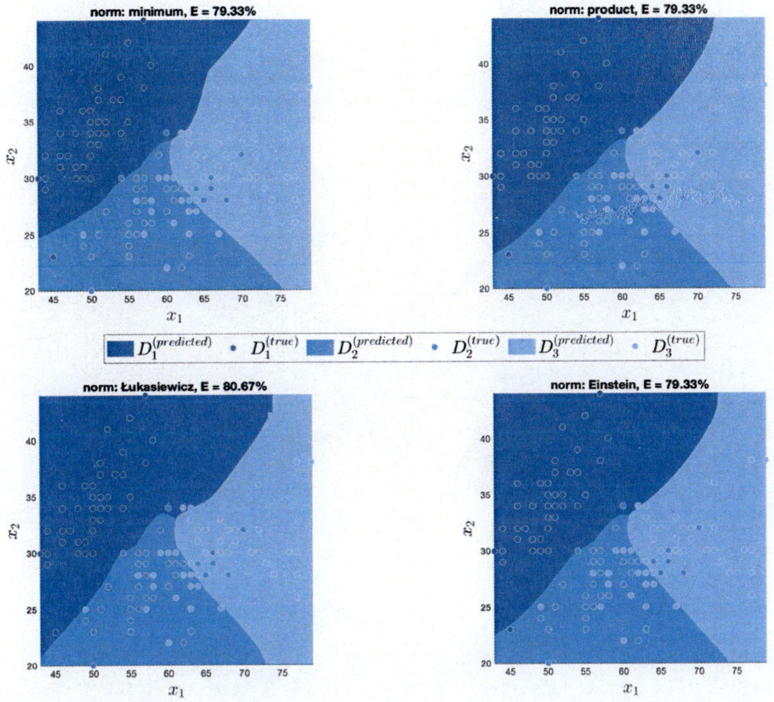

Fig. 4. Decision support according to four different t-norms within inclusion measure calculation for two features of Iris data set. Three areas indicate final decision which is supported according to the highest belief value. Each result is related to different classification quality E.

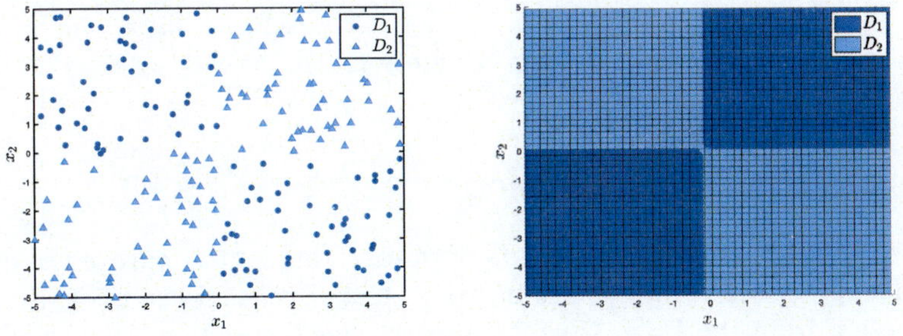

Fig. 5. Presentation of XOR data (left diagram) and decision area required for a perfect classification

This time, decision boundaries for different t-norms are not viewed hence the final decision is the same. Instead, detailed shape of belief values for two dimensional space is investigated. We can see on Fig. 6. that the product norm (second diagram from the top) shapes the most regular belief value. According to the data, we can see that interesting area is located in the middle of domain.

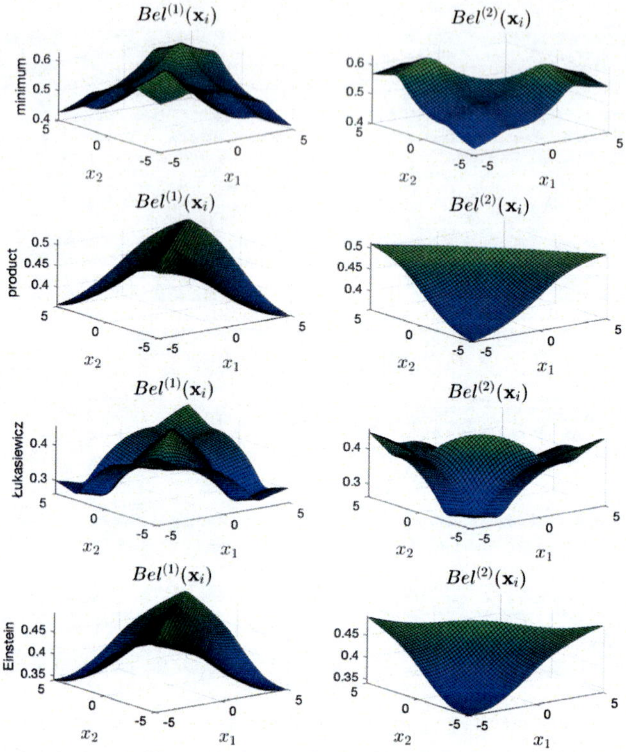

Fig. 6. Differences between shapes of the calculated belief function results according to four examined triangle operations, minimum, product, Łukasiewicz and Einstein.

For this area only the minimum norm influence low belief values, which is quite undesired to lower the final decision certainty in this crucial area. The rest of norms allow keeping high belief values.

4 Discussion and Conclusions

This paper presents preliminary results of research aimed to improve the quality of DSS. The proposed solutions are based on fuzzy sets and use the generalized belief function from the DST. Current research has focused on a modified computation of fuzzy set inclusions. These sets help to model imprecise information that is inferred with knowledge described by the fuzzy focal elements. So far, one of the things mentioned has not been replaced with its fuzzy model. Therefore, checking the influence of the t-norm operator on the inclusion result, the result of the belief function and the final result of the decision support is an important element of improving the DSS.

Out of several triangular operations, only the minimum function has been used so far. Based on the research results, it was noticed that triangular operations, in particular the Łukasiewicz's, product and Einstein's norms, due to their influence on the entire decision support process, require further attention. In the conducted research, the triangular operation verified for the calculation of inclusions is itself also used for: aggregation of the input information matching to the rule premise, the stage of BPA and building the knowledge base from fuzzy focal elements. Therefore, the starting point for further studies is the need to examine the use of the above-mentioned triangular operations also for these aspects of designing DSS. Łukasiewicz's t-norm, according to the previous point, seems to have the best influence on the inclusion operation. Nevertheless, its final evaluation must be made on the basis of analyzing all stages of building and using the DSS.

Acknowledgments. This research is financed from the statutory activities of the Faculty of Automatic Control, Electronics and Computer Science of the Silesian University of Technology.

References

1. Arrieta, A.B., et al.: Explainable Artificial Intelligence (XAI): Concepts, taxonomies, opportunities and challenges toward responsible AI. Inf. Fusion **58**, 82–115 (2020)
2. Atanassov, K., Pasi, G., Yager, R.: Intuitionistic fuzzy interpretations of multi-criteria multi-person and multi-measurement tool decision making. Int. J. Syst. Sci. **36**(14), 859–868 (2005)
3. Behounek, L., Dankova, M.: Variable-domain fuzzy sets - part II: apparatus. Fuzzy Sets Syst. **380**, 19–43 (2020)
4. Dempster, A.P.: Upper and lower probabilities induced by a multivalued mapping. Ann. Math. Stat. **38**(2), 325–339 (1967)
5. Deschrijver, G., Kerre, E.E.: On the relationship between some extensions of fuzzy set theory. Fuzzy Sets Syst. **133**(2), 227–235 (2003)
6. Deschrijver G., Kerre E.E.: Triangular norms and related operators in L*-fuzzy set theory. In: Klement E.P., MesiarLogical, R. (eds), Algebraic, Analytic and Probabilistic Aspects of Triangular Norms, pp. 231–259, Elsevier Science B.V. (2005)
7. Huang, B., Wu, W.-Z., Yan, J., Li, H., Zhou, X.: Inclusion measure-based multi-granulation decision-theoretic rough sets in multi-scale intuitionistic fuzzy information tables. Inf. Sci. **507**, 421–448 (2020)
8. Li, F., Jin, C., Zhang, X., Wang, Y., Liu, X.: Attribute importance measurement method based on data coordination degree. Knowledge-Based Systems **192**, 105359 (2020)
9. Madrid, N., Ojeda-Aciego, M.: Functional degrees of inclusion and similarity between L-fuzzy sets. Fuzzy Sets Syst. **390**, 1–22 (2020)
10. Mieszkowicz-Rolka A., Rolka L.: Variable precision fuzzy rough set model with linguistic labels. In: IEEE International Conference on Fuzzy Systems (FUZZ-IEEE), Glasgow, United Kingdom (2020)
11. Pękala B., Rak E., Kosior D., Mrukowicz D., Bazan J.G.: Application of similarity measures with uncertainty in classification methods. In: IEEE International Conference on Fuzzy Systems (FUZZ-IEEE), Glasgow, United Kingdom (2020)
12. Pawlak, Z.: Rough Sets: Theoretical Aspects of Reasoning About Data. Kluwer Academic Publisher, Norwell, MA (1992)

13. Porebski, S., Straszecka, E.: Using fuzzy numbers for modeling series of medical measurements in a diagnosis support based on the dempster-shafer theory. In: Rutkowski, L., Scherer, R., Korytkowski, M., Pedrycz, W., Tadeusiewicz, R., Zurada, J. (eds) Artificial Intelligence and Soft Computing, ICAISC 2018. Lecture Notes in Computer Science, vol. 10842. Springer, Cham (2018)
14. Pękala, B., Dyczkowski, K., Grzegorzewski, P., Bentkowska, U.: Inclusion and similarity measures for interval-valued fuzzy sets based on aggregation and uncertainty assessment. Inf. Sci. **547**, 1182–1200 (2021)
15. Rajati, M.R., Mendel, J.M.: Advanced computing with words using syllogistic reasoning and arithmetic operations on linguistic belief structures. In: IEEE International Conference on Fuzzy Systems (FUZZ-IEEE), Hyderabad (2013)
16. Sarabi-Jamab, A., Araabi, B.N.: An information-based approach to handle various types of uncertainty in fuzzy bodies of evidence. PLOS ONE **15**(1), e0227495 (2020)
17. Ogawa, H., Fu, K.S., Yao, J.T.P.: An inexact inference for damage assessment of existing structures. Int. J. Man-Mach. Stud. **22**(3), 295–306 (1985)
18. Singh, V., Yadav, S.P., Mesiar, R.: Information measures in atanassov's intuitionistic fuzzy environment and their application in decision making. IEEE Trans. Fuzzy Syst. **28**(11), 2905–2917 (2020)
19. Shafer, G.: A Mathematical Theory of Evidence. Princeton University Press (1976)
20. Straszecka, E.: Combining uncertainty and imprecision in models of medical diagnosis. Inf. Sci. **176**(20), 3026–3059 (2006)
21. Yang, J.-B., Liu, J., Wang, J., Sii, H.-S., Wang, H.-W.: Belief rule-base inference methodology using the evidential reasoning approach-RIMER. IEEE Trans. Syst., Man, Cyber. - Part A: Syst. Human. **36**(2), 266–285 (2006)
22. Yu, G.: Relationships between fuzzy approximation spaces and their uncertainty measures. Inf. Sci. **528**, 181–204 (2020)
23. Zadeh, L.A.: Possibility theory and soft data analysis. In: Cobb, L., Thrall, R. M., (eds) Mathematical Frontiers of the Social and Policy Sciences, pp. 69–129. Westview Press (1981)

Certain Aspects of the OFNBee Algorithm Operation for Different Fuzzifiers

Dawid Ewald[1](✉), Huber Zarzycki[2], and Jacek M. Czerniak[1]

[1] Kazimierz Wielki University, Bydgoszcz, Kujawsko-Pomorskie, Poland
dawidewald@ukw.edu.pl

[2] General Tadeusz Kościuszko Military Academy of Land Forces in Wrocław, Wrocław, Poland
hubert.zarzycki@awl.edu.pl

Abstract. The article describes a new hybrid method for bee optimisation using the arithmetic of ordered fuzzy numbers. The aim of this combination is to more realistically represent the decision-making mechanisms of bees, commonly referred to as waggle dance. The OFN notation allows to more accurately transfer the information between individuals, which contributes to the improvement of the optimisation algorithm. The article describes the fuzzy functions necessary for the new method to work, but also for the very use of OFN in other algorithms. Since there are no fuzzing methods available in the literature, a few such methods are proposed in this article. The new method also utilises the defuzzification functions, and the two most well-known of them were used for the experiment, namely Golden Ratio and Mandala Factor. The last section is a description of the new method and the results achieved by OFNBee for selected functions: Bukin N.6, Cross-in-tray, Drop Wave, Eggholder, Levy and Holder Table.

1 Introduction

To properly select one of the artificial intelligence methods, we should carefully consider the problem to be solved by such a tool [7,8]. In this context, swarm intelligence allows us to use the collective action of respective individuals to solve a large group of optimisation problems. There are many nature-inspired swarm algorithms [15–17]. The majority of such algorithms is based on the social behaviour of ants, termites or bees [19,21]. Still, some of proposed solutions operate on the basis of the mechanisms known in colonies of:

- bacteria [4],
- mites [3],
- chickens [18].

Optimisation algorithms based on the behaviour of insects or animals are an attempt to utilise natural mechanisms for optimization. However, the difficulty of recording the behaviour of living organisms in algorithms leads to simplifications.

K. T. Atanassov et al. (Eds.): IWIFSGN 2020/BOS/SOR 2020, LNNS 338, pp. 241–256, 2022.
https://doi.org/10.1007/978-3-030-95929-6_19

Although effective, such action limits the possibilities of these methods. So it seems reasonable to use the mechanisms of fuzzy logic, which describes real-world phenomena in a more natural way. Given how bee optimisation algorithms have been developing and where their inspiration comes from, we should consider combining more natural fuzzy number arithmetic with bee algorithms. Selected fuzzy logic issues and selected elements of ordered fuzzy numbers are presented in this paper below.

2 OFN

The history of ordered fuzzy numbers (OFN) begins in 2002. Witold Kosinski, together with Piotr Prokopowicz and Dominik Slezak, focused on eliminating the flaws of classic algebra of fuzzy numbers. The drawbacks in question are mainly the fact that by doing several operations on given L-R numbers, the obtained numbers may be too fuzzy, meaning less useful. This entails great computational complexity and the inability to apply backward inference. The authors of the OFN also wanted to develop arithmetic that would enable operations on both triangular and trapezoidal numbers. Let us begin with the definition of OFN.

Definition 1 [9,13,14]. An ordered fuzzy number is understood as a triple $A = (R; \mu_A; s_A)$, where $\mu_A : R \longrightarrow [0;1]$ is the membership function of the number, whereas $s_A \in \{-1; 0; 1\}$ denotes orientation. Fuzzy numbers can be divided into two groups in terms of orientation:

- OFN with positive orientation - direction of the OX axis increase
- OFN with negative orientation - direction opposite to the OX axis increase.

The graphical representation of ordered fuzzy number A is the pair $A = (C_A; s_A)$. Where $C_A \subseteq R^2$ is a curve described as follows:

$$C_A = (l_A, p_A) \times \{0\} \cup [1_A^-, 1_A^+] \times \{1\} \cup up_A \cup down_A \quad (1)$$

where:

- $up_A, down_A \subseteq R^2$ and sa are graphs of monotonic functions;
- s_A means orientation;
- C_A is a membership curve.

Functions $f_A; g_A$ correspond respectively $up_A, down_A \subseteq R^2$ (Fig. 1)

$$up_A = \{(f_A(y), y) : y \in [0,1]\} \quad (2)$$

$$down_A = \{(g_A(y), y) : y \in [0,1]\} \quad (3)$$

Orientation corresponds to the order of f_A and g_A graphs.

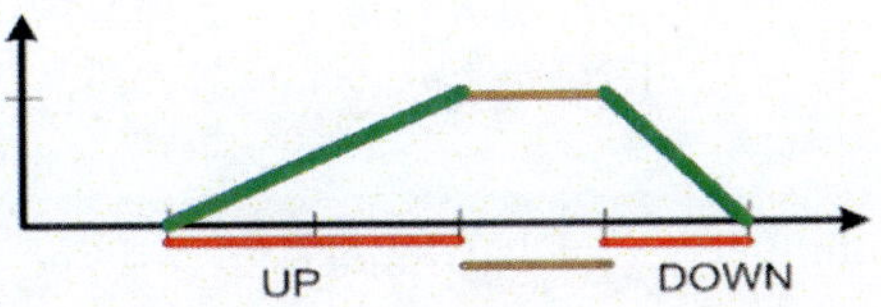

Fig. 1. OFN

The graphic interpretation of the two fuzzy numbers and real number $_0$ is shown in the figure below (Fig. 2).

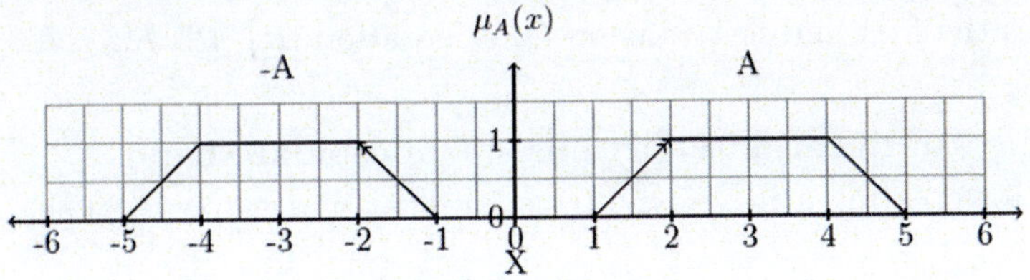

Fig. 2. Opposites A and (-A)

2.1 Fuzzy Operators

Fuzzy methods must be defined for the new method to work properly [23]. They are essential for the operation of OFNBee. The choice of method may have an impact on the algorithm results.

The Golden Ratio

The inverse golden ratio method is based on the concept of the golden ratio coefficient derived from the Fibonacci sequence. In historical terms, we can see terms such as harmonic division, golden proportion or divine proportion.

Currently, the best known interpretation of the golden ratio is to split a section into two parts so that the ratio of the longer part to the shorter one is the same as the ratio of the entire section to the longer part, as shown in Fig. 3.

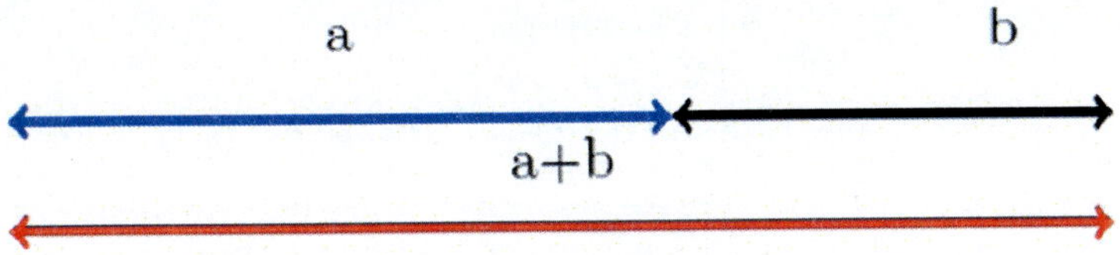

Fig. 3. The golden ratio

$$GRf = \begin{cases} R1 = number - 2.4721, \\ R4 = number + 1.5278. \end{cases} \quad (4)$$

where:
number – means the number we want to make an ordered fuzzy number as per the
$R1$, $R4$ – is the first and last value in OFN notation $[R1, R2, R3, R4]$.

$$GRf = \begin{cases} R1 = A + 2.4721 \\ R4 = A - 1.5278 \end{cases} \quad (5)$$

where:
number – means the number we want to make an ordered fuzzy number as per the
$R1$, $R4$ – is the first and last value in OFN notation $[R1, R2, R3, R4]$.

beeT Type Fuzzification Functions - Trapezoidal Shape
The rules for creating a beeT type fuzzy number in a trapezoidal shape are as follows:

$$beeTf\alpha\beta\gamma\delta = \begin{cases} R1 = number, \\ R2 = number + \mid number * \beta \mid, \\ R3 = number + \mid number * \gamma \mid, \\ R4 = number + \mid number * \delta \mid, \end{cases} \quad (6)$$

where:
$R1, R2, R3, R4$ – subsequent values of the ordered fuzzy number,
α – determines the number orientation and has the values:

- p – positively oriented number – $[R1, R2, R3, R4] \longrightarrow [R1, R2, R3, R4]$
- n – negatively oriented number – $[R4, R3, R2, R1] \longrightarrow [R4, R3, R2, R1]$.

number – means the number we want to make an ordered fuzzy number as per the beeT type,
β, γ, δ – values for the level of blur.

beeT Type Fuzzification Functions - Right Triangle Shape
The rules for creating a beeT type fuzzy number in a right triangle shape are as follows:

$$beeTf\alpha\beta = \begin{cases} R1 = liczba, \\ R2 = liczba, \\ R3 = liczba, \\ R4 = liczba + \mid liczba * \beta \mid, \end{cases} \quad (7)$$

where:
$R1, R2, R3, R4$ – subsequent values of the ordered fuzzy number,
α – determines the number orientation and has the values:

- p – positively oriented number – $[R1, R2, R3, R4] \longrightarrow [R1, R2, R3, R4]$
- n – negatively oriented number – $[R4, R3, R2, R1] \longrightarrow [R4, R3, R2, R1]$.

number – means the number we want to make an ordered fuzzy number as per the beeT type,
β – values for the level of blur.

beeM Type Fuzzification Functions - Trapezoidal Shape

The rules for creating a beeM type fuzzy number in a trapezoidal shape and positive orientation are as follows:

$$beeMTp\beta\gamma = \begin{cases} R1 = number, \\ R2 = number, \\ R3 = number + \mid number * \beta \mid, \\ R4 = number + \mid number * \gamma \mid, \end{cases} \tag{8}$$

where:
$R1, R2, R3, R4$ – subsequent values of the ordered fuzzy number,
α – determines the number orientation and has the values:

- p – positively oriented number – $[R1, R2, R3, R4] \longrightarrow [R1, R2, R3, R4]$
- n – negatively oriented number – $[R4, R3, R2, R1] \longrightarrow [R4, R3, R2, R1]$.

number – means the number we want to make an ordered fuzzy number as per the beeT type,
β, γ – values for the level of blur.

beeM Type Fuzzification Functions - Trapezoidal Right Triangle Shape

The rules for creating a beeT type fuzzy number in a right triangle shape are as follows:

$$beeMTrp\beta\gamma = \begin{cases} R1 = number, \\ R2 = number, \\ R3 = number, \\ R4 = number + \mid number * \beta \mid, \end{cases} \tag{9}$$

where:
$R1, R2, R3, R4$ – subsequent values of the ordered fuzzy number,
α – determines the number orientation and has the values:

- p – positively oriented number – $[R1, R2, R3, R4] \longrightarrow [R1, R2, R3, R4]$
- n – negatively oriented number – $[R4, R3, R2, R1] \longrightarrow [R4, R3, R2, R1]$.

number – means the number we want to make an ordered fuzzy number as per the beeT type,
β – values for the level of blur.

3 Defuzzification Functions Dedicated to OFN

A very important feature of OFN is the number orientation; this extra value distinguishes ordered numbers from other solutions. That is why it is so important to use orientation-sensitive defuzzification function. The first work on defuzzification functions was done by W. Kosinski [10–12,22]. His work was continued in numerous articles by W. Dobrosielski. The functions proposed by W. Dobrosielski are described further in the paper. In the conducted experiment, an important feature of defuzzification functions is the sensitivity to OFN orientation, and therefore only methods that meet this condition are described here.

Golden Ratio Defuzzification Operator
The idea of golden ratio is described in the chapter titled Inverse Golden Ratio. This section presents the Golden Ratio (GR) defuzzification function. The method allows us to determine the real value of fuzzy number A according to the formula (9) [5]:

$$GR = \frac{min(supp(A)) + |supp(A)|}{\Phi} \quad where\ \Phi = 1,618033998875\ldots \tag{10}$$

where:

- GR is the defuzzification operator,
- $supp(A)$ means support of fuzzy number A in universe X.

Mandala Factor
Another operator used in the work on the new method is the Mandala Factor operator [1,2] proposed by J. Czerniak. A mandala is an image composed of coloured grains of sand by Buddhist monks. The essence of these works is the harmonious combination of a square and a circle, where the circle symbolises heaven, infinity. The square is interpreted as a symbol of the inner sphere associated with the earth and man. The whole arrangement begins and ends at the point where both shapes are connected. The inspiration by grains of sand in mandalas lies at the heart of Mandala Factor. Given the trapezoidal OFN number A presented in Fig. 4, the contour marked by the sides of the OFN number and the OX axis should be filled with virtual sand grains, as shown in Fig. 5. Then a rectangle is built and filled with virtual sand grains. The filling means pouring sand vertically in the columns until it is exhausted. The non-fuzzy value of the number is in the point above which the last of filled column ends (Fig. 6). The mathematical notation is as follows [2]:

$$MF(A) = \begin{cases} c + r\ , & dla\ (A)\ positively\ oriented\ number \\ c - r\ , & dla\ (A)\ negatively\ oriented\ number \end{cases} \tag{11}$$

where:

$$r = \frac{1}{d-c}\int_c^d x\,dx - \frac{c}{d-c}\int_c^d dx + \frac{f}{f-e}\int_f^e dx \\ -\frac{1}{f-e}\int_e^f x\,dx + \int_d^e dx \tag{12}$$

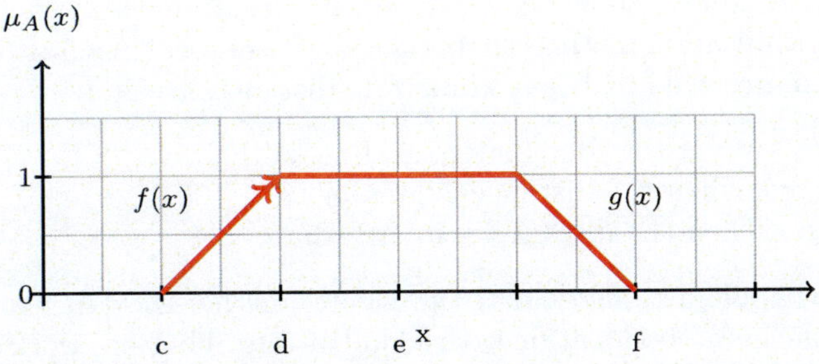

Fig. 4. Number OFN A

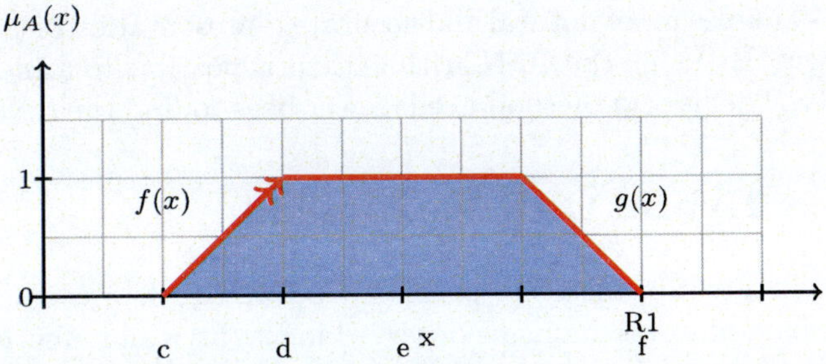

Fig. 5. Visualization of the Mandala Factor operation - step one

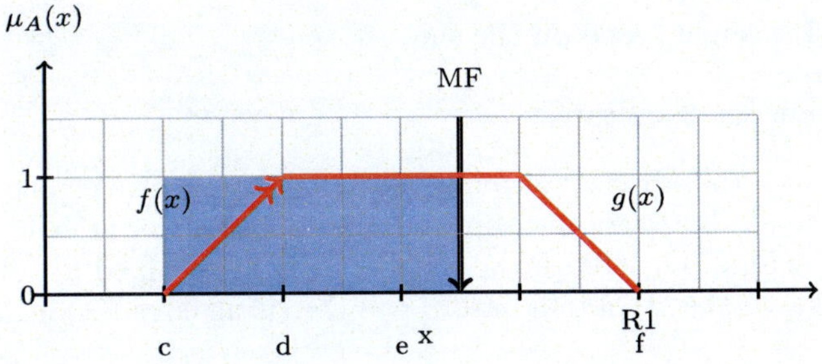

Fig. 6. Visualization of the Mandala Factor operation - step two

4 A New Optimisation Method and Its Application

The new OFNBee optimisation method was created by combining ordered fuzzy numbers with bee optimisation. The result is presented further in the article. The entomological research on the bee dance and the amount of information conveyed in it quickly demonstrated that the current bee optimisation algorithms interpret this issue very superficially. These algorithms focus on the group work of bees rather than on the exchange of information between individuals. In this case, unique data might be lost or the search process may take longer. The use of ordered fuzzy numbers to describe how bees exchange information has made it possible to create a new optimisation method that is close to its natural inspiration. There are three types of dance used by bees to illustrate distances to food:

- round dance - when the food source is close to the hive,
- waggle dance - when the distance is greater than 100 m,
- sickle dance - when the distance is intermediate.

The research by apologists also lead to the conclusion that the dance process is how bees agree on the new food location with a kind of vote. The herd work of bees gives excellent results in finding the best source of food. Acting collectively and sharing information, bees perfectly optimise their capabilities and resources. Exchange of information is crucial here. Orientation - the specific feature of fuzzy numbers in OFN notation - enables more natural and accurate representation of the intelligent behaviour of bees. By using the OFN arithmetic it is possible to achieve solutions of a new quality. The hybrid method needs fewer bees to find the best solution.

5 The New Hybrid OFNBee Method

OFN notation in bee optimisation seems to be a completely natural way to describe the behavioural mechanisms observed in the hive and mentioned above. Such mechanisms are presented in the new method through dedicated fuzzy operators. The input data is information carried by a single bee, i.e.:

- the direction of food,
- the navigation angle based on the sun,
- flight length,
- food abundance in the source.

Figure 7 shows the OFN describing the information provided by the scout bee: aF - the angle between the sun and the food source, sF - food amount, dF - distance to food. Ordered fuzzy number A is determined as follows. First, support(A) is set as the trapezoid base. Then the rising edge f(x) is drawn at the

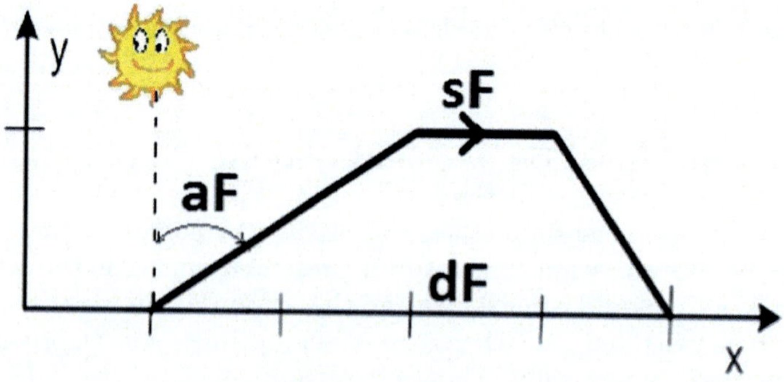

Fig. 7. Graphical interpretation of the OFN number in OFNBee

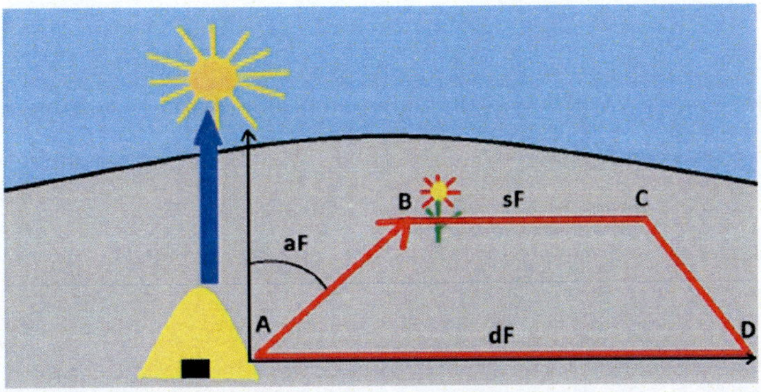

Fig. 8. Displaying the OFN number in OFNBee

angle of $90° - aF$. The second trapezoid base is set aside from the intersection point of function f(x) z y = 1. Finally, the falling edge function g(x) connects the two ends of the trapezoid bases, as shown in Fig. 8. The operation of the algorithm can be divided into several stages. In the first step, points are selected at random from the solution space. Such points form the tables of current solutions. The function value is calculated for each table element. Worker bees create a new table based on the solution table. The values for the new points are calculated and sorted so that the best score is the first item in the table. The next stage is the "bee dance", i.e. the exchange of information on the best sources of food (the best solution). This step begins with checking if the new source of food is better than the best remembered source. If not, the selected fuzzy operator is used to fuzzify the value with negative orientation; otherwise the orientation is positive. This operation is done for each worker bee. Then we use the arithmetic of ordered fuzzy numbers to calculate the probability that the bees choose the solution according to formula 13. The orientations of numbers depend on whether the nectar value at the point is greater (positive orientation) or lower (negative orientation) than the current highest value. This step involves defuzzification of probaOFN value through the use of one of the selected defuzzification functions. The next stage is to create a neighbourhood. The first step is to randomly select values from 0 to 1 and compare them with the selection probability recorded for all function values. If the random value is lower than the selection probability calculated for a given function value, the neighbourhood is created as per the procedure in the OFNBee Algorithm. However, when the selected number is greater than the selection probability, the value at the point is copied. At the end of the stage, the function value is calculated at the new points. The table is then sorted. This process is repeated in cycles until the predetermined number of epochs is reached.

$$probaOFN = \frac{maksOFNf_p\beta\gamma}{\sum_{i=1}^{30} OFNf_ip\beta\gamma} \tag{13}$$

gdzie:

- $probaOFN$ – probability that the source is selected,
- $maksOFNf_p\beta\gamma$ – currently the highest amount of food (positively oriented fuzzy number),
- $\sum_{i=1}^{30} OFNf_ip\beta\gamma$ – fuzzy sum of all food values in the table.

6 Configuration of OFNBee Algorithm Parameters

The new OFNBee algorithm has been implemented in R language. The OFN-bee algorithm includes several configuration options. For proper operation it is necessary to specify the fuzzification and defuzzification operators and the population size when running the algorithm. Fuzzification and defuzzification operators were tested in the first part of the experiment. Each of the fuzzy functionals was launched with one of the two defuzzification functionals, and then the results for both were compared. Each algorithm run was repeated 30 times for all combinations of fuzzification and defuzzification operators. The new OFN-Bee algorithm proved its efficiency in the previous experiment. But another six commonly known and used test functions were selected to prove its effectiveness, which are usually seen when testing new algorithms [6,20]:

- Bukin N.6,
- Cross-in-tray,
- rop Wave,
- Eggholder,
- Levy,
- Holder Table.

These functions were chosen because the minimum value is different from zero except for Bukin N.6 and Levy functions. Thanks to the ability to generate ABC algorithm results, R language enables more precise time and accuracy analysis. Such features allowed us to better compare the two algorithms. Using algorithms written in the same language and run on the same hardware, we get the most reliable conditions for comparing both solutions.

7 Results of Experiments on Mathematical Functions

This part of the experiment provides a compilation of results for the best of the traditional ABC bee algorithms using OFNBee in the form of graphs. The purpose of selecting subsequent functions is to check if the algorithm is working properly. The values from 0 to 25 on the OX axis are obtained using the relevant beeT fuzzy functions according to: for GR defuzzification functions:

1. beeTfp010207,
2. beeTfn010205,
3. beeTfn010203,

4. beeTfn010405,
5. beeTrfn05,
6. beeTrfn07,
7. beeTrfn11,
8. beeTrRfn051,
9. beeTrRfn0711,
10. beeTrRfp0913.
11. beeTrRfn010.

and for the MF defuzzification functions:

12. beeTfp010207,
13. beeTfn010205,
14. beeTfn010203,
15. beeTfn010405,
16. beeTrfn05,
17. beeTrfn07,
18. beeTrfn11,
19. beeTrRfn051,
20. beeTrRfn0711,
21. beeTrRfp0913.
22. beeTrRfn010.

The last value marked on the graphs is the result of the ABC algorithm (red triangle). All functions were run 30 times for each combination of fuzzification

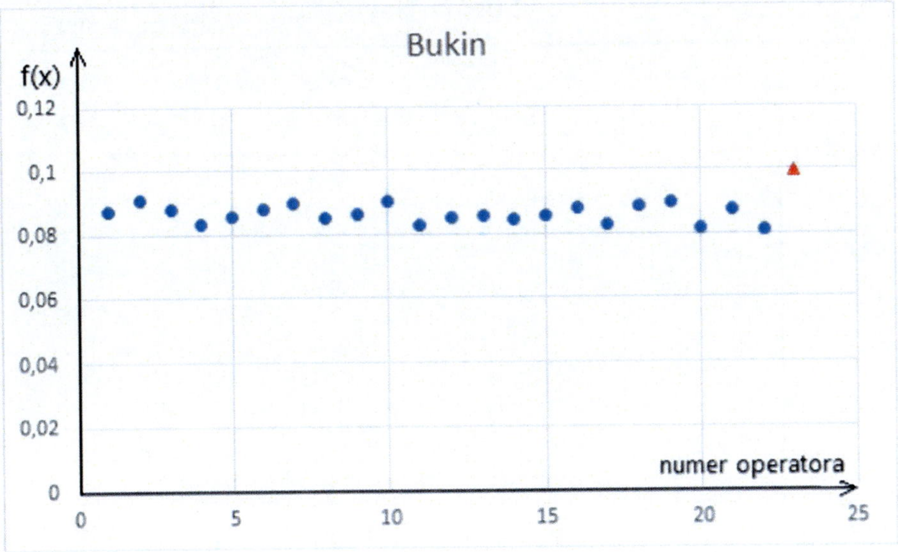

Fig. 9. Results of 30 independent OFNBee runs for the Bukin function. Blue points mean subsequent OFNBee values when using various fuzzification and defuzzification functions. The red triangle marks the best result of the ABC algorithm.

and defuzzification functions, and the calculated average is marked with a blue dot on the graph.

As can be seen in Fig. 9 for the Bukin N.6 function, in the case of all fuzzification and defuzzification operators, the new method achieves better results than ABC. A similar situation can be seen in Fig. 10 for the Cross-in-tray function, where the new method achieves results better than ABC for all fuzzification and defuzzification operators. At the same time, the fuzzification and defuzzification operators gave the same result for this function. In the case of the Eggholder function, the new method turned out to be much better for all fuzzification and defuzzification operators, as shown in Fig. 11. The results were similar for the Levy function (Fig. 12). The ABC algorithm marked with a red triangle indicates that OFNBee performs much better in this function. For the Holder Table function, the result in Fig. 13 demonstrates that the new method achieves exactly the expected point regardless of selected fuzzification and defuzzification operators.

8 Discussion of Results

Bukin N.6 function Fig. 9 we can see that the results for the new hybrid OFNBee method are closer to the expected 0 than in the ABC algorithm. In case of the Cross-in-tray function, the global minimum value is $-2{:}06261$. The results of both algorithms for the function are shown in Graph 1.10. We can see that all configurations of the new method achieve better results than the ABC algorithm.

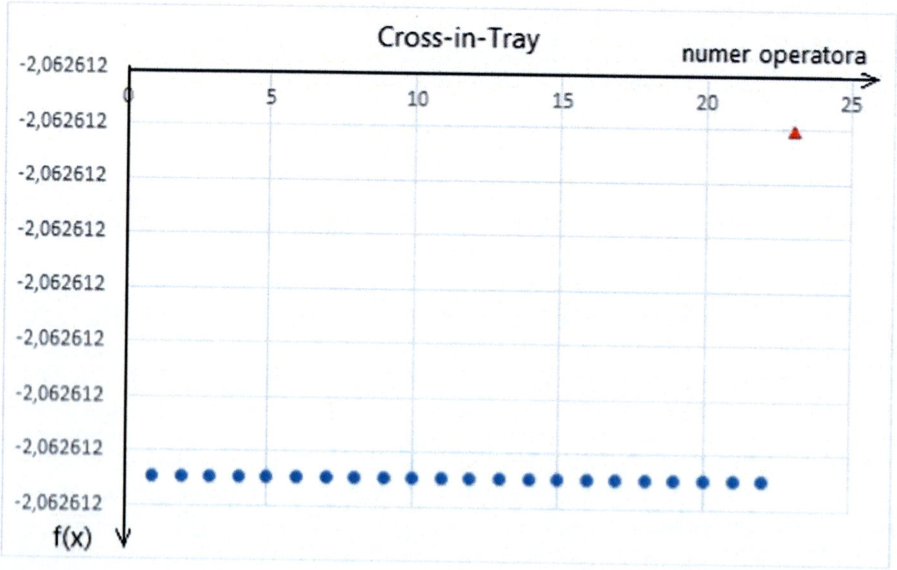

Fig. 10. Results of 30 independent OFNBee runs for the Cross-in-tray function. Blue points mean subsequent OFNBee values when using various fuzzification and defuzzification functions. The red triangle marks the best result of the ABC algorithm

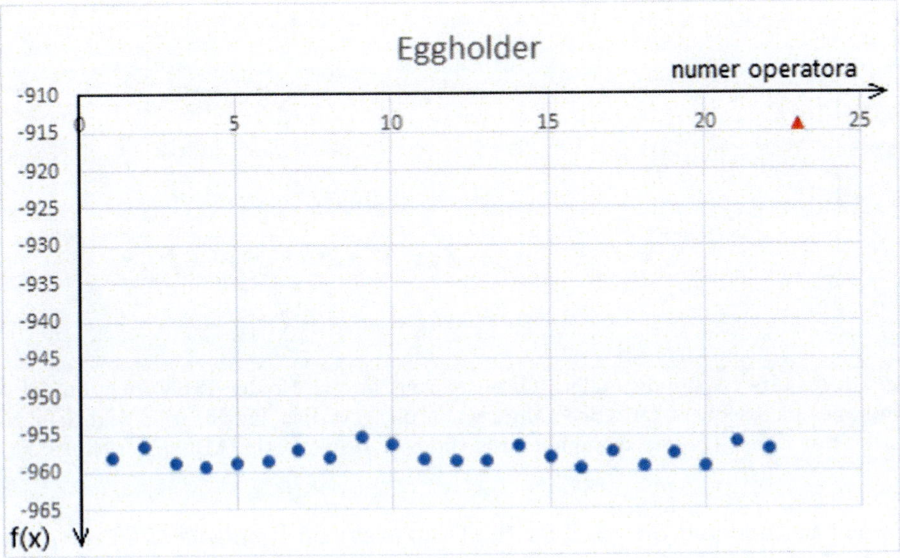

Fig. 11. Results of 30 independent OFNBee runs for the Eggholder function. Blue points mean subsequent OFNBee values when using various fuzzification and defuzzification functions. The red triangle marks the best result of the ABC algorithm

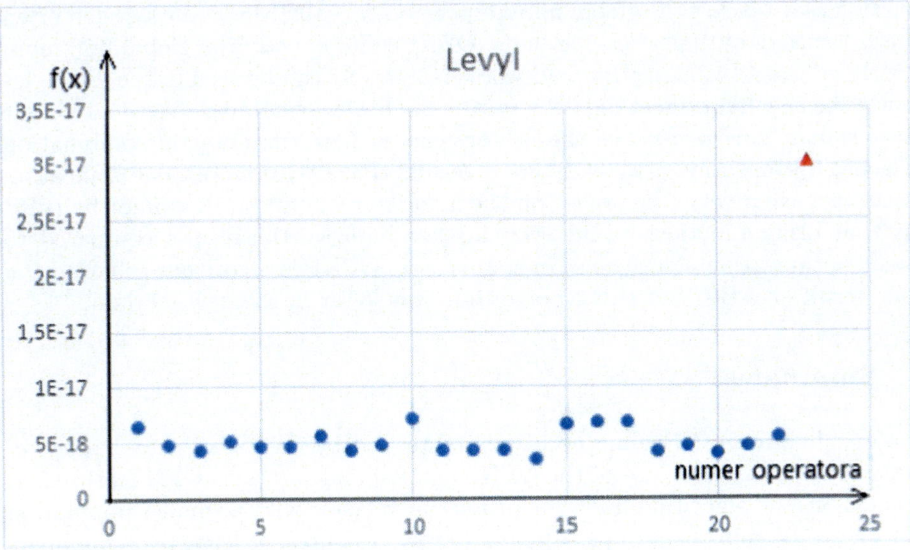

Fig. 12. Results of 30 independent OFNBee runs for the Levy function. Blue points mean subsequent OFNBee values when using various fuzzification and defuzzification functions. The red triangle marks the best result of the ABC algorithm

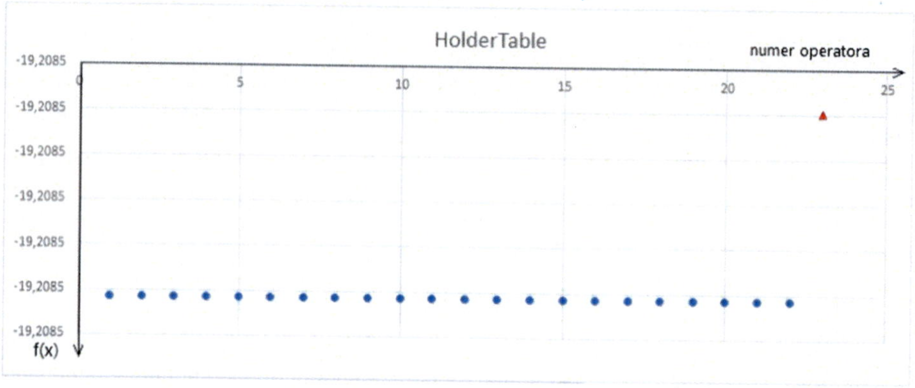

Fig. 13. Results of 30 independent OFNBee runs for the Holder Table function. Blue points mean subsequent OFNBee values when using various fuzzification and defuzzification functions. The red triangle marks the best result of the ABC algorithm

The new method and the ABC for the Drop Wave function generated the same results of -1, which reaches the global minimum at point -1. Therefore, it should be considered that the new hybrid optimisation method is in this case as effective as ABC. Eggholder achieves the minimum at point $-959{:}6407$. In Fig. 11, the ABC algorithm results are worse than for OFNBee. Also in the case of the Levy function, which reaches the global minimum at point 0, the new OFNBee method got better results - Graph 12. An interesting case is the Holder Table function as it reaches the global minimum at point $-19{:}2085$, and its design often causes issues in optimisation methods. OFNBee hit exactly the global minimum point, and the ABC algorithm had worse results. As can be seen in the examples above, the new hybrid method achieves results better or comparable to ABC, but never worse. This is because the use of ordered fuzzy numbers in combination with bee optimisation gives very good results. Fuzzification and defuzzification functions have a very big impact on the algorithm operation. It cannot be ruled out that other functions could even further improve the results compared to classic optimisation algorithms. In some cases, OFNBee algorithm achieved the same result as ABC, but the average time was lower in the new method.

9 Summary

The new hybrid OFNBee method is characterised by the use of three groups of bees, but thanks to the use of the OFN notation and the basic OFN feature, i.e. orientation, the algorithm finds a solution much faster with a smaller population size - fewer bees in the groups. Thanks to their features, ordered fuzzy numbers can very well reflect the sense of decision-making in the hive during the dance of real bees.

The new method utilises the fuzzification and defuzzification operators, and the algorithm result is influenced by a combination of fuzzification and defuzzification operators. Also, it cannot be ruled out that the selection of operators in terms of the problem to be solved may have an impact on the OFNBee results.

References

1. Czerniak, J.M.: Zastosowania skierowanych liczb rozmytych w wybranych algorytmach optymalizacji rojowej (2019)
2. Czerniak, J.M., Dobrosielski, W.T., Filipowicz, I.: Comparing fuzzy numbers using defuzzificators on OFN shapes. In: Prokopowicz, P., Czerniak, J.M., Mikolajewski, D., Apiecionek, L., Slezak, D. (eds.) Theory and Applications of Ordered Fuzzy Numbers. A Tribute to Professor Witold Kosinski. Studies in Fuzziness and Soft Computing, pp. 99–132. Springer, Cham (2017). https://doi.org/10.1007/978-3-319-59614-3_6
3. Czerniak, J.M., Zarzycki, H.: Artificial Acari optimization as a new strategy for global optimization of multimodal functions. J. Comput. Sci. **22** (2017)
4. Das, S., Biswas, A., Dasgupta, S., Abraham, A.: Bacterial foraging optimization algorithm: theoretical foundations, analysis, and applications. In: Abraham, A., Hassanien, A.E., Siarry, P., Engelbrecht, A. (eds.) Foundations of Computational Intelligence. Studies in Computational Intelligence, vol. 3, pp. 23–55. Springer, Heidelberg (2009). https://doi.org/10.1007/978-3-642-01085-9_2
5. Dobrosielski, W.T., Szczepański, J., Zarzycki, H.: A proposal for a method of defuzzification based on the golden ratio–GR. In: Atanassov, K.T., et al. (eds.) Novel Developments in Uncertainty Representation and Processing. AISC, pp. 75–84. Springer, Cham (2016). https://doi.org/10.1007/978-3-319-26211-6_7
6. Eberhart, R.C., Shi, Y.: Comparing inertia weights and constriction factors in particle swarm optimization. In: Proceedings of the 2000 Congress on Evolutionary Computation, vol. 1, pp. 84–88. IEEE (2000)
7. Galas, K.: Drive unit as a replacement for the platform. Stud. Mater. Appl. Comput. Sci. (ISSN 1689-6300) **12**(1), 10–14 (2020)
8. Izukl, B., Piechowiak, M.: The impact of ant colony optimization parameters on the connections efficiency in networks. Stud. Mater. Appl. Comput. Sci. (ISSN 1689-6300) **12**(2), 4–9 (2020)
9. Kosiński, W., Markowska-Kaczmar, U.: On evolutionary approach for determining defuzzyfication operator. In: Proceedings of the International Multiconference on Computer Science and Information Technology, pp. 93–101 (2006)
10. Kosinski, W.: On defuzzyfication of ordered fuzzy numbers. In: Rutkowski, L., Siekmann, J.H., Tadeusiewicz, R., Zadeh, L.A. (eds.) Artificial Intelligence and Soft Computing - ICAISC 2004. LNCS, vol. 3070, pp. 326–331. Springer, Heidelberg (2004). https://doi.org/10.1007/978-3-540-24844-6_46
11. Kosinski, W., Piasecki, W., Wilczynska-Sztyma, D.: On fuzzy rules and defuzzification functionals for Ordered Fuzzy Numbers. In: Proceedings of AI-Meth 2009 Conference. AI-METH Series, Gliwice, November 2009, pp. 161–178 (2009)
12. Kosinski, W., Wilczynska-Sztyma, D.: Defuzzification and implication within ordered fuzzy numbers. In: Fuzzy Systems (FUZZ), 2010 IEEE International Conference on Computational Intelligence, pp. 1–7. IEEE (2010)
13. Kosiński, W., Prokopowicz, P.: Algebra liczb rozmytych. Matematyka Stosowana **5**, 37–63 (2004)

14. Kosiński, W., Prokopowicz, P., Slezak, D.: On algebraic operations on fuzzy numbers. In: Intelligent Information Processing and Web Mining: Proceedings of the International IIS: IIPWM 2003 Conference held in Zakopane, pp. 353–362 (2003)
15. Kwaśnicka, H.S.: Multiobjective particle swarm optimization using fuzzy logic (2011)
16. Kwaśnicka, H.S., Markowska-Kaczmar, U., Kwasnicka, H.: Metody inspirowane natura w zastosowaniach (2011)
17. Lukowski, J.: Logical description of a combinatorial system by the binary representation method. Stud. Mater. Appl. Comput. Sci. (ISSN 1689-6300) **11**(1), 10–12 (2019)
18. Meng, X., Liu, Y., Gao, X., Zhang, H.: A new bio-inspired algorithm: chicken swarm optimization. In: Tan, Y., Shi, Y., Coello, C.A.C. (eds.) Advances in Swarm Intelligence. LNCS, vol. 8794, pp. 86–94. Springer, Cham (2014). https://doi.org/10.1007/978-3-319-11857-4_10
19. Pham, D.T., Negm, M., Otri, S.: Using the bees algorithm to solve a stochastic optimization problem. In: 4th International Virtual Conference on Intelligent Production Machines and Systems (IPROMS), Whittles, Dunbeath, Scotland, pp. 454–461 (2008)
20. Pohlheim, H.: Examples of objective functions. Retrieved **4**(10), 2012 (2007)
21. Polberg, S., Paprzycki, M., Ganzha, M.: Developing intelligent bots for the diplomacy game. In: Ganzha, M., et al. (eds.) Proceedings of the 2011 Federated Conference on Computer Science and Information Systems, Los Alamitos, CA. IEEE CS Press, pp. 589–596 (2011)
22. Wilczyńska-Sztyma, D., Wielki, K.: Direction of research into methods of defuzzification for ordered fuzzy numbers. In: XII International PhD Workshop OWD 2010, 23–26 October 2010, July 2019
23. Zadeh, L.: Fuzzy sets. Inf. Control **8**(3), 338–353 (1965). https://www.sciencedirect.com/science/article/pii/S001999586590241X

Engineering Applications

A Hybrid System for Intelligent Waste Measurement

Łukasz Apiecionek[1(✉)], Robert Palka[2], Mateusz Biedziak[2], Maciej Tomczak[2], Łukasz Blechman[2], and Dawid Ewald[1]

[1] Casimir the Great University, Bydgoszcz, Poland
lapiecionek@ukw.edu.pl
[2] TELDAT Sp. z o.o. sp. k., Bydgoszcz, Poland

Abstract. The technological growing progress increases the demand for various goods, which generates significantly a greater amount of wastes. Properly management of the waste is a very important aspect. To increase the effectiveness and the accuracy of this process, it is necessary to implement a relevant system for their control and monitoring. The National Regulation of the Minister of Environment from 14th of December 2016 "on recycling levels, preparation for reuse and recovery by other methods of other municipal waste fractions" indicates that in following years they must meet the increasing recycling rates. Eventually until 2020 they should attain the level of 70%. To achieve that, it is necessary to improve Waste Management System. The only possible, relatively simple and low-cost solution, which would improve management of waste, is an efficient and precise solution for weighing and cataloguing the type of the collected waste from a given location. In this article, authors show a hybrid system for the intelligent waste measurement.

Keywords: Hybrid system · Intelligent system · Waste measurement

1 Introduction

The technological growing progress increases the demand for various goods, which generates significantly a greater amount of waste. Properly management of the waste is a very important aspect. To increase the effectiveness and the accuracy of it, is necessary to implement a relevant system for their control and monitoring. The National Regulation of the Minister of Environment from 14th of December 2016 "on recycling levels, preparation for re-use and recovery by other methods of other municipal waste fractions" indicates that in following years they must meet the increasing recycling rates. Finally, in 2020, they must be at the level of 70%. Moreover, the control of waste transport from their collection to the municipal installation, will also be required [1,2,12]. To achieve that, it is necessary to improve Waste Management System. Current systems do not give such a possibility and are very expensive in implementation. This is mainly due to the fact that the devices are inaccurate because of their poor adjustment too difficult and dynamically changing external conditions during their usage (e.g. ambient temperature, weighing methods - insufficient elimination of measurement errors,

K. T. Atanassov et al. (Eds.): IWIFSGN 2020/BOS/SOR 2020, LNNS 338, pp. 259–269, 2022.
https://doi.org/10.1007/978-3-030-95929-6_20

connected to the lack of leveling of the garbage vehicles used for waste collection during the measurements and using single-point measurements). According to the survey among entities involved in municipal waste collection, the price was the main obstacle in implementing earlier such a system. The second factor was the measurement inaccuracy. The only possible, relatively simple and low-cost solution, which would improve management of waste, is the efficient and precise solution for weighing and cataloguing the type of collected waste from a given location. Later in this article, authors show the hybrid system for the intelligent waste measurement.

2 Waste Management

According to the data collected by the Central Statistical Office, 12 485 thousand tons of municipal waste was produced in Poland in 2018, which is 4.3% more than in 2017. This gives an average of 325 kg of collected waste per capita. The highest amounts were in the Lower Silesia voivodeship to which the city of Wroclaw has definitely contributed - 394 kg per capita. The lowest amounts were in Swietokrzyskie voivodeship - 201 kg.

The smallest increase was noted in mixed packaging waste - 15 kg (14 kg in 2017), plastics - 9 kg (8 kg in 2017) and paper and cardboard - 7 kg (6 kg in 2017). The total of collected glass waste per capita increased by 8.3% regard to 2017 and is at the level of 13 kg. A substantial increase in the amount of biodegradable waste has also been noted - 26 kg (23 kg in 2017). A similar situation applied to the large size waste and this was the biggest increase from 11.5 kg to 14 kg (21.7%) [3, 13, 14]. Figure 1 shows a structure of separately collected municipal waste by waste fraction.

As shown in the abovementioned data, the amount of the generated waste is increasing significantly. This is because of the continuous technological progress, which generates an increased need for goods. A constant growth results in increasing the financial outlays in the context of the control of waste management, contributing to a decrease in the savings of the public sector. To correct the current situation, it is necessary to implement an efficient and precise solution for weighing, cataloguing and monitoring the type of collected waste from a given location. According to the survey among the entities involved in municipal waste collection (Sanitras Sp. z o.o. and REMONDIS Sp. z o.o.) the main obstacle in implementing such a system earlier, was its price (100% of the answers).

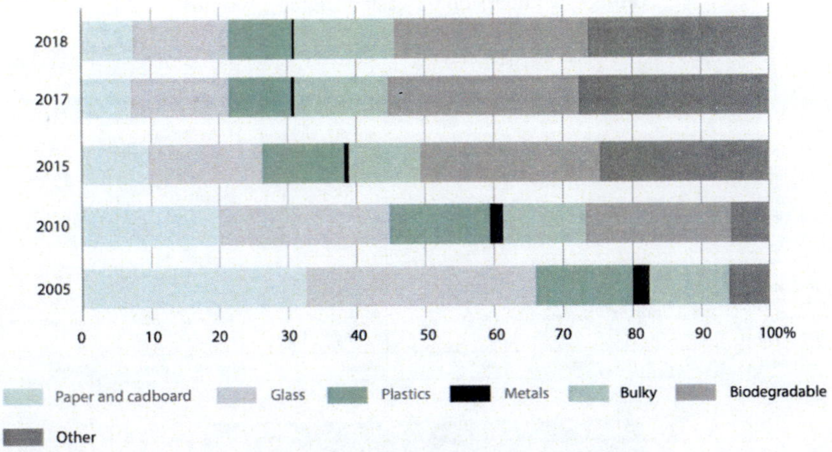

Fig. 1. Structure of separately collected municipal waste by waste fraction [3]. Source: Environment 2019, Central Statistical Office

3 Hybrid System for Intelligent Waste Measurement Method

Nowadays on the market are competitive systems for weighing municipal waste. These products differ to each other in their parameters. This is a very important factor that affects the ease of use and the possibility of using the system in any area and in any weather conditions. That is why the TELDAT company decided to compare its solution with competitive systems. The list of parameters of six solutions and their comparison is presented in Table 1. The analysis of the parameters of the existing solutions has shown that all the competitive solutions designed for weighing municipal waste do not meet users' requirements. One of the key parameters, which proves it, is not taking into account the vehicle's tilt during measurements or its too low compensation. Garbage trucks perform their tasks in different conditions. Therefore, it is necessary to take into account obstacles such as steep driveways or high kerbs.

In order to create a hybrid and intelligent solution made for efficient and precise weighing, cataloguing and monitoring of the collected waste, TELDAT company developed a project named "Innovative, complex and precise system for dynamic, fast and automatic waste weighing, managing and monitoring with the use of an innovative hybrid measurement method", co-financed by the National Centre for Research and Development. The goal of the project was to develop devices able to measure the weight in the Y(b) accuracy class, monitor and record it precisely and, in the future, to account the waste producers, basing on the weight of different types of waste.

Hybrid system for intelligent waste measurement includes: sensors layer, conversion layer, computation layer and result layer. The structure of above-mentioned system is shown on Fig. 2. The first layer (sensors layer) is responsible for automatic measurement of waste weight taking into consideration the vehicle deflection. In this layer there are hydraulic block with the oil pressure sensor, connection block and a beam

Table 1. Comparison of waste weighing systems

Feature	TELDAT	Elte GPS [4]	Progress Wagotronic [5]	XTrack [6]	C-Trace [7]	Globtrak [8]
Measurement accuracy	Y(b) class	Y(b) class	III-IIII OIML class	2 kg for $0.7\ m^3$ containers 5 kg for $1.1\ m^3$ containers	2 kg for two-wheeled containers 5 kg for four-wheeled containers	± 2 kg
Real measurement accuracy	YES	NO	NO	NO	NO	NO
Operating temperature range	$-30 \div +65\ ^\circ C$	No data	$-25 \div +60\ ^\circ C$	$-10 \div +40\ ^\circ C$	$-10 \div +40\ ^\circ C$	$-20 \div +50\ ^\circ C$
Real operating temperature range	$-30 \div +65\ ^\circ C$	No data	No data	No data	$-20 \div +60\ ^\circ C$	No data
Tilt compensation	25°	None	10°	None	No tilt message	None

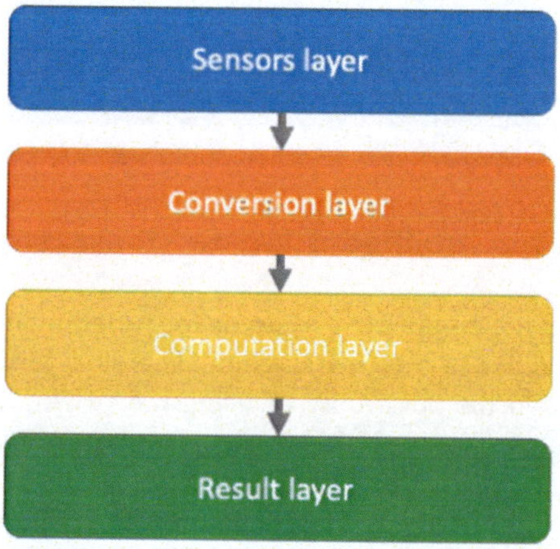

Fig. 2. Structure of hybrid system for intelligent waste measurement. Source: own study

with load cell sensors. The hydraulic block branches the oil flow through the hydraulic system, which gives the possibility to measure the oil pressure in a given point of the position of the piston accurately. The measurement beam with two load cell sensors, performs the bin weight measurement. Sensors used to build the system are consistent with the requirements of the OIML R60 standard and which were correctly performing the measurements in various environmental conditions and in a every place of use. The measurement error related to the low quality of the devices could be avoided thanks to the use of this type of sensor.

The vehicle deflection was not yet taken into consideration in the existing solutions on a larger scale, which results in the incorrectness of the measurements. Therefore, in the second layer (conversion layer) the measurement error related to the deflection of the construction was eliminated by the use of the result correction based on the developed algorithms. Moreover, the conversion layer is responsible for converting data from sensors to one common format.

In the next step, the converted data are transferred to the computation layer. This layer is using the Central Computational Unit which is responsible for reading and processing data from the sensors, as well as controlling the solenoid valves.

The task of the result layer is transfer finished results from the computation layer to the special database. Results are also showing on the Result Display Unit.

Devices included in the hybrid system for intelligent waste measurement are shown on Fig. 3. The aforementioned database is part of the Municipal Waste Management System. This system includes server software (databases, data access interface) and client software dedicated to export companies (e.g. municipal companies) and Local Government Units. It is supposed to cooperate with Communication Modules working on garbage trucks and allow processing and presentation of reports to all users. The system interface is shown in Fig. 4.

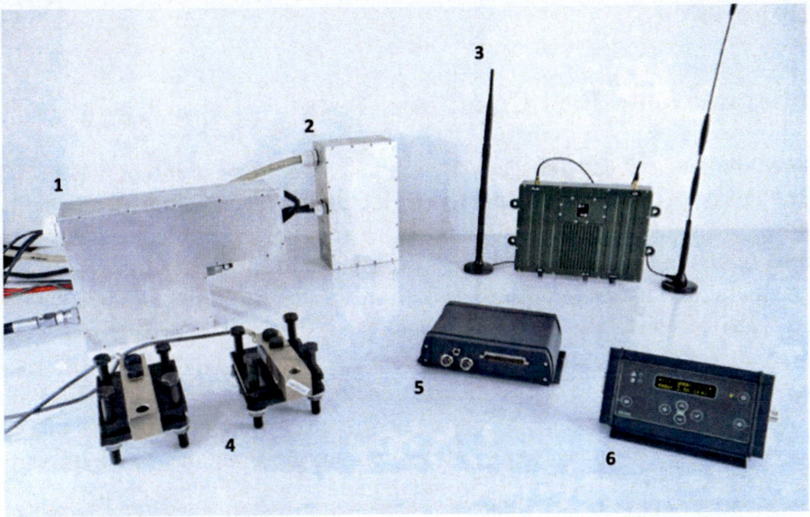

Fig. 3. Devices included in the system (1 - hydraulic block, 2 - connection block, 3 - Communication Module, 4 - load cell sensors, 5 - Central Computational Unit, 6 - Result Display Unit). Source: own study

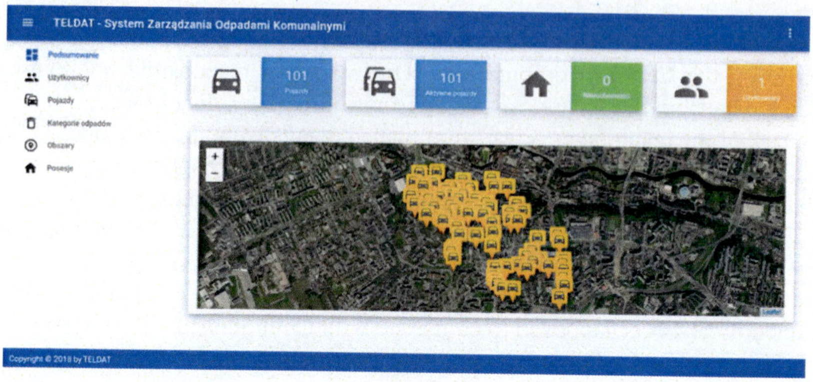

Fig. 4. Devices included in the system (1 - hydraulic block, 2 - connection block, 3 - Communication Module, 4 - load cell sensors, 5 - Central Computational Unit, 6 - Result Display Unit). Source: own study

4 Laboratory and Real Conditions Tests

To achieve the data needed for analysis and potential changes in construction, laboratory tests were performed, with the use of a climatic chamber, dust chamber, water tank and mechanical shaker. Tests included the resistance to different environmental conditions (temperature range for devices working in difficult conditions), adaptation to fast temperature changes (a quick pass from the temperature below zero to above zero and again to below zero, in order to check the resistance of the electronic circuit to condensation), dustproofs, water-resistance and resistance of vibration and shock.

Temperature changes are one of the main factors influencing the operation of electronic devices, therefore climate tests are one of the most important stages of strength testing. On the basis of many years of research and observations, the authors of the publication "Temperature tests of electronic modules" proved that every 10 °C increase of devices temperature can cause twice as much damage to the silicon connectors of power transistors. An increase of 15 °C can cause damage to capacitors and 35 °C to resistors. On the other hand, if the temperature changes from 27 °C to 70 °C, it reduces the strength of soldering connections [9]. Such situations are not acceptable when the devices are designed to operate in difficult and changing environmental conditions. According to IBM data, when temperature increase only by 5°, this turns into a 10–15% increase in failure rate [10]. TELDAT has decided to carry out tests to ensure full climatic resistance of the equipment to various climatic conditions.

The test consisted in checking devices resistance in temperature range from −30 °C to +65 °C with a 5° jump. The test of adaptation to fast temperature changes consisted in a quick transition from positive to negative temperature and vice versa, to check the resistance of electronic systems to condensation. Figure 5 shows prototype devices placed in a climate chamber. During the test of resistance to different environmental conditions devices worked correctly, i.e.: they correctly measured weight, processed data from sensors, displayed the results and faultless data transmission. After the analysis of the collected results, the range of temperature of use was determined from −30 °C

Fig. 5. Climatic chamber tests and vibration/shock tests. Source: own study

to +65 °C. Moreover, test of adaptation to fast temperature changes confirmed that the condensation not effect on operation of the devices.

Many manufacturers of mobile devices use different types of protection of components against water. However, it does not always operate properly. Dustproofs and water resistance have a huge impact on the operation of electronic devices, in difficult conditions [11].

The waterproofness test consisted in checking resistance of devices to flooding (IPx7 classification) in a water tank. The device submerged to 1-m depth. The dustproofs test consisted in checking the resistance of the communication module to dust (IP6x classification) in a dust chamber for 8 h. The above tests were carried out in accordance with the Polish Standard PN EN 60529:2003. Figure 6 shows devices during water and dust tightness tests. After tests, personnel visually checked condition of devices by opening their casings. Water and dust did not get inside enclosures, so components were not exposed to water and dust. Test confirmed devices tightness classification.

The resistance to vibration and shock test consisted in checking the resistance of devices to vibrations occurring in actual operation conditions. The test was carried out on the PN-EN 60068-2-6 standard. The Central Computational Unit and Result Display Unit were subjected to 20 cycles of mechanical sine wave operation with an amplitude of not less than 1 g and frequency range of 12.5–25 Hz with sweeping. The Communication Module was subjected to cycles of mechanical sine wave with amplitude 1.5–5 g and frequency range 12.5–50 Hz. Devices were vibrated in three mutually perpendicular axes. At the end checked mechanical damage and conducted a functional check of devices. The Fig. 7 shows devices during resistance of vibration and shock test. The resistance of vibration and shock test confirmed that mechanical vibrations did not cause damages to devices. Moreover, the devices retained full metrological capacity. It was confirmed by performing a series of 10 measurements using

Fig. 6. Devices during water and dust tightness tests. Source: own study

standard weights. Table 2 shows comparison of results obtained before and after resistance of vibration and shock test. This confirmed the metrological capability of equipment because errors are within limits of the standard. In order to verify the ability of the devices to work in different environmental conditions, the tests were performed in the outside laboratories, including: Environmental Test Laboratory in the Military Communication Institute in Warsaw, Laboratory of Vibroacoustics, Shock Resistance and Magnetic Fields in the Research and Development Maritime Technology Centre in Gdynia and the Research Laboratory in the BOSMAL Automotive Research and Development

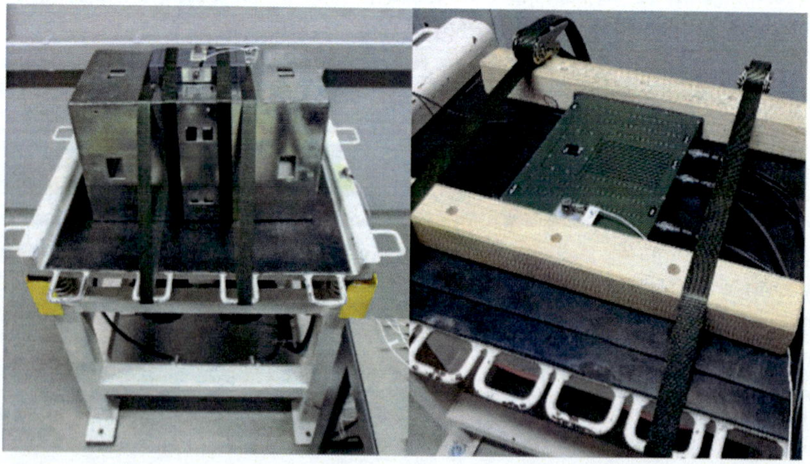

Fig. 7. Devices during resistance of vibration and shock test. Source: own study

Table 2. Comparison of weighing results before and after resistance of vibration and shock test

No.	Weigh [kg]	Voltage before shaking [V]	Voltage after shaking [V]
1	10	4,067	4,145
2		4,046	4,231
3		4,024	4,12
4		4,039	4,121
5		4,027	4,108
6		4,013	4,109
7		4,017	4,095
8		4,015	4,109
9		4,013	4,109
10		4,014	4,098

Institute in Bielsko-Biała. Tests included the adaptation to dynamically changing temperature (quick pass from the temperature below 0 to above 0 and again to below 0, in order to check the resistance of the electronic circuit to condensation), resistance to different environmental conditions (temperature ranges for devices operating in difficult environmental conditions - form $-30\,^{\circ}\mathrm{C}$ to $+65\,^{\circ}\mathrm{C}$), resistance to vibration and shock, water and dust resistance. Devices received positive test results, which confirmed their ability to work in different environmental conditions and in every place of use.

A real-life scenario with the use of a vehicle equipped with a dust-free construction has also been performed. The tests included checks on the test area, as well as in 10 different locations, selected having regard to the unevenness and slope of the terrain (Fig. 8). Reference weights, which were made especially for these tests, were used. According to the previously developed test plan, 10 measurements of each weight were undertaken at each location. The algorithms used correlated the existing inequalities and obtained satisfactory results. Table 3 shows results of a series of 10 measurements for 10 kg and 80 kg reference weights.

Obtained results were satisfying due to a small error of maximum 1 kg.

This type of system is intended to join precise, dynamic, fast and automatic measurements with waste management and their monitoring. This solution has to have a competitive price, as well as the accuracy of the measurements in the Y(b) class in different environmental conditions and in every place of use.

Fig. 8. Tests in real environment. Source: own photo

Table 3. Results of a series of 10 measurements for 10 kg and 80 kg reference weights

No.	Weigh [kg]	Calculated weight [kg]	Error [kg]
1	10	9	1
2		10	0
3		10	0
4		11	1
5		10	0
6		10	0
7		9	1
8		9	1
9		10	0
10		10	0
No.	Weigh [kg]	Calculated weight [kg]	Error [kg]
1	80	81	1
2		80	0
3		80	0
4		80	0
5		79	1
6		80	0
7		80	0
8		80	0
9		81	1
10		80	0

5 Conclusions

This project made it possible to acquire knowledge, based on which it can be concluded that it is possible to develop a system for the vehicles with dust-free construction, able to weigh municipal waste taking into consideration the deflection of the vehicle. Currently the deflection is not taken into account in the available solutions on a larger scale, which results that the results of the measurements are improper. The conducted tests proved that in order to eliminate the error from the vehicle deflection, it is necessary to develop an algorithm counteracting the resultant error. The use of the OIML R60 compliant sensors will assure that the measurements are correct in all environmental conditions and in every place of use. Moreover, the results of tests confirmed the resistance of the solution to different environmental conditions, fast temperature changes, dust, water, vibration and shock.

This publication is the result of the developed project named "Innovative, complex and precise system for dynamic, fast and automatic waste weighing, managing and monitoring with the use of an innovative hybrid measurement method", co-financed by the National Centre for Research and Development.

References

1. The National Regulation of the Minister of Environment from 14th of December 2016 on recycling levels, preparation for re-use and recovery by other methods of other municipal waste fractions, p. 3 (2016)
2. Law of 13th September 1996 on keeping the municipalities clean and in order, p. 51 (1996)
3. Central Statistical Office: Environment. Warsaw 2019, pp. 153–158 (2019)
4. Company homepage Elte GPS. https://www.eltegps.pl. Accessed 23 Nov 2020
5. Company homepage Progress Wagotronic. http://wagi.pl. Accessed 23 Nov 2020
6. Company homepage XTrack. http://xtrack.com/pl. Accessed 24 Nov 2020
7. Company homepage C-Trace. https://www.c-trace.de/pl-PL/home. Accessed 24 Nov 2020
8. Company homepage Globtrak. https://www.globtrak.pl. Accessed 24 Nov 2020
9. Cwirko, J.O., Cwirko, R.O.: Badania temperaturowe modułów elektronicznych. Biuletyn WAT **LVII**(2), 133–142 (2008)
10. Vilsbeck, Ch.: IDE - praca ciagła. PC World Komputer (2003)
11. Hirsz, M.: Stopnie ochrony zapewniane przez obudowy (IP), KDBEx, pp. 1–8
12. Aleksandra, M.: Selected methods of artificial intelligence implemented in C/C++ programming language. Stud. Mater. Appl. Comput. Sci. **11**(2), 13–18 (2019). ISSN 1689-6300
13. Rojek, I., Krolikowski, Z., Marciniak, M.: Obstructive sleep apnea, microcontroller, database, Web application, mobile application. Stud. Mater. Appl. Comput. Sci. **11**(2), 15–21 (2019). ISSN 1689-6300
14. Bednorz, I.: Monitoring of network equipment and reporting through Spectrum and Performance Management applications from CA Technologies, vol. 12, issue number 2, pp. 10–15 (2019)

Method for Multiplication Transmission Interfaces

Łukasz Apiecionek[1(✉)], Robert Palka[2], Mateusz Biedziak[2], Maciej Tomczak[2], Łukasz Blechman[2], and Dawid Ewald[1]

[1] Casimir the Great University in Bydgoszcz, Bydgoszcz, Poland
lapiecionek@ukw.edu.pl
[2] TELDAT Sp. z o.o. sp. k., Bydgoszcz, Poland

Abstract. In the era of the universal Internet access, the popularity of mobile devices increases. They are used to the very broad extent and offer more and more innovative functionalities. The number of mobile devices and users grows up every year. In 2019 recorded 6.8 billion users and 13 billion mobile devices. In turn, it is estimated that through continuous growth, in 2023 the number of users will reach 7.3 billion and mobile devices - 16.8 billion. With the popularity of increase of mobile solutions, including smartphones and tablets, the possibilities of their use are extending. Current trends are focused on the miniaturization and the maximum mobility of solutions, while increasing the performance of components. Nowadays, many organizations operate globally. Their employees operate in large areas and often in different locations. Therefore, very the important thing is to provide secure, the efficient and effective communication by using all available transmission interfaces. The equipment used e.g.: in security and defence sectors is exposed to various types of danger and unusual situations. Therefore, it is necessary that the devices used in these sectors should meet the required standards and be resistant to climate change and extremely difficult operating conditions. In this article, authors show modern solution of mobile devices, which meet resistant to extremely difficult operating conditions and use the method for multiplication transmission interfaces.

Keywords: Mobile devices · Transmission interfaces · Laboratory tests · Real conditions tests

1 Introduction

In the era of the universal Internet access, the popularity of mobile devices increases. They are used to the very broad extent and offer more and more innovative functionalities. The number of mobile devices and users grows up every year. Mobile Statistics Report 2019–2023, created by THE RADICATI GROUP presents data on the number of active users and mobile devices, including smartphones and tablets in the whole world. In 2019 recorded 6.8 billion users and 13 billion mobile devices. In turn, it is estimated that through continuous growth, in 2023 the number of users will reach 7.3 billion and mobile devices - 16.8 billion [1]. Chart 1 shows the number of mobile devices in the whole world between 2019–2023.

K. T. Atanassov et al. (Eds.): IWIFSGN 2020/BOS/SOR 2020, LNNS 338, pp. 270–279, 2022.
https://doi.org/10.1007/978-3-030-95929-6_21

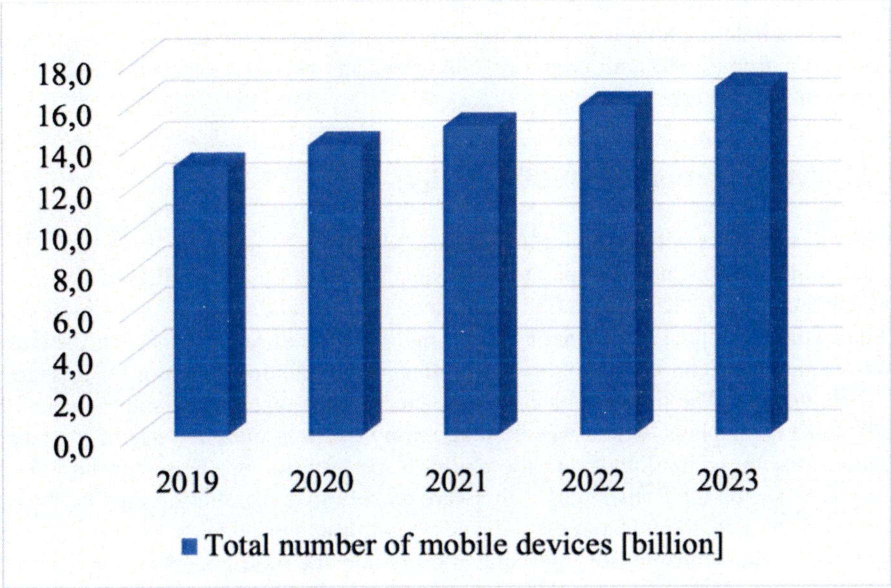

Fig. 1. The number of mobile devices in the whole world between 2019–2023. Source: THE RADICATI GROUP, INC., Mobile Statistics Report 2019–2023, A Technology Market Research Firm, Londyn 2019, s. 2–3

With the popularity increase of mobile solutions, including smartphones and tablets, the possibilities of their use are extending. Current trends are focused on miniaturization and maximum mobility of solutions, while increasing the performance of components. Nowadays, many organizations operate globally. Their employees operate in large areas and often in different locations. Therefore, very important thing is to provide the secure, efficient and effective communication by using all available transmission interfaces. The equipment used e.g. in security and defence sectors is exposed to various types of danger and unusual situations. Therefore, it is necessary that the devices used in these sectors should meet the required standards and be resistant to climate change and extremely difficult operating conditions. Employees who perform their duties in office, production hall or on battlefield need an access to the data exchange regardless of the prevailing conditions. Any malfunction in the functioning of a device, caused by not adapting to prevailing threats, exposes the safety of the employee. To prevent this kind of situations, mobile devices must be resistant to sudden climate change and demanding the operation environment. However, their existing functionality must remain unchanged. Therefore, it is very important that current users of mobile solutions are not forced to have the certain usage restrictions [8–10].

Intel company in the publication Deploying Tablets Safely in Manufacturing to Boost Productivity, described series of tests that were performed to identify the business, safety and ergonomics requirements for using mobile devices, including tablets, in a production environment. Ruggedized devices may protect against falling and damage, which could result in data loss and increase production costs [2]. Therefore, later in this article, authors will show modern solution of mobile devices, which meet resistant to extremely difficult operating conditions and use the method for multiplication transmission interfaces.

2 Modern Solution of Mobile Devices

In order to create a solution of mobile devices that provide secure, efficient and effective data transmission and will operate in all weather conditions the TELDAT company conducts the project: "Creating a family of innovative tablet devices with the increased resistance to mechanical and climatic conditions, allowing the effective use of the latest data transmission technologies" which is co-financed by National Centre for Research and Development. The project objective is to create three types of mobile tablets with 5″, 8″ and 12″ displays, which will be featured by the high mobility, use of the latest data transmission technologies and the ability to be used in very harsh terrain conditions. Currently, on the Polish market there are no solutions like this, created by Polish company, that meet all criteria of the defense standards and provide the secure, efficient and effective data transmission by using the method for multiplication transmission interfaces. Figure 1 shows prototypes of mobile tablets.

Fig. 2. Prototypes of mobile tablets (from the left: 5″, 8″ and 12″). Source: own study

Analysts, constructors and specialists from the Documentation and Production Preparation Workshop, carried out an analysis to select the structural elements and equipment components, that the developed configuration could ensure the increased resistance to mechanical and climatic conditions. Table 1 presents the technical specification of prototype mobile tablets. Displays used in the production of prototype devices enable to operate in high sunshine conditions and also have an automatic backlight adjustment function by using a built-in light sensor. The touch panel is made in capacitive technology with the Multi-Touch function, so it enables the use of tablets in special gloves. Moreover, the screens have an Anti-Glare coating, which disperses the light falling on them eliminating the so-called mirror effect. All devices have been equipped with two cameras, on both sides of a case, which ensures to record images and make e.g. videoconferences. Mobile tablets are also equipped with a built-in battery a capacity indicator which enables checking the state of a battery charge, also during the process of storing devices. All three devices are equipped with a set of connectors and ports, i.e.: RS232, USB, PS/2. Additionally, 5″ and 8″ devices use accelerometer and magnetometer. The 12″ device has additionally a cryptographic and SD card reader. Mobile tablets are operating with Microsoft Windows 10 Professional in 64-bit version.

Table 1. Technical specification of prototype mobile tablets

Components	Type of device		
	Tablet 5″	Tablet 8″	Tablet 12″
Processor	Intel Atom E3930, 1.30 GHz	Intel Atom E3930, 1.30 GHz	Intel Core i7 7600U, 2.80 GHz
RAM	8 GB	8 GB	8 GB
Graphic	Intel HD Graphics, 1 GB	Intel HD Graphics, 1 GB	Intel HD Graphics 620, 1 GB
Memory	SSD 32 GB	SSD 240 GB	SSD 120 GB
Display	5″, 800×480 pixels	8.4″, 1024×768 pixels	12.1″, 1024×768 pixels
Battery	19 Wh	78 Wh	156 Wh

3 Method for Multiplication Transmission Interfaces

To provide the secure, efficient and effective communication, TELDAT company is operating on architecture for the simultaneous use of all available transmission interfaces in the mobile devices presented in the above chapters of this article. Figure 2 shows a scheme of the solution. In current solutions of mobile devices, the data transmission can be ensured by using technologies such as: 3G, LTE or Wi-Fi (802.11 standard). These solutions choose only one and the best available transmission interface. It may lead to the data loss when the connection is broken. Only after reconnecting the possibility of the data transmission is available [3]. The situation of a temporary connection loss is unacceptable in security and defence sectors. To provide the stable and uninterrupted network access in mobile tablets, the TELDAT company is planning to develop and implement method for multiplication transmission interfaces.

The method for multiplication transmission interfaces consists in sending the data by all available transmission interfaces simultaneously. Thus, transmission does not block the bandwidth of interfaces. Moreover, disclosing of all transmitting information makes more difficulty, because data transmission is divided into several transmission interfaces. If an enemy intercept data from one interface, the rest of data is safe. Figure 3 shows a scheme of method for multiplication transmission interfaces. His solution will be application-independent by separating transmission layer from software layer. Moreover, it will ensure:

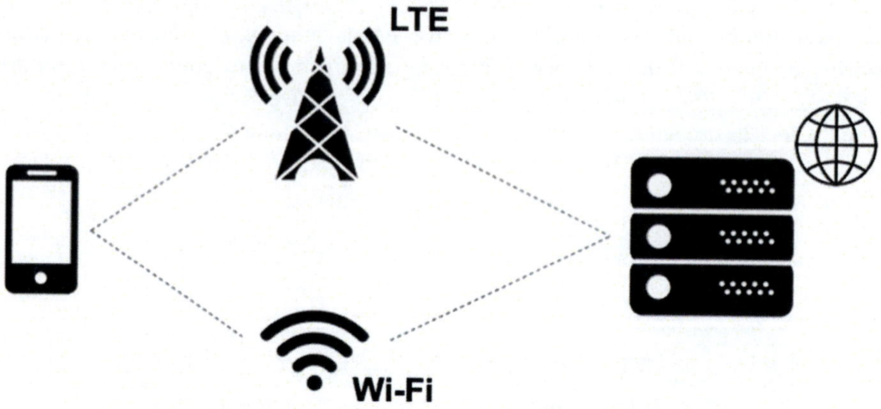

Fig. 3. Scheme of simultaneous use of all available transmission interfaces. Source: Ł. Apiecionek, M. Sobczak, W. Makowski, T. Vince, Multi Path Transmission Control Protocols as a security solution, 2015 IEEE 13th International Scientific Conference on Informatics, Poprad 2015, p. 29

- increased stability of data transmission – communication will be maintained regardless of a disappearance of individual interfaces signal;
- increase security of transmission – data will be transmitted through different paths, which means that interception one of them will not allow to intercept all information [4];
- increase in the data transfer rate – data will be transferred over all currently available interfaces, which will cause the effective bandwidth to be equal to the sum of the available interfaces.

Mobil tablets equipped with the method for multiplication transmission interfaces may allow to use of mobile technologies:

- for monitoring of events and systems, e.g.: in a protection of property and in monitoring of high-risk production lines (transmission of data, information, video images);
- for monitoring of the status of operated devices in Police, Fire Brigade, Ambulance Service and Polish Army [5];
- for cooperating with systems designed for public services and organizational units – ensure the stable and fast communication in Crisis Management Systems (e.g.: Police, Fire Brigade, Medical Rescue, Border Guard);
- in systems dedicated to the Armed Forces by increasing the reliability and speed of data transmission where human life and safety depend on it;
- in systems designed for all companies and institutions operating in large areas and at different locations - ensure stable and secure information exchange.

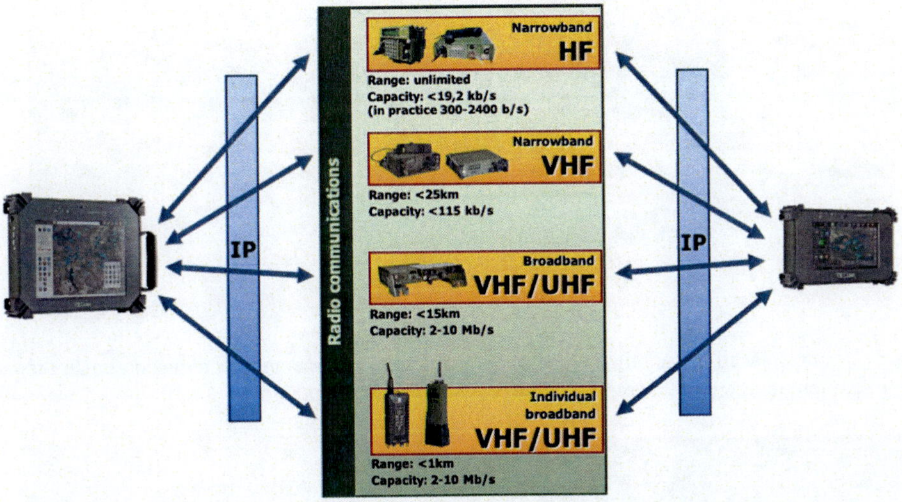

Fig. 4. Scheme of method for multiplication transmission interfaces. Source: own study

4 The Laboratory and Real Conditions Tests

To meet all rigorous criteria of defence standards and other specialist requirements, mobile devices has been tested in the laboratory. The plan assumed tests of:

- resistance to reduced ambient temperature, rapid temperature changes and increased ambient temperature;
- resistance to increased humidity;
- resistance to sinusoidal vibrations;
- resistance to falls;
- waterproofness.

Adverse temperature conditions, in the particular increased ambient temperature may increase a probability of failure of electronic devices and their components [6]. Therefore, the TELDAT engineers decided to verify the behavior of devices and their components during tests of resistance to reduced ambient temperature, rapid temperature changes and increased ambient temperature. Tests procedures were carried out in accordance with NO-06-A107:2005 and NO-06-A103:2005 defence standards. Tests assumed to place mobile devices connected to the power supply in the climate chamber and then check their operation at the $-50\,^{\circ}\mathrm{C}$ for 12 h, $-40\,^{\circ}\mathrm{C}$ for 4 h and $+60\,^{\circ}\mathrm{C}$ for 5 h. The next step was turn off mobile tablets and keep them for 6 h at $+70\,^{\circ}\mathrm{C}$. Figure 4 shows devices placed in the climate chamber during this test. Devices were operating correctly during tests. Moreover, after completion of tests, structural elements and components had not been damaged.

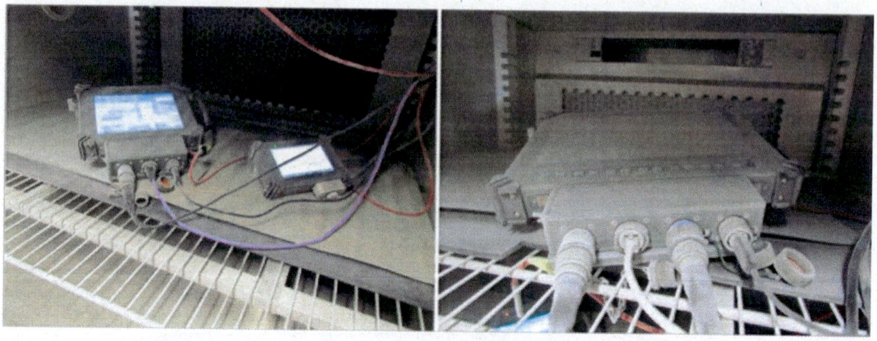

Fig. 5. Devices placed in the climate chamber during tests of resistance to reduced ambient temperature, rapid temperature changes and increased ambient temperature

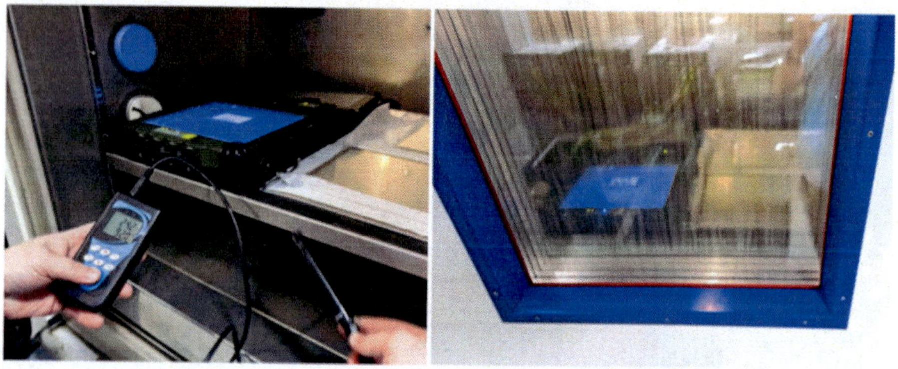

Fig. 6. Devices placed in the climate chamber during test of resistance to increased humidity. Source: own study

The second test was to verify the devices resistance to the increased humidity. The test procedure, like the first one, was carried out in accordance with NO-06-A107:2005 and NO-06-A103:2005 defence standards. The test assumed to place mobile devices connected to the power supply in the climate chamber and then check their operation at +35 °C with 80% a humidity for 5 h. The next step was to turn off the devices and keep them at +70 °C for 6 h. The Fig. 5 shows devices placed in the climate chamber during a test of resistance to the increased humidity. The damages of structural elements and components after a completion of the test had not been observed. This proved the resistance of mobile tablets to the increased environmental humidity.

The next test was to verify the devices resistance to sinusoidal vibrations from 12.5 Hz up to 50 Hz, with an overload up to 5g in three x/y/z planes. The test assumed to place devices on the mechanical vibration exciter and then to verify a sinusoidal vibrations influence on the devices' operation. Figure 6 shows mobile tablets during the test of resistance to sinusoidal vibrations.

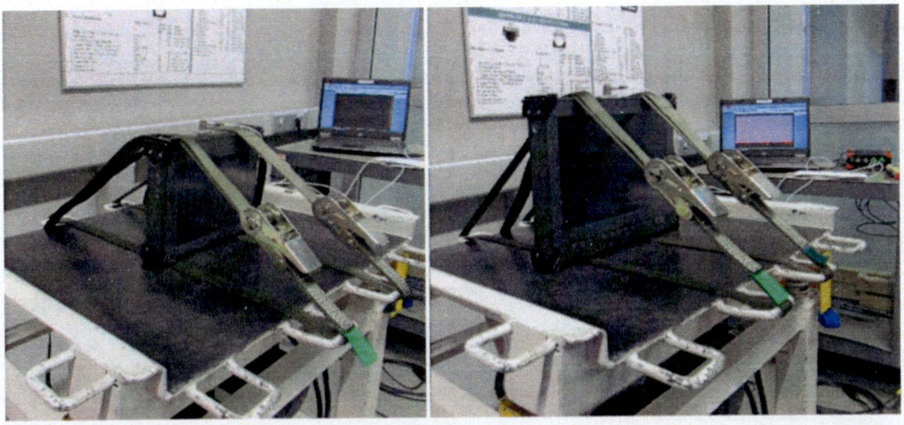

Fig. 7. Devices placed on the mechanical vibration exciter during test of resistance to sinusoidal vibrations. Source: own study

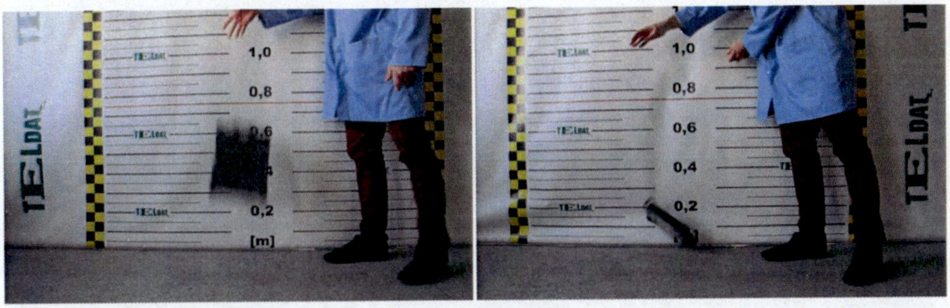

Fig. 8. Test of resistance to falls. Source: own study

During the test, the difference in equipment performance was not exceeding 20%. Moreover, after the test completion, no mechanical damage to the structure was observed. Test confirmed devices' resistance to sinusoidal vibrations of the frequency from 12.5 Hz to 50 Hz with the overload up to 5 g.

The another laboratory test was a fall resistance test. The aim of the test was to check resistance of mobile tablets to falls from 0.75 m. The test was carried out in accordance with the NO-06-A103:2005 defence standard. The Fig. 7 shows the test of resistance to falls. After completion of test, the construction elements and components of devices were thoroughly inspected. This inspection did not show any mechanical damage. This clearly confirmed the resistance of devices to the effects of a fall from a height of 0.75 m and meeting the requirements of standard NO-06-A103:2005 to this extent.

Housings of all mobile devices are equipped with functional elements such as: buttons, switch, cryptographic and SD card readers, cameras and communication ports. All these elements should be carefully sealed, because they are the most vulnerable for the water access to the interior of the devices [7]. Figure 8 shows the potential water access points to a typical mobile device. Therefore, it was decided to carry out

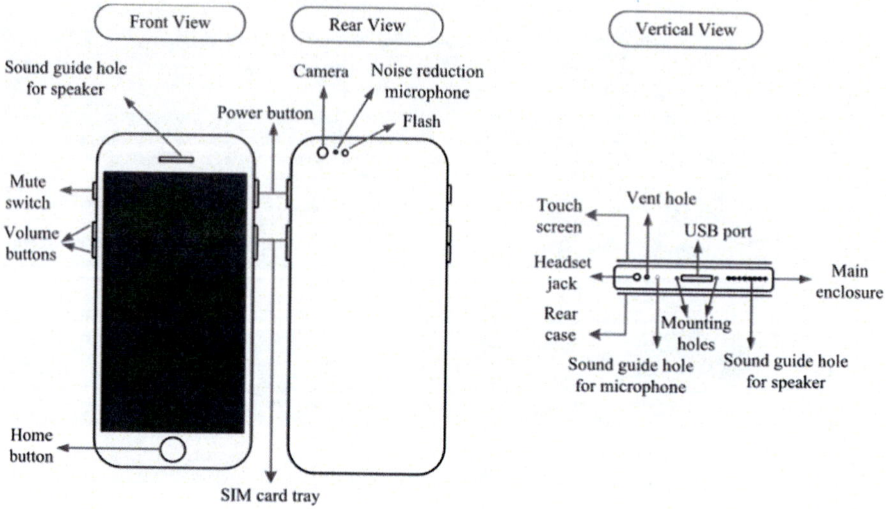

Fig. 9. Potential water access points to a typical mobile device. Source: Yu Q., Xiong R., Li C., Pecht M. G., Water Resistant Smartphone Technologies, IEEE Access vol. 7, 2019, s. 42757–42773

Fig. 10. Devices placed in the water tank during the test of waterproofness. Source: own study

a waterproofness test of devices to check and ensure their actual resistance. Test assumed the immersion of mobile tablets in water tank to the depth of 1m for 2h. The Fig. 9 shows devices placed in the water tank. After the test, the personnel visually checked condition of devices by opening their casings. Water did not get inside enclosures, so components were not exposed to water. Moreover, devices were operating correctly during this test. This confirmed devices' water resistance (Fig. 10).

5 Conclusions

The implementation of this project has confirmed that it is possible to create mobile devices - including tablets - which are resistant to very difficult conditions such as low and high temperature, mechanical damage, falls from high or flooding. This has been confirmed by tests under laboratory conditions. Moreover, it is crucial to provide the secure, efficient and effective communication. So the TELDAT company created an architecture for simultaneous use of all available transmission interfaces. Thus, transmission does not block the bandwidth of interfaces. Moreover, a disclose all transmitting information makes more difficulty, because the data transmission is divided into several transmission interfaces. If an enemy intercepts data from one interface, the rest of data is safe. Currently, the company is working on the fourth stage of project.

This publication is the result of the developed project named "Creating a family of innovative tablet devices with the increased resistance to mechanical and climatic conditions, allowing the effective use of the latest data transmission technologies", which is co-financed by National Centre for Research and Development.

References

1. The Radicati Group, Inc.: Mobile Statistics Report 2019–2023, A Technology Market Research Firm, London, pp. 2–3 (2019)
2. Intel IT: Deploying Tablets Safely in Manufacturing to Boost Productivity, Mobility Tablets, USA, pp. 1–9 (2013)
3. Apiecionek, Ł., Sobczak, M., Makowski, W., Vince, T.: Multi path transmission control protocols as a security solution. In: 2015 IEEE 13th International Scientific Conference on Informatics, Poprad, p. 29 (2015)
4. Apiecionek, Ł: Limiting energy consumption by decreasing packets retransmissions in 5G network. Mob. Inf. Syst. **2017**, 3 (2017)
5. Apiecionek, Ł., Palka, R., Brazkiewicz, P., Biedziak, M.: Mozliwosci zastosowania sieci 4G do monitorowania urzadzen eksploatowanych w straży pozarnej, Innowacyjne technologie w straży pożarnej, red. J. Roguski, CNBOP-PIB, Warsaw, pp. 259–269 (2018)
6. Cwirko, J., Cwirko, R.: Badania temperaturowe modułów elektronicznych. Biuletyn WAT **LVII**(2), 133–142 (2008)
7. Yu, Q., Xiong, R., Li, C., Pecht, M.G.: Water resistant smartphone technologies. IEEE Access **7**, 42757–42773 (2019)
8. Aleksandra, M.: Selected methods of artificial intelligence implemented in C/C++ programming language. Stud. Mater. Appl. Comput. Sci. **11**(2), 13–18 (2019). ISSN: 1689-6300
9. Rojek, I., Krolikowski, Z., Marciniak, M.: Obstructive sleep apnea, microcontroller, database, web application, mobile application. Stud. Mater. Appl. Comput. Sci. **11**(2), 15–21 (2019). ISSN: 1689-6300
10. Bednorz, I.: Monitoring of network equipment and reporting through spectrum and performance management applications from CA technologies. Stud. Mater. Appl. Comput. Sci. **12**(2), 10–15 (2019)

Application of Game Method for Modelling to Locate a Forest Fire Ignition Point in the Presence of Wind

Veselina Bureva[1], Evdokia Sotirova[1], Peter Vassilev[2], Vassia Atanassova[2], Olympia Roeva[2](✉), Krassimir Atanassov[2], and Hristo Tsakov[3]

[1] Laboratory of Intelligent Systems, "Prof. Dr. Assen Zlatarov" University, "Prof. Yakimov" Blvd., 8010 Burgas, Bulgaria
{vbureva,esotirova}@btu.bg
[2] Department of Bioinformatics and Mathematical Modelling, Institute of Biophysics and Biomedical Engineering, Bulgarian Academy of Sciences, Acad. G. Bonchev Str., bl. 105, 1113 Sofia, Bulgaria
olympia@biomed.bas.bg, krat@bas.bg
[3] Forest Research Institute – Bulgarian Academy of Sciences, 132, "St. Kliment Ohridski" Blvd., 1756 Sofia, Bulgaria
hristotsakovbg@abv.bg

Abstract. In the paper an investigation related to forest fires spread in the presence of wind is presented. The process is illustrated by a real example of a forest fire in the Haskovo region and Yambol region of Bulgaria. The Game Method for Modelling (GMM) is used to determine the forest fire ignition point from the constructed map of fire damages. The GMM is applied to the map containing the initial configuration of the observed area to simulate the fire. In the end the result is compared with the final configuration to determine similarities between them.

Keywords: Forest fire · Game method for modelling · Ignition point

1 Introduction

The standard Conway's Game of Life (CGL, see, e.g., [8, 11–15, 27, 28]) has a "universe" which is an infinite two-dimensional orthogonal grid of square cells, each of which is in one of two possible states, alive or dead, or (as an equivalent definition) in the square there is an asterisk, or there is not. The first situation corresponds to the case when the cell is alive and the second corresponds to the case when the cell is dead.

In [1], an extension of the CGL, called Game Method for Modelling (GMM) is described.

In this research work we investigate the possibility to identify fire ignition point in the case, when there is wind. Our approach uses the GMM and it is a continuation of the research from [7], where the process flows without the presence of wind. The process is illustrated by a real example of a forest fire in the Haskovo region and Yambol region of Bulgaria in August 2007.

K. T. Atanassov et al. (Eds.): IWIFSGN 2020/BOS/SOR 2020, LNNS 338, pp. 280–293, 2022.
https://doi.org/10.1007/978-3-030-95929-6_22

The burned area includes the lawns with dense vegetation, dry grasses and shrubs between the villages Melnica and Srem. The terrain of the district is defined as hilly to semi-mountainous. The climate of the area is temperate-continental, with a Mediterranean influence, which leads to higher temperatures. In general, forest fires in the region are divided into crown fires – 24% and surface fires – 76%. In this area, fire activity peaked in the years 2007 and 2020. The Haskovo region and Yambol region are marked as areas with high risk of fire according to the rankings of Ranking of forest fire risk areas (Table 1) [25].

2 Game Method for Modeling

Game Method for Modeling (GMM) is introduced [1] as an extension of Conway's Game of Life. In a series of papers, the theory and applications of the GMM have been presented [2–7, 9, 10, 16–24, 26]. The standard GMM has a "universe" presented as infinite two-dimensional orthogonal grid of cells. Each cell has coordinates. Depending on the modeled area the set of cells can contain different values. The water resources are marked using letter "*R*". The letter "*S*" is assigned for stones. According to the homogeneity of the grassy and forested area the number from 1 to 9 is written. The number 1 indicates the presence of some grasses, shrubs and trees. The number 9 indicates the presence of very dense grassy areas and forests. Depending on the intensity of vegetation in the area, the numbers 2 to 8 are used.

Let's have a finite area of vegetation in the area of the fire, but before it occurs. We assume that the area before the fire is called *initial configuration*. An *initial configuration* is the combination of every ordered set of $(n+2)$-tuples with an initial component being the number of the object; the second, third, etc., until the $(n+1)$-st being its coordinates; and the $(n+2)$-nd being its respective symbol from the set S. Let's have the same finite area of vegetation in the area of the fire, but after it occurs. We assume that the area after the fire is called *final configuration*. The *final configuration* is in the same format but after an iterative procedure of modification within a given number of iterations, where the rules from A are being applied. An *elementary step* in the transformation of the model is the single application of a rule from A over a given configuration K and that will be denoted by $A(K)$. In this sense, if K is the initial configuration, and L is the final configuration obtained from K in the course of multiple application the rules from A, then configurations $K_0, K_i, \ldots, K_m$ will exist, for which:

$$\begin{aligned} K_0 &= K, \\ K_{i+1} &= A_1(K_i), \text{ for } 0 < i < m-1, \\ K_m &= L. \end{aligned} \tag{1}$$

Table 1. Rankings of the regions in Bulgaria by risk of forest fire [25].

№	Region	Density of fires	Actual flammability of the forest territory	Integral indicator of risk	Areas of I cl. at %	Degree of risk	Average annual percentage of burning of forest area
1	Vidin	0.17	11.92	2.030	6.7	High	1.19
2	Lovech	0.22	7.61	1.674	22.9	High	0.76
3	Vratsa	0.27	4.64	1.253	5.6	High	0.46
4	Pleven	0.22	4.71	1.036	8.4	High	0.47
5	Haskovo	0.14	7.17	1.004	22.0	High	0.72
6	Yambol	0.10	6.98	0.698	14.5	High	0.70
7	St. Zagora	0.10	3.55	0.355	25.3	High	0.36
8	Sofia Province	0.14	2.15	0.301	23.2	High	0.22
9	Kyustendil	0.18	1.65	0.297	29.1	Average	0.16
10	Montana	0.09	2.58	0.232	41.6	Average	0.26
11	Plovdiv	0.17	1.17	0.199	33.3	Average	0.12
12	Pazardzhik	0.15	1.27	0.190	38.1	Average	0.13
13	Pernik	0.16	1.14	0.182	40.4	Average	0.11
14	Dobrich	0.25	0.64	0.16	18.2	Average	0.06
15	Burgas	0.09	1.66	0.149	22.0	Average	0.17
16	V. Tarnovo	0.10	1.14	0.114	7.46	Average	0.11
17	Gabrovo	0.09	1.23	0.111	32.4	Average	0.12
18	Sliven	0.11	0.92	0.101	16.5	Average	0.09
19	Razgrad	0.14	0.59	0.083	9.10	Low	0.06
20	Sofia-town	0.15	0.49	0.073	31.9	Low	0.049
21	Silistra	0.08	0.75	0.060	8.30	Low	0.08
22	Blagoevgrad	0.09	0.50	0.045	58.0	Average	0.05
23	Shumen	0.09	0.39	0.035	11.5	Low	0.04
24	Varna	0.22	1.10	0.242	3.90	Average	0.11
25	Smolyan	0.09	0.23	0.021	41.9	Low	0.02
26	Kardzhali	0.12	0.10	0.012	67.0	Average	0.01
27	Ruse	0.08	0.12	0.010	6.00	Low	0.01
28	Targovishte	0.05	0.17	0.008	7.40	Low	0.02

The algorithm for applying the GMM is described in details in [1].

3 Application of Game Method for Modelling to Locating the Forest fire's Ignition Point

If some rule determines that symbol B must be changed to symbol C, we will further denote this fact by $B \to C$.

The rules for changing the symbols during the fire:

A.1. $R \to R$;
A.2. $S \to S$;
A.3. $n \to n - 1$, for $n \in [1, 9]$;
A.4. $1 \to 0$.

In the example below, R denotes a river, S – denotes stones (rocky area), the digits represent the quantity of wood mass (fuel) per unit square.

On the first step, we take a field of the forested area in the form of a grid with the above denotations (configuration K_b, 'b' with the meaning "before the fire"). On the next step, we take the same field but with configuration K_a after the fire has burnt out. Then, we construct a configuration $D_0 = K_b - K_a$. This configuration contains as elements digits corresponding to the quantity of burnt wood mass. Using the above rules in a stepwise manner, from configuration D_0 we subsequently obtain configurations $D_1, D_2, \ldots$. This process of generating configurations finishes when only few squares remain in the field containing the number 1. These squares are the potential ignition points of the fire. For each of them, we determine its coordinates and after that for each of them we consider sequentially the initial configuration K_b in which fire occurs at the point with the respective coordinates. On this configuration we apply the following rules:

B.1. $R \to R$;
B.2. $S \to S$;
B.3. $0 \to 0$;
B.4. On the initial time-step, the fire starts from a fixed cell containing digit n ($1 > n > 9$) that represents the density of the trees in the forest. On the second time-step for the same cell $n \to (n - 1)$. On the third time-step for the same cell

$$(n-1) \to \begin{cases} 0, & \text{if } n = 1 \\ n-2, & \text{if } n > 1 \end{cases}. \tag{2}$$

B.5. At the same moment all neighbouring cells of a cell with fire change their digits according to the rule (2);
B.5. At the next moment all cells with fire change their digits based on rule (2) and all their neighbouring cells in which there is no fire yet also change their digits according to rule (2);

B.6. Depending of the wind the cells activate the neighbourhood cells according to different scope of the field. In the investigated case we have wind from the top side. Therefore, the cells in the bottom part of the map are burnt by 3 cells in the direction of the fire at each step (Fig. 1b). The cells in the upper side activate the neighbourhood cells with the standard step of 1 (Fig. 1a).
B.7. The process continues until all cells in the region contain only digit 0. In the opposite case, go to Step **B.5.**

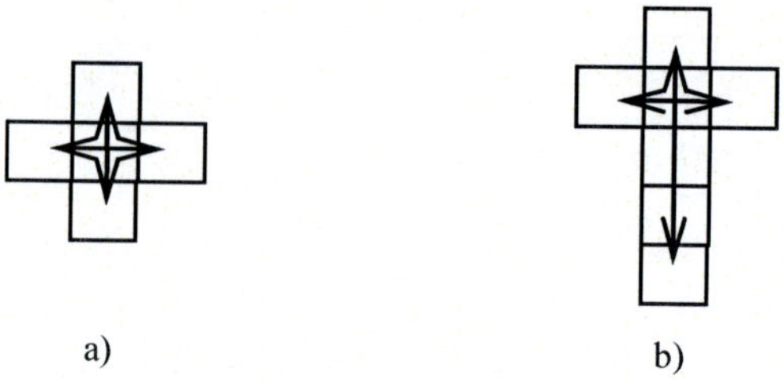

Fig. 1. a) Standard step. b) Step with wind

As a result, for each potential ignition point with coordinates (p, q) we obtain a final configuration $K_f(p, q)$ and we compare each of them to the configuration K_a. The one that resembles the most K_a in shape and value of digits corresponds to the most likely origin of fire in the Haskovo region and Yambol region in Bulgaria.

The area (Haskovo region and Yambol region in Bulgaria) of forest fire in 2007 is presented on the Fig. 2. The forest fire is concentrated in the vegetation between Srem and Melnica villages. The fire spreads southwards and at a certain point in time the flames are high and cross over the Tundzha River. The area south of the Tundzha River has been affected by a previous fire. There was a third forest fire west of the Tundzha River. The region between Srem and Melnica villages and Tundzha River will be investigated to locate the fire ignition point. The opposite process of forest fire spread is simulated using GMM.

For the investigation purpose a grid of cells is applied to the maps. The border is added to highlight the fire region. The border is visualized with black cells. The river is marked using blue cells containing the letter "*R*". In the other cells depending on the forest intensity the numbers from 1 to 9 are written. The number 9 means big part of forest and bushes while the number 1 is assigned for small area of forest and bushes. The cells containing the number 0 are cells damaged from the fire. These cells are completely burned (Fig. 3).

Therefore, the grid of cells representing the investigated area of forest fire spread will be analyzed. The initial configuration for GMM is presented on Fig. 4. It contains mostly places with slightly pronounced vegetation. Several cells do not contain vegetation. These cells have been damaged by previous fires. The Tundzha River is visualized on the left side of the selected area.

During most of the time of the burning the wind was coming from the north as it shown on Fig. 1b. Thus, we accept that the wind has not changed direction. It had been a strong wind, so our assumption is that the southward propagation of the fire is thrice as fast as in the other directions.

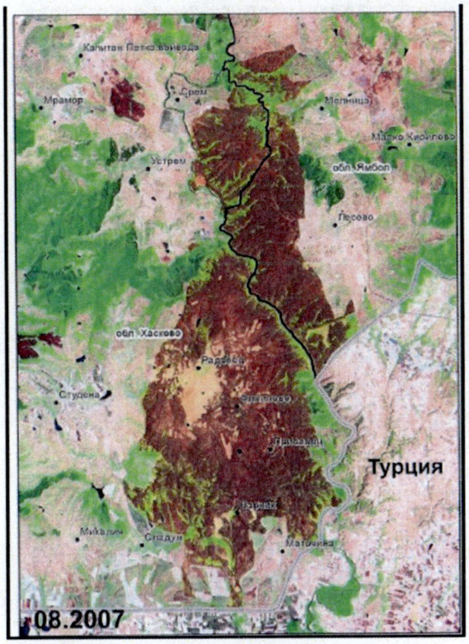

Fig. 2. The investigated region of forest fire

As can be seen from Fig. 2, the fire has spread to the south of river Tundzha but that has occurred after the fire to the north and east of the river had reached it. Since our goal is to point out possible sources of origins (ignition points) for the fire, we do not consider the fire to the south of the river.

The grid of cells with final configuration of the burned area is presented on Fig. 5. After the forest fire small part of vegetation is available. The burned area is represented using the number 0. The constructed area with fire damages is presented on the Fig. 6. The grid of cells is calculated as subtraction of the initial configuration and the final configuration. The result contains the fire damage.

In the next steps the algorithm iterates following the opposite fire direction. We start from the two points that are the upper border and bottom border of the fire to identify the ignition point. Then we iterate the procedure opposite to fire. The step of the algorithm starting from the **upper border of the fire** burns the vegetation with step of 1. The step of the algorithm starting from the starting from the **bottom side** of the fire burns the vegetation area with the step 3 against to the wind. The iterations are presented on Figs. 7, 8, 9, 10, 11, 12, 14 and 15.

The final configuration with fire damages from Fig. 6 is compared with constructed configuration from Fig. 19 and Fig. 20. The digits of the cells in the grids are identical or similar. The difference of the cell values is around 1 or 2. The differences are mainly in areas with more vegetation. The values in the cells of the grids located at the fire

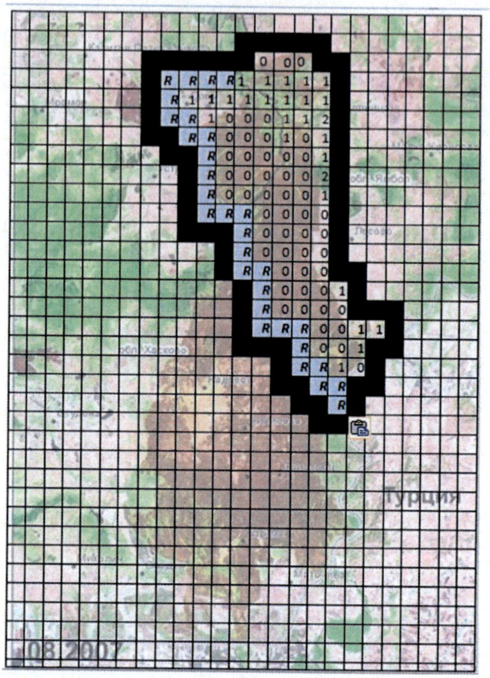

Fig. 3. Grid of cells for the investigated area of forest fire around Haskovo region and Yambol region in 2007

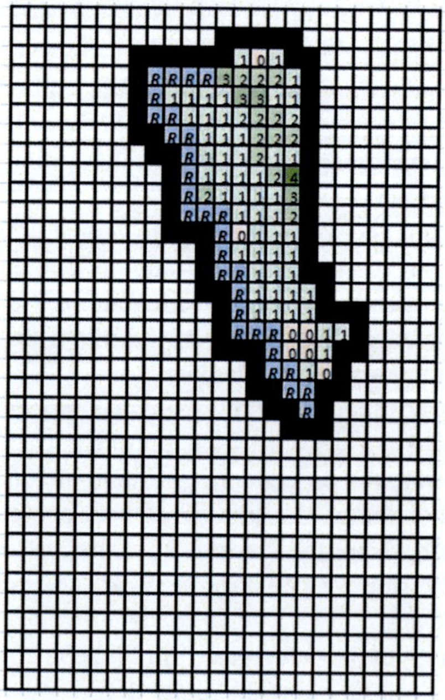

Fig. 4. Initial configuration

boundaries are almost the same. Therefore, slight discrepancies in the degree of burnt areas are mainly around the point of occurrence of the fire. The constructed model includes the presence of wind and almost homogeneous terrain of fire spread.

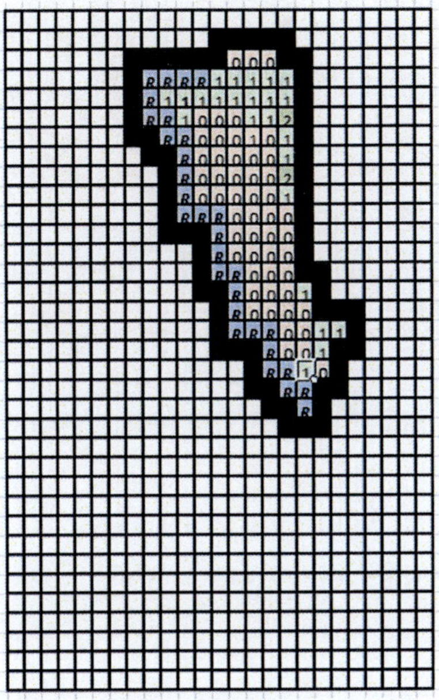

Fig. 5. Final configuration

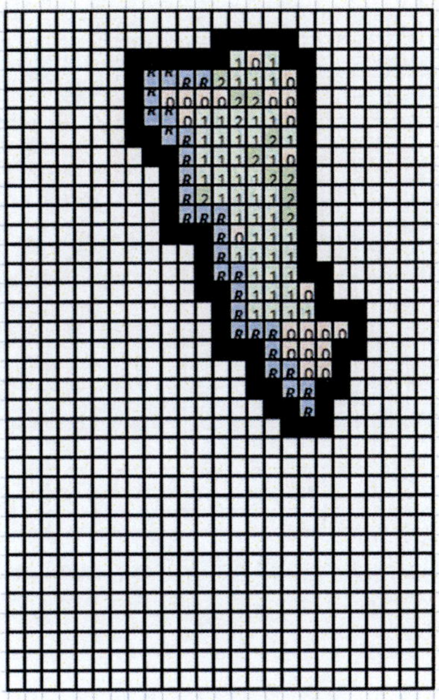

Fig. 6. Fire damages

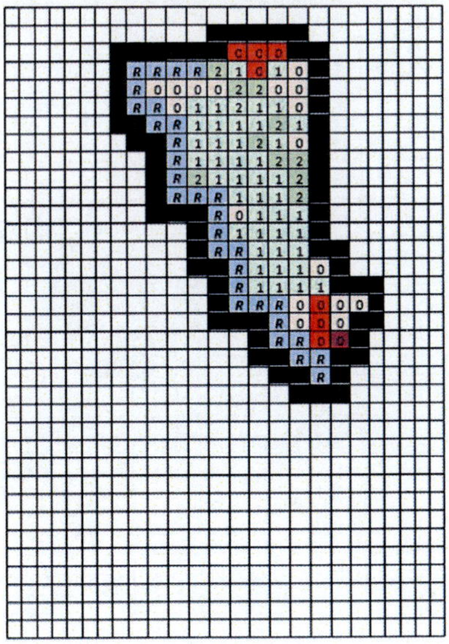

Fig. 7. Reversed fire process - iteration 1

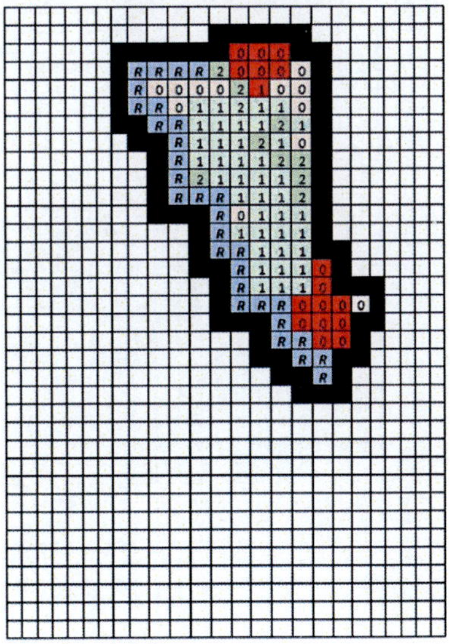

Fig. 8. Reversed fire process - iteration 2

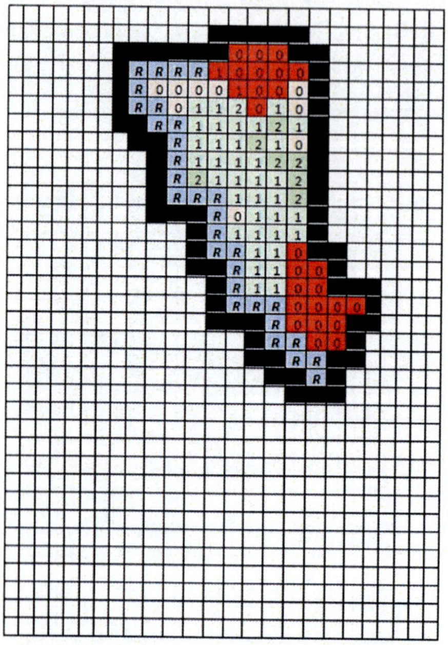

Fig. 9. Reversed fire process - iteration 3

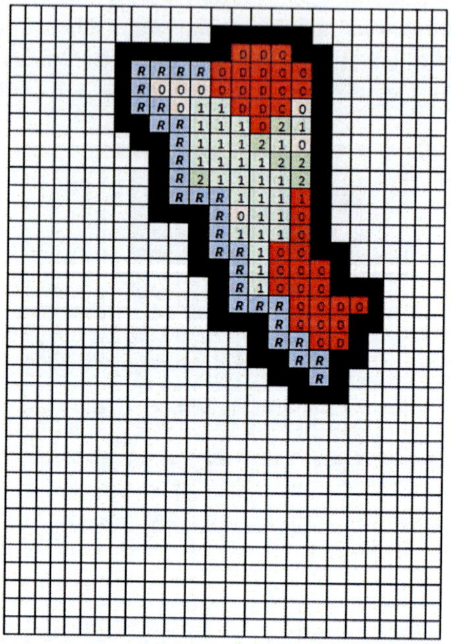

Fig. 10. Reversed fire process - iteration 4

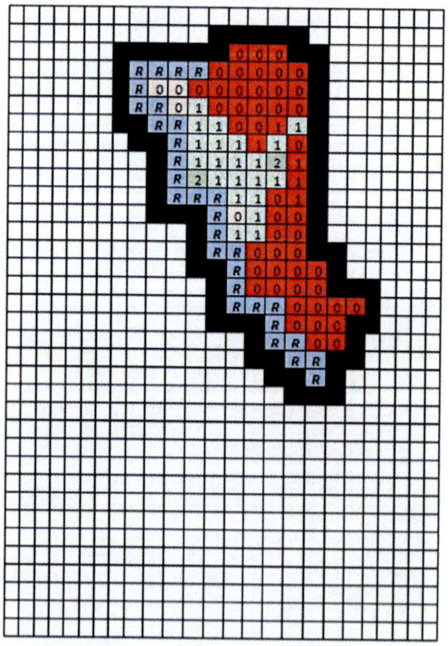

Fig. 11. Reversed fire process - iteration 5

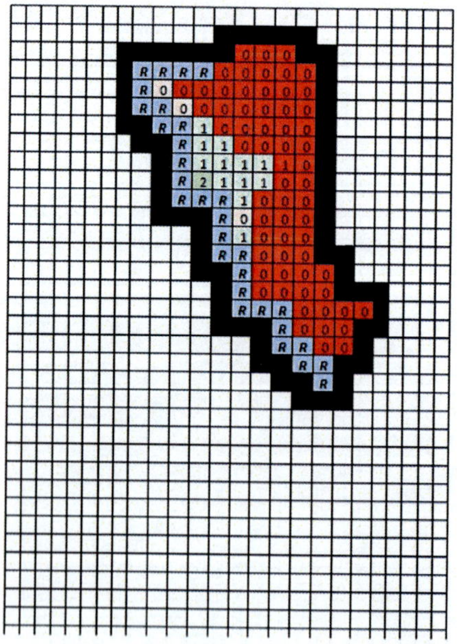

Fig. 12. Reversed fire process - iteration 6

The forest fire ignition point is identified in the 8 iteration. The flames are initiated from the region close to the Tundzha river.

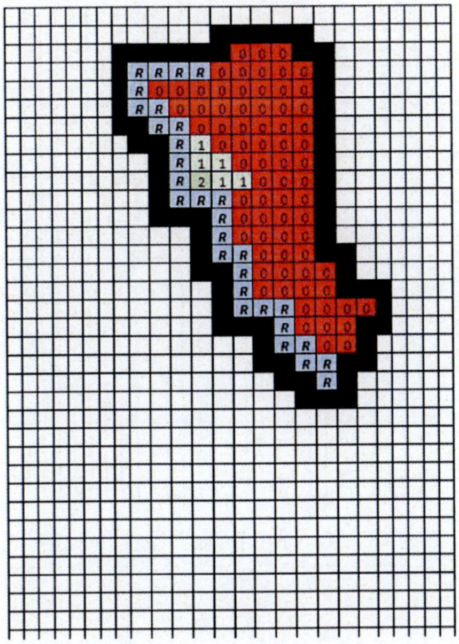

Fig. 13. Reversed fire process - iteration 7

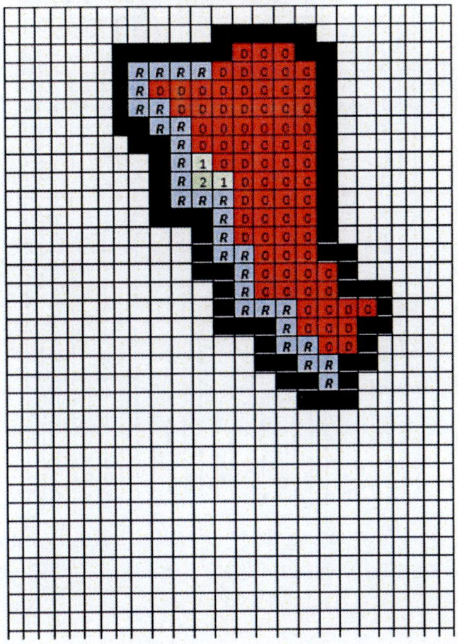

Fig. 14. Reversed fire process - iteration 8

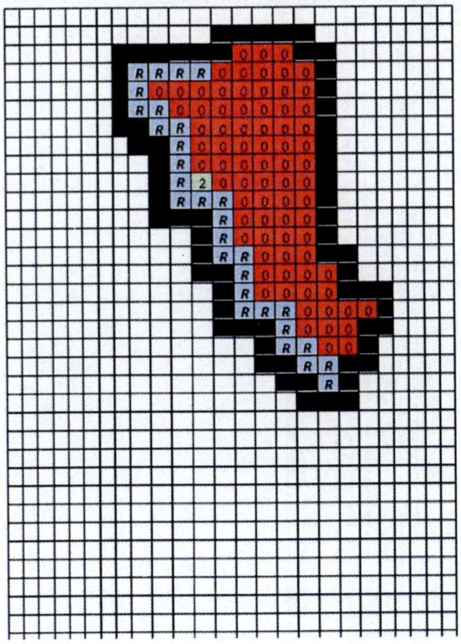

Fig. 15. Reversed fire process - iteration 9

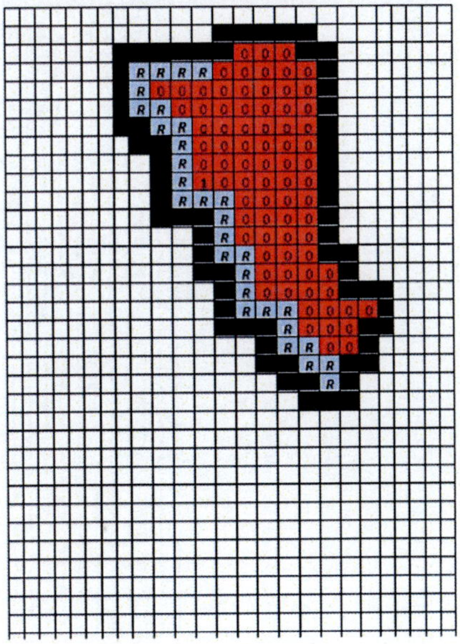

Fig. 16. Reversed fire process - iteration 10

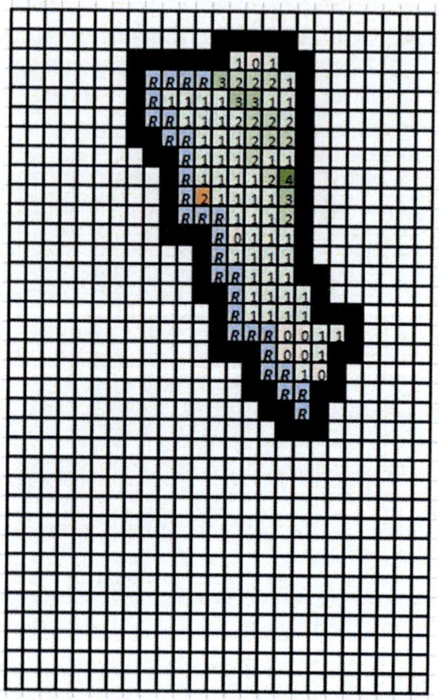

Fig. 17. Starting point of fire as a result of the reversed fire simulation

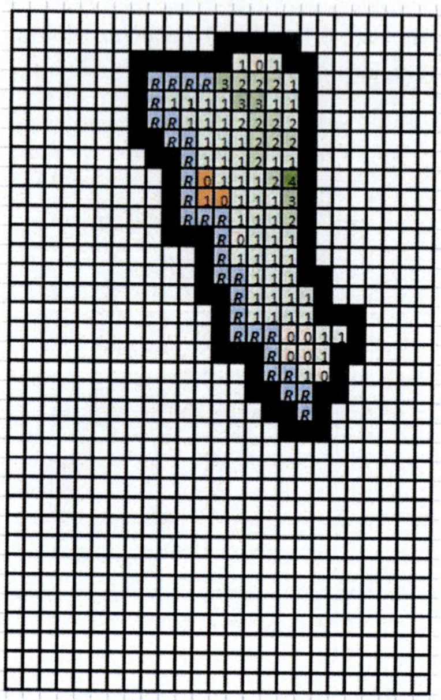

Fig. 18. Forward fire process - iteration 1

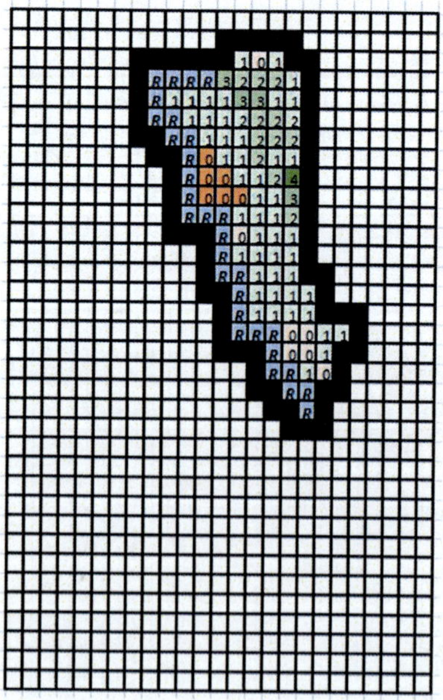

Fig. 19. Forward fire process - iteration 2

The application of the GMM over a selected area determines one possible ignition point for locating the fire (Fig. 16). This point will be considered as starting point for modeling the fire behavior (Fig. 17). The cell of fire ignition point is mapped to the initial configuration. Therefore, we start the fire simulation (Figs. 17, 18 and 19).

The application of GMM over the map of the forest fire around Haskovo region and Yambol region locates one ignition point. Mapping the obtained results over the initial configuration of the forest fire we find they agree with the climatic factors taken into account.

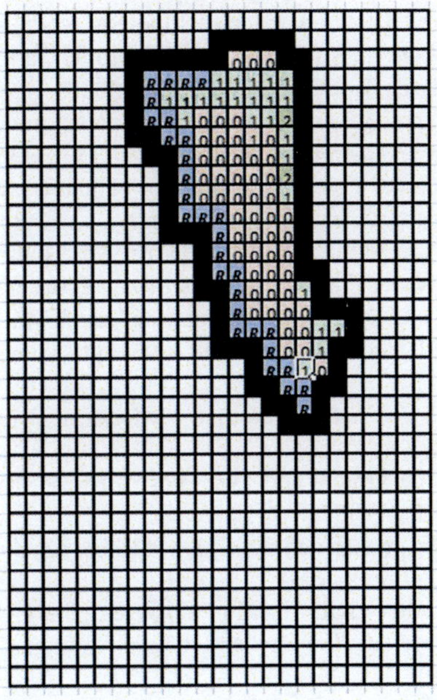

Fig. 20. Final configuration as a result of the forward fire simulation

4 Conclusion

Investigation related to forest fire in the Haskovo region and Yambol region of Bulgaria is presented. The GMM is applied to determine the forest fire ignition point. Therefore the ignition point is used to simulate the fire. The final configuration of the fire is compared with the configuration from simulation of the fire to determine similarities between the fire damage. The similarities between the cases present the right way of the fire flow.

Acknowledgments. The authors are grateful for the support provided by the Project DN-16-6/11.12.2017 "*Integrated Approach for Modeling of the Forest Fires Spread*", funded by the Bulgarian National Science Fund, Bulgarian Ministry of Education and Science.

References

1. Atanassov, K.: Game Method for Modelling. Prof. M. Drinov Academic Publishing House, Sofia (2011)
2. Atanassova, L., Atanassov, K.: Intuitionistic fuzzy interpretations of conway's game of life. In: Dimov, I., Dimova, S., Kolkovska, N. (eds.) NMA 2010. LNCS, vol. 6046, pp. 232–239. Springer, Heidelberg (2011). https://doi.org/10.1007/978-3-642-18466-6_27
3. Atanassova, L., Atanassov, K.: On a game-method for modelling with intuitionistic fuzzy estimations: part 1. In: Lirkov, I., Margenov, S., Waśniewski, J. (eds.) LSSC 2011. LNCS, vol. 7116, pp. 182–189. Springer, Heidelberg (2012). https://doi.org/10.1007/978-3-642-29843-1_20
4. Atanassova, L., Atanassov K.: On a game-method for modeling with intuitionistic fuzzy estimations. Part 2. In: Sabelfeld, K., Dimov, I. (eds.) Monte Carlo Methods and Applications, pp. 29–38. Walter de Gruyter GmbH, Berlin/Boston (2013)
5. Atanassov, K., Atanassova, L., Sasselov, D.: On the combinatorial game-method for modelling in astronomy. Compt. Rend. Acad. Bulg. Sci. **47**(9), 5–7 (1994)
6. Atanassov, K., Sotirova, E.: On some applications of game method for modeling, part 4: interpretations of some stochastic processes. Proc. Jangjeon Math. Soc. **16**(3), 301–310 (2013)
7. Bureva, V., Atanassova, L., Atanassov, K., Delkov, A.: Application of the game method for modelling for locating the wildfire ignition point. In: Proceedings of the 4th International Conference on Numerical and Symbolic Computation Developments and Applications, Porto, Portugal, 11–12 April 2019, pp. 397–413. ISEP – Instituto Superior de Engenharia do Porto (2019)
8. Deutsch, A., Dormann, S.: Cellular Automaton Modeling Biological Pattern Formation. Birkhauser, Boston (2005)
9. Dimitrov, D.: Modelling the growth and dynamics of forest stands by game-method. Adv. Model. Anal. **2**(1–2), 11–22 (1998)
10. Dimitrov, D.: Modelling the growth and dynamics of forest stands by extended game-method. Adv. Model. Anal. **4**(1–2), 7–21 (1999)
11. Eigen, M., Winkler, R.: Ludus Vitalis. Erwin Poell, Heidelberg (1973)
12. Gardner, M.: The Unexpected Hanging and Other Mathematical Diversions. Simon & Schuster, New York (1969)
13. Gerhardt, M., Schuster, H.: Anregungen Zellulare Automaten Simulieren Reale Vorgange. Magazinur Computer Technik (2), 44–50 (1992)
14. Nowak, M.: Evolutionary Dynamics. The Belknap Press of Harvard University Press, Cambridge (2006)
15. Otto, S., Day, T.: A Biologist's Guide to Mathematical Modeling in Ecology and Evolution. Princeton University Press, Princeton (2007)
16. Sotirova, E., Atanassov, K., Fidanova, S., Velizarova, E., Vassilev, P., Shannon, A.: Application of the game method for modelling the forest fire perimeter expansion. Part 1: a model fire intensity without effect of wind. In: Proceedings of IFAC Workshop on Dynamics and Control in Agriculture and Food Processing, Plovdiv, 13–16 June 2012, pp. 159–163 (2012)
17. Sotirova, E., Atanassov, K., Fidanova, S., Velizarova, E., Vassilev, P., Shannon, A.: Application of the game method for modelling the forest fire perimeter expansion. Part 2: a model fire intensity with effect of wind. In: Proceedings of IFAC Workshop on Dynamics and Control in Agriculture and Food Processing, Plovdiv, 13–16 June 2012, pp. 165–169 (2012)
18. Sotirova, E., Atanassov, K., Fidanova, S., Velizarova, E., Vassilev, P., Shannon, A.: Application of the game method for modelling the forest fire perimeter expansion. Part 3: a model of the forest fire speed propagation in different homogenous vegetation types. In: Proceedings of IFAC Workshop on Dynamics and Control in Agriculture and Food Processing, Plovdiv, 13–16 June 2012, pp. 171–174 (2012)

19. Sotirova, E., et al.: Hexagonal Game Method model of forest fire spread with intuitionistic fuzzy estimations. Notes Intuitionistic Fuzzy Sets **19**(3), 73–80 (2013)
20. Sotirova, E., Dimitrov, D., Atanassov, K.: On some applications of game method for modelling part 1: forest dynamics. Proc. Jangjeon Math. Soc. **15**(2), 115–123 (2012)
21. Sotirova, E., Dobrinkova, N., Atanassov, K.: On some applications of game method for modeling. Part 2: development of forest fire. Proc. Jangjeon Math. Soc. **15**(3), 335–342 (2012)
22. Sotirova, E., Sotirov, S., Atanassova, L., Atanassov, K.T., Bureva, V., Doukovska, L.: Game method for modelling with intuitionistic fuzzy rules. In: Atanassov, K.T., et al. (eds.) IWIFSGN 2016. AISC, vol. 559, pp. 153–168. Springer, Cham (2018). https://doi.org/10.1007/978-3-319-65545-1_15
23. Sotirova, E., Sotirov, S., Dimitrov, A., Atanassov, K.: On some applications of the game method for modelling: part 3: simulation of oil transformation in marine environment. Proc. Jangjeon Math. Soc. **16**(2), 293–300 (2013)
24. Sotirova, E., Velizarova, E., Fidanova, S., Atanassov, K.: Modeling forest fire spread through a game method for modeling based on hexagonal cells. In: Lirkov, I., Margenov, S., Waśniewski, J. (eds.) LSSC 2013. LNCS, vol. 8353, pp. 321–328. Springer, Heidelberg (2014). https://doi.org/10.1007/978-3-662-43880-0_36
25. Surveying and mapping the risk of forest fires in the country (in Bulgarian). http://www.iag.bg/data/docs/Ocenka_i_kartografirane_na_risk_ot_gorski_pozhari.pdf
26. Velizarova, E., Sotirova, E., Atanassov, K., Vassilev, P., Fidanova, S.: On the game method for the forest fire spread modelling with considering the wind effect. In: 2012 Proceedings of 6th IEEE International Conference on Intelligent Systems, pp. 216–220 (2012). Art. No. 6335219
27. Velten, K.: Mathematical Modeling and Simulation. Wiley-VCH Verlag GmbH & Co. KgaA, Weinheim (2009)
28. http://en.wikipedia.org/wiki/Conway%27s_Game_of_Life

An Alternative Fuzzy Logic Inference Method on the Example of a PID Controller

Wojciech T. Dobrosielski(✉)

Kazimierz Wielki University, J.K. Chodkiewicza 30 Street,
85-064 Bydgoszcz, Kujawsko-Pomorskie, Poland
wdobrosielski@ukw.edu.pl

Abstract. This paper presents a different approach to the inference process of a fuzzy controller. An important trait of the approach is the use of an OFN (ordered fuzzy numbers) property and an arithmetic apparatus. The solutions of inference processing are based on the work by Mamdani or Takagi-Sugeno, which share a common feature, a mechanism of a rule base. The approach postulated in this work does not feature a classic distribution; it only applies characteristic OFN properties using an arithmetic apparatus. The problem of system control is presented as a mathematical formula in which OFN are substituted. The approach simplifies the operations processed in the controller, an important element for which is the mathematical model of the process discussed in the paper. The results are interpreted within the area of interest of the given system. As an example, a discussion of the possible solutions of a PID controller for OFN is provided.

Keywords: Fuzzy logic · Fuzzy controller · Ordered fuzzy numbers OFN · PID

1 Introduction

The problem of uncertainty or ambiguity of described real-life models can be solved with many theories, among which is fuzzy logic. The theory postulated in [40] by Zadeh became a mathematical tool to solve this and similar problems. The application of a fuzzy set as $\mu_A(x)$, degree of membership x in set A, which is characterized as a pair set.

$$A = \{(x, \mu(x))\} \ \forall x \in X \tag{1}$$

A literature review focused on fuzzy logic revealed another step which consists in using a new definition of $L - R$ fuzzy number interval range. This solution was introduced by Dubois and Prade [6].

$$A = \mu_A(x) = \begin{cases} L(\frac{m-x}{\alpha}), & \text{for } x \leq m \\ 1, & \text{for } m \leq x \leq m \\ R(\frac{x-m}{\beta}), & \text{for } x \geq m \end{cases} \tag{2}$$

K. T. Atanassov et al. (Eds.): IWIFSGN 2020/BOS/SOR 2020, LNNS 338, pp. 294–307, 2022.
https://doi.org/10.1007/978-3-030-95929-6_23

A modification of the [40] theory assumes that an $L - R$ fuzzy number is characterized by two decreasing functions, $L(x)$ and $R(x)$, which are called the functions of shape. The original theory was restructured due to the constraints which demonstrate that performance of operations is quite complex. This complexity mainly stems from that operations are performed both on the support and the degrees of membership. As it turned out, this proposal, albeit improved in comparison to its original form, does not solve all issues. While basic arithmetic operations were simplified, inconvenience lingers. The freedom of arithmetic operations on fuzzy numbers introduces a certain inaccuracy to a system. The inaccuracy is the expansion of the support, like the resultant function of membership. Reference literature provides different names of this inaccuracy, like intervals of imprecision, expansion of imprecision, or enhancement of fuzziness with the number of performed operations and the lack of solution to linear equations in a set of fuzzy numbers. A consequence of imprecision expansion on fuzzy numbers is that the result of the operation $A + (-A)$ is not an element neutral to addition, but a fuzzy zero instead of a real zero. Another inconvenience of classic arithmetic of fuzzy numbers is the lack of backward chaining. Before the essence of this problem is explained, note that forward chaining can be expressed as follows:

$$A + B = C, \tag{3}$$

with A, B, C being fuzzy numbers.

Respectively, A and B are premises of the given C of a conclusion. If the completion is known or a specific end of an action is expected, backward chaining can be used in the system of interest. This is the inference most useful in expert models for which facts are sought which fulfil the specific conclusion. Respectively, this kind of inference is expressed with the following Eq. (4).

$$C - B = A, \tag{4}$$

with A, B, C being fuzzy numbers, like: A is the known fact, B is the fact sought after, and C is the assumed result. Considering backward chaining with the use of fuzzy number arithmetic, a conclusion is drawn that Eq. 4 should have a completely different form. The correct form is $C - A \neq B$, and the sought fact being fuzzy number B is not the same variable as it was in Eq. (3). Hence it is evident that basing on a part of the inputs and the result, it is not possible to trace back the course of the whole event of interest.

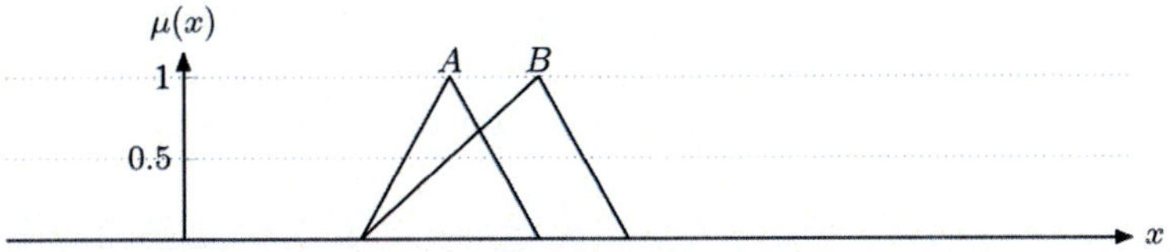

Fig. 1. Fuzzy numbers A, B as addition operand

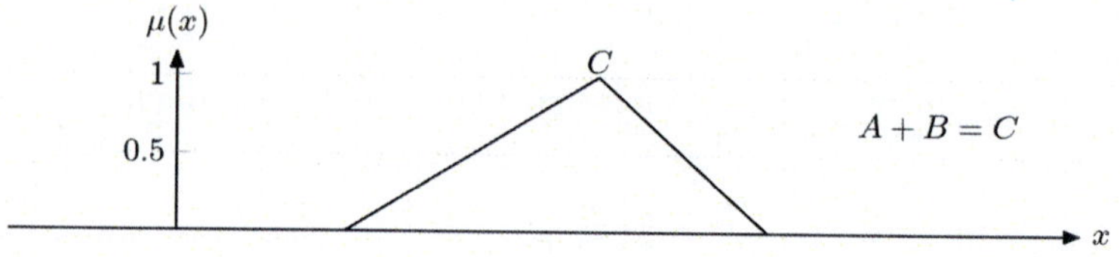

Fig. 2. Fuzzy number C as the addition result or an inference conclusion

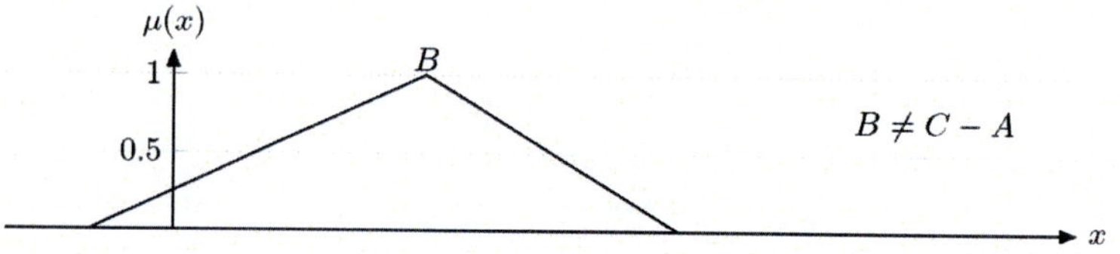

Fig. 3. Example of the missing backward chaining; the resulting fuzzy number B is different than B in Fig. 1

Another modification of existing fuzzy logic solutions was the theory of OFN (ordered fuzzy numbers). The OFN theory allows solving the inconvenienced discussed here. The generalization of Zadeh's postulate proposed in [17–21] in the form of an OFN makes algebraic operations consistent with real numbers, feature the distributive law for positive numbers, commutation of addition, and existence of elements neutral to addition and multiplication; hence, the algebraic operations satisfy the properties of real-numbers algebra. The redefining of classic fuzzy sets, where Zadeh argues that a fuzzy number is an organised pair, expands the definition with an organised pair of functions. Hence, the OFN proposed by Kosiński is defined as follows:

Definition 1. OFN A is an organised pair of functions

$$A = (f_A, g_A) \tag{5}$$

with $f_A, g_A : [0, 1] \to R$ are continuous functions.

Considering the specific elements of Definition (1), f is the part *up* (ascending), while g is the part *down* (descending). Assuming that both functions are strictly monotone, an inverse function exists, f_A^{-1} and g_A^{-1}. From the condition of continuity it is inferred that both parts are limited intervals and expressed as these limits: $UP_A = (l_A, 1_A^-)$, $DOWN_A = (1_A^+, p_A)$. Regarding [13], it is possible to show that the intervals are ordered. A supplement to the definition of the limits is that the edges are mapped to characteristic points of an OFN:

$$l_A = f_A(0), 1_A^- = f_A(1), 1_A^+ = g_A(1), p_A = g_A(0). \tag{6}$$

The limits are real number and do not need to be specific. This means that they do not need to satisfy condition $l_A \leq 1_A^-$ and $1_A^+ \leq p_A$. A graphical interpretation is shown in Fig. 4. To fully produce an analogy to the function of membership

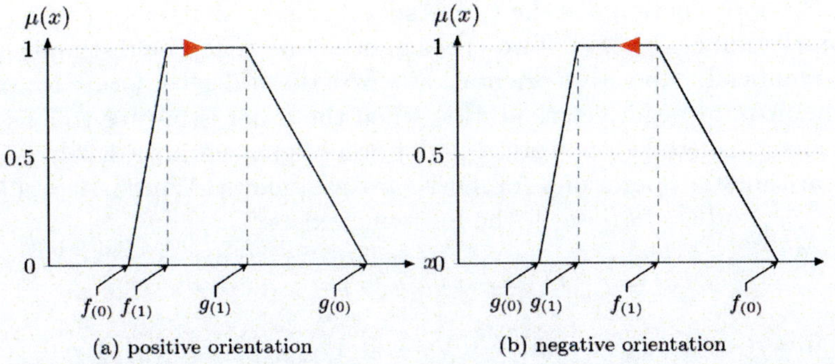

Fig. 4. OFN in two versions of order and with characteristic points

μ_A, constant function 1 must be introduced which, together with intervals UP and $DAWON$, forms a single interval defined as follows for R.

The introduced OFN definition shows that it is an expansion of classic fuzzy numbers, otherwise known as convex numbers. A graphical interpretation of the OFN and its reference to convex numbers is shown in Fig. 5.

Hence, the 'order' is a property resulting from the very definition (1) of OFN, where the order of function pairs f and g is important so that the function is

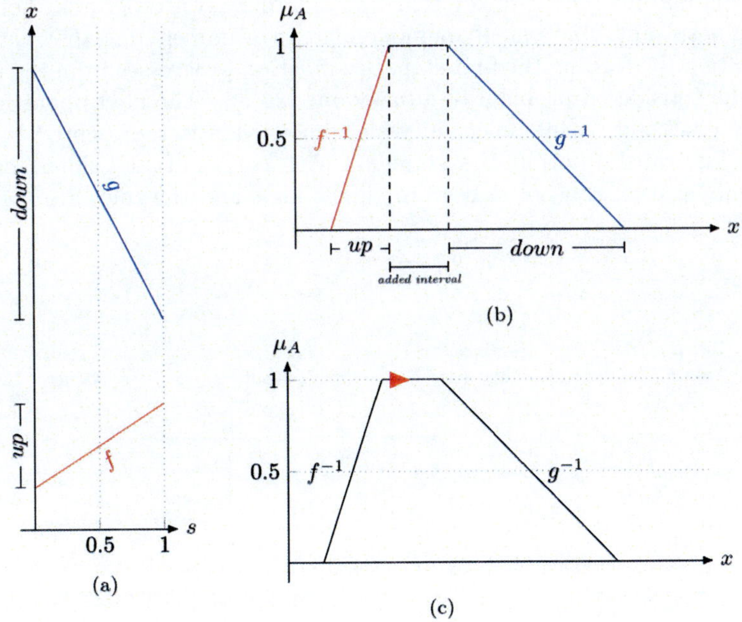

Fig. 5. Example (a) of an OFN and the interpretation (b), (c) referring to a convex fuzzy number

the origin f and function g is the end. As a consequence, number (f_A, g_A) is not identical to number (g_A, f_A). Two types of orientation can be distinguished. An ordered number has positive orientation when the order conforms to axis OX of real numbers or negative orientation when the order is inverse to axis OX of real numbers. A detailed interpretation of order is shown in [1,4,29].

The arithmetic operations for OFN are as follows. Where $A = (f_A, g_A)$, $B = (f_B, g_B)$ and $C = (f_C, g_C)$, the following exists:

$$f_c(x) = f_A(x) \star f_B(x) \wedge g_c(x) = g_A(x) \star g_B(x) \tag{7}$$

where sign $\star$ denotes $+, -, \cdot$ and $\div$ if $\forall y \in [0,1]\,, f_B(y) \neq 0, g_B(y) \neq 0$

The operations defined so are analogous to operations on real number. Especially for any OFN A:

$$A - A = A + (-1) \cdot A = 0 \tag{8}$$

and

$$A \cdot A^{-1} = A \cdot \frac{1}{A} = 1 \tag{9}$$

Using the foregoing traits of OFN, it is possible to easily overcome the limitations of classic convex fuzzy sets.

2 Background

Fuzzification and inference are key elements of a fuzzy controller and their product is the resultant function of membership, an argument for the method of defuzzification. Regarding the first step, fuzzification progresses from the n acute space of the considered problem to a fuzzy one [24,32]. The next operating step of a fuzzy controller is inference, in which knowledge is represented by rules. The rules for transformation of statements with concomitant axioms forms an inference apparatus, and not only so in classic logic and the rules are applied in fuzzy logic.

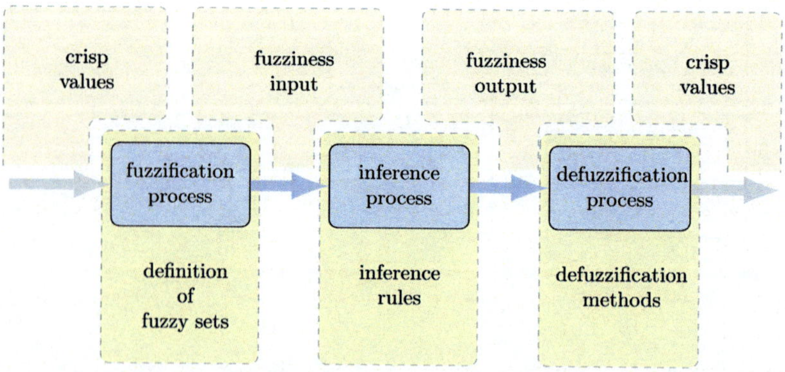

Fig. 6. Structure of a fuzzy controller

In both logics, premises exist which justify conclusion. By this reasoning, inferences are indirectly generated from premises, and the 'recipe' for achieving a decision is called an 'implication'. The operator of consequence in fuzzy logic is detailed in [7,8,34]. As stated by the authors of the cited works, a number of implication families can be distinguished; the most often implications are based on t-norms, like S-implications, formed by a suitable use of a t-conorm and R-implications as t-norms [2]. The methods of inferring imperfect knowledge to generate a proper resultant formula for fuzzy logic is based on application of deductive inference, like the generalized modus ponens, as shown in Table 1.

Table 1. Fuzzy inference rule of *modus ponens*

Premise	x is A'
Implication	**IF** x is A **THEN** y is B
Inference	y is B'

For methods for determination of the resultant function of membership, the rules of Larsen, Łukasiewicz, minimum and maximum (the Zadeh rule), and Mandami can be used. In a formal approach, the recipe for inferences will first be the determination of the degree of fulfilment of premises in the specific rule. The aggregation of the specific rule and the calculation of the degrees of conclusion activation based on a specific implication enable conducting the last step of inference. Combination of the parts into a whole of the previous step with s-norm operators provides resultant functions of membership; this process is called 'accumulation'. The foregoing theory of fuzzy control is called a Mamdani fuzzy system and detailed in [25,27]. In this proposal, both predecessor and successor are linguistic terms [41]. If the conclusion of a given rule features functions, it is a Takagi-Sugeno fuzzy system [37,38]. The main difference between the fuzzy systems by Mamdani and Takagi-Sugeno is that in the former system, operations are done on fuzzy sets, while the other system facilitates description of non-linear systems.

The application of OFN (ordered fuzzy numbers) is a completely different approach to fuzzy control inference. Unlike in the solutions by Mamdani or Takagi-Sugeno, OFN uses a mathematical formula only. It can be generally said that control of a process is analytically presented in the form of algebraic, differential, or derivative equations. To present this unique approach in an inference unit, it is prudent to use an example of a PID (proportional-integral-derivative) controller. It is known that the PID algorithm is one of the most mature and widely applied control strategies in industrial control solutions [15] and the other possible control solutions like [3,10,23,31,33]. The applied coupling of a loop-back with a proportional, integral, and derivative term calculates the subsequent corrective settings as shown in Fig. 8.

$$u(t) = K_P e(t) + K_I \int_0^t e(t')\, dt' + K_D \frac{de(t)}{dt} \tag{10}$$

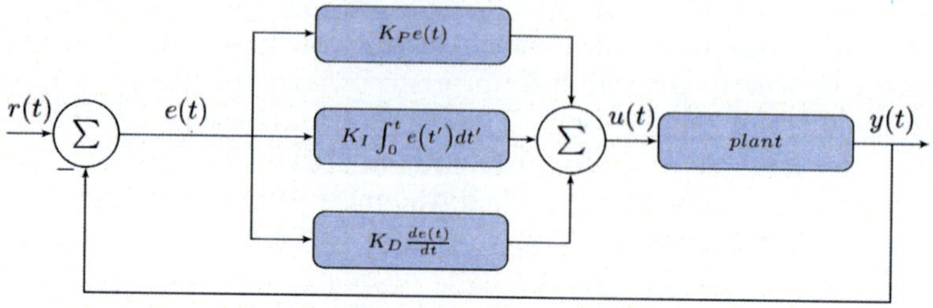

Fig. 7. Diagram of a PID controller

with: $e(t) = r(t) - y(t)$ error signal between reference $r(t)$ and system output $y(t)$, where the units are named accordingly: KP - *proportional*, KI - *integral*, and KD - *derivative.*

It is known that the design and use of a PID controller is a quite complex task, especially if fuzzy solutions are involved [14,28,30]. The concomitant problems include complexity of structures and dimensionality. The authors point out that the systems feature poor dynamics and performance of controlled objects. If OFN are used, a PID controller is limited to the mathematical formula alone (10), which is nested in the controller's inference unit. This solution is possible only by that the OFN arithmetic behaves in the same way with real numbers.

3 Solutions

Irrespective of the domain a control problem is solved in, the reasoning can be simplified to a system shown in Fig. 8. A typical fuzzy controller discussed in [11] and [26,35,36], and its derivations [12,22,39], are solutions with a good performance in solving of non-linear systems.

A self-setting fuzzy PID controller is one of the many proposals. A characteristic of the solutions is the addition of a fuzzy unit to a conventional PID system. The fuzzy term processes control error value e and control error increment $\frac{de}{t}$ into automatic gain control of K_p, K_i, and K_i.

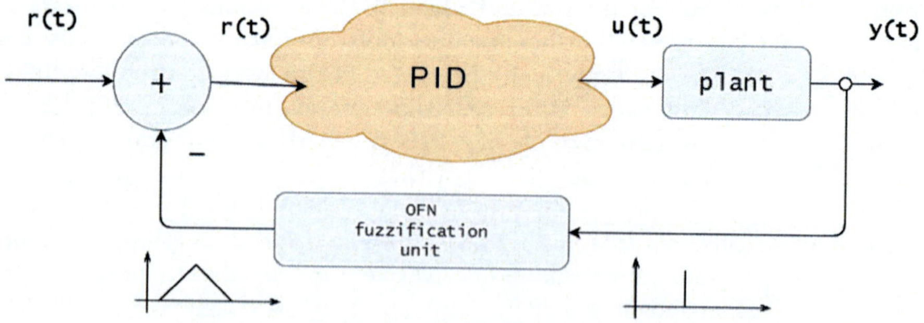

Fig. 8. General case of a PID controller due to the IN/OUT

The proposal concerns a discussion of the possible solutions of a PID controller for OFN. In the previous section, the reasons favouring the use of OFN were explained. The reasons are particularly beneficial in PID controllers. The controller input receives a set value which, together with the object's output value in the adder unit, is interpreted as the error value.

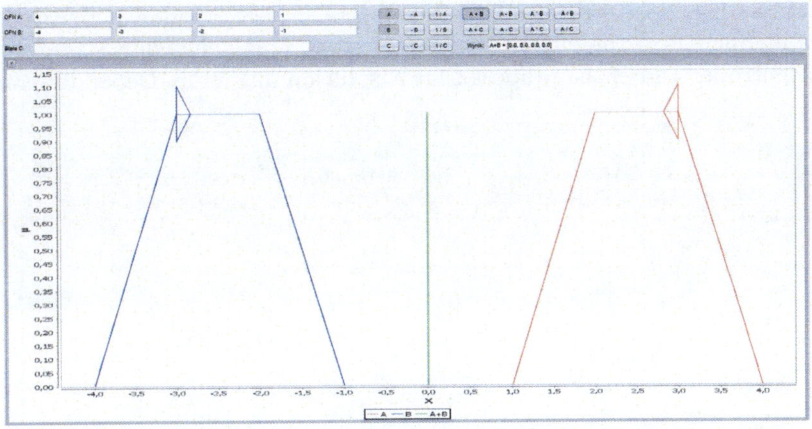

Fig. 9. Case interpretation with error $e = 0$

The discussion of a solution for OFN begins with the product, the object error as a comparison of inputs: the set value and the output value in the adder unit are key elements of a PID controller. Three operating states of the controller are classified for the purposes of this section. The first state is a situation where an OFN is input to the controller as variable x. The OFN can, in many instances, be identical to a singleton with a conventional value of 10 units of measure. For example, it is a commanded setting for a solar cell axis of rotation. Given that the controlled system achieves the value, its output is variable y with a value of 10. To solve a PID controller for OFN, a transformation from 'measured value' $y(t)$ to an OFN is applied in a 'OFN fuzzification unit'. The unit operates by defuzzying and acute value into OFN. The expected product is error $e = 0$, or that the system reaches set value x. For the interpretation of OFN, this description is illustrated in Figs. 3a, b, and c. As shown in Fig. 9 above, there are two OFN. OFN $y = [-1, -2, -3, -4]$ is the response of the controlled object (the measured value), and OFN $r = [1, 2, 3, 4]$ is the set value. The resulting error of the system is zero and calculated with Eq. (11).

$$e(t) = r(t) - y(t) \tag{11}$$

The behaviour of OFN is exceptionally well demonstrated in this example. This applies to the properties like arithmetic operations which result in real zero, and the effect of orientation on the process. As shown in Fig. 1, the interpretation of orientation of an OFN is the role of error in the system. Here, the set value and the response approach a state in which error is $e = 0$, which is obvious for a system approaching a set value. The discussed OFN vary between one another, where OFN r is a positive OFN value with negative orientation and OFN y is a negative OFN value with positive orientation. Error e can be interpreted as an OFN in this form: $[0, 0, 0, 0]$.

In other examples, the error is different from zero, which results in errors more or less than zero. If the error is less than zero, the measured value increases. If the error value increases, the output value decreases. This behaviour is typical of a classic controller and of the proposed OFN solution in Fig. 10. Depending on the

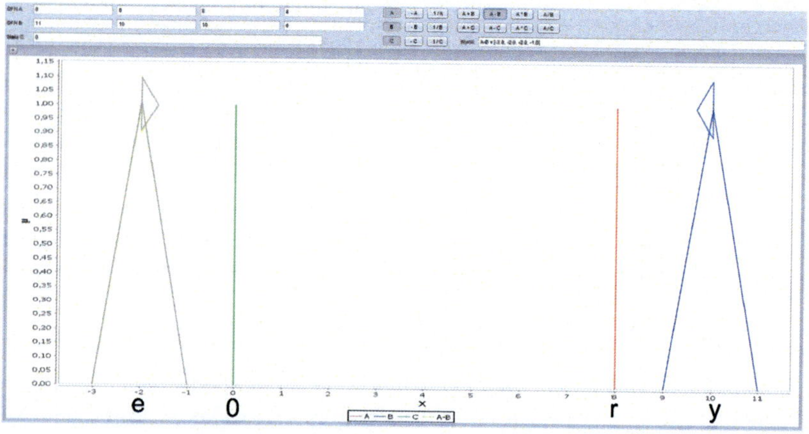

Fig. 10. If error $e(t) < 0$, output $y(t)$ increases. Constant set value $r(t)$, and zero value, "0"

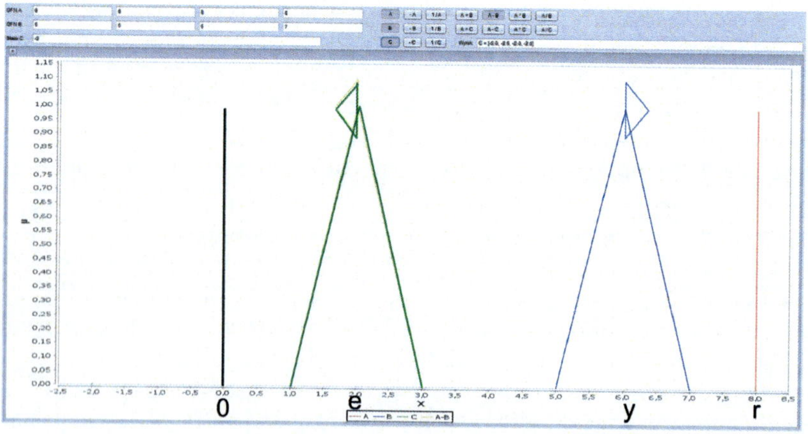

Fig. 11. If error $e(t) > 0$, output $y(t)$ decreases. Constant set value $r(t)$, and zero value, "0"

assumed scheme, a fuzzy PID controller can be represented by specific versions of links between the proportional, integral, and derivative units. Here, the units will be discussed for the application of OFN only. First controllers mainly included a proportional unit, which acted as an amplifier. In formal terms, it can be represented by a product of error and constant K_p. The arithmetic operation of multiplication for OFN is shown in Fig. 12. It is formalized as Eq. (12).

$$P_{out} = K_p * error \tag{12}$$

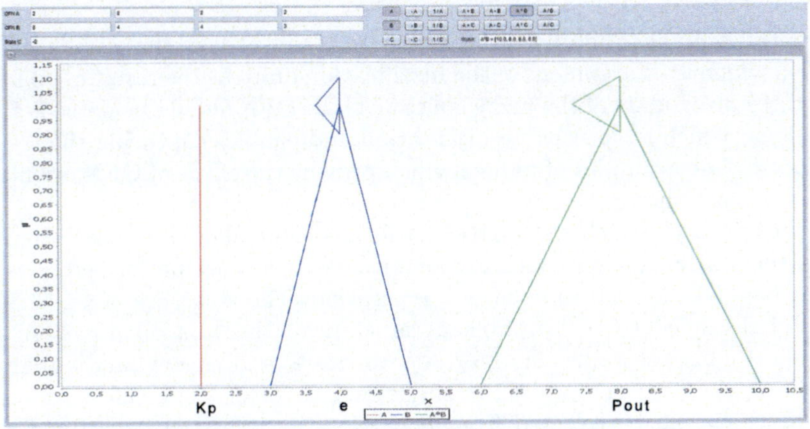

Fig. 12. Operation of the proportional unit for OFN

In this example, the multiplication of constant K_p by the fuzzy number produces a new fuzzy number. The kernels of OFN and the constant value behave as real number arithmetic operators. The example of multiplication with OFN proves that a proportional system can be implemented with the use of OFN arithmetic. The resulting gain, especially a too high value of gain, can cause the system to overshoot. The proportional unit is a two-fold challenge. The first inconvenience is the non-zero error and oscillation. To eliminate these complications, two additional units were added to the controller. The integral unit solves the problem of non-zero error. Integration is based on the response to the quantity and duration of *error* to reduce the *error* value to zero. This is done by a mathematical formula (13) which calculates the area of the error curve in defined time intervals.

$$I_{out} = K_i * (errorSum + (error + lastError) * \frac{\Delta Time}{2}) \tag{13}$$

The last of the units of the PID controller is the derivative unit, which solves the problem of oscillation damping. The higher the oscillation is, the stronger its damping is. In other terms, the derivative unit algorithm improves the dynamics of the system and responds to the slope of equation $\frac{\Delta e}{\Delta t}$. When considering the same unit with OFN, the equation desirable for calculations can be expressed as shown in (14).

$$D_{out} = K_d * \left(\frac{error - lastError}{\Delta Time}\right) \tag{14}$$

By performing the operations of the derivative unit with Eq. (14) for OFN not unlike in other units of the fuzzy PID controller, it is possible for the controller to produce inference. To summarize the results which will affect the controlled object settings, Eq. (15) is used.

$$u(t) = P_{out}(t) + I_{out}(t) + D_{out}(t) \tag{15}$$

The control value of the object resulting from Eq. (15) is the resultant fuzzy function, which is an argument of the fuzzification unit. As the result of operation of the three fuzzy units of the fuzzy controller, the OFN should be defuzzied with a functional which is sensitive to orientation and enables capturing the trend of the process. The examples of papers which present the effect of OFN orientation on inference are shown in [5,9,42].

In the approach which is devoid of a rule base and where inference is reduced to a mathematical rule on which arithmetic operations are performed, it is necessary to apply a mechanism of OFN generation. The structure of OFN should be especially fitted to the variable parameters of the controlled system. The selection of constant values K_P, K_I, K_D warrants a brief explanation. In this description, the constant values are singletons; however, they can be described as OFN which can be selected from many existing methods, like the *Ziegler-Nichols* method.

4 Conclusions

This paper presents a different approach to fuzzy logic control process. A novel feature is the lack of fuzzy rules, unlike in classic fuzzy controllers. By application of OFN postulates, especially arithmetic, it is possible to swiftly perform mathematical operations. Application of a specific process formula with OFN can simplify operations in the fuzzy controller. This paper presents the classic solutions of PID control with fuzzy implementations, and a proposal for OFN is discussed. Proportional, integral, and derivative units are presented for OFN. The paper discusses the origin of error in three main cases. The situations include an error equal to zero, an error more than a set value, and an error less than a set value. With the resultant function of membership generated, a transition to a defuzzied (acute) value is performed to command the controlled object. The impact of the resulting error value and variable derivatives is explained in the figures. The conclusions from the experiment are mainly that the application of OFN is desirable given that it employs the mathematical formula of the process to be solved. It will be necessary to develop an engineering solution to enable an in-depth analysis of behaviour of a PID controller for OFN. The development is planned for further research into the solution discussed in this paper.

References

1. Apiecionek, Ł., Czerniak, J.M., Dobrosielski, W.T., Ewald, D.: New proposal of fuzzy observation of DDoS attack. In: Atanassov, K. et al. (eds.) Uncertainty and Imprecision in Decision Making and Decision Support: Cross-Fertilization, New Models and Applications. IWIFSGN 2016. AISC, vol. 559, pp. 240–250. Springer, Cham (2018). https://doi.org/10.1007/978-3-319-65545-1_22
2. Baczynski, M., Jayaram, B.: Fuzzy Implications. Studies in Fuzziness and Soft Computing Series, Springer-Verlag, Heidelberg (2008). https://doi.org/10.1007/978-3-540-69082-5
3. Chodak, J., Heiner, M.: Spike–as a supporting tool for a model parameters optimization via branched simulations. In: Algorithms Tools Petri Nets, p. 18 (2019)
4. Czerniak, J.M., Zarzycki, H., Dobrosielski, W.T., Szczepański, J.: Application of OFN notation in the fuzzy observation of WIG20 index trend for the period 2008–2016 (2018). In: Atanassov, K., et al. (eds.) Uncertainty and Imprecision in Decision Making and Decision Support: Cross-Fertilization, New Models and Applications. IWIFSGN 2016. AISC, vol 559, pp. 190–201. Springer, Cham. https://doi.org/10.1007/978-3-319-65545-1_18
5. Dobrosielski, W.T., Szczepański, J., Zarzycki, H.: A proposal for a method of defuzzification based on the golden ratio–GR. In: Atanassov, K., et al. (eds.) Novel Developments in Uncertainty Representation and Processing. AISC, vol. 401, pp. 75–84. Springer, Cham (2016). https://doi.org/10.1007/978-3-319-26211-6_7
6. Dubois, D., Prade, H.: Operations on fuzzy numbers. Int. J. Syst. Sci. **9**(6), 613–626 (1978)
7. Dubois, D., Prade, H.: Fuzzy logics and the generalized modus ponens revisited. Cybern. Syst. **15**(3–4), 293–331 (1984)
8. Dubois, D.J.: Fuzzy Sets and Systems: Theory and Applications, vol. 144. Academic Press, New York (1980)
9. Ewald, D., Zarzycki, H., Apiecionek, Ł, Czerniak, J.M.: Ordered fuzzy numbers applied in bee swarm optimization systems. J. Univ. Comput. Sci. **26**(11), 1475–1494 (2020)
10. Fecko, B., Vince, T., Pushkar, K., Maliakova, M.: Design and realization of general purpose movement actuator. In: 2020 IEEE Problems of Automated Electrodrive. Theory and Practice (PAEP), pp. 1–4 (2020)
11. Harris, C.J., Husband, T., Brown, M., Moore, C.G.: Intelligent Control: Aspects of Fuzzy Logic and Neural Nets, vol. 6. World Scientific, London (1993)
12. He, S.Z., Tan, S., Xu, F.L., Wang, P.Z.: Fuzzy self-tuning of PID controllers. Fuzzy Sets Syst. **56**(1), 37–46 (1993)
13. Kaucher, E.: Fundamentals of Numerical Computation (Computer-Oriented Numerical Analysis), Chap. Interval Analysis in the Extended Interval Space IR, pp. 33–49. Springer, Vienna (1980). https://doi.org/10.1007/978-3-7091-8577-3_3
14. Khan, A.A., Rapal, N.: Fuzzy PID controller: design, tuning and comparison with conventional PID controller. In: 2006 IEEE International Conference on Engineering of Intelligent Systems, pp. 1–6 (2006)
15. Ang, K.H., Chong, G., Li, Y.: PID control system analysis, design, and technology. IEEE Trans. Control Syst. Technol. **13**(4), 559–576 (2005)
16. Kosinski, W., Piechór, K., Prokopowicz, P., Tyburek, K.: On algorithmic approach to operations on fuzzy numbers. In: Methods of Artificial Intelligence in Mechanics and Mechanical Engineering, pp. 95–98 (2001)

17. Kosiński, W., Prokopowicz, P.: Algebra of fuzzy numbers. Math. Applicanda **32** (2004)
18. Kosiński, W., Prokopowicz, P., Ślęzak, D.: Drawbacks of fuzzy arithmetics - new intuitions and proposition
19. Kosiński, W., Prokopowicz, P., Ślęzak, D.: On algebraic operations on fuzzy numbers. In: Kłopotek, M.A., Wierzchoń, S.T., Trojanowski, K. (eds.) Intelligent Information Processing and Web Mining. Advances in Soft Computing, vol. 22, pp. 353–362. Springer, Heidelberg (2003). https://doi.org/10.1007/978-3-540-36562-4_37
20. Kosiński, W., Prokopowicz, P., Ślęzak, D.: Ordered fuzzy numbers. Bull. Pol. Acad. Sci. **51**(3), 327–338 (2003)
21. Kosinski, W., Tyburek, K.: On soft computing and modelling. Image Process. Commun. **11**(1), 71–82 (2006)
22. Lin, Y., Ye, X.L.: Cascade fuzzy self-tuning PID control for the liquid-level control of double water-tank. In: Advanced Materials Research, vol. 383, pp. 207–212. Trans Tech Publications (2012)
23. Liu, Y., Ma, X., Qiao, W.: Key factors analysis of arctic navigation security based on fuzzy AHP (2019)
24. Lotfi, A., Tsoi, A.C.: Learning fuzzy inference systems using an adaptive membership function scheme. IEEE Trans. Syst. .Man Cybern. Part B (Cybern.) **26**(2), 326–331 (1996)
25. Mamdani, E.H.: Application of fuzzy algorithms for control of simple dynamic plant. Proc. Inst. Electr. Eng. **121**(12), 1585–1588 (1974)
26. Mamdani, E.H.: Application of fuzzy algorithms for control of simple dynamic plant. In: Proceedings of the Institution of Electrical Engineers, vol. 121, pp. 1585–1588. IET (1974)
27. Mamdani, E., Assilian, S.: An experiment in linguistic synthesis with a fuzzy logic controller. Int. J. Man-Mach. Stud. **7**(1), 1–13 (1975). http://www.sciencedirect.com/science/article/pii/S0020737375800022
28. Mann, G.K.I., Hu, B.-G., Gosine, R.G.: Analysis of direct action fuzzy PID controller structures. IEEE Trans. Syst. Man Cybern. Part B (Cybern.) **29**(3), 371–388 (1999)
29. Marszałek, A., Burczyński, T.: Ordered fuzzy candlesticks. In: Prokopowicz, P., Czerniak, J., Mikołajewski, D., Apiecionek, Ł., Ślęzak, D. (eds.) Theory and Applications of Ordered Fuzzy Numbers. Studies in Fuzziness and Soft Computing, vol. 356, pp. 183–194. Springer, Cham (2017). https://doi.org/10.1007/978-3-319-59614-3_10
30. Mizumoto, M.: Realization of PID controls by fuzzy control methods. Fuzzy Sets Syst. **70**(2), 171 – 182 (1995). http://www.sciencedirect.com/science/article/pii/016501149400215S. Modern Fuzzy Control
31. Molnár, J., Gans, ., Slavko, O.: Design and implementation self-balancing robot. In: 2020 IEEE Problems of Automated Electrodrive. Theory and Practice (PAEP), pp. 1–4 (2020)
32. Patyra, M.J., Grantner, J.L., Koster, K.: Digital fuzzy logic controller: design and implementation. IEEE Trans. Fuzzy Syst. **4**(4), 439–459 (1996)
33. Qiao, W., Liu, Y., Ma, X.: A methodology to evaluate human factors contributed to maritime accident by mapping fuzzy FT into ANN based on HFACS. Ocean Eng. **197**, 106892 (2020)
34. Smets, P., Magrez, P.: Implication in fuzzy logic. Int. J. Approximate Reasoning **1**(4), 327–347 (1987)

35. Åström, K.: Intelligent tuning. IFAC Proceedings Volumes 25(14), 359–370 (1992), https://www.sciencedirect.com/science/article/pii/S1474667017507612, 4th IFAC Symposium on Adaptive Systems in Control and Signal Processing 1992, Grenoble, France, 1-3 July
36. Åström, K., Hägglund, T., Hang, C., Ho, W.: Automatic tuning and adaptation for PID controllers - a survey. Control Eng. Pract. **1**(4), 699–714 (1993). https://www.sciencedirect.com/science/article/pii/096706619391394C
37. Sugeno, M., Kang, G.: Structure identification of fuzzy model. Fuzzy Sets Syst. **28**(1), 15–33 (1988). http://www.sciencedirect.com/science/article/pii/0165011488901133
38. Takagi, T., Sugeno, M.: Fuzzy identification of systems and its applications to modeling and control. IEEE Trans. Syst. Man Cybern. SMC **15**(1), 116–132 (1985)
39. Yang, Y., Bian, H.: Design and realization of fuzzy self-tuning PID water temperature controller based on plc. In: 2012 4th International Conference on Intelligent Human-Machine Systems and Cybernetics, vol. 2, pp. 3–6. IEEE (2012)
40. Zadeh, L.: Fuzzy sets. Inf. Control **8**(3), 338–353 (1965)
41. Zadeh, L.A.: Outline of a new approach to the analysis of complex systems and decision processes. IEEE Trans. Syst. Man Cybern. SMC **3**(1), 28–44 (1973)
42. Zarzycki, H., Dobrosielski, W., Apiecionek, Ł., Vince, T.: Center of circles intersection, a new defuzzification method for fuzzy numbers. Bulletin of the Polish Academy of Sciences. Technical Sciences **68**(2) (2020)

Data Quotation Period and the Quality of ANN Learning in the Polish Power Exchange Model

Dariusz Ruciński(✉)

Computer Science Institute, UPH - University of Natural Sciences and Humanities, 3-Maja 54, 08-110 Siedlce, Poland
dariusz.rucinski@uph.edu.pl

The work contains the results of research on the impact of the length of data quotations on the Day-Ahead Market and thus the number of teaching and testing pairs on the learning quality of the Artificial Neural Network as a model of the Polish Power Exchange. The main target of the research was to find out the quality of Artificial Neural Network reproduction as multidimensional approximation of the Polish Power Exchange Day Ahead Market model. The data covers the entire period of functioning of the Polish stock exchange (2002–2019 first half of the year), hence it was necessary to make a decision as to the number of teaching and testing pairs in order to achieve such accurate neural model to reflect the operation of the real system. The research was carried out in the MATLAB and Simulink environment with the use of our own m-file library.

1 Introduction

The modeled object is the Polish Electricity Exchange (PPE) for business processes related to the Day-Ahead Market (DAM). The input data included 24 quantities concerning the volume of electricity supplied and sold on the DAM [MWh] in particular hours of the day, and the output values, respectively, obtained in particular hours of the day, were average weighted with the volume of electricity sold [PLN/MWh]. The figures used in the modeling relate to six months, i.e. 181 days (from 01/01/2015 to 30/06/2015). Previous papers of author listed in references presents the neural model of the Day-Ahead Market (DAM) system of the Polish Power Exchange (PPE) along with its evolutionary modifications and quantum inspirations, implemented in the form of a Perceptron Artificial Neural Network (ANN) model taught using the data recorded on January 1 2015 to June 30, 2015 (semi-annual period).

There were doubts as to whether the period of the data adopted in the experiment is too short, which may cause difficulties in making a reliable experimental analysis, and there was a need to check the functioning of the model on the data for subsequent periods in order to make a reliable analysis of the practical usefulness of the obtained results. including the repeatability of the obtained result, which was the DAM of PPE neural model. In order to obtain the neural model of the PPE DAM system, the impact of the length of the period of data recorded

K. T. Atanassov et al. (Eds.): IWIFSGN 2020/BOS/SOR 2020, LNNS 338, pp. 308–326, 2022.
https://doi.org/10.1007/978-3-030-95929-6_24

on the Day-Ahead Market and thus the number of training and testing pairs on the quality of the Artificial Neural Network was examined in a manner that did not raise any doubts in the selection of the data length used in the modeling experiment. This aim of the research is related to the search for the quality of the ANN reproduction of the multidimensional approximation of the DAM model. For this purpose, the experiment used figures from the entire period of operation of the DAM, i.e. from 2002 to 2019 (the first half of the year).

Due to the relatively large set of data used in this case to teach ANN, i.e. 17.5 years (previously only a half-year period was used), an attempt was made to divide the data into different research periods in order to determine the quality of the approximation in the ANN learning process, e.g. the least squares method is influenced by data with a sufficiently long training cycle, that is, correspondingly large amounts of training and testing file data. At this point, therefore, the problem of the multiplicity of the set of training and testing pairs, i.e. the number of input vectors of such a length, that the learned neural model accurately reflects the operation of the real system, i.e. universal, i.e. a substitute diagram of the real system, which can be used in carrying out various types of tests on the model, so that there is no need to conduct them on the real system, including forecasting average weighted with the volume of electricity prices in particular hours of the day. An additional goal is to check whether the number of training and testing pairs depends on the specificity and properties of the PPE DAM system, whether it depends on the quality and quantity of training and testing data. Therefore, apart from the results of the research on learning the Perceptron ANN, the study also shows the results of the research on the impact of the number and quality of training and testing pairs taken from the PPE DAM on the quality of the neural model.

The research was carried out in the MATLAB and Simulink environment using the developed own m-file library and functions available in the Neural Network Toolbox (Guide for MATLAB 2019). The R^2 coefficient of determination and the mean square error MSE (see: Ruciński 2018) were used as a measure of the model quality. It was assumed that the model is considered correctly learned for a given number of training pairs and testing pairs, if the value of R^2 and the mean square error MSE do not differ more than their acceptable measured values from the highest value of R^2 and the lowest MSE obtained for all tested models.

The examined neural model concerns the setting of prices [PLN] on the PPE DAM depending on the volume of electricity supplied and sold in particular hours of the day [MWh]. Modeling market decision problems for the balancing market or modeling electricity prices on the electricity market is not a completely new issue, but so far, not so much a model of a neural system has been searched for, but rather, inter alia, a specific model of decision making in the balancing market (see: Toczyłowski et al. 2010) or price shaping (see: Arabas, Adamowicz 2003), or a price model depending on different input quantities and adopted modeling methods (see: Fijorek et al. 2010; Malko 1995; Conejo et al. 2005; Weron 2006) or research on the quality of electricity load prediction (see: Ciechulski, Osowski 2014).

2 The Problem of the Number of Learning Pairs

The subject of the work is the selection of the quantity of learning data. If there is a set of $\{u_i\}_1^q$ input data and $\{d_i\}_1^q$ of expected data (the so-called teacher), then $\{u_i, d_i\}_1^q$ is a set of training pairs. Thus, the question arises as to what size of the learning pairs set is appropriate for a particular Artificial Neural Network, so that it does not over-train or under-train (see: Osowski 2013; Zurada et al. 1994). One of the proposals was presented by Vapnik and Czerwonienkis, named as the VC dimension (see: Blumer 1989), which is a measure of the capacity, power of a family of classification functions. The problem investigated by Vapnik and Czerwonienkis was to determine how large the training data set should be for the classifying function to work without error, eg having a binary classification function, two possible results can be obtained: d ∈ {0.1} The functions that implement the result d (so-called dichotomy) can be described as the set Z F(u, w), where the vector w belongs to the space W, and the function F belongs to the m - dimensional space R mapping the space 0,1, which can be written as:

$$Z = \{F(u, w) : w \in W, F : R^m W\{0, 1\}\} \tag{1}$$

where the set of input vectors u must also be m-dimensional, which can be written as follows:

$$L = \{u_i \in R_m, i = 1, 2, ..., q\} \tag{2}$$

In the case of a two-dimensional space, the input set can be divided into two subsets $L_0\, and\, L_1$:

$$F(u, w) = \begin{Bmatrix} 0 \text{ for } u \in L_0 \\ 1 \text{ for } u \in L_1 \end{Bmatrix} \tag{3}$$

If $\Delta_z(L)$ represents the number of unique classifying functions in a two-dimensional space, implemented by ANN, and $\Delta_z(l)$ represents the maximum cardinality of the set $\Delta_z(L)$, then for $\|L\| = 1$ it can be concluded that L can be divided by Z on $\Delta_z(L) = 2^L$ ways. The VC dimension allows to estimate the degree of ANN learning, which is used in the statistical learning theory (see: Koronacji, Ćwik 2006). There is a relationship between the number of teaching pairs and the VC dimension. The number of free parameters is proportional to the VC dimension. For networks with a threshold activation function, the VC dimension is:

$$O(wlogw), \tag{4}$$

where: w - number of free parameters (weights).
For a Multi-Layer Perceptron (MLP) network with a sigmoidal activation function (see: Żurada et al. 1996), the VC dimension is:

$$O(w^2), \tag{5}$$

which shows the advantage of MLP networks due to their greater "capacity" than other ANNs.

3 Characteristics of the MLP

The Perceptron model is one of the most popular ANN tools used to classify linearly separated patterns, i.e. linear regression consisting in finding the straight line best suited to the points representing the data, which is illustratively shown in Fig. 1 (see: Patterson, Gibson 2020; Żurada et al. 1996).

The work of Rosenblatt (1957) gives evidence that in the case of linearly separated patterns, a single-layer perceptron is sufficient for the separability of functions. For example, logic gates AND and OR, can be separated by a line in two-dimensional space, but not XOR (for its implementation, a two-layer Artificial Neural Network is needed due to the fact that the values 0 and 1 are diagonally on the first layer. Assuming that the cost of the function amounting to C will be understood as the difference between the expected value and the value obtained from ANN, and therefore in this case it will depend on the independent variable which is weight w and will have the form C(w). Assuming further that the function C(w) is differentiable, the optimal solution is to obtain such vector $w^\star$, that:

$$C(w^\star) \leq C(w), \tag{6}$$

where: $C(w^\star)$ is a minimum of the cost. In such a case $\Delta C(w^\star) = 0\Delta$, where gradient for m-dimensional vector is equal:

$$\Delta = \left[\frac{\partial}{\partial w_1}, \frac{\partial}{\partial w_2}, \ldots, \frac{\partial}{\partial w_n}\right]^T \tag{7}$$

from which:

$$\Delta C(w^\star) = \left[\frac{\partial C}{\partial w_1}, \frac{\partial C}{\partial w_2}, \ldots, \frac{\partial C}{\partial w_n}\right]^T \tag{8}$$

In order for the cost gradient to be equal zero, it is necessary to find out a local minimum (for a derivative equal to zero), it is done in subsequent iterations by differentiating the cost function and changing the weights according to a specific criteria:

$$C(w(n+1)) < C(w(n)), \tag{9}$$

where: n - next iteration number. Optimization can be done in a few ways e.g.: 1) by the fastest gradient descent method, then:

$$G = \Delta C(w), \tag{10}$$

$$w(n+1) = w(n) + \eta G(n), \tag{11}$$

$$\Delta w(n) = w(n+1) - w(n) = -\eta(G(n). \tag{12}$$

The value to found out is the cost of iteration C(w(n + 1)) which is found based on the Taylor series expansion around w(n). Taking the first derivative into account, the series takes the form:

$$\begin{aligned} C(w(n+1) &\approx C(w(n)) + G(w(n))^T \Delta w(n) \\ &= C(w(n) - \eta G(w(n)^T G(n) = C(w(n) - \eta \|G(n)\|^2 \end{aligned} \tag{13}$$

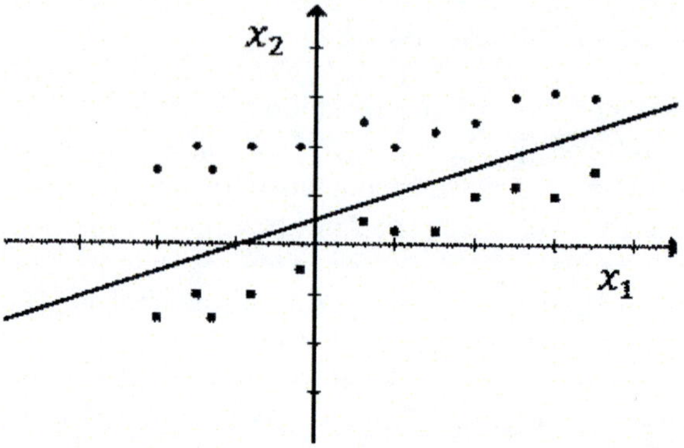

Fig. 1. An example of the classification of linearly separated learning patterns using ANN of the Perceptron type with a single neuron (linear regression). Marking: x_1, x_2 axis of input vector values in two-dimensional space. Source: (see: Patterson, Gibson 2020; Żurada et al. 1996).

And since both the value of the gradient norm and the cost of the nth iteration are always positive, the cost of the n + 1 iteration will be smaller, and the scale of change will depend on the parameter $0 < \eta < 1$. 2) Newton's method, which using the second derivative of the Taylor series for the cost function:

$$\begin{aligned} \Delta C(w(n) &= C(w(n+1)) - C(w(n)) \\ &\approx G(n)^T \Delta w(n) + \tfrac{1}{2}\Delta w(n)^T H(n) \Delta w(n), \end{aligned} \tag{14}$$

where: $\frac{1}{2}\Delta w(n)^T H(n)\Delta w(n)$ (the second term of the dependence (14) – is the second derivative of the cost function called, $\mathbf{H}$(n) – Hessian:

$$\mathbf{H} = \begin{bmatrix} \frac{\partial^2 C(w)}{\partial w_1^2}, & \frac{\partial^2 C(w)}{\partial w_1 \partial w_2}, & \ldots, & \frac{\partial^2 C(w)}{\partial w_1 \partial w_m} \\ \frac{\partial^2 C(w)}{\partial w_2 \partial w_1}, & \frac{\partial^2 C(w)}{\partial w_2^2} & \ldots, & \frac{\partial^2 C(w)}{\partial w_2 \partial w_m} \\ \ldots, & \ldots, & \ldots, & \ldots \\ \frac{\partial^2 C(w)}{\partial w_m \partial w_1}, & \frac{\partial^2 C(w)}{\partial w_m \partial w_2}, & \ldots, & \frac{\partial^2 C(w)}{\partial w_m^2} \end{bmatrix} \tag{15}$$

The minimum cost will be achieved if its increment is zero. By differentiating the Eq. (14) according to the variable $\Delta w(n)$ we get:

$$\Delta C(w(n)) = G(n) + \mathbf{H}(n)\Delta w(n) = 0 \tag{16}$$

where does it follow that:

$$\Delta w(n) = \mathbf{H}^{-1}(n)G(n) \tag{17}$$

wherein:

$$w(n+1) = w(n) + \Delta w(n) = w(n) - \mathbf{H}^{-1}(n)G(n) \tag{18}$$

From equation (18), and more specifically from its second term, it follows that Newton's method allows for asymptotic convergence of the cost function provided that the matrix **H** is positive for every n. Equation (15) uses the differentiation of the function according to the vector, where the function C(w) is a function of the actual values of the parameter w, which is the m-dimensional vector $[w_1, w_2, ..., w_m]$. If the function is differentiated according to this vector then:

$$\frac{\partial C(w)}{\partial w} = \left[\frac{\partial C(w)}{\partial w_1}, \frac{\partial C(w)}{\partial w_2}, \frac{\partial C(w)}{\partial w_3}, ..., \frac{\partial C(w)}{\partial w_m}\right] \tag{19}$$

The result is a vector consisting of the derivatives for each dimension. If the function is a form of an internal product, then there is a relationship:

$$f(w) = u^T w = \Sigma_{k=1}^{m} u_k w_k \tag{20}$$

then

$$\frac{\partial f(w)}{\partial w_k} = u_k \tag{21}$$

that is the result of the function differentiation will be a vector u. For quadratic form functions we get:

$$f(w) = w^T R w = \Sigma_{k=1}^{m} \Sigma_{j=1}^{m} w_k r_{kj} w_j \tag{22}$$

where:

R is the matrix of real values. In turn, for $k = j$ the matrix values are w^2, and if the function is differentiated:

$$\frac{\partial f(w) \partial w_k}{=} \Sigma_{j=1}^{m} r_{kj} w_j = \mathbf{Rw} \tag{23}$$

then the result is a matrix with dimensions m × m. The matrix **R** is positive if for any vector **a** $\in R^m$ the expression $a^T Ra > 0$ is true (see: Kaczorek et al. 1998). The positive matrices have the following properties: for positive R matrices there are inverse R − 1 matrices and the eigenvalues are such that for det (R) = 0 they are all positive.

4 Research Characteristics

In the case under consideration, the PPE DAM is modeled as a system (see: Mielczarski 2000; Tchórzewski 2013). The data for training and testing of ANN was collected from the DAM from the entire period of PPE operation, 01/07/2002 to 30/06/2019 (see: Towarowa Giełda Energii S.A. 2020). This period was divided into periods of different lengths of samples, as is shown in Table 1. The result of ANN learning was a neural model in the form of a PPE system equivalent scheme. Learning the ANN of the PPE DAM neural model required the preliminary preparation of data obtained from the DAM PPE (e.g. their normalization), i.e. both the input values for 24 variables being the volume of electricity supplied and sold in particular hours of the day, as well as the output values for 24 variables being weighted average of prices obtained in each hour of the day for the supplied and sold electricity was normalized.

Table 1. Possibilities of training ANN with the use of data from periods with different possible lengths of quotations on the PPE DAM. Source: own author elaboration.

Period	Number of learning subsets
Month	204
Quarter	68
Half-year	34
Year	17
2 years	8
3 years	5
4 years	4
5 years	3
6 years in steps of 1 year	12
7 years in steps of 1 year	11
8 years in steps of 1 year	10
9 years in steps of 1 year	9
10 years in steps of 1 year	8
15 years in steps of 1 year	3
The entire period under study, 01/07/2002–30/06/2019	1

5 Data Used in Calculations and Market Characteristics

The data used for the neural modeling was obtained from electricity quotations on the PPE DAM. The input data for the model (u) was the volume of delivered and sold electricity expressed in MWh, and as the output data for the model (s) - average prices weighted by the volume of delivered and sold ee in each hour of the day, i.e. 0–1, 1–2, 2–3, etc. (see: Towarowa Giełda Energii SA 2020; Ruciński 2018). Quotations on the PPE DAM are held every day, contracts are concluded 24 or 48 h before the electricity is supplied to the customer. Contracts are concluded for individual delivery hours, while quotes are held according to a specific schedule, where each PPE participant has his own account (wallet) and can submit any number of orders (see: Mielczarski 2000; Miller, Bućko 2014).

Each order specifies, inter alia, the code of the forward instrument of a given type of order, type of order (purchase or sale), portfolio for which the order is placed, volume (amount of energy purchased or sold), etc. Information about transactions is published on the PPE website and they were the source of data for conducted analysis (see: Towarowa Giełda Energii SA 2020; Mielczarski 2000).

6 Tools and Method of Modelings

The MATLAB and Simulink environment (see: Guide for MATLAB 2019) was selected for the ANN design and learning process from among other similar tools that can be used in the design of neural models, such as SPHINX, SAS or SAP. Technical considerations prevailed in the selection process, because the above-mentioned calculations are dedicated to this type of calculation. Neural Network Toobox environment selected for neural modeling and its successor Deep Learning Toolbox (Paterson 2020) and Matlab language, as well as the fact that it is in the latest version by the Institute of Informatics of UPH in Siedlce. The MATLAB and Simulink environment gives great opportunities to support research using the Neural Network Toolbox (NNT) and Deep Learning Toolbox (DLT) libraries, along with the possibility of using a very high-level language such as Matlab. The process of designing a neural model in the MATLAB and Simulink environment using the NNT and DLT functions has been reduced to the following steps:

Step1. Designing the ANN architecture, in this case using the feedforwardnet (hiddenSizes, trainFcn) function, which has two optional arguments: hiddenSizes (number of neurons in hidden layers) and trainFcn (ANN training function).
Step2. Selection of the neuron activation function for individual layers, in this case the functions were respectively: for the first layer of neurons - tansig and for the second layer of neurons - purelin
Step3. Determining the arguments of the neti for retrieving input and output values (so-called training pairs), i.e. for retrieving input data, output data and selected ANN training parameters, including methods of initialization of initial weights and bias values, while re-initialization of weights is possible using the neti = init (net) function. The modeling results for the MSE error obtained for the Perceptron Artificial Neural Network (PANN) are shown in Table 2, the determination coefficient R2 error results - in Table 3, and the selected error courses for various learning subsets for the MSE error - in Table 4 and for the R2 in Table 5. The method of designing and teaching PSSN has been presented in detail in the author's earlier works (see: Ruciński 2018; Ruciński 2017; Ruciński 2016, Tchórzewski, Ruciński 2016).

Table 2. Modeling results for the MSE error value obtained for the Perceptron ANN. Source: own author elaboration.

The obtained MSE values for the appropriate length of modeling periods			
The length of the quotation period on the PPE DAM	Maximum	Minimum	Mean
Month	0,024845	2,49E-12	0,000727
Quarter	0,006832	7,29E-05	0,000907
Half-year	0,006224	0,000209	0,001114
Year	0,005502	0,000326	0,001377
2 years	0,00408	0,00044	0,001615
3 years	0,003139	0,00064	0,001338
4 years	0,003612	0,000603	0,001552
5 years	0,002895	0,000579	0,001517
6 years in steps of 1 year	0,002526	0,000707	0,001429
7 years in steps of 1 year	0,00235	0,000773	0,001462
8 years in steps of 1 year	0,002179	0,000919	0,001501
9 years in steps of 1 year	0,00262	0,000851	0,001595
10 years in steps of 1 year	0,002515	0,000931	0,001651
15 years in steps of 1 year	0,002027	0,001756	0,001922
Entire period (10 trials)	0,002098	0,001624	0,001904

Table 3. Modeling results concerning the R^2 error value obtained for the Perceptron Artificial Neural Network. Source: own author elaboration.

The obtained R^2 values for the appropriate length of modeling periods			
The length of the quotation period on the PPE DAM	Maximum	Minimum	Mean
Month	0,910676	0,13284	0,671464
Quarter	0,883035	0,260872	0,664489
Half-year	0,847599	0,20746	0,622259
Year	0,787481	0,395959	0,60331
2 years	0,74061	0,46056	0,567373
3 years	0,655004	0,492299	0,5669
4 years	0,662723	0,467276	0,545216
5 years	0,560887	0,516093	0,539543
6 years in steps of 1 year	0,654797	0,368147	0,533565
7 years in steps of 1 year	0,638559	0,462133	0,536493
8 years in steps of 1 year	0,641572	0,434553	0,516764
9 years in steps of 1 year	0,577501	0,437375	0,507126
10 years in steps of 1 year	0,537034	0,472314	0,501819
15 years in steps of 1 year	0,484907	0,442602	0,45778
Entire period (10 trials)	0,474964	0,411678	0,446714

Table 4. Selected charts of MSE values for the periods listed in Table 2. Source: own author elaboration.

Course of **maximum MSE errors** for subsequent periods from one month, through quarter, half-year, year, 2 years, up to the entire period of PPE operation under study

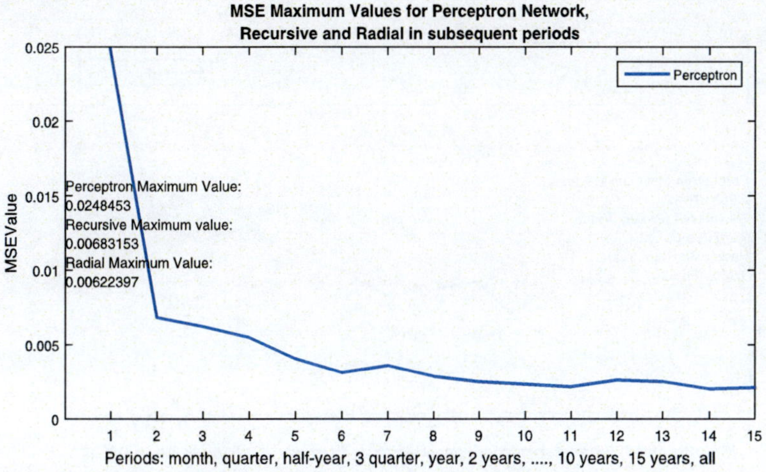

Course of **mean MSE errors** for subsequent periods from one month, through quarter, half-year, year, 2 years, up to the entire period of PPE operation under study

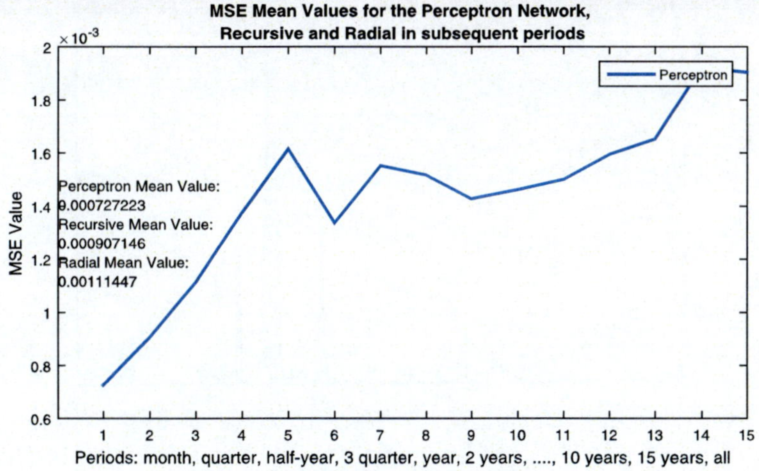

(*continued*)

Table 4. (*continued*)

Course of **minimal MSE errors** for subsequent periods from one month, through quarter, half-year, year, 2 years, up to the entire period of PPE operation under study
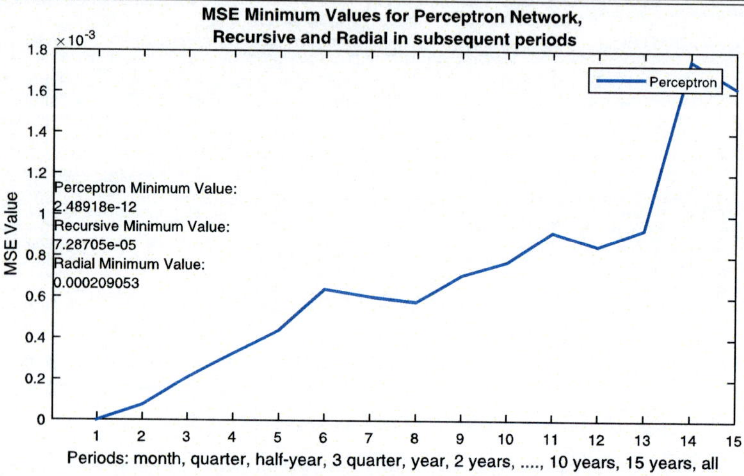

Course of **MSE errors for monthly periods**, taking into account the successive number of periods
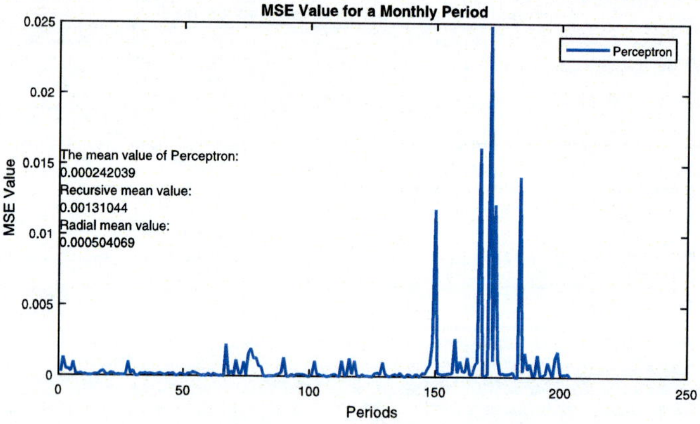

(*continued*)

Table 4. (*continued*)

Course of **MSE errors for quarter periods**, taking into account the successive number of periods

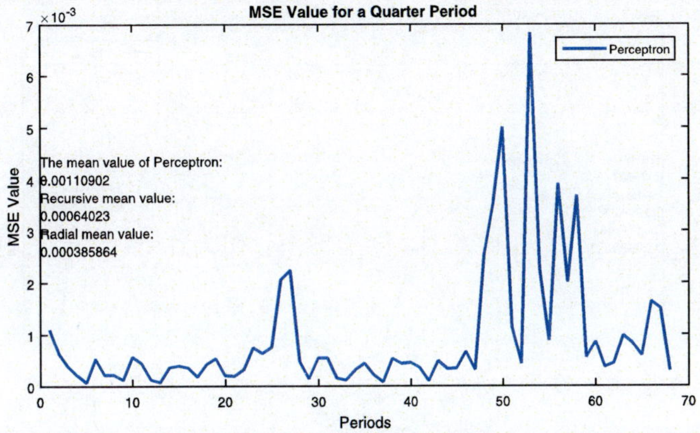

Course of **MSE errors for half year periods**, taking into account the successive number of periods

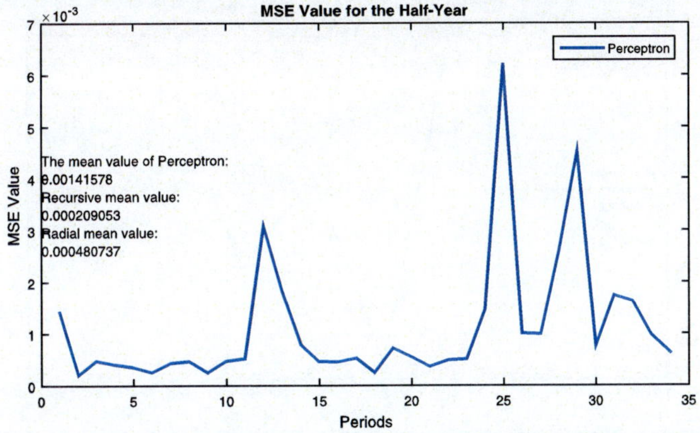

(*continued*)

Table 4. (*continued*)

Course of **MSE errors for year periods**, taking into account the successive number of periods

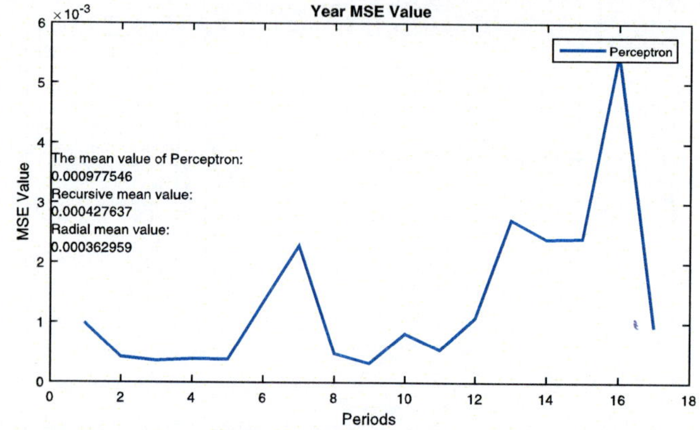

Course of **MSE errors for the entire period** of operation of the PPE DAM, with 10 data attempts

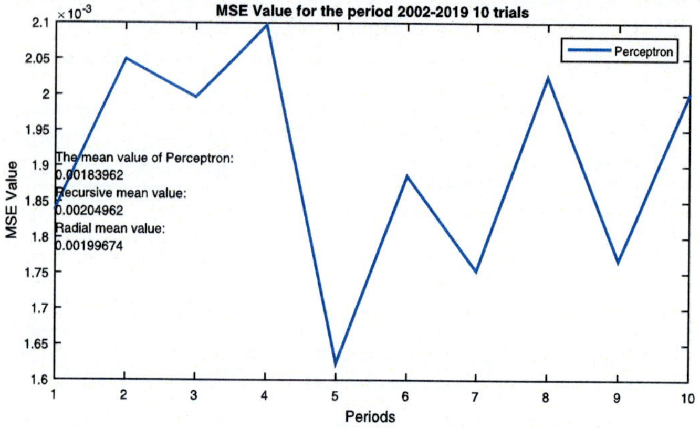

Table 5. Selected graphs concerning the values of the regression coefficient R^2 for the periods specified in Table 3. Source: own author elaboration.

Course of **maximum** R^2 **errors** for subsequent periods from one month, through quarter, half-year, year, 2 years, up to the entire period of PPE operation under study

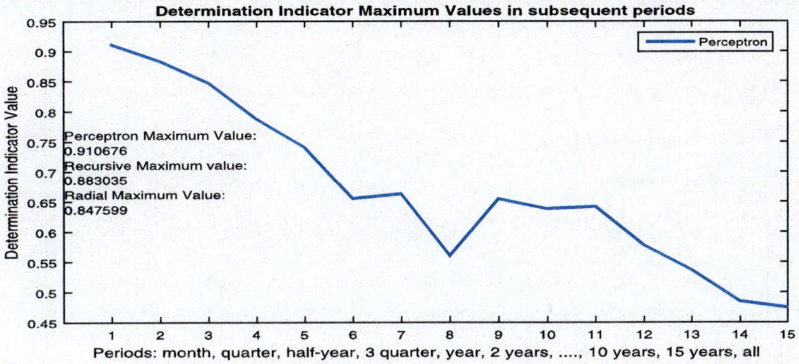

Course of **mean** R^2 **errors** for subsequent periods from one month, through quarter, half-year, year, 2 years, up to the entire period of PPE operation under study

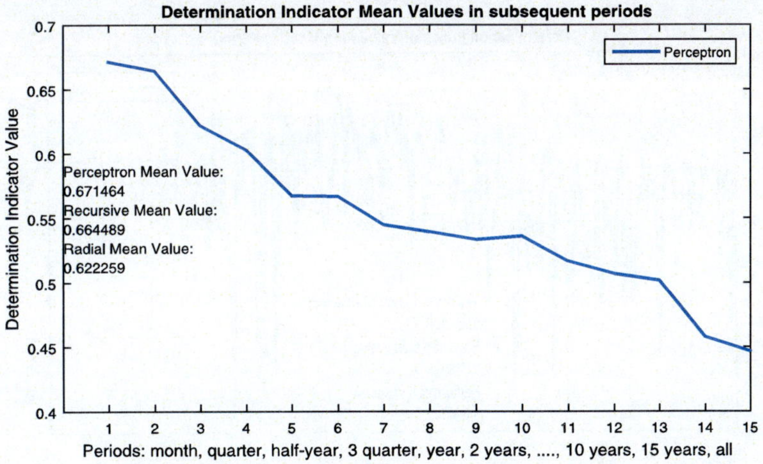

(*continued*)

Table 5. (*continued*)

Course of **minimal R^2 errors** for subsequent periods from one month, through quarter, half-year, year, 2 years, up to the entire period of PPE operation under study
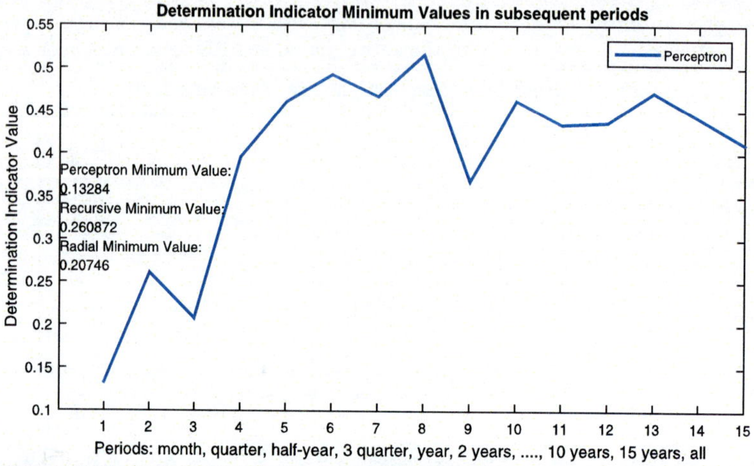
Course of R^2 **errors for monthly periods**, taking into account the successive number of periods
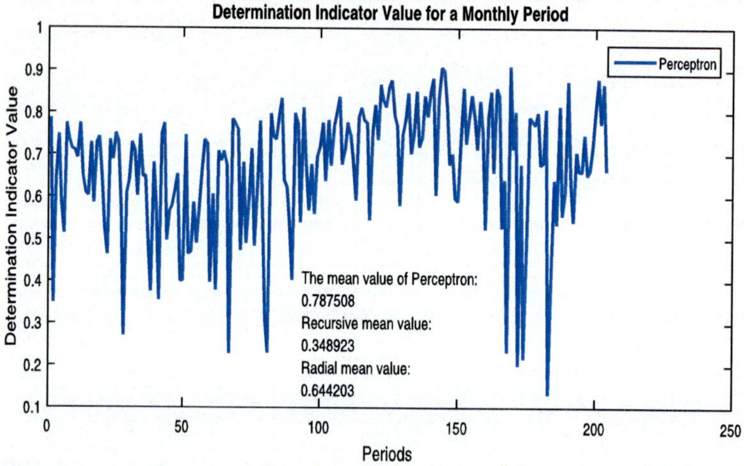

(*continued*)

Table 5. (*continued*)

Course of R^2 **errors for quarter periods**, taking into account the successive number of periods
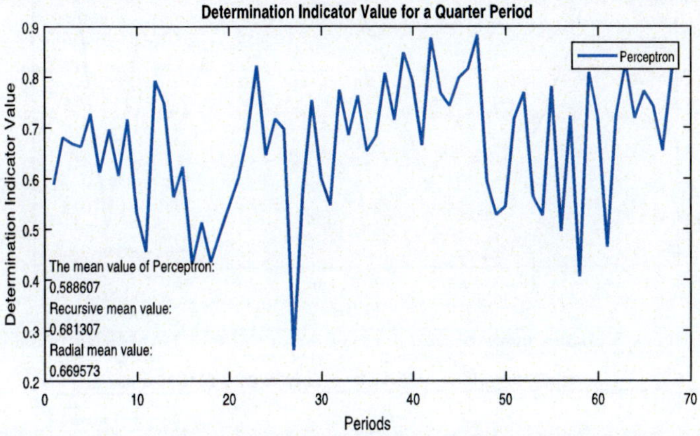
Course of R^2 **errors for half year periods**, taking into account the successive number of periods
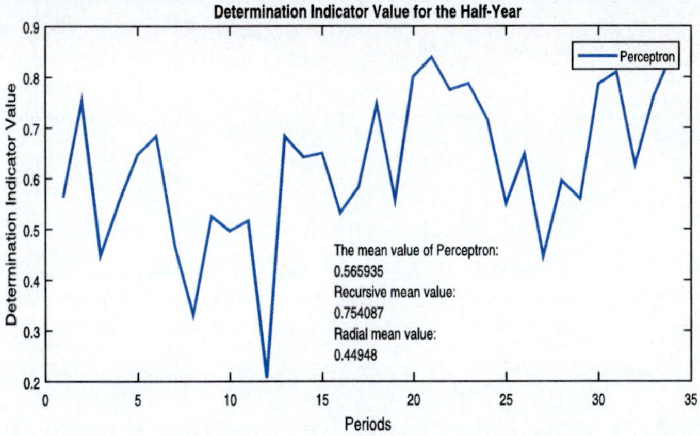

(*continued*)

Table 5. *(continued)*

Course of R^2 **errors for year periods**, taking into account the successive number of periods

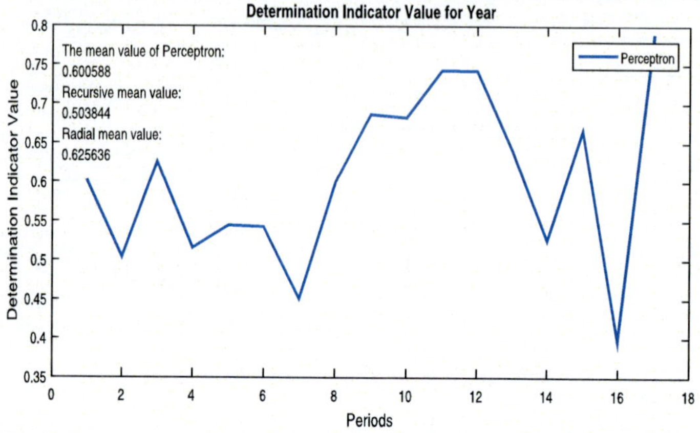

Course of R^2 **errors for the entire period** of operation of the PPE DAM, with 10 data attempts

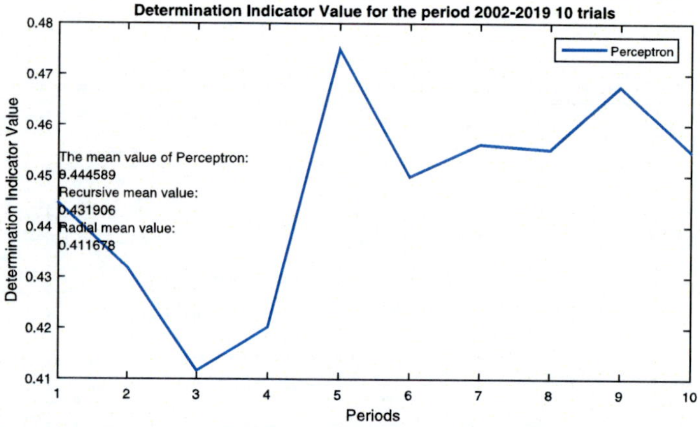

7 Conclusions and Directions for Further Research

1. As a result of the research, the results were obtained concerning the impact of the length of data quotations on the Day-Ahead Market (DAM) and thus the number Please, do not put anything in the header Nor in the footer of teaching and testing pairs on the quality of learning the Artificial Neural Network (ANN) as a model of the Polish Power Exchange (PPE).
2. Moreover, research results were obtained concerning the influence of various ANN learning parameters, including the number of layers and the number of neurons in the layers, as well as various activation functions, and learning rules on the quality of the DAM model.

3. Many detailed research results were obtained, which in particular include the results of the analysis of changes in the MSE index and for R2 for different periods of the training sequences, and these indicators indicated better results for shorter intervals, which were, for example: for the MSE error for a period of:

month: error value was in the range 0.0245483 to 2.489 * 10–12,
year: error value was 0.0550218 0.000326423,
10 years: error value ranged from 0.00209772 to 0.00162369,

for the determination index R^2 for the period equal to:

month: the R^2 value was in the range 0.13284 to 0.910676,
year: the R^2 value was in the range 0.395959 to 0.787481,
10 years: the R2 value was in the range 0.472314 to 0.537034.

4. It was noted, inter alia, that the quality of the model is determined by the lowest possible MSE error value and the highest possible value of the coefficient of determination R2. for the average value for each period of the above-mentioned the ratios were respectively for the period of:

month: MSE = 0.000727, coefficient of determination R^2 = 0.671464,
year: MSE = 0.00137711, coefficient of determination R^2 = 0.60331,
10 years: MSE = 0.00165136, coefficient of determination R2 = 0.501819,

which indicates that as the number of training pairs increased for the respective period: quarter, half-year, year, 10-year, etc., the MSE error and regression error rates deteriorated.

5. It was found that selecting the optimal period of neural modeling of the DAM of PPE, the two factors mentioned and tested should be taken into account, i.e. the value of the MSE error and the value of the regression error expressed with the determination index R^2, which turned out to be a very important element of qualitative research, apart from the stability and accuracy of the obtained neural model.

6. An attempt to develop a synthetic measure of the number of samples and thus the length of the measuring period may be the subject of further targeted and in-depth research on the above-mentioned range.

General Remark: Based on the conducted neural modeling for subsequent periods of data recording in the years 2002–2019, the first half of the year (of various lengths), there were no differences in the quality of the measured indicators, both MSE and R^2, obtained for individual time intervals specially selected for this purpose, i.e. incl. month, quarter, half-year, year, etc., regardless of whether these were data from the beginning of the PPE period in 2002, or they are closer to the current times, including the first half of 2019. On the other hand, it was noted that the differences are visible in the aspect of the length of the training set itself, i.e. different results were obtained for the month, quarter, half-year, year, etc., which was tried to be shown in this article.

References

Arabas, J., Adamowicz, L.: Prognozowanie zapotrzebowania a rynek energii. CIRE (2003). www.cire.pl
Guide for MATLAB, Simulink, Neural Network Toolbox. MathWorks, USA (2019)
Malko, J.: Wybrane zagadnienia prognozowania w elektroenergetyce. OW PWr, Wrocław (1995)
Kaczorek, T.: Wektory i macierze w automatyce i elektronice. WNT, Warszawa (1998)
Koronacki, J., Ćwik, J.: Statystyczne systemy uczące się, 327 p. AOW EXIT, Warszawa (2006)
Mielczarski, W.: Rynki energii elektrycznej. Wybrane aspekty techniczne i ekonomiczne, 308 p. ARE, Warszawa (2000)
Osowski, S.: Sieci neuronowe do przetwarzania informacji, 422 p. OWPW, Warszawa (2013)
Paterson, J., Gibson, A.: Deep Learning. Praktyczne wprowadzenie. Hellion, Warszawa (2020)
Rosenblatt, F.: The perceptron: a perceiving and recognizing automaton (project PARA) report (85-460-1). Cornell Aeronautical Laboratory, Inc. (1957)
Tchórzewski, J.: Rozwój systemu elektroenergetycznego w ujęciu teorii sterowania i systemów. OW PWr., Wrocław (2013)
Weron, R.: Modeling and Forecasting. Electricity loads and Prices: A Statistical Approach. Wiley, Hoboken (2006)
Żurada, J., et al.: Sztuczne Sieci Neuronowe. Wydawnictwo Naukowe PWN, Warszawa (1996)
Blumer, A., et al.: Learnability and the Vapnik-Chervonenkis dimension. J. ACM **36**(4), 929–965 (1989)
Ciechulski, T., Osowski, S.: Badanie jakości predykcji obciążeń elektroenergetycznych za pomocą sieci neuronowych SVM, RBF i MLP. Przegląd Elektrotechniczny **90**(8), 148–151 (2014)
Conejo, A.J., et al.: Forecasting electricity prices for a day-ahead pool-based electric energy market. Int. J. Forecast. **21**(3), 435–462 (2005)
Fijorek, K., at al.: Prognozowanie cen energii elektrycznej na Rynku Dnia Następnego metodami data mining. Rynek Energii (12), 1–8 (2010)
Miller, A., Bućko, P.: Zastosowanie Sztucznych Sieci Neuronowych do Prognozowania Cen na Giełdzie Energii. Zeszyty Naukowe Wydziału Elektrotechniki i Automatyki Politechniki Gdańskiej, Nr. 40 (2014)
Ruciński, D.: Modelowanie neuronalne cen na Towarowej Giełdzie Energii Elektrycznej wspomagane algorytmem ewolucyjnym oraz inspirowane obliczeniami kwantowymi, rozprawa doktorska pod kierunkiem dr hab. inż. Jerzego Tchórzewskiego, prof. UPH w Siedlcach, IBS PAN, Warszawa (2018)
Ruciński, D.: The neural modeling in chosen task of Electric Power Stock Market. Studia Informatica. Systemy i technologie informacyjne **21**(1–2), 63–83 (2017)
Ruciński, D.: Neural-evolutionary modeling of polish electricity power exchange. In: EPNet 2016, pp. 1–6. Katedra Energoelektryki PWr, Institute of Power Engineering and Control Systems of Lviv Polytechnic National University, Ukraine and O/Wrocławski SEP, XPlore Digital Library (2016)
Toczyłowski, E., et al.: Modelowanie Rynkowych Problemów Decyzyjnych na Rynku Energii Elektrycznej. Rynek Energii - 12/2010 (2010)
Towarowa Giełda Energii S.A. (2020). www.tge.pl. Accessed 21 Jan 2020
Zurada, J., et al.: Sensitivity analysis for minimization of input dimension for feedforward neural networks. In: Proceedings of IEEE International Symposium on Circuits and Systems, vol. 6, pp. 447–50 (1994)

3D Software Implementation of the Game Method for Modelling Forest Fires in MyGL Software Tool

Nikolay Ikonomov[1], Pencho Marinov[2], Peter Vassilev[3], Olympia Roeva[3](✉), Dafina Zoteva[3,4], Vassia Atanassova[3], and Krassimir Atanassov[3]

[1] Institute of Mathematics and Informatics, Bulgarian Academy of Sciences, Acad. G. Bonchev Street, bl. 8, 1113 Sofia, Bulgaria
nikonomov@math.bas.bg

[2] Institute of Information and Communication Technologies, Bulgarian Academy of Sciences, Acad. G. Bonchev Street, bl. 25 A, 1113 Sofia, Bulgaria

[3] Institute of Biophysics and Biomedical Engineering, Bulgarian Academy of Sciences, Acad. G. Bonchev Street, Bl. 105, Sofia 1113, Bulgaria
olympia@biomed.bas.bg

[4] Sofia University "St. Kliment Ohridski", 15 Tsar Osvoboditel Blvd., 1504 Sofia, Bulgaria

Abstract. A preliminary version of the 3D visualization software tool MyGL (My OpenGL Game of Life) v2.0, which is a continuation of the previously developed software tool MkBGFire for creating abstract game models of forest fires, is considered in the present paper. The standard John Conway's Game of Life uses two states for each cell in the model: 0 (dead) and 1 (alive). To enable simulation of forest fires, here, three states are used: 0 (normal cell or pre-fire state), 1 (cell currently on fire), 2 (cell has burned out). The transition between 0 and 1 is "heating the vegetation to ignition point". It reflects the number of time steps it takes for the cell to start burning. The transition between 1 and 2 is "burning out the vegetation". This is the number of time steps required for the cell to burn out completely. The OpenGL technology is used for 3D visualization of the cells. The primary focus is on cube cells, while hexagonal prisms have been considered for backward compatibility. Three algorithms are currently implemented. The feasibility of the MyGL software is illustrated with an example of a forest fire scenario.

Keywords: MyGL software · Game method for modelling · Forest fires

1 Introduction

Modelling and simulating the spread of forest fires is a particularly topical issue today as manifested by the extended fire season across the world and the

K. T. Atanassov et al. (Eds.): IWIFSGN 2020/BOS/SOR 2020, LNNS 338, pp. 327–337, 2022.
https://doi.org/10.1007/978-3-030-95929-6_25

devastating wildfires in Southern Europe, beyond the Arctic Circle and in the Australian bushlands, to name a few. To deal with the negative impact of fires, new and improved tools for modelling and predicting the fire propagation are required. A recent study on the main topics of forest fire behaviour and fuel model research are presented in [5]. The potential performance in a real emergency scenario of a new data driven system SmartQFIRE is considered in [7]. Authors in [6] applied a two-way coupling algorithm for simulation of forest fire spread and proved the approach as a better option for prediction the fire size and the shape of a fire. Different mathematical models have been developed to simulate the forest fires behaviour, and to predict the parameters of the fire spread, like direction and intensity. One of them, the Game Method for Modelling (GMM) [1] was inspired by Conway's Game of Life, a planar cellular automata simulation, in which simple rules give rise to surprisingly complex behaviours.

In contrast with Conway's model, which alternates (possibly in repeating cycles) between the states of "dead" and "alive" (namely, 0→1→0 or 1→0→1), the considered GMM model only goes from the state 0 (not burning) to 1 (burning) to 2 (burned out), i.e. 0→1→2, where it stops.

The landscape where the modelled forest fire is located is represented in the plane as a finite (or in the abstract case an infinite) grid tessellated by square or hexagonal cells. This and similar approaches have gained popularity in the last decade due to the ease of use and their relative simplicity.

A simple field fire simulation, where the GMM has been employed, is proposed in [3]. The field is represented by hexagonal cells, and no wind or slope are considered in the model.

In [8] the authors propose the MkBGFire software for forest fire spread simulations developed with regards to the particularities of Bulgarian vegetation. The MkBGFire is based on the paradigm of the Game Method for Modelling [1]. In particular, a square or hexagonal grid is used. The software is based on a mathematical model, and vector models and cellular automata models are implemented. The MkBGFire software uses the OpenGL technology for 2D and 3D visualization of the cells of the game model [8].

In this paper, a new - extended - version of the MkBGFire software is discussed. The implemented 2D algorithm is the same as the one in MkBGFire software, and a 3D algorithm is added to the new version. Both software applications are available at https://forestfires.info/software/.

The so described 3-dimensional case of GMM will be applied for modelling different processes. For example, applications of the GMM for modelling forest dynamic and forest fires are described for the 2-dimensional case in [11–15]. All these can be re-written for the 3-dimensional case which will be essentially close to a real situation.

2 3D Visualization Software MyGL

The three-dimensional model uses a four-dimensional matrix matA [X] [Y] [Z] [T], where X is the width, Y is the depth (corresponding to the 2D model), Z is the height of the modelled vegetation, and T is the time step.

The 3D algorithm can be described as follows. In order to simulate the fire, the following three states are used: 0 (normal, or pre-fire, cell), 1 (cell currently on fire) and 2 (cell has burned out). Each cell in the matrix has three main parameters: $fuel$ material, $resistance$ to fire and *state* (all of them are integer values). Cells marked as ignition points have $resistance = 0$ and $state = 1$. At each iteration of the algorithm, each cell of the matrix is sequentially updated. When a cell has $state = 1$, then its 6 neighbours (cube) and 8 neighbours (hexagonal prism) are considered.

For each cell with $state = 1$ (cell on fire), the $fuel$ is reduced by one point, and when the $fuel$ reaches 0, then $state = 2$. If $state = 0$ and $fuel$ ¿ 0, then $resistance$ is decreased by one point. If $resistance$ reaches 0, then $state = 1$. If $state = 2$, the cell is no longer considered - it is completely burned out.

The full list of cell parameters is as follows: $fuel$, $resistance$ to fire, *state*, $flag$, *block type*, $fire\ start\ mark$ (ignition point). The $flag$ is used to ensure that at each time step any cell is modified only once, since the matrix processing starts from location (0, 0, 0), and is processed by width, then depth, finally height.

The main window of the MyGL software tool is shown in Fig. 1.

It is divided into two panels. The parameters necessary for starting a simulation and their default/ initial values are listed on the left side of the main window. They are explained in details below. The preview of the current terrain and the fire propagation during a simulation can be observed on the right side panel of the main window.

The initial parameters for setting a simulation, as well as the description of the cell and model characteristics are summarized in Table 1.

Table 1. Simulation parameters, cell characteristics and model characteristics

Simulation parameters	
Parameter	**Meaning**
eyeX, eyeY, eyeZ, locX, locY, locZ	Show the coordinates of the position of the terrain
Camspeed	Shows the camera speed when moving through the terrain
Rotangle	Indicates the rotation angle when rotating the terrain
X, Y, Z	Define the size of the terrain (default/minimum of 10, with 10 increment)
Cube/Prism	Changes the cell type for the terrain.
Cell characteristics	
Parameter	**Meaning**
K1Fuel	Shows the time steps necessary for the cell to burn out completely
K2Resist	Shows the time steps necessary for the cell to start burning
K3State	Three states are used, in order to enable simulation of forest fires: 0 (normal cell), 1 (cell currently on fire), 2 (cell has burned out)
K4Flag	The flag ensures that at each time step any cell is modified only once
Block	There are several block types defined that allow to draw a very simple forest layout with cubes/prisms Default block characteristics are as follows: Air: $fuel = -1$, $resistance = -1$, $state = -1$; Water: $fuel = -1$, $resistance = -1$, $state = -1$; Rock: $fuel = -1$, $resistance = -1$, $state = 0$; Leaves: $fuel = 2$, $resistance = 1$, $state = 0$; Bush: $fuel = 10$, $resistance = 5$, $state = 0$; Trunk: $fuel = 20$, $resistance = 10$, $state = 0$
Start	Marks the cells as ignition points
Result	Saves changes to the blocks.
Model characteristics	
Parameter	**Meaning**
Algorithm	Three algorithms are implemented. • NGL (3D): Some of the theoretical foundations of the approach used in NGL are considered in [2] • STF (2D) and PGM (2D): The algorithms are discussed in [4]
Windspeed/Windangle	The parameters are used only for STF and PGM algorithms
Rka, Rkb, R09, R10, i30	These five parameters (Rka, Rkb, R09, R10, i30) are used by the algorithm PGM

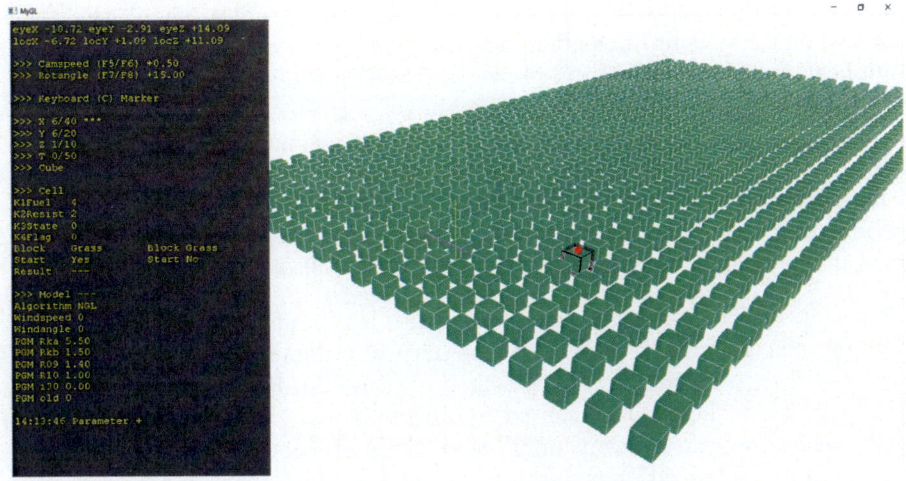

Fig. 1. Starting screen of MyGL

3 Implemented Algorithms in MyGL Software

3.1 Short Remarks on the Game Method for Modelling

Let a set of symbols S and an n-dimensional simplex (in the sense, e.g., of [9]) comprising of n-dimensional cubes (when $n = 3$, a 3-dimensional grid of cubes) or other types of simplexes called *cells* be considered.

It is assumed that material points (referred as *objects*) can be found in some of the centres of the n-dimensional cells.

The GMM-grid can be either finite or infnite. In the first case, for i-th dimension of the grid there is a natural number g_i that corresponds to the number of the sequential cells of the grid in the present dimension. Therefore, when the n-dimensional GMM-grid is finite, there is a vector $\langle g_1, g_2, ..., g_n \rangle$ of the lenghts of its sides. In general, finite grids will be used.

A set of rules $\mathcal{A}$ are also considered as follows:

1) Rules for the motion of the objects along the vertices of the simplex;
2) Rules for the interactions among the objects, e.g., when they are collected in one cell.

Let the rules from the i-th type be denoted as i-rules, where $i = 1, 2$.

When $S = \{\perp, *\}$, the standard Conway's Game of Life is obtained, where $\perp$ symbolizes an empty symbol.

An *initial configuration* is a set of (ordered) $(n + 2)$-tuples with an initial component being the number of the object; the second, third, etc., until the $(n+1)$-st components are its coordinates; and the $(n+2)$-nd – its corresponding symbol from S.

A *final configuration* is the ordered set of $(n + 2)$-tuples having the above form and being a result of the modifications that occur during a certain number of applications of the rules from $\mathcal{A}$ over a (fixed) initial configuration.

The single application of a rule from $\mathcal{A}$ over a given configuration K is called an *elementary step* in the transformation of the model and is denoted by $A_1(K)$, where $A_1 \in \mathcal{A}$.

For some initial configuration the new configurations are obtained in a stepwise manner. Some constructive criteria for stopping the process should be determined. For example, such a condition may be the following.

1. The rules of the GMM are applied over the initial configuration and its derivates for exactly s iterations, where the natural number s is fixed in advance.
2. A predefined configuration is obtained on the GMM-grid. For example, if a process of interaction between some objects is modelled, the process should stop when the grid contains only one of these objects.
3. A previous configuration is obtained on the GMM-grid, i.e., the process oscillates. This criterion is applicable for deterministic processes.

In this sense, if K is an initial configuration and L is a final configuration derived from K through multiple applications of the rules from $\mathcal{A}$, then interim configurations $K_0, K_1, ..., K_m$ will exist, for which

$$K_0 = K,$$

$$K_{i+1} = A_{i+1}(K_i), \quad \text{for } 0 \leq i \leq m - 1,$$

$$K_m = L$$

(the equality "=" is used in the sense of identity of the configurations) and this will be denoted by

$$L = \mathcal{A}(K) \equiv A_m(A_{m-1}(...A_1(K)...)),$$

where $A_1, A_2, ..., A_m \in \mathcal{A}$.

Obviously, the form of each configuration will influence the form of the GMM-grid.

Some Forms of the 3-Dimensional GMM Cells

While in the 2-dimensional case of the GMM, the cells can have forms of an equilateral triangle, square or an equilateral hexagon, in the 3-dimensional case of the GMM, the cells can have the forms from Figs. 2a and 2b.

In the two forms of cells, each fixed cell C has neighbouring cells that share common faces with C.

When an object moves from the cell C to a neighbouring cell, it has 6 possible directions in case of the cell's form from Fig. 2a, while in the case of Fig. 2b the possibilities are 8.

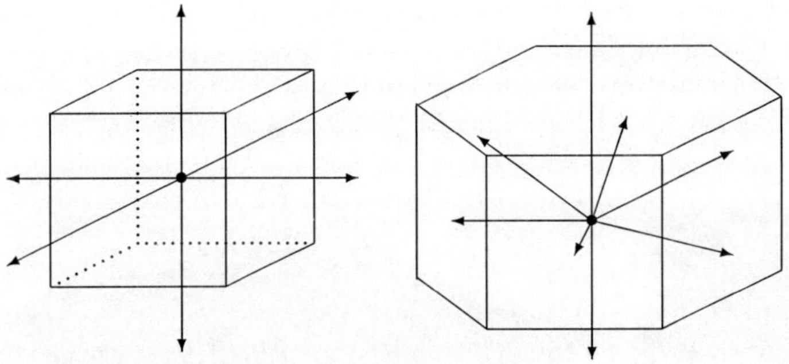

(a) Cube

(b) Hexagonal Prism

Fig. 2. Cells in 3-dimensional case of the GMM

More complex is the case in Fig. 3. Now, the cell C has four higher and four lower neighbouring cells, while the vertical neighbouring cells are also four.

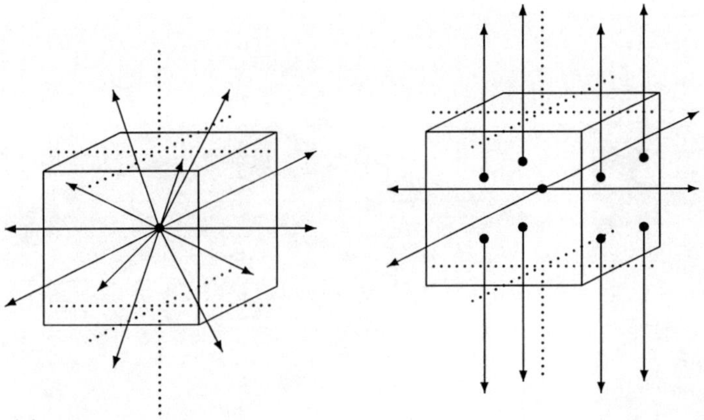

(a)

(b)

Fig. 3. Cell with four neighbouring cells on each side

Let these cell be called 1-neighbouring cells for C. The 1-neighbouring cells for the 1-neighbouring cells of C will be 2-neighbouring cells about C, etc.

4 Field Fire Spread Simulation Scenarios Using MyGL Software

Using an example of forest fire scenario, the feasibility of the MyGL software is illustrated.

As can be seen in Fig. 4, fire starts from the cube marked with a red dot. On the field to the left of the river there are two groves and two places with shrubs. On the 10th time step, the grass around the bush on the left is burning (red blocks), part of it has already burned (black blocks) (Fig. 5).

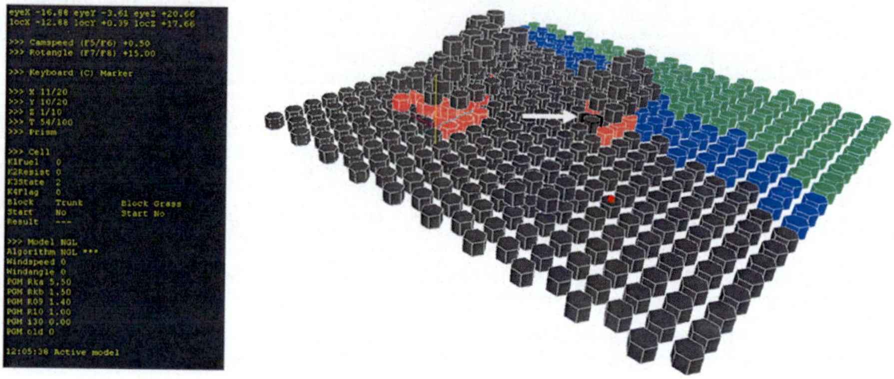

Fig. 4. Fire simulation: start point of the fire

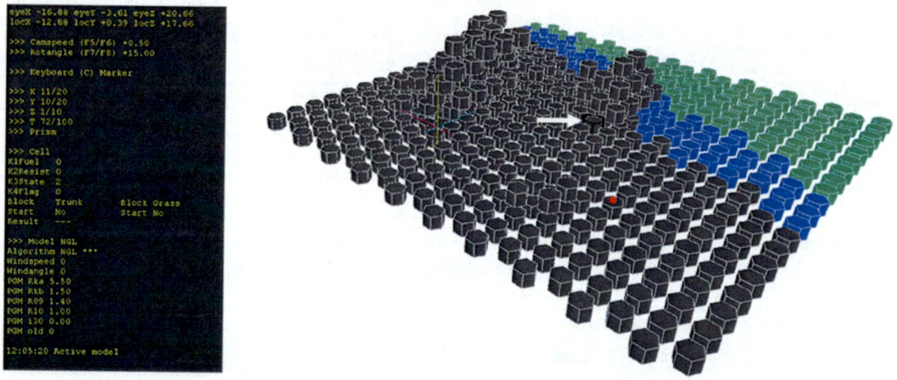

Fig. 5. Fire simulation at time step 10

The characteristics of any block of the terrain at each step of the simulation can be observed as the cursor is placed on the given block. For example, the marked block in Fig. 5 (part of bush) shows that $fuel = 9$ (the initial $fuel$ for a bush type cell is 10) and $state = 1$. At each subsequent iteration, the $fuel$ of this cell decreases. At time step 20, this part of the bush has burned out.

A simulation of the fire is illustrated in Fig. 6 for eight different time steps.

(a) Time step = 20

(b) Time step = 30

(c) Time step = 32

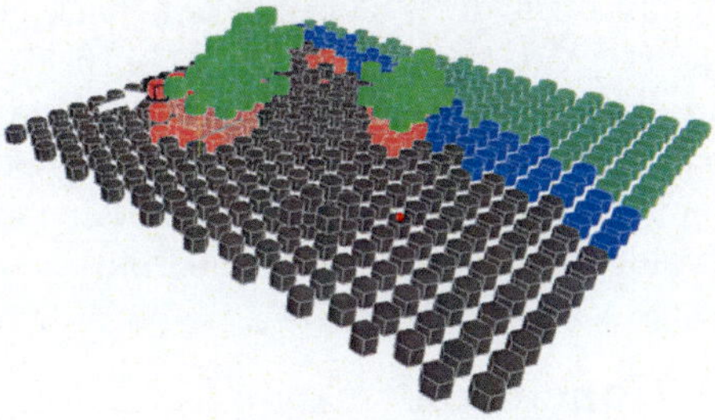

(d) Time step = 40

(e) Time step = 45

(f) Time step = 50

(g) Time step = 54

(h) Time step = 72

Fig. 6. Fire simulation at different time steps

At time step 20, the fire reaches the two groves, the lower parts of the first bush have already burned out. At time step 30, the second bush is burning. The marked red cube (part of the bush) shows $fuel = 8$ and $state = 1$. At the same time, some of the tree trunks are already burning. At time step 32, the marked red cube (trunk) shows $fuel$ value of 16 (the initial $fuel$ for a trunk cell is 20) and $state = 1$. At time step 40, the leaves of the grove, located farther away from the river, are already burning, while all the grass has burned out. The marked red cube (leaves) shows $fuel = 1$ (the initial $fuel$ for a leaves cell is 2) and $state = 1$.

At time step 45, the leaves of the trees from the grove, located near the river, start burning. The lower standing leaves of the trees from the other grove have already burnt out. The marked black cube shows $fuel = 0$ and $state = 2$. At time step 50, the tree leaves in both groves have already burned out ($fuel = 0$ and $state = 2$ of the marked cube), while all tree trunks are currently burning. At time step 54, some trunks have already burned out. The marked black cube shows $fuel = 0$ and $state = 2$. Finally, at time step 72, all the vegetation on the left side of the river has burned out.

5 Conclusion

An extended version of the software MkBGFire (developed for forest fire spread simulations specific to the Bulgarian vegetation), named MyGL, is demonstrated in the present paper. The MyGL software is based on the paradigm of the Game Method for Modelling, which itself is an extension of the famous Conway's Game of Life, implementing vector models and cellular automata models. A 3D modelling and visualization algorithm is added to the new version. A simple example is considered and presented in a stepwise manner to illustrate the MyGL software. The fire spread is simulated over an illustrative terrain containing as basic elements a river, two small groves and two bushes (near and far from the river). The value changing rules and the model parameters are initialized with regards to the specifics of the vegetation of Bulgaria.

For future work, the authors intend to further develop the current model, which currently considers flat terrains. In order to improve the modelling capacity of the MyGL software, simulations with sloped terrain will be considered, as well as the introduction of additional parameters for wind, which is a crucial factor to fire spread necessary to take into account for more realistic simulations.

Acknowledgments. This work is supported by the Bulgarian National Science Fund Project DN16/6 "Integrated approach for modelling of the forest fires spread".

References

1. Atanassov, K.: Game Method for Modelling. Marin Drinov Academic Publishing House, Sofia (2011)

2. Atanassova, L., Atanassov, K., Angelova, N.: Short remark on 3-dimensional game method for modelling. In: Atanassov, K.T., Atanassova, V., Kacprzyk, J., Kaluszko, A., Krawczak, M., Owsinski, J.W., Sotirov, S., Sotirova, E., Szmidt, E., Zadrozny, S. (eds.) IWIFSGN 2018. AISC, vol. 1081, pp. 379–386. Springer, Cham (2021). https://doi.org/10.1007/978-3-030-47024-1_35
3. Fidanova S., Marinov P.: Field fire simulation applying hexagonal game method. In: Dimitrov V., Georgiev, V., (Eds.) Proceedings of the 7th International Conference on Information Systems and Grid Technologies, pp. 201–209 (2013)
4. Fidanova, S., Marinov, P.: The impact of slope on fire spread simulation. Environ. Eng. Manage. J. **15**(3), 505–510 (2016)
5. Güngöroglu, C.: Forest fire studies on fire behaviour: key topics and their importance. In: 14th International Combustion Symposium (INCOS2018), 25-27 April 2018, pp. 113–116 (2019)
6. Lopes, A.M.G., Ribeiro, L.M., Viegas, D.X., Raposo, J.R.: Simulation of forest fire spread using a two-way coupling algorithm and its application to a real wildfire. J. Wind Eng. Ind. Aerodyn. **193**, 103967 (2019)
7. Rios, O., Valero, M.M., Pastor, E., Planas, E.: A data-driven fire spread simulator: validation in Vall-llobrega's fire. Front. Mech. Eng. (2019). https://doi.org/10.3389/fmech.2019.00008
8. Roeva, O., et al.: MkBGFire software – an example of game modelling of forest fires in Bulgaria. In: Atanassov, K.T., Atanassova, V., Kacprzyk, J., Kaluszko, A., Krawczak, M., Owsinski, J.W., Sotirov, S., Sotirova, E., Szmidt, E., Zadrozny, S. (eds.) IWIFSGN 2018. AISC, vol. 1081, pp. 387–397. Springer, Cham (2021). https://doi.org/10.1007/978-3-030-47024-1_36
9. Kuratovski, K.: Topology. Academic Press, New York (1966)
10. Atanassov, K.: On game-method for modelling. Adv. Stud. Contem. Math. **22**(2), 189–207 (2012)
11. Sotirova, E., Atanassov, K., Fidanova, S., Velizarova, E., Vassilev, P., Shannon, A.: Application of the game method for modelling the forest fire perimeter expansion. Part 1: a model fire intensity without effect of wind. In: IFAC Workshop on Dynamics and Control in Agriculture and Food Processing, Plovdiv, 13–16 June 2012, pp. 159–163 (2012)
12. Sotirova, E., Atanassov, K., Fidanova, S., Velizarova, E., Vassilev, P., Shannon, A.: Application of the game method for modelling the forest fire perimeter expansion. Part 2: a model fire intensity with effect of wind. In: IFAC Workshop on Dynamics and Control in Agriculture and Food Processing, Plovdiv, 13–16 June 2012, pp. 165–169 (2012)
13. Sotirova, E., Atanassov, K., Fidanova, S., Velizarova, E., Vassilev, P., Shannon, A.: Application of the game method for modelling the forest fire perimeter expansion. Part 3: a model of the forest fire speed propagation in different homogenous vegetation types. In: IFAC Workshop on Dynamics and Control in Agriculture and Food Processing, Plovdiv, 13–16 June 2012, pp. 171–174 (2012)
14. Sotirova, E.N., Dimitrov, D.D., Atanassov, K.T.: On some application of game-method for modelling. Part. 1: forest dynamics. In: Proceedings of the Jangjeon Mathematical Society, vol. 15, no. 2, pp. 115–123 (2021)
15. Sotirova, E.N., Dimitrov, D.D., Atanassov, K.T.: On some application of game method for modelling. Part 2: Development of forest fire. In: Proceedings of the Jangjeon Mathematical Society, vol. 15, no. 2, pp. 335–342 (2012)

Balancing Predictive Relevance of Ligand Biochemical Activities

Marek Pecha[1,2](✉)

[1] Department of Applied Mathematics, FEECS, VŠB-TU Ostrava, Ostrava, Czechia
[2] The Czech Academy of Sciences, Institute of Geonics, Ostrava, Czechia
marek.pecha@vsb.cz

Abstract. In this paper, we present a technique for balancing predictive relevance models related to supervised modelling ligand biochemical activities to biological targets. We train uncalibrated models employing conventional supervised machine learning technique, namely Support Vector Machines. Unfortunately, SVMs have a serious drawback. They are sensitive to imbalanced datasets, outliers and high multicollinearity among training samples, which could be a cause of preferencing one group over another. Thus, an additional calibration could be required for balancing a predictive relevance of models. As a technique for this balancing, we propose the Platt's scaling. The achieved results were demonstrated on single-target models trained on datasets exported from the ExCAPE database. Unlike traditional used machine techniques, we focus on decreasing uncertainty employing deterministic solvers.

1 Ligand-Based Supervised Biochemical Activity Modelling

In silico computational supervised modelling based on ligand profiles is widely used as a complementary method in the pharmaceutical industry; specifically in the early stage of drug design and development to obtain an indication of off-target interactions. The aim is to train a model that generalizes the biological activities of ligands, i.e. small biochemical molecule, ion, or protein, by reusing information from former *in vitro* laboratory experiments. A common approach to predicting activities of ligands on biological targets, e.g. genes, is to exploit a panel of structure-analysis models such as quantitative structure-activity relationship (QSAR) with one model per a biological target. Then, using statistical learning, we train a classification or regression model used to predict categorical (active or inactive) or numerical values, e.g. probability of reaction, for a new unseen ligand, respectively. The suggested strategies on how to train the models are the single and multi-target schemes. In our text, we focus on a simplified single-target modelling that is a baseline for more general approaches, e.g. the multi-target ones mentioned above.

Several machine learning methods are typically employed for constructing such models. Currently, the most popular ones are Deep Neural Network (DNNs) and Support Vector Machines (SVMs). Despite the fact that Deep Learning (DL) is getting popular in recent years, SVMs are still applicable. One of the most advantages of the SVMs approach is that they find learning functions maximizing geometric margins,

K. T. Atanassov et al. (Eds.): IWIFSGN 2020/BOS/SOR 2020, LNNS 338, pp. 338–348, 2022.
https://doi.org/10.1007/978-3-030-95929-6_26

unlike generally composed functions produced by DNNs. A benefit of this additional implicit (in contrast with DNNS) requirement is that reducing generalization error leading to preventing overfitting of a model and improving its robustness in terms of a bias-variance tradeoff. Respecting later, SVM seems to be appropriate as a technique for modelling biochemical activity, which exploits ligand profiles, for the reason that a multivariance among samples is typically small. This fact is arising from the nature of in vitro experiments because laboratories test biochemical activities of quite similar ligands to a particular target commonly than testing ligands belonging to entirely different groups.

However, SVMs have a serious drawback; they are sensitive to imbalanced datasets, outliers and multicollinearies among training samples, which could be a cause of preferencing one group over another. Therefore, we propose to use Platt scalling for an additional model calibration, which is based on transforming the SVM classification model output into a posterior probability distribution by fitting logistic regression model to SVM raw prediction scores. This calibrating technique is practically used for reducing impact of overfitting to predictor mainly caused by training data. After training calibrated models, we demonstrate balanced predictive relevance of these models by converting them to label prediction using an optimal threshold.

This text is organized as follows. In Sect. 2, the SVM formulations with a relaxed-bias term are presented. A calibration technique based on the Platt scaling is introduced in Sect. 3. Numerical experiments are presented in Sect. 4.

2 Support Vector Machines for Relaxed-Bias Classification

SVMs belong to conventional machine learning techniques, and they are practically used for both classification and regression. Unlike the DL underlying architecture, SVMs could be considered as the single perceptron problems that find the learning functions maximizing the geometric margins. Therefore, we can explain the qualities of a learning model and the underlying solver behaviour straightforwardly. In this paper, we focus only on the linear C-SVM for classification, in which a misclassifications error term is penalized by a user-specified penalty C. We denote the C-SVM as SVM for simplification in further text.

SVM was originally designed by [1] as a supervised binary classifier, i.e. a classifier that decides whether a sample falls into either Class A or Class B by means of a model determined from already categorised samples in the training phase of the classifier. Let us denote the training data as an ordered sample-label pairs such that

$$T := \{(\mathbf{x}_1, y_1), (\mathbf{x}_2, y_2), \ldots, (\mathbf{x}_m, y_m)\}, \tag{1}$$

where m is the number of samples, $\mathbf{x}_i \in \mathbb{R}^n$, $n \in \mathbb{N}$, is the i-th sample and $y_i \in \{-1, 1\}$ denotes the label of the i-th sample, $i \leq m$. The linear SVM solves the problem of training the classification model in the form of a so-called maximal-margin hyperplane:

$$H = \langle \mathbf{w}, \mathbf{x} \rangle + \hat{b}, \tag{2}$$

where $\mathbf{w}$ is a normal vector of hyperplane H and $\hat{b} = \frac{b}{\|\mathbf{w}\|}$ determines the offset of the hyperplane H from the origin along its normal vector $\mathbf{w} \in \mathbb{R}^n$. In the cases of the relaxed-bias classification, we do not consider $\hat{b}$ in a classification model, however we include it into the problem by means of augmenting the vector $\mathbf{w}$ and each sample $\mathbf{x}_i$ with an additional dimension so that $\widehat{\mathbf{w}} \leftarrow \begin{bmatrix} \mathbf{w} \\ B \end{bmatrix}$, $\widehat{\mathbf{x}}_\text{i} \leftarrow \begin{bmatrix} \mathbf{x}_i \\ \gamma \end{bmatrix}$, and $\gamma \in \mathbb{R}^+$ is a user defined variable, typically set to 1.

Let $p \in \{1, 2\}$ for purposes related to our application, then the problem of finding hyperplane $\widehat{H} = \langle \widehat{\boldsymbol{w}}, \widehat{\boldsymbol{x}} \rangle$ can be formulated as a constrained optimization problem in the following primal formulation:

$$\underset{\widehat{\boldsymbol{w}},\, \widehat{\xi}_i}{\arg\min}\ \frac{1}{2}\langle \widehat{\boldsymbol{w}}, \widehat{\boldsymbol{w}} \rangle + \frac{C}{p}\sum_{i=1}^{n} \widehat{\xi}_i^p \ \text{ s.t. } \begin{cases} y_i \langle \widehat{\boldsymbol{w}}, \widehat{\boldsymbol{x}}_i \rangle \geq 1 - \widehat{\xi}_i, \\ \widehat{\xi}_i \geq 0 \ \text{ if } p = 1,\ i \in \{1, 2, \ldots, n\}, \end{cases} \tag{3}$$

where $\widehat{\xi}_i = \max\{0, 1 - y_i \langle \widehat{\boldsymbol{w}}, \widehat{\boldsymbol{x}}_i \rangle\}$ is the hinge loss function releated to augmented samples $\widehat{\boldsymbol{x}}_i$. Essentially, the hinge loss function quantifies error between predicted and correct classification of sample $\widehat{\boldsymbol{x}}_i$. The variable $C \in \mathbb{R}^+$ is a penalty parameter that penalizes misclassification error. Generally, a higher value of C increases the importance of minimising the hinge loss functions ξ_i and, on the other hand, causes maximizing $\|\mathbf{w}\|$, i.e. minimizing the width of margin, leading to poor generalization capabilities of a classification model. The goal is to find a reasonable value of C such that a resulting model balances the robustness and performance tradeoff.

Further, we can say in general, the minimizer associated with formulation (3) corresponding to an optimal rotation of separating hyperplane $\widehat{H}$ in one-dimension higher feature-space $\mathbb{R}^{n+1}$ than the original feature space has.

To reduce a number of unknowns, we can dualize the primal formulation (3) using the Lagrange duality so that, for both $p = 1$ and $p = 2$, they result in:

$$\underset{\boldsymbol{\alpha}}{\arg\min}\ \frac{1}{2}\boldsymbol{\alpha}^T \boldsymbol{Y}^T \boldsymbol{K} \boldsymbol{Y} \boldsymbol{\alpha} - \boldsymbol{\alpha}^T \mathbf{e} \ \text{ s.t. } \mathbf{o} \leq \boldsymbol{\alpha} \leq C\mathbf{e}, \tag{4}$$

$$\underset{\boldsymbol{\alpha}}{\arg\min}\ \frac{1}{2}\boldsymbol{\alpha}^T \left(\boldsymbol{Y}^T \boldsymbol{K} \boldsymbol{Y} + C^{-1}\boldsymbol{I} \right) \boldsymbol{\alpha} - \boldsymbol{\alpha}^T \mathbf{e} \ \text{ s.t. } \mathbf{o} \leq \boldsymbol{\alpha}, \tag{5}$$

respectively. $\boldsymbol{K} \in \mathbb{R}^{m \times m}$ is a matrix of inner products called the Gram matrix such that $\boldsymbol{K} := \boldsymbol{X}^T\boldsymbol{X}$, $\boldsymbol{X} = [\mathbf{x}_1, \mathbf{x}_2, \ldots, \mathbf{x}_m]$ is data matrix of the training samples, $\mathbf{y} = [y_1, y_2, \ldots, y_m]^T$ is vector of corresponding labels, $\boldsymbol{Y} = diag(\mathbf{y})$, and $\mathbf{o} \in \mathbb{R}^m$, $\mathbf{e} \in \mathbb{R}^m$ denote a zero-vector and an all-ones vector, respectively. In general, the matrix $\boldsymbol{K}$ is symmetric positive semi-definite (SPS) of a rank

$$\text{rank}(\boldsymbol{K}) = \min\{\min\{m, n\}, N_g\}, \tag{6}$$

where N_g maximum number of linearly independent training samples, m and n are a numbers of training samples and their features, respectively.

Comparing the dual SVM-QP formulations, specifically, $\ell 1$-loss (4) and $\ell 2$-loss (5), we can see that they differ in the forms of the related Hessians and constraints. While the Hessian is generally an SPS matrix in a case of (4), the Hessian related to the formulation (5) is regularized by means of the matrix $C^{-1}\boldsymbol{I}$. This can provide a better convergence rate for (5) and an associated optimization problem could be more stable.

On the other hand, $\ell 1$-loss SVM could produce a more robust model in the sense of performance score, because using a linear sum of ξ_i leads to catching the outliers during a training phase of a classifier.

Further, for obtaining a solution of the original primal problem, we introduce dual to primal reconstruction formula as follows:

$$\widehat{\boldsymbol{w}} = \widehat{\boldsymbol{X}\boldsymbol{Y}}\boldsymbol{\alpha}, \tag{7}$$

Using the reconstructed normal vector $\widehat{\boldsymbol{w}}$, we can set the decision rule:

$$\operatorname{sgn}\left(\langle \widehat{\boldsymbol{w}}, \widehat{\boldsymbol{x}}_i \rangle\right) = \begin{cases} +1 \ldots \; \widehat{\boldsymbol{x}}_i \in \text{Class A}, \\ -1 \ldots \; \widehat{\boldsymbol{x}}_i \in \text{Class B}. \end{cases} \tag{8}$$

In sense of equivalence of solutions, we can easily show a connection between the classification models associated with standard model (2) and relaxed-bias formulations. Let us write the separating hyperplane equation in a component-wise form such that:

$$\widehat{H} := \langle \widehat{\boldsymbol{w}}, \widehat{\boldsymbol{x}} \rangle = \underbrace{w_1 x_1 + w_2 x_2 + \cdots + w_m x_m}_{=\langle \mathbf{w}, \mathbf{x} \rangle} + \underbrace{B\gamma}_{=:b}, \tag{9}$$

which is equivalent to (2). However, the bias term b is incorporated into the regularization term in sense of Tikhonov regularization, the resulting model could slightly differ from this attained by standard (non-relaxed) formulations. On the other side, we have not to deal with equality constraints, which are appear in the standard dual formulations, and sometimes, they are reason why solvers could diverge.

3 Model Calibration

Calibrating a classification model refers to a special type of statistical inference that transforms a uncalibrated output (raw prediction), particularly, decision function $f(\mathbf{x})$, to a probability of class membership $P(class \mid input)$. Commonly, the calibration is required when we need to adjust robustness of a classification model, balances of class preference and provides a cost-sensitive classification. In this section, we pay attention to an estimation of the probability employing a well-known calibration technique called Platt's scaling – introduced by [7]. Commonly, this technique is known as Platt's calibration in the machine learning communities.

An idea beyond this technique is based on fitting a parametric form of sigmoid-shaped function that maps the uncalibrated SVM output to the posterior probability $P(y = 1 \mid \mathbf{x}) \approx P_{A,B}(y = 1 \mid \mathbf{x})$, where

$$P_{A,B}(y = 1 \mid \mathbf{x}) = \frac{1}{1 + \exp\left(Af(\mathbf{x}) + B\right)}. \tag{10}$$

The parameters A, B determine the slope of the sigmoidal curve and lateral displacement, respectively, and they are practically fitted using maximum likelihood estimation

(MLE). In order to the relaxed-bias classification mentioned in Sect. 2, we assume that the raw SVM output

$$f(\mathbf{x}) := \widehat{H}\left(\widehat{\boldsymbol{x}} \leftarrow \begin{bmatrix} \mathbf{x} \\ \gamma \end{bmatrix}\right) = \langle \widehat{\boldsymbol{w}}, \widehat{\boldsymbol{x}} \rangle \tag{11}$$

is proportional to the log odds of positive samples in the model (10).

In the original paper, Platt suggested to use an additional training set, i.e. a calibration set, for training calibration curve on output of general instance-based SVM to avoid incorporating bias failures, i.e. cases when $1 - y_i f_i > 0$, on functional-margin $\gamma = |1|$. Let us denote such dataset as an ordered set:

$$T_{CA} := \{(f_1, y_1), (f_2, y_2), \ldots (f_l, y_l)\}, \tag{12}$$

where l is a number of the calibration samples, f_j is estimate of $f(\mathbf{x}_j)$ for $j \in \{1, 2, \ldots, l\}$. On the other hand, when an optimal model performance is attained in a reasonably small value of the penalty C, e.g., in real-world applications employing linear SVMs, see [7], and data is well-behaved, bias on margin failures usually become small. Therefore, it often possible to simply fit the sigmoid on the training dataset.

To prevent model overfitting, Platt proposed additional transformation of binary labels y_j to target probabilities t_j such that $t_j = \frac{N_p+1}{N_p+2}$ iff $y = +1$, or $t_j = \frac{1}{N_n+2}$ iff $y = -1$, where N_p and N_n are numbers of positive and negative calibration samples, respectively.

The best parameter setting (A^*, B^*) is determined by minimizing negative log likehood (cross-entropy error function) on calibration data so that:

$$(A^*, B^*) = \arg\min_{A,B} - \sum_{j=1}^{l} \left[\, t_j \log(p_j) + (1 - t_j)\log(1 - p_j) \,\right], \tag{13}$$

where $p_j = \frac{1}{1+\exp(Af_j+B)}$. To solve (13), an author in [7] proposed to use Levenberg-Marquardt (LM) algorithm [3]. Unfortunately, a technique for updating a damping factor for LM introduced by Platt causes that solver could not converge to a minimum of (12). It is discussed in [4]. To avoid issue arising from a damping factor associated with the LM method, the authors suggested the Newton method with backtracking line-search.

Though the proposed approach is favourable due to its simplicity, the trust-region methods are more robust. Since we focus on training robust predictors in this paper, we exploit the Newton method with trust region in all numerical experiments presented in Sect. 4.

4 PermonSVM

The PermonSVM package is a part of the PERMON toolbox designed for usage in a massively parallel distributed environment containing hundreds or thousands computational cores. Technically, it is an extension of the core PERMON package called PermonQP, from which it inherits basic data structures, initialization routines, build system, and utilizes computational and QP transformation routines, e.g. normalization of an objective function, dualization, etc. Programmatically, core functionality of PERMON toolbox is written on the top of the PETSc framework, follows the same design and coding style, making it easy-to-use for anyone familiar with PETSc. It is usable on all main operating systems and architectures consisting of smartphones through laptops to high-end supercomputers.

PermonSVM supports distributed parallel (through MPI) reading from formats like SVMLight, HDF5, PETSc binary file formats, more than 4 problem formulations of classification problem, two types of parallel cross-validation, namely *k*-fold and stratified *k*-fold cross-validation. The resulting QP-SVM problem with implicitly represented Hessian, in which Gram matrix $\boldsymbol{X}^T\boldsymbol{X}$ is not assembled, is proceeded by solvers provided by the PermonQP package or the PETSc framework. Unlike standard machine learning libraries, PERMON toolbox provides interface functions to change underlying QP-SVM solver, monitoring and tweaking the algorithms. In Code 1, we present an example of a usage PermonSVM API. Our libraries are developed as an open-source project under the BSD 2-Clause Licence.

```
MPI_comm      comm = PETSC_COMM_WORLD;
SVM           svm;
PetscViewer v;

char          f_train[PETSC_MAX_PATH_LEN] = "cnr1.h5";
char          f_test[PETSC_MAX_PATH_LEN] = "cnr1.t.h5";

TRY( SVMCreate(comm,&svm) );
TRY( SVMSetType(svm,SVMPC) ); /* Platt calibration type */
TRY( SVMSetFromOptions(svm) );

TRY( PetscViewerHDF5Open(comm,f_train,FILE_MODE_READ,&v) );
TRY( SVMLoadTrainingDataset(svm,viewer) );
TRY( PetscViewerDestroy(&v) );

TRY( PetscViewerHDF5Open(comm,f_train,FILE_MODE_READ,&v) );
TRY( SVMLoadCalibrationDataset(svm,viewer) );
TRY( PetscViewerDestroy(&v) );

TRY( PetscViewerHDF5Open(comm,f_test,FILE_MODE_READ,&v) );
TRY( SVMLoadTestDataset(svm,v) );
TRY( PetscViewerDestroy(&v) );

TRY( SVMSetHyperOpt(svm,PETSC_TRUE) );
TRY( SVMSetNfolds(svm,3) );

TRY( SVMTrain(svm) );
TRY( SVMTest(svm) );

TRY( SVMDestroy(&svm) );
```

Code 1. Example of calling PermonSVM API.

5 Numerical Experiments

In this section, we analyze numerical experiments related to balancing predictive relevance of the single-target relaxed-bias classification model using the Platt's Calibration technique, which we introduced in Sect. 2 and Sect. 3, respectively. We benchmark this approach on datasets associated with biochemical activities of ligands on 4 biological targets, namely abl1 (Abelson murine leukemia viral oncogene homolog 1 protein), adora2a (Adenosine A_2A receptor), cnr1 (cannabinoid receptor type 1), and cnr2 (cannabinoid receptor type 2). These datasets were exported from the ExCAPE database, which was developed by [8]. While it is possible to calibrate a model on the same dataset, on which the model was trained, see Sect. 3, it could be problematic to decide if a bias of an uncalibrated model is small enough. Therefore, we split training samples into the training and calibration datasets. After training models, we evaluate their performance on the test dataset using precision, sensitivity, and area under the curve receiver operating characteristic (AUC) performance scores. The input datasets were divided into training, calibration and test datasets such that they consist of 640, 200, 160 ligands, respectively, and a ratio of active and inactive ones is sufficiently preserved. Characteristics associated with these datasets are summarized in Table 1.

For training uncalibrated classification models, we choose the best penalty C_{BE} from the set $\widehat{C} = \{2^p, p \in \{-7, -6, \ldots, 6, 7\}\}$ algorithmically employing the hyper-parameter optimization (HyperOpt) by means of grid-search combined with stratified 3-fold cross validation (CV). The value of the best penalty C_{BE} is selected so that accumulated related precision and sensitivity during CV are maximized. All components of the initial guess $\mathbf{x}_0$ are set to $0.99 * C$, proposed in [6]. The relative norm of projected gradient being smaller than 1e−1, discussed in [5], is used as stopping criterion for the MPRGP (Modified Proportioning and Reduced Gradient Projection) algorithm, see [2], in all presented experiments. The expansion step-length α is fixed and determined such as $\alpha = 1.95/\|\boldsymbol{A}\|_2$, where $\|\boldsymbol{A}\|_2 = \sqrt{\lambda_{max}\left(\boldsymbol{A}^T\boldsymbol{A}\right)}$, where $\boldsymbol{A}$ denotes the Hessian matrix associated with (4) and (5).

Table 1. The characteristics of training, calibration and test dataset related to abl1, adora2a, cnr1, cnr2 biological targets.

Target (dataset)	#ligands		
	#active	#incative	$\sum$
abl1 (training)	312 (48.75%)	328 (51.25%)	640
abl1 (calibration)	92 (46.00%)	108 (54.00%)	200
abl1 (test)	81 (50.62%)	79 (49.38%)	160
adora2a (training)	343 (53.59%)	297 (46.40%)	640
adora2a (calibration)	105 (52.50%)	95 (47.50%)	200
adora2a (test)	95 (59.38%)	65 (40.62%)	160
cnr1 (training)	392 (61.25%)	248 (38.75%)	640
cnr1 (calibration)	123 (61.50%)	77 (38.50%)	200
cnr1 (test)	110 (68.75%)	50 (31.25%)	160
cnr2 (training)	405 (63.28%)	235 (36.72%)	640
cnr2 (cablibration)	127 (63.50%)	73 (36.50%)	200
cnr2 (test)	112 (70.00%)	48 (36.50%)	160

Using PETSc implementation of the Newton method without preconditioning with default setting, the sigmoid-shaped calibration function is computed by minimizing cross-entropy (13) on calibration data. Since the Newton method converges quickly to optimal solution $\mathbf{x}^*$ when vector $\mathbf{x}$ is close enough to $\mathbf{x}^*$, [7] proposed initial guesses for parameters of sigmoid such that $A_0 = 0$ and $B_0 = \log \frac{l^+ + 1}{l^- + 1}$, where l^+ and l^- denote numbers of active and inactive samples associated with the calibration dataset T_{CA}. To avoid numerical difficulties or catastrophic cancellations that could arise from evaluation $1 - p_i$, where p_i is close to 1, we evaluate (10) by using $\frac{\exp(-Af(\mathbf{x})-B)}{1+\exp(-Af(\mathbf{x})-B)}$ when $Af(\mathbf{x}) + B \geq 0$ else we use (10). This numerical improvements were proposed by [4]. Other numerical obstacles could arise from evaluating Hessian

$$H = \begin{bmatrix} \sum_{i=1}^{l} f_i^2 p_i (1 - p_i) & \sum_{i=1}^{l} f_i p_i (1 - p_i) \\ \sum_{i=1}^{l} f_i p_i (1 - p_i) & \sum_{i=1}^{l} p_i (1 - p_i) \end{bmatrix} \tag{14}$$

associated with cross-entropy function (13). Thus, we replace the term $(1 - p_i)$ by means of $\frac{1}{1+\exp(-Af(\mathbf{x})-B)}$ when $Af(\mathbf{x}) + B \geq 0$, else $\frac{\exp(Af(\mathbf{x})+B)}{1+\exp(Af(\mathbf{x})+B)}$, see [4]. Since the Hessian (14) is SPS in general [4], we regularize it by the matrix $\sigma \boldsymbol{I}$, where $\sigma = 1\text{e}{-12}$ in our experiments.

Instead of stochastic optimization, which commonly used in the machine learning community, our used solvers are deterministic in these experiments. They pass all training samples in one iteration. Therefore we consider terms epoch and iteration as identical in this text. In other words, we do not take into account a batch of a training dataset during the training phase of the classifier. By this, we obtain strictly settled and reproducible training pipelines, unlike employing DNNs or other traditionally used techniques. On the other hand, deterministic solvers suffer on their cost in the sense of computational resources. Thus, training predictors can take longer than in the case of stochastic optimization. At this expense, we reduce uncertainty during the training process, which could be crucial for some scientific application, e.g. ones related to the pharmaceutical industry.

After calibrating a classification model, we convert probabilities to label prediction using optimal threshold (thr.), i.e. $y = 1$ if $p >$ thr. to demonstrate the balanced class predictive relevance of calibrated models. The optimal threshold is determined using grid-search so that absolute value of the difference of precision score (Pre.) and sensitivity (Sen.) on the test dataset is minimized, and F1 score must be greater than 0.50, i.e. predictive ability of model must be better than random.

Because of all datasets are too small to utilize more than one processor core, all experiments were run on 1 MPI process pinned to a processor core. In all presented experiments, we utilized the same node of the ANSELM supercomputer at IT4Innovations. Evaluations of performance scores are summarized in Table 2 and Table 3 for uncalibrated models and models after calibration, respectively.

Table 2. abl1, adora2a, cnr1, cnr2 uncalibrated single-target models: evaluation of performance scores of models trained with parameter C_{BE} determined by means of HyperOpt, particularly, grid-search combined with 3-fold cross-validation. We report precision (Pre.), sensitivity (Sen.), F1 score, and AUC as qualitative metrics of a model performance for the both $\ell 1$-loss and $\ell 2$-loss SVM.

Target	Loss	Uncalibrated model				
		C_{BE}	Pre. [%]	Sen. [%]	F1	AUC
abl1	$\ell 1$	2^{-6}	71.60	65.17	0.68	0.66
	$\ell 2$	2^{-5}	69.14	60.22	0.64	0.64
adora2a	$\ell 1$	2^{-6}	70.53	82.72	0.76	0.74
	$\ell 2$	2^{-7}	70.53	83.75	0.77	0.74
cnr1	$\ell 1$	2^{-6}	90.00	82.50	0.86	0.77
	$\ell 2$	2^{-6}	87.27	81.36	0.84	0.74
cnr2	$\ell 1$	2^{-6}	83.93	82.46	0.83	0.72
	$\ell 2$	2^{-5}	86.61	82.91	0.85	0.74

Looking at performance scores presented in Table 2, we can see that $\ell 1$-loss SVM outperforms $\ell 2$-loss in overall performance scores (F1 and AUC) in order to abl1 and cnr1 datasets, while $\ell 2$-loss SVM provides slightly better models for adora2a and cnr2 datasets. As we mentioned in Sect. 2, $\ell 1$-loss SVM commonly produces better quality models. However, we have to take into an account that we relax bias b in our approaches. Therefore the models are relaxed as well, which could be a cause of these unexpected results in the sense of performance score of models.

Table 3. abl1, adora2a, cnr1, cnr2 calibrated single-target models: quality of models in probabilistic sense (Brier score) and performance scores in sense of binary classification on test datasets related to models, which are converted from target probabilities to labels, using the optimal threshold (Thr.). Results are presented for the both $\ell 1$-loss and $\ell 2$-loss SVM.

Target	Loss	Calibrated model				
		Brier score	Binary classification			
			Thr.	Pre. [%]	Sen. [%]	AUC
abl1	$\ell 1$	0.1105	0.52	65.43	64.63	0.64
	$\ell 2$	0.0947	0.54	60.49	60.49	0.60
adora2a	$\ell 1$	0.1280	0.41	78.95	78.95	0.74
	$\ell 2$	0.1222	0.44	78.95	78.95	0.74
cnr1	$\ell 1$	0.0905	0.63	83.64	83.64	0.74
	$\ell 2$	0.0710	0.58	83.64	83.64	0.74
cnr2	$\ell 1$	0.0889	0.53	83.04	83.04	0.72
	$\ell 2$	0.0611	0.53	83.93	83.93	0.73

Analysing predictive relevance of uncalibrated models, we can see that active ligands are preferred in cases of models related to abl1, cnr1, and cnr2, on the other hand, inactive ligands are prefered for adora2a dataset. To balance predictive relevance, we perform model calibration. After this, we can see in Table 3, the models trained using $\ell 2$-loss seem to be better calibrated by comparing Brier scores for all cases than ones related to the $\ell 1$-loss SVM. This could simple consequence of underlying model robustness. Even we use relaxed approach, the $\ell 1$-loss SVM still tries to produce a more robust model than the $\ell 2$-loss SVM, since using a linear sum of hinge loss functions instead of a sum of squared hinge loss functions leads to slightly better catching the outliers.

Table 4. abl1, adora2a, cnr1, cnr2 biological targets: elapsed time related to training of models including HyperOpt and calibration.

Loss	Elapsed time [s] (HyperOpt + Training + Calibration)			
	abl1	adora2a	cnr1	cnr2
$\ell 1$	2.15	2.61	1.95	2.57
$\ell 2$	1.38	1.86	1.57	1.58

Thus, calibrating models related to the $\ell 2$-loss SVM has significant impact than in a case of the $\ell 1$-loss SVM; as we can see, predictive relevances of classes are well balanced. Moreover, from the Table 4, we can observe speedups 1.56 (abl1), 1.40 (adora2a), 1.24 (cnr2), and 1.62 (cnr1) in order to using the $\ell 2$-loss SVM against the $\ell 1$-loss SVM.

However, calibrating models could cause a deterioration of overall model performance determined by means of AUC as we can see in Table 4. Specifically, the over performance scores of models decrease by 1% to 4% in the cases of the abl1 (both loss-type models), cnr1 ($\ell 1$-loss model) and cnr2 ($\ell 2$-loss model). In order to models related to adora2a target, $\ell 1$-loss and $\ell 2$-loss models associated with cnr1 and cnr2, respectively, AUC scores are same. Since models were trained, calibrated and tested on different datasets, we can consider the calibrated models have the same overall performance score in the sense of AUC as their related uncalibrated models. Comparing the quality of the remaining calibrated and uncalibrated models is application-specific. In some application, they could be considered as models of same quality. For more strict quality merits, the models can be considered that differ significantly.

From achieved results, it seems that it is better to train models using the $\ell 2$-loss SVM that are not such robust as in the case of the $\ell 1$-loss SVM and, then, perform their calibration. Moreover, we can obtain a better convergence rate by employing this approach. We observe speedup up to 1.62 in case of training model on the cnr2 dataset.

6 Conclusion

In this paper, we focused on a problem dealing with balancing predictive relevance of single-target models trained using SVMs. This calibration could be required since

SVMs is sensitive to imbalanced datasets, outliers and high multicorrelation among training samples.

Regarding calibration improvements of models, we observe that an additional calibration works significantly better for models trained using the $\ell 2$-loss SVM with relaxed-bias from achieved results. It seems this could be a consequence of that models are not such robust as in the case of the $\ell 1$-loss SVM. Moreover, we achieve speedup up to 1.62 by means of the approach based on $\ell 2$-loss SVM. On the other hand, calibrating models could cause a deterioration of overall model performance as we saw in the presented numerical experiments. Therefore, it makes sense calibrating models just for critical applications, e.g. biochemical modelling presented in this paper, where balanced predictive relevance is required.

Since we achieved some unexpected results in the sense of model performance scores, which are probably caused relaxing the bias term of the hyperplane H, we are going to focus on calibrating model trained to employ training based on full-formulation dual formulation of SVM, i.e. with equality constraint. Further, we are going to test another calibration technique, e.g. the isotonic regression.

Acknowledgments. The author acknowledges the grant programme "Support for Science and Research in the Moravia-Silesia Region 2017" (RRC/10/2017), financed from the budget of the Moravian-Silesian region; and the Grant of SGS No. SP2020/84, VSB - Technical University of Ostrava. This result was also produced support of long-term conceptual development of the research organization of the Institute of Geonics of the Czech Academy of Sciences, RVO: 68145535. The author would like to thank a reviewer for the constructive feedback as well.

References

1. Cortes, C., Vapnik, V.: Support-vector networks. Mach. Learn. **20**(3), 273–297 (1995). https://doi.org/10.1007/BF00994018
2. Dostál, Z.: Optimal Quadratic Programming Algorithms, with Applications to Variational Inequalities. Springer Optimization and Its Applications, vol. 23. Springer, New York (2009). https://doi.org/10.1007/b138610
3. Levenberg, K.: A method for the solution of certain non-linear problems in least squares. Q. Appl. Math. **2**(2), 164–168 (1944)
4. Lin, H.T., Lin, C.J., Weng, R.C.: A note on Platt's probabilistic outputs for Support Vector Machines. Mach. Learn. **68**(3), 267–276 (2007)
5. Pecha, M., Hapla, V., Horák, D., Čermák, M.: Notes on the preliminary results of a linear two-class classifier in the PERMON toolbox. In: AIP Conference Proceedings, vol. 1978 (2018)
6. Pecha, M., Horák, D.: Analyzing l1-loss and l2-loss support vector machines implemented in PERMON Toolbox. In: Zelinka, I., Brandstetter, P., Trong Dao, T., Hoang Duy, V., Kim, S. (eds.) AETA 2018 - Recent Advances in Electrical Engineering and Related Sciences: Theory and Application. AETA 2018. LNEE, vol. 554, pp. 13–23. Springer, Cham (2020). https://doi.org/10.1007/978-3-030-14907-9_2
7. Platt, J.: Probabilistic outputs for support vector machines and comparisons to regularized likelihood methods. Adv. Large Margin Classif. **10**(3), 61–74 (1999)
8. Sun, J., et al.: ExCAPE-DB: an integrated large scale dataset facilitating big data analysis in chemogenomics. J. Cheminformatics **9**(1), 1–9 (2017)

Modelling the Process of Production of Automotive Gasoline by the Use of Generalized Nets

Danail Stratiev[1], Dafina Zoteva[1,2], Dicho Stratiev[3], and Krassimir Atanassov[1,4](✉)

[1] Department of Bioinformatics and Mathematical Modelling, Institute of Biophysics and Biomedical Engineering, Bulgarian Academy of Sciences, Acad. G. Bonchev Str., Block 105, 1113 Sofia, Bulgaria
danail.stratiev@gmail.com, krat@bas.bg

[2] Faculty of Mathematics and Informatics, Sofia University "St. Kliment Ohridski", 5 "James Bourchier" Blvd, 1164 Sofia, Bulgaria

[3] LUKOIL, Neftochim Burgas, Burgas, Bulgaria
stratiev.dicho@neftochim.bg

[4] Intelligent Systems Laboratory, Prof. Asen Zlatarov University, 1, "Prof. Yakimov" Blvd, 8010 Burgas, Bulgaria

Abstract. A generalized net model of the process of production of different grades of automotive gasoline is constructed. The processes in Lukoil Neftohim Burgas, Bulgaria refinery are considered as an example. The model can be used for synchronization and optimization of the process. It is simulated in GN IDE (Generalized Net Integrated Development Environment) – a platform independent software for constructing and simulating generalized nets models.

Keywords: Automotive gasoline · Generalized Net · Model

1 Introduction

Generalized Nets (GNs) [3,4] were defined as an extension of Petri Nets of Carl-Adam Petri. Now, GNs are used as a tool for modelling parallel processes in different areas of science, knowledge and practice – from artificial intelligence to medicine, from economics to chemistry, etc. [1].

In order to support and enhance the possibilities of application of the GNs, a platform independent software tool has been developed. Generalized Net Integrated Development Environment (GN IDE) [2,7] allows construction and simulation of the functioning of GN-models by implementing the latest versions of the algorithms for a transition and a GN functioning defined in the GNs theory. GN IDE has been successfully employed in the simulation of the functioning of

Danail Stratiev is PhD student at the Institute of Biophysics and Biomedical Engineering.

K. T. Atanassov et al. (Eds.): IWIFSGN 2020/BOS/SOR 2020, LNNS 338, pp. 349–365, 2022.
https://doi.org/10.1007/978-3-030-95929-6_27

a GN-model that compares different mathematical models of *E. coli* fed-batch fermentation process [8], a GN-model of wastewater treatment processes [10] and a GN-model of the processes of loading and transportation of raw materials on open construction cites [25].

The first attempts to apply GNs for modelling and simulation of processes in a petrochemical plant were published in [5,9]. The constructed GN-models follow the idea of hierarchical models expanded by the union or the composition of simpler ones. The simplified global model describes the general movement of fuels, petrochemicals and plastics, produced and processed in the plant, taking into consideration the demands for these goods and the influence of local and global factors (marketing situation, regulations, dynamics of prices, etc.).

Modelling and simulation of the complex processes in a petroleum refinery is a powerful tool to optimize their performance and improve the refinery's profitability [6,11–23,26–33]. Most models applied in petroleum refining employ linear programming (LP) technique.

In the present paper, the capabilities of GNs are employed to model the production of different grades of automotive gasoline, for example, in Lukoil Neftohim Burgas, Bulgaria refinery. Several refining processes run in parallel producing different automotive gasoline components: reformate, hydrotreated fluid catalytic cracking (HTFCC) gasoline, MTBE (methyl-tertiary butyl ether), alkylate, hydrotreated straight run naphtha, C5-cut, isobutane, and some additional components like bio-ethanol, and imported MTBE (nine components) which are bought from the market depending on the gasoline grades demand and the ordered gasoline grade specifications. The functioning of the constructed GN-model is simulated in GN IDE.

The paper is organized in the following way. The description of the constructed GN-model is presented in Sect. 2. Section 3 contains specifics of the simulation of its functioning in GN IDE. Concluding remarks are given in Sect. 4.

2 Generalized Net Model of Production of Four Automotive Gasoline Grades (A-95H; A-95 B-9; A-98H; A-98H B-9) from Nine Gasoline Components in the Lukoil Neftohim Burgas Refinery

The production of gasoline in a refinery is a typical parallel process in which the gasoline components are produced by several refining processes like catalytic reforming of heavy naphtha; FCC of vacuum gas oil; production of MTBE by reaction of iso-butene, made in the FCC unit, with methanol; production of alkylate by reacting the C4 alkenes made in the FCC unit with iso-butane; production of C5-cut made in the central refinery gas fractionation unit. All these gasoline components, along with the imported MTBE and bio-ethanol, are blended in accordance with predefined recipes into final marketable automotive gasoline grades. Figure 1 is an illustration of the properties of the nine gasoline components studied in this work and the produced thereof four gasoline grades.

The GN-model contains 21 transitions, 63 places and 13 types of tokens (see Fig. 2). The first group of nine tokens are:

α-tokens that represent HTFCC gasoline;
β-tokens that represent Reformate;
γ-tokens that represent Alkylate;
δ-tokens that represent Hydrotreated Straight Run Light Naphtha (HTLSRN);
ε-tokens that represent C5 cut;
ζ-tokens that represent MTBE;
η-tokens that represent MTBE import;
θ-tokens that represent Bio ethanol;
ι-tokens that represent i-butane.

They stay permanently in places $l_{11}, l_{13}, l_{15}, l_{17}, l_{19}, l_{21}, l_{23}, l_{25}, l_{27}$, respectively, with the initial and current characteristic

"current quantity of the respective gasoline component".

Each one of them is split to two tokens if there is a request of the respective raw material - the original one and a new one $\alpha', \beta', ...$ that enters place $l_{10}, l_{12}, l_{14}, l_{16}, l_{18}, l_{20}, l_{22}, l_{24}, l_{26}$, respectively, with the characteristic

"necessary quantity of the respective gasoline component".

The second group of four tokens are:

κ-tokens that represent A-95H;
λ-tokens that represent A-95H B9;
μ-tokens that represent A-98H;
ν-tokens that represent A-98H B9.

They stay permanently in places $l_{58}, l_{60}, l_{62}, l_{64}$, respectively, with an initial and current characteristic

"current quantity of the respective gasoline grade: A-95H; A-95H B-9; A-98H; A-98H B-9".

When $\alpha, \beta, ...$ tokens enter each of the places $l_{58}, l_{60}, l_{62}, l_{64}$, they unite with the token staying permanently in the respective place and it changes the value of its current characteristic.

When there is a necessity for the respective gasoline grade (A-95H; A-95H B-9; A-98H; A-98H B-9 - four gasoline grades), tokens $\kappa, \lambda, \mu, \nu$ split into two tokens - the original one and a new one $\kappa', \lambda', ...$ that enters place $l_{57}, l_{59}, l_{61}, l_{63}$, respective with a characteristic

"necessary quantity of the respective gasoline grade".

The GN-transitions have the following forms.

$$Z_1 = \langle \{l_1, l_{11}\}, \{l_{10}, l_{11}\}, r_1 \rangle$$

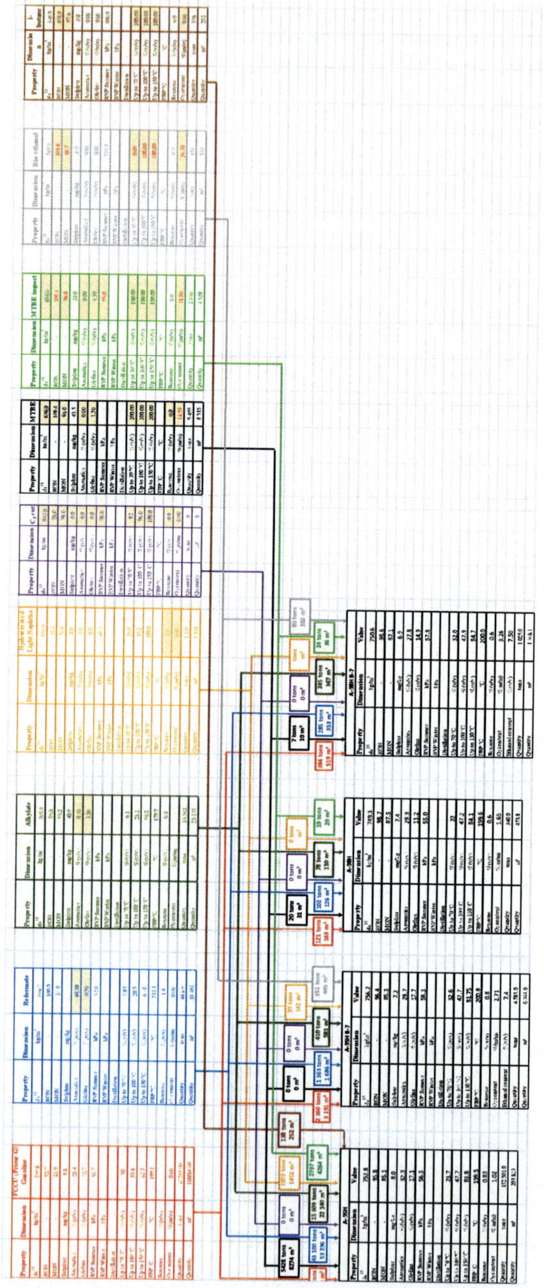

Fig. 1. Production scheme of four automotive gasoline grades made by nine components in a petroleum refinery by nine gasoline components

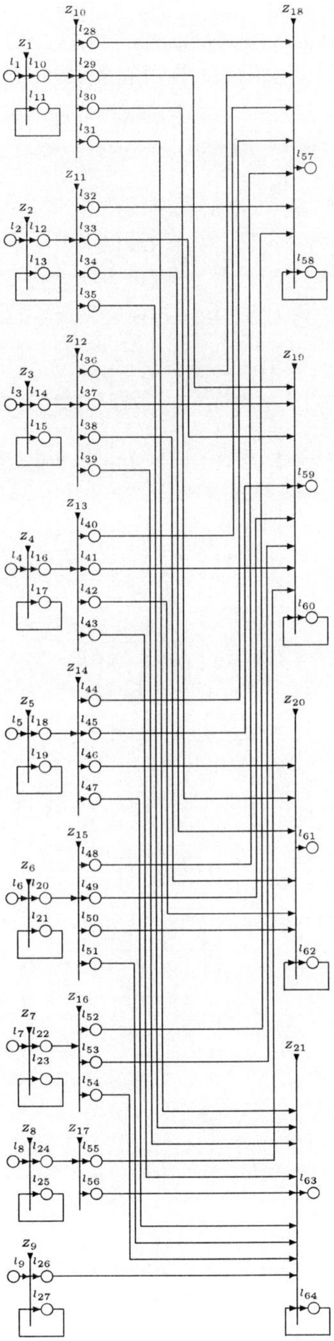

Fig. 2. The generalized net

where

$$r_1 = \begin{array}{c|cc} & l_{10} & l_{11} \\ \hline l_1 & false & true \\ l_{11} & W_{11,10} & true \end{array}$$

where
$W_{11,10}$ = "There is a necessity of $a_{1,1} + a_{2,1} + a_{3,1} + a_{4,1}$ tons/cubic meters HTFCC gasoline".

$$Z_2 = \langle \{l_2, l_{13}\}, \{l_{12}, l_{13}\}, r_2 \rangle$$

where

$$r_2 = \begin{array}{c|cc} & l_{12} & l_{13} \\ \hline l_2 & false & true \\ l_{13} & W_{13,12} & true \end{array}$$

where
$W_{13,12}$ = "There is a necessity of $a_{1,2} + a_{2,2} + a_{3,2} + a_{4,2}$ tons/cubic meters Reformate".

$$Z_3 = \langle \{l_3, l_{15}\}, \{l_{14}, l_{15}\}, r_3 \rangle$$

where

$$r_3 = \begin{array}{c|cc} & l_{14} & l_{15} \\ \hline l_3 & false & true \\ l_{15} & W_{15,14} & true \end{array}$$

where
$W_{15,14}$ = "There is a necessity of $a_{1,3} + a_{2,3} + a_{3,3} + a_{4,3}$ tons/cubic meters Alkylate".

$$Z_4 = \langle \{l_4, l_{17}\}, \{l_{16}, l_{17}\}, r_4 \rangle$$

where

$$r_4 = \begin{array}{c|cc} & l_{16} & l_{17} \\ \hline l_4 & false & true \\ l_{17} & W_{17,16} & true \end{array}$$

where
$W_{17,16}$ = "There is a necessity of $a_{1,4} + a_{2,4} + a_{3,4} + a_{4,4}$ tons/cubic meters HTLSRN".

$$Z_5 = \langle \{l_5, l_{19}\}, \{l_{18}, l_{19}\}, r_5 \rangle$$

where

$$r_5 = \begin{array}{c|cc} & l_{18} & l_{19} \\ \hline l_5 & false & true \\ l_{19} & W_{19,18} & true \end{array}$$

where
$W_{19,18}$ = "There is a necessity of $a_{1,5} + a_{2,5} + a_{3,5} + a_{4,5}$ tons/cubic meters C5 cut".

$$Z_6 = \langle \{l_6, l_{21}\}, \{l_{20}, l_{21}\}, r_6 \rangle$$

where

$$r_6 = \begin{array}{c|cc} & l_{20} & l_{21} \\ \hline l_6 & false & true \\ l_{21} & W_{21,20} & true \end{array}$$

where
$W_{21,20}$ = "There is a necessity of $a_{1,6} + a_{2,6} + a_{3,6} + a_{4,6}$ tons/cubic meters MTBE".

$$Z_7 = \langle \{l_7, l_{23}\}, \{l_{22}, l_{23}\}, r_7 \rangle$$

where

$$r_7 = \begin{array}{c|cc} & l_{22} & l_{23} \\ \hline l_7 & false & true \\ l_{23} & W_{23,22} & true \end{array}$$

where
$W_{23,22}$ = "There is a necessity of $a_{1,7} + a_{2,7} + a_{3,7} + a_{4,7}$ tons/cubic meters MTBE import".

$$Z_8 = \langle \{l_8, l_{25}\}, \{l_{24}, l_{25}\}, r_8 \rangle$$

where

$$r_8 = \begin{array}{c|cc} & l_{24} & l_{25} \\ \hline l_8 & false & true \\ l_{25} & W_{25,24} & true \end{array}$$

where
$W_{25,24}$ = "There is a necessity of $a_{1,8} + a_{2,8} + a_{3,8} + a_{4,8}$ tons/cubic meters Bio ethanol".

$$Z_9 = \langle \{l_9, l_{27}\}, \{l_{26}, l_{27}\}, r_9 \rangle$$

where

$$r_9 = \begin{array}{c|cc} & l_{26} & l_{27} \\ \hline l_9 & false & true \\ l_{27} & W_{27,26} & true \end{array}$$

where
$W_{27,26}$ = "There is a necessity of $a_{1,9} + a_{2,9} + a_{3,9} + a_{4,9}$ tons/cubic meters i-butane".

$$Z_{10} = \langle \{l_{10}\}, \{l_{28}, l_{29}, l_{30}, l_{31}\}, r_{10} \rangle$$

where

$$r_{10} = \begin{array}{c|cccc} & l_{28} & l_{29} & l_{30} & l_{31} \\ \hline l_{10} & W_{10,28} & W_{10,29} & W_{10,30} & W_{10,31} \end{array}$$

where
$W_{10,28}$ = "There is a necessity of $a_{1,1}$ tons/cubic meters HTFCC gasoline",
$W_{10,29}$ = "There is a necessity of $a_{2,1}$ tons/cubic meters HTFCC gasoline",
$W_{10,30}$ = "There is a necessity of $a_{3,1}$ tons/cubic meters HTFCC gasoline",
$W_{10,31}$ = "There is a necessity of $a_{4,1}$ tons/cubic meters HTFCC gasoline".

$$Z_{11} = \langle \{l_{12}\}, \{l_{32}, l_{33}, l_{34}, l_{35}\}, r_{11} \rangle$$

where

$$r_{11} = \begin{array}{c|cccc} & l_{32} & l_{33} & l_{34} & l_{35} \\ \hline l_{12} & W_{12,32} & W_{12,33} & W_{12,34} & W_{12,35} \end{array}$$

where
$W_{12,32}$ = "There is a necessity of $a_{1,2}$ tons/cubic meters Reformate",
$W_{12,33}$ = "There is a necessity of $a_{2,2}$ tons/cubic meters Reformate",
$W_{12,34}$ = "There is a necessity of $a_{3,2}$ tons/cubic meters Reformate",
$W_{12,35}$ = "There is a necessity of $a_{4,2}$ tons/cubic meters Reformate".

$$Z_{12} = \langle \{l_{14}\}, \{l_{36}, l_{37}, l_{38}, l_{39}\}, r_{12} \rangle$$

where

$$r_{12} = \begin{array}{c|cccc} & l_{36} & l_{37} & l_{38} & l_{39} \\ \hline l_{14} & W_{14,36} & W_{14,37} & W_{14,38} & W_{14,39} \end{array}$$

where
$W_{14,36}$ = "There is a necessity of $a_{1,3}$ tons/cubic meters Alkylate",
$W_{14,37}$ = "There is a necessity of $a_{2,3}$ tons/cubic meters Alkylate",
$W_{14,38}$ = "There is a necessity of $a_{3,3}$ tons/cubic meters Alkylate",
$W_{14,39}$ = "There is a necessity of $a_{4,3}$ tons/cubic meters Alkylate".

$$Z_{13} = \langle \{l_{16}\}, \{l_{40}, l_{41}, l_{42}, l_{43}\}, r_{13} \rangle$$

where

$$r_{13} = \begin{array}{c|cccc} & l_{40} & l_{41} & l_{42} & l_{43} \\ \hline l_{16} & W_{16,40} & W_{16,41} & W_{16,42} & W_{16,43} \end{array}$$

where
$W_{16,40}$ = "There is a necessity of $a_{1,4}$ tons/cubic meters HTLSRN",
$W_{16,41}$ = "There is a necessity of $a_{2,4}$ tons/cubic meters HTLSRN",
$W_{16,42}$ = "There is a necessity of $a_{3,4}$ tons/cubic meters HTLSRN",
$W_{16,43}$ = "There is a necessity of $a_{4,4}$ tons/cubic meters HTLSRN".

$$Z_{14} = \langle \{l_{18}\}, \{l_{44}, l_{45}, l_{46}, l_{47}\}, r_{14} \rangle$$

where

$$r_{14} = \begin{array}{c|cccc} & l_{44} & l_{45} & l_{46} & l_{47} \\ \hline l_{18} & W_{18,44} & W_{18,45} & W_{18,46} & W_{18,47} \end{array}$$

where
$W_{18,44}$ = "There is a necessity of $a_{1,5}$ tons/cubic meters C5 cut",
$W_{18,45}$ = "There is a necessity of $a_{2,5}$ tons/cubic meters C5 cut",

$W_{18,46}$ = "There is a necessity of $a_{3,5}$ tons/cubic meters C5 cut",
$W_{18,47}$ = "There is a necessity of $a_{4,5}$ tons/cubic meters C5 cut".

$$Z_{15} = \langle \{l_{20}\}, \{l_{48}, l_{49}, l_{50}, l_{51}\}, r_{15} \rangle$$

where

$$r_{15} = \begin{array}{c|cccc} & l_{48} & l_{49} & l_{50} & l_{51} \\ \hline l_{20} & W_{20,48} & W_{20,49} & W_{20,50} & W_{20,51} \end{array}$$

where
$W_{20,48}$ = "There is a necessity of $a_{1,6}$ tons/cubic meters MTBE",
$W_{20,49}$ = "There is a necessity of $a_{2,6}$ tons/cubic meters MTBE",
$W_{20,50}$ = "There is a necessity of $a_{3,6}$ tons/cubic meters MTBE",
$W_{20,51}$ = "There is a necessity of $a_{4,6}$ tons/cubic meters MTBE".

$$Z_{16} = \langle \{l_{22}\}, \{l_{52}, l_{53}, l_{54}\}, r_{16} \rangle$$

where

$$r_{16} = \begin{array}{c|ccc} & l_{52} & l_{53} & l_{54} \\ \hline l_{22} & W_{22,52} & W_{22,53} & W_{22,54} \end{array}$$

where
$W_{22,52}$ = "There is a necessity of $a_{1,7}$ tons/cubic meters MTBE import",
$W_{22,53}$ = "There is a necessity of $a_{2,7}$ tons/cubic meters MTBE import",
$W_{22,54}$ = "There is a necessity of $a_{4,7}$ tons/cubic meters MTBE import".

$$Z_{17} = \langle \{l_{24}\}, \{l_{55}, l_{56}\}, r_{17} \rangle$$

where

$$r_{17} = \begin{array}{c|cc} & l_{55} & l_{56} \\ \hline l_{24} & W_{24,55} & W_{24,56} \end{array}$$

where
$W_{24,55}$ = "There is a necessity of $a_{2,8}$ tons/cubic meters Bio ethanol",
$W_{24,56}$ = "There is a necessity of $a_{4,8}$ tons/cubic meters Bio ethanol".

$$Z_{18} = \langle \{l_{26}, l_{28}, l_{32}, l_{36}, l_{40}, l_{44}, l_{48}, l_{52}, l_{58}\}, \{l_{57}, l_{58}\}, r_{18} \rangle$$

where

$$r_{18} = \begin{array}{c|cc} & l_{57} & l_{58} \\ \hline l_{26} & false & true \\ l_{28} & false & true \\ l_{32} & false & true \\ l_{36} & false & true \\ l_{40} & false & true \\ l_{44} & false & true \\ l_{48} & false & true \\ l_{52} & false & true \\ l_{58} & true & true \end{array}.$$

$$Z_{19} = \langle \{l_{29}, l_{33}, l_{37}, l_{41}, l_{45}, l_{49}, l_{55}, l_{60}\}, \{l_{59}, l_{60}\}, r_{19} \rangle$$

where

$$r_{19} = \begin{array}{c|cc} & l_{59} & l_{60} \\ \hline l_{29} & false & true \\ l_{33} & false & true \\ l_{37} & false & true \\ l_{41} & false & true \\ l_{45} & false & true \\ l_{49} & false & true \\ l_{55} & false & true \\ l_{60} & true & true \end{array}.$$

$$Z_{20} = \langle\{l_{30}, l_{34}, l_{38}, l_{42}, l_{46}, l_{50}, l_{53}, l_{62}\}, \{l_{61}, l_{62}\}, r_{20}\rangle$$

where

$$r_{20} = \begin{array}{c|cc} & l_{61} & l_{62} \\ \hline l_{30} & false & true \\ l_{34} & false & true \\ l_{38} & false & true \\ l_{42} & false & true \\ l_{46} & false & true \\ l_{50} & false & true \\ l_{53} & false & true \\ l_{62} & true & true \end{array}.$$

$$Z_{21} = \langle\{l_{31}, l_{35}, l_{39}, l_{43}, l_{47}, l_{51}, l_{54}, l_{56}, l_{64}\}, \{l_{63}, l_{64}\}, r_{21}\rangle$$

where

$$r_{21} = \begin{array}{c|cc} & l_{63} & l_{64} \\ \hline l_{31} & false & true \\ l_{35} & false & true \\ l_{39} & false & true \\ l_{43} & false & true \\ l_{47} & false & true \\ l_{51} & false & true \\ l_{54} & false & true \\ l_{56} & false & true \\ l_{64} & true & true \end{array}.$$

3 Simulation of the Constructed GN-Model in GN IDE

3.1 Description of the GN-Model in a GN XML File

The presented GN-model of the production of automotive gasoline grades in Lukoil Neftohim Burgas, Bulgaria refinery is simulated in GN IDE (Fig. 3).

Each GN-model simulated in GN IDE has to be descried in a GN XML file. The structure and the hierarchy of the elements in such a file are strictly defined by a GN XML schema [2,7].

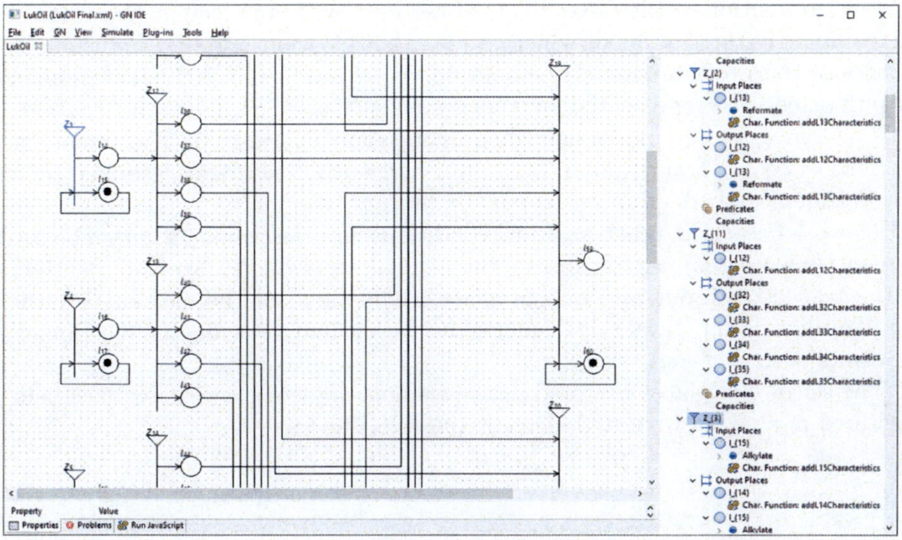

Fig. 3. The GN-model of the production of automotive gasoline grades in GN IDE

```
<?xml version="1.0" ?>
<gn xmlns="http://www.clbme.bas.bg/GN" name="LukOil"
    time="256" timeStart="0" timeStep="1" language="JavaScript" root="true">
    <transitions>
        <transition id="Z1" name="Z_{1}" priority="0" startTime="0"
            lifeTime="-1" positionX="100" positionY="100" sizeY="100" >
        <transition id="Z10" name="Z_{10}" priority="1" startTime="0"
            lifeTime="-1" positionX="220" positionY="50" sizeY="200" >
            <inputs>
                <input ref="l10">
                    <arc>
                        <point positionX="150" positionY="130"/>
                        <point positionX="220" positionY="130"/>
                    </arc>
                </input>
            </inputs>
            <outputs>
            <predicates>
                <predicate input="l10" output="l28">W10_28</predicate>
                <predicate input="l10" output="l29">W10_29</predicate>
                <predicate input="l10" output="l30">W10_30</predicate>
                <predicate input="l10" output="l31">W10_31</predicate>
            </predicates>
        </transition>
```

Fig. 4. Part of the GN XML file describing the GN-model of the production of automotive gasoline grades

Part of the GN XML file with the description of the GN-model of the production of automotive gasoline grades presented in Sect. 2 is shown in Fig. 4.

According to the specifications of the GN XML schema, the GN XML file consists of one main element (root) $\langle gn \rangle$, associated with the GN itself. The four elements $\langle transitions \rangle$, $\langle places \rangle$, $\langle tokens \rangle$ and $\langle functions \rangle$, corresponding to the sets of transitions, places, tokens, as well as the predicate and characteristic functions of the GN-model, are embedded in the element $\langle gn \rangle$.

Each transition of the GN-model is presented with an element ⟨*transition*⟩, the compulsory attributes of which are its identification number and the coordinates of the position where it should be visualized in GN IDE. Its input and output places are represented by the nested elements ⟨*inputs*⟩ and ⟨*outputs*⟩ with references to the places described in the element ⟨*places*⟩. The visual connection between a transition and its input and output places is described by an ⟨*arc*⟩ element embedded in an ⟨*input*⟩ or ⟨*output*⟩ element.

The predicates in the transition conditions are described in an embedded element ⟨*predicates*⟩. Attributes of the element ⟨*predicate*⟩ are the identifiers of the respective input and output places. The value of the element is either the identifier of the JavaScript function to be evaluated or one of the Boolean constants *true* or *false*.

The set of GN-places is stored in the element ⟨*places*⟩, each GN-place being described in a separate nested element ⟨*place*⟩ (Fig. 5).

```xml
<place id="l10" name="l_{10}" priority="0" capacity="-1" merge="false"
    positionX="150" positionY="130" char="addL10Characteristics"/>

<place id="l11" name="l_{11}" priority="2" capacity="-1" merge="true"
    positionX="150" positionY="180" char="addL11Characteristics"/>
```

Fig. 5. Part of the description of the GN-places in the GN XML file

Important attributes for each GN-place are the attributes *merge* and *char*. The value of the *merge* attribute is *true* when merging of tokens is allowed in this particular place. Otherwise, $merge = false$. The attribute *char* contains the identifier of the characteristic function which sets new characteristics to the tokens entering this place.

The GN-tokens are described in the element ⟨*tokens*⟩ (Fig. 6).

```xml
<tokens>
    <token id="alpha" name="HTFCC" priority="1" host="l11" entering="0" leaving="-1">
        <char name="HTFCC_current_quantity" type="number" history="1">1000000</char>
    </token>
    <token id="beta" name="Reformate" priority="1" host="l13" entering="0" leaving="-1">
        <char name="Reformate_current_quantity" type="number" history="1">1000000</char>
    </token>
    <token id="gamma" name="Alkylate" priority="1" host="l15" entering="0" leaving="-1">
    <token id="delta" name="HTLSRN" priority="1" host="l17" entering="0" leaving="-1">
    <token id="epsilon" name="5 cut" priority="1" host="l19" entering="0" leaving="-1">
    <token id="zeta" name="MTBE" priority="1" host="l21" entering="0" leaving="-1">
    <token id="eta" name="MTBE import" priority="1" host="l23" entering="0" leaving="-1">
    <token id="theta" name="Bio Ethanol" priority="1" host="l25" entering="0" leaving="-1">
    <token id="iota" name="i-Butane" priority="1" host="l27" entering="0" leaving="-1">
    <token id="kappa" name="A-95H" priority="1" host="l58" entering="0" leaving="-1">
    <token id="lambda" name="A-95H B9" priority="1" host="l60" entering="0" leaving="-1">
    <token id="mu" name="A-98H" priority="1" host="l62" entering="0" leaving="-1">
    <token id="nu" name="A-95H B9" priority="1" host="l64" entering="0" leaving="-1">
</tokens>
```

Fig. 6. Description of the GN-tokens in the GN XML file

An identifier, priority, position and a list of initial characteristics is given for each GN-token.

The simulated production process is not continuous as in the original GN-model, described in Sect. 2. For the purpose of the simulation, it is assumed that the quantities of the raw input components necessary for the production of the automotive gasoline grades are limited. Therefore, the input places l_1, l_2, l_3, l_4, l_5, l_6, l_7, l_8 and l_9 are omitted in the simulated version of the GN-model. Instead, as an initial characteristic of the tokens in places l_{11}, l_{13}, l_{15}, l_{17}, l_{19}, l_{21}, l_{23}, l_{25} and l_{27}, the "current quantity" of the respective input raw component is set to a big enough positive number.

The description of the GN-model in a GN XML file ends with the element $\langle functions \rangle$ with the definitions of all predicate and characteristic functions in JavaScript (Fig. 7).

```
<functions><![CDATA[

  var substance_quantities =
  [
      [ 84420, 2360, 121, 384 ],
      [ 43100, 1363, 102, 285 ],
      [ 15609, 410, 78, 245],
      [ 1009, 99, 0, 0 ],
      [ 0, 0, 0, 0 ],
      [ 5428, 0, 20, 7 ],
      [ 2797, 0, 19, 24 ],
      [ 0, 351, 0, 80 ],
      [ 138, 0, 0, 0 ]
  ];

  function W11_10()
  {
      var token_for_transfer = gn.places["l11"].tokens[0];

      if( typeof token_for_transfer.chars["HTFCC_current_quantity"] === 'undefined' )
          return false;

      var HTFCC_current_quantity = token_for_transfer.chars["HTFCC_current_quantity"].value;

      var necessary_quantity =
          substance_quantities[0][0] + substance_quantities[0][1] +
          substance_quantities[0][2] + substance_quantities[0][3];

      return (HTFCC_current_quantity > necessary_quantity) && emptyPlace("l10");
  }
```

Fig. 7. Part of the predicate functions of the GN-model, defined in the GN XML file

The predicate and the characteristic functions of a GN-model need to be defined in a rather formal way so that the functioning of the GN-model to be simulated in GN IDE. In this particular case, they are also subject of interpretation.

Each of the automotive gasoline grades is produced according to a predefined recipes. The particular quantities of the input raw components required to obtain the four final products are described in the matrix *substance_quantities* of size 9 (raw components) $\times$ 4 (final products). The predicate function

$$W_{11,10} = \text{"There is a necessity of } a_{1,1} + a_{2,1} + a_{3,1} + a_{4,1} \text{ tons/cubic meters HTFCC gasoline"}$$

is then interpreted in the following way. The total necessary quantity of the HTFCC gasoline is calculated. If it does not exceed the available quantity of the raw component (the characteristic "current quantity" in the corresponding

α-token) and the output place l_{10} is empty, then the predicate is evaluated as *true*.

The rest of the predicate and characteristic functions of the GN-model are adapted in a similar way.

Other important details often omitted in the general description of a GN-model are the priorities of the GN-transitions, places and tokens, as well as the types of the GN-transitions. However, they are essential when a simulation is performed. Therefore, in order to simulate the accurate functioning of the GN-model, different priorities of the places are considered and some of the transition types are determined. Those transitions, whose types are not defined, are activated on each step of the GN functioning.

3.2 Results of the Simulation of the GN-Model in GN IDE

Upon starting the simulation in GN IDE, the tokens enter their designated input places.

On the first step of the GN functioning, the transitions Z_1, Z_2, ..., Z_9 are activated. Since no transition type is specified for these transitions, they will be activated on each step of the GN functioning. The relevant predicate functions are evaluated to determine if the tokens in places l_{11}, l_{13}, ..., l_{27} can be transferred.

When a token is transferred from the input place l_{11} to the output place l_{10}, the characteristic function of the corresponding output place is called (Fig. 8).

```
function addL10Characteristics( token )
{
    if( typeof token.chars["HTFCC_current_quantity"] === 'undefined' )
        return;

    token.name = "HTFCC";

    var new_quantity =
        substance_quantities[0][0] + substance_quantities[0][1] +
        substance_quantities[0][2] + substance_quantities[0][3];

    token.chars["HTFCC_current_quantity"].value = new_quantity;
}
```

Fig. 8. A characteristic function, defined in the GN XML file, associated with the l_{10} place

The characteristic "current quantity" of the token in place l_{10} is updated with the transferred necessary quantity of the raw component. At the same time, the available quantity of this raw component (the token in place l_{11}) is decreased by the same value.

Similar actions affect the rest of the activated transitions.

Transitions Z_{10}, Z_{11}, ..., Z_{17} are not activated on this first step, since there are no tokens in their input places.

The transition Z_{18} (Z_{19}, Z_{20} and Z_{21} likewise) can be activated when there are tokens in the input places l_{26}, l_{28}, l_{l32}, ..., l_{48}, l_{52}, or there is a token only in the input place l_{58}. Therefore, the transition is activated on this first step. However, no token is transferred to the output place l_{57}. The corresponding predicate is evaluated as $false$, since the quantity of the final product is still 0.

On the second step of the GN functioning all transitions are activated. Z_{10}, Z_{11}, ..., Z_{17} have higher priorities than the rest of the GN-transitions. Upon their activation, the tokens in input places l_{10}, l_{12}, ..., l_{24}, respectively, are split into the necessary number of tokens each of which with the requested quantity of the corresponding input raw component.

When there are tokens in the input places of the transitions Z_{18}, Z_{19}, Z_{20} and Z_{21}, they are activated. The tokens in their input places are merged in l_{58}, l_{60}, l_{62} and l_{64}, respectively. These actions represent the production of the final products – the four automotive gasoline grades.

On the next time step the functioning of these transitions is performed in two steps. First, each of the tokens in l_{58}, l_{60}, l_{62} and l_{64} is split into two different tokens. One of the resulting tokens is transferred to the corresponding output place (l_{57}, l_{59}, l_{61} and l_{63}, respectively), while the other stays in l_{58}, l_{60}, l_{62} and l_{64}, respectively. The characteristic "current quantity" of the final product in the second token is then updated. On the second step of the functioning of these transitions, the tokens in their input places are again merged in l_{58}, l_{60}, l_{62} and l_{64}, respectively. New quantities of the final products are produced.

The production process continues until at least of the input raw components is exhausted.

The results of the performed simulation of the GN-model in GN IDE is a verification of the model itself.

4 Conclusion

The present GN-model describes the processes of production of four different grades of automotive gasoline observed in Lukoil Neftohim Burgas, Bulgaria refinery. It can be used for synchronization and optimization of these processes in a real process management and automation system. The functioning of the GN-model is simulated in GN IDE.

In near future, this model will be extended by incorporating the gasoline blending model developed in [24]. Simulations of different scenarios including disruptions in the processes of gasoline component production, which have occurred in the real life, will be further performed. Evaluation of the best economical course of action in such situations will be discussed. This GN-model can be used as a tool for assessment of the development of unwanted events and the impact they could have on the environment and on the profitability of petroleum refining and the economics of the district this refinery is located.

Acknowledgements. This paper has been partially supported by the Bulgarian National Science Fund under the Grant Ref. No. Ref. No. DN 02/10 "New Instruments for Knowledge Discovery from Data and their Modelling".

References

1. Alexieva, J., Choy, E., Koycheva, E.: Review and bibliography on generalized nets theory and applications. In: Choy, E., Krawczak, M., Shannon, A., Szmidt, E. (eds.) A Survey of Generalized Nets Raffles KvB Monograph, vol. 10, pp. 207–301(2007)
2. Angelova, N., Todorova, M., Atanassov, K.: GN IDE: implementation, improvements and algorithms. Comptes rendus de l'Academie Bulgare des Sciences **69**(4), 411–420 (2016)
3. Atanassov, K.: Generalized Nets. World Scientific, Singapore, London (1991)
4. Atanassov, K.: On Generalized Nets Theory. "Prof. M. Drinov" Academic Publishing House, Sofia (2007)
5. Atanassov, K., Kacprzyk, A., Skenderov, V., Kyuchukov, A.: Principle generalized net models of the activity of a petrochemical combine. In: Proceedings of the Eighth International Workshop on Generalized Nets, Sofia, 26 June 2007, pp. 38–41 (2007)
6. Castillo, P., Castro, P., Mahalec, V.: Global optimization of nonlinear blend-scheduling problems. Engineering **3**(2), 188–201 (2017)
7. Dimitrov, D.G.: GN IDE: a software tool for simulation with generalized nets. In: Proceedings of the Tenth International Workshop on Generalized Nets, pp. 70–75 (2009)
8. Dimitrov, D., Roeva, O.: Comparison of different mathematical models of an E. coli fed-batch cultivation process using a generalized net model. In: Proceedings of the 13th International Workshop on Generalized Nets, London, UK, pp. 15–23 (2012)
9. Dimitrova, S., Dimitrova, L., Kolarova, T., Petkov, P., Atanassov, K., Christov, R.: Generalized net models of the activity of NEFTOCHIM petrochemical combine in Bourgas. In: Atanassov, K. (ed.) Applications of generalized nets. World Scientific Publ. Co., Singapore, pp. 208–213 (1993)
10. Georgieva, V., Angelova, N., Roeva, O., Pencheva, T.: Simulation of parallel processes in wastewater treatment plant using generalized net integrated development environment. Comptes rendus de l'Academie bulgare des Sciences, Tome **69**(11), 1493–1502 (2016)
11. Hou, Y., Wu, N., Zhou, M., Li, Z.: Pareto-optimization for scheduling of crude oil operations in refinery via genetic algorithm. IEEE Trans. Syst. Man Cybern. **47**(3), 517–530 (2017)
12. Jalanko, M., Mahalec, V.: Supply-demand pinch based methodology for multi-period planning under uncertainty in components qualities with application to gasoline blend planning. Comput. Chem. Eng. **119**, 425–438 (2018)
13. Jia, Z., Ierapetritou, M.: Mixed-integer linear programming model for gasoline blending and distribution scheduling. Ind. Eng. Chem. Res. **42**(4), 825–835 (2003)
14. Joly, M., Moro, L., Pinto, J.: Planning and scheduling for petroleum refineries using mathematical programming. Braz. J. Chem. Eng. **19**(2), 207–228 (2002)
15. Reisig, W., Rozenberg, G. (eds.): Lectures on Petri Nets II: Applications: Advances in Petri Nets. Lecture Notes in Computer Science, Springer, Heidelberg (1998). https://doi.org/10.1007/3-540-65307-4
16. Li, Z., Ierapetritou, M.: Production planning and scheduling integration through augmented Lagrangian optimization. Comput. Chem. Eng. **34**(6), 996–1006 (2010)
17. Li, J., Xiao, X., Boukouvala, F., et al.: Data-driven mathematical modeling and global optimization framework for entire petrochemical planning operations. AIChE J. **62**(9), 3020–3040 (2016)

18. Li, F., Yang, M., Du, W., et al.: Development and challenges of planning and scheduling for petroleum and petrochemical production. Front. Eng. Manage. **7**, 373–383 (2020). https://doi.org/10.1007/s42524-020-0123-3
19. Mouret, S., Grossmann, I., Pestiaux, P.: A new Lagrangian decomposition approach applied to the integration of refinery planning and crude-oil scheduling. Comput. Chem. Eng. **35**(12), 2750–2766 (2011)
20. Neiro, S., Pinto, M.: A general modeling framework for the operational planning of petroleum supply chains. Comput. Chem. Eng. **28**, 871–896 (2004)
21. Shah, N., Sahay, N., Ierapetritou, M.: Efficient decomposition approach for large-scale refinery scheduling. Ind. Eng. Chem. Res. **54**(41), 9964–9991 (2015)
22. Siamizade, M.: Global optimization of refinery-wide production planning with highly nonlinear unit models. Ind. Eng. Chem. Res. **58**(24), 10437–10454 (2019)
23. Simao, L., Dias, D., Pacheco, M.: Refinery scheduling optimization using genetic algorithms and cooperative coevolution. In: IEEE Symposium on Computational Intelligence in Scheduling. Honolulu, HI, 1–5 April 2007, pp. 151–158 (2007)
24. Stratiev, D., et al.: Evaluation of accuracy of available in literature gasoline blending models to predict octane numbers of gasoline blends. Petrol. Sci. Technol. **35**(11), 1146–1153 (2017). https://doi.org/10.1080/10916466.2017.1312445
25. Tzanov, V., Zoteva, D., Atanassova, V., Atanassov, K.: Generalized net model of processes of loading and transportation of raw materials of open construction site. In: Proceedings of the 16th Workshop on Generalized Nets and Data Mining, 10 February 2018, Sofia, Bulgaria, pp. 36–40. ISSN 1313-6860
26. Wu, N., Chu, F., Chu, C., Zhou, M.: Hybrid petri net modeling and schedulability analysis of high fusion point oil transportation under tank grouping strategy for crude oil operations in refinery. IEEE Trans. Syst. Man Cybern. Part C (Appl. Rev). **40**(2), 159–175 (2010)
27. Wu, N., Zhou, M., Chu, F.: A petri net-based heuristic algorithm for realizability of target refining schedule for oil refinery. IEEE Trans. Autom. Sci. Eng. **5**(4), 661–676 (2008). https://doi.org/10.1109/TASE.2008.916737
28. Yang, Y., Barton, P.: Integrated crude selection and refinery optimization under uncertainty. AIChE J. **62**(4), 1038–1053 (2016)
29. Yang, J., Gu, H., Rong, G.: Supply chain optimization for refinery with considerations of operation mode changeover and yield fluctuations. Ind. Eng. Chem. Res. **49**(1), 276–287 (2010)
30. Yang, Y., Vayanos, P., Barton, P.: Chance-constrained optimization for refinery blend planning under uncertainty. Ind. Eng. Chem. Res. **56**(42), 12139–12150 (2007)
31. Zhang, J., Zhu, X., Towler, G.: A level-by-level debottlenecking approach in refinery operation. Ind. Eng. Chem. Res. **40**(6), 1528–1540 (2001)
32. Zhao, H., Ierapetritou, M., Shah, N., Rong, G.: Integrated model of refining and petrochemical plant for enterprise-wide optimization. Comput. Chem. Eng. **97**, 194–207 (2017)
33. Zhao, H., Rong, G., Feng, Y.: Effective solution approach for integrated optimization models of refinery production and utility system. Ind. Eng. Chem. Res. **54**(37), 9238–9250 (2015)

Social, Medical and Health Care Applications

Students' Forward Thinking – An Analysis of Cultural Divides Through the 'Reverse Clustering'

Jan W. Owsiński[1(✉)], Cristian Ciurea[2], Florin Gh. Filip[3], and Jarosław Stańczak[1]

[1] Systems Research Institute, Newelska 6, Warsaw, Poland
{Jan.Owsinski,Jaroslaw.Stanczak}@ibspan.waw.pl
[2] The Bucharest University of Economic Studies, Piata Romana 6, Bucharest, Romania
cristian.ciurea@ie.ase.ro
[3] Romanian Academy, Bucharest, Romania
ffilip@acad.ro

Abstract. The paper presents an analysis of data collected about the educational track of university students, using a "reverse clustering" approach. The results are interpreted in the context of a crowdsourcing effort made by authors and the respondents of the questionnaire, in order to extract relevant data for the countries involved in the study. The results obtained and reported here indicate that the cultural profiles of the respondents have a great influence regarding the responses, associated with the educational and professional perspectives of the students involved.

Keywords: Crowdsourcing · Decision-making · Reverse clustering · Cultural divides

1 Introduction

This paper reports on a study, oriented at gaining knowledge through a dispersed community, using web-based tools. A questionnaire-based attempt was made of inquiring into the opinions of university students, concerning the ways they are taught, the potential for change, and their own hypothetical participation in such a change, as well as their, similarly hypothetical, contribution to it. In this paper we concentrate on the analysis of the responses obtained to the questionnaire from the point of view distinctions, brought into the respective response profiles, by the cultural backgrounds and their differences. For this purpose, we use a special technique, which we refer to as "reverse clustering".

K. T. Atanassov et al. (Eds.): IWIFSGN 2020/BOS/SOR 2020, LNNS 338, pp. 369–380, 2022.
https://doi.org/10.1007/978-3-030-95929-6_28

The here reported study is being carried out in the framework of the bilateral research project between the Polish and Romanian Academies[1]. This project is focused on the multi-person decision processes, with special stress on the use of the novel, IT enabled technologies and methodologies. It was deemed appropriate to perform, in the framework of this project, a limited, but valid experimental study, which would potentially constitute an introduction to a broader effort, undertaken in the future.

This technique tries to reconstruct an initially given partition of the set of observations, using the clustering algorithms. Depending upon the character of this given partition and its relation to the data at hand on the set of observations, the reconstruction may be more or less successful, indicating various important aspects of both the partition given and the data set, as well as the potentially different partition ultimately obtained from the exercise.

We outline, first, the study itself, and its planned course, as well as objectives and methods used. Then we characterise in a general manner the results obtained. In the subsequent section we describe the "reverse clustering" approach, to then give the report on the results, important from the point of view of the analysis, undertaken here. The paper ends with conclusions and outline for continuation of the study.

2 The Study Intended

2.1 The General Characteristics of the Responses

The present paper refers to a questionnaire-based study, launched in March 2020 in Romania and Poland. As of the time of this writing, i.e. middle of July 2020, some 50 responses have been obtained to the questionnaire. The questionnaire contains altogether 61 distinct items, organised into, roughly, seven domains. The responses obtained provide quite a variety of characteristics, first of all in terms of respondents. Thus, in particular,

- 5 university-level schools are represented,
- Not less than 7 directions of study are represented,
- Respondents belong to 4 nationalities
- The sample contains both intramural and extramural students
- The respondents currently are at all the years of study, from 1 to 4
- The gender balance among respondents is almost perfect.
- The age of respondents ranges from 18 to 37.

Therefore, without discussing the "representativeness", being here, of course, not really feasible, we can say that the sample acquired is highly interesting.

[1] The project, entitled "Multiparticipant cooperative decision making: consensus building vs. crowdsourcing-based decisions", is implemented by the Systems Research Institute of the Polish Academy of Sciences and the Bucharest University of Economic Sciences in the framework of bilateral cooperation between the Academies of the two countries.

It should also be noted that – given the timing of the present writing and the coronavirus pandemic – the issues of novel and technology-enabled ways of providing educational services become a highly important, if not outright urgent matter.

2.2 The Questionnaire and the Overview of Responses

As mentioned, a questionnaire-based survey was launched, addressed at university-level students in the schools on the Romanian and Polish sides (a Reader, interested in a broader account on the questionnaire and the related part of the study, is referred to [CO19] and [OCF20]). The substantive sense of the inquiry is related to the way university education in the respective study directions is carried out, the tools that are and can be used, the potential changes to be introduced and the hypothetical contribution from students. So, the questionnaire contains the following groups of items:

- Basic data on the respondents (the school, the study direction, the intra- or extramural studies, the year of study, nationality, gender, age)
- Expected (preferred) nature of employment after studies
- Attitude towards studying (attending lectures and exercises, superfluous lectures, relation to life problems, excess theory in the curriculum, teamwork of students)
- Use of the web (university website, use by lecturers, use by oneself – and for what purposes, use of social media)
- Learning and teaching and the kinds of resources used (with a suggestion of an "appropriate mix" of various kinds of media to be used)
- Necessity of changes in teaching, with indication of particular domains (lectures, lecture materials, exercises,...) and the potential tools or instruments
- Student contribution to the potential change (can students contribute? how? do you have your own idea?).

Let us add that the questionnaire was formulated in English language, while, apparently, at least some of the responding students have definite problems with the language, and this could be a source of some – hopefully marginal – errors in responding. The questionnaire was made accessible online, using Google Forms, and is available at: https://forms.gle/kJsWg2aGW5hLqTU99.

We summarise below the statistics regarding the gender, nationality and university of students having responded to the questionnaire:

Gender:

- Female: 49%
- Male: 51%

Nationality:

- Romanian: 51%
- Polish: 43%

- Ukrainian: 4%
- Other: 2%

School:

- Warsaw School of Information Technologies (WIT): 49%
- Bucharest University of Economic Studies: 40%
- Polytechnic University of Bucharest: 6%
- Babes Bolyai University: 2%
- National University of Political Studies, Bucharest: 2%

Against this background of basic statistics, let us quote some of the distribution characteristics of responses to specific items in the questionnaire. Thus, for instance, one of the questions asked whether a student spent more than 10 h per week on own study, and the following distribution of responses was obtained:

- Students of Polish nationality, spending less than 10 h per week on own study: 9
- Students of Romanian nationality, spending less than 10 h per week on own study: 9
- Total for students spending less than 10 h per week on own study: 18 (38%)

Now, regarding those who study at least 10 h per week, the rough distribution was:

- Students of Polish nationality: 11,
- Students of Romanian nationality: 15,
- Total for those spending 10 h and more on own study: 29 (62%).

It might be interesting to know what prospective occupations the students responding expected, hopefully, to be involved in after graduation. Thus, the positive responses (summing up to more than the total of respondents) were as follows:

- Programmer: 31
- Designer/developer: 34
- Manager: 21
- Project manager: 23
- Clerk: 8
- Self-employed/business person: 29
- Sales person: 9
- Other: 14.

This appears to be a typical perspective for the current conditions on the labour market, linking wishes with pragmatic assessment of the situation. Finally, to sum up, the responses to the question as to whether the students have their own idea on the tools and methods for improving the teaching process and its outcomes were: only 34% YES, and 66% NO. Even worse, of those responding YES, quite some, when asked for concrete (even simple) concepts, gave no answers.

Concerning the question "Do you have an own idea of a tool/method to improve the teaching process and its outcomes?" only 32% of respondents said "Yes", which represents a very small percentage considering the innovative potential of the young generation. Then, as concerns the question "Can students contribute to the change?" all the respondents answered positively, with indicated values from 3 to 5 out of the interval 0–5 (0 meaning here not at all, up to 5, meaning absolutely yes).

Against this background, it seemed to be interesting to try to analyse the connections between the response profiles (as indicated – more than 60 items) and the presumed cultural differences (nationalities, schools, etc.). This analysis was done with the use of the "reverse clustering" approach. The analysis of the cultural divides was meant to precede the subsequent stage of the study project, namely a more widely conceived questionnaire, involving –potentially and hypothetically – a more concrete input from the respondents in terms of "crowdsourcing".

3 The "Reverse Clustering" Approach

This approach has been developed and used by the representatives of the Polish part of the project team, as reported in, e.g., [OKO17,OOS17,OSZ18,OSO21]. The approach consists in solving of the following, very specific optimisation task:

Assume we dispose of a data set, denoted X, composed of n observations x_i, $i = 1,\ldots,n$, each of which is characterised by a vector of m values, x_{ik}, $k = 1,\ldots,m$. At the same time, we also dispose of a partition of this set of observations, denoted P_A, meaning, actually, that the set of observations (or of their indices i, this set being denoted I, $I = \{1,\ldots,i,\ldots,n\}$) is divided up into a certain number, say p_A, of subsets, call them A_q, $q = 1,\ldots,p_A$, in such a manner that $\bigcup_q A_q = I$ and $A_q \cap A'_q = \varnothing$, that is – the partition P_A is exhaustive and the subsets (clusters), forming it, are disjoint. Thus, we deal with some data set and with a partition of it. Without any further specifications, the task is to determine (to find) a clustering procedure that will produce, when applied to the data set X, a partition P_B that is as close to P_A as possible (this partition P_B being composed of the subsets, or clusters, B_q, $q = 1,\ldots,p_B$). This amounts to minimising some sort of dissimilarity metric $D(P_A, P_B)$ by searching in the space of the available clustering procedures and their parameters.

The metric $D(P_A, P_B)$, which is quite natural, and which is actually largely used in the approach outlined, is the classical Rand index of similarity of two partitions. This is a very simple and powerful construct, based on the concept of having pairs of observations, indicated by their number from the set I, either in the same or in different clusters (subsets) in the two compared partitions. Thus, for any two partitions of the same set (here: I) we have four counts of such pairs:

a – the number of pairs of observations which are in the same cluster in both partitions (no matter how we designate this cluster), b – the number of pairs of observations, which are in the same cluster in one partition, but in different clusters in the second partition, c – the number of pairs that are in different clusters in one partition, but are in the same cluster in the second partition, and, finally, d – the number of pairs of observations that are in different clusters in both partitions. Of course,

$$a + b + c + d = \frac{1}{2} * n * (n - 1) \quad (1)$$

since they must add to the total number of observation pairs considered. Now, we can either maximise the quotient $(a+d)/(a+b+c+d)$, meaning similarity of the two partitions, or minimise the quotient $(b+c)/(a+b+c+d)$, meaning the dissimilarity of the two partitions. There is no qualitative difference to the use of any of these two. On the other hand, there are statistical considerations, which motivate some modifications of these two criteria, related to the properties of the distribution of values of these criteria, and leading to some modifications of their (simple) form, which we shall not go into here.

Concerning the search in the space of the clustering procedures and their parameters, a number of pragmatic assumptions have to be made, related to:

- The classes of clustering algorithms accounted for (in the practical applications we dealt with these were the hierarchical agglomeration algorithms, the k-means family, and a density-based algorithm, usually represented by DBSCAN).
- The parameters of the clustering methods – which, in general, are quite different, not only as to their character and meaning, but also as to their number. Regarding the hierarchical agglomeration algorithms, such as single or complete linkage etc., a natural choice of the parameterisation is the classical Lance-William formula and its coefficients; concerning the k-means family the essential parameter is the number of clusters; concerning the density-based algorithms – usually the local distance threshold and the number of neighbours.
- Another important parameter is constituted by the distance definition, which is very important for some of the algorithms (like k-means), but is irrelevant for many others; in our implementation of the approach we simply used the exponent in the Minkowski formula for distance between observations, $d(x_i, x_j)$.
- Finally, it was assumed that the partitions obtained may depend upon the choice of variables among the m values, characterising each observation; thus, either weights of variables (between 0 and 1) were subject to optimisation, or, simply, their choice.

All these parameters (choice of algorithm class, parameters of the algorithm, distance definition, variable weights or choice) can be said to form a vector (of variable length), denoted by Z. If so, the task at hand is to minimise

$D(P_A, P_B)$ by searching in the space of Z's, the Z corresponding to the minimum of $D(P_A, P_B)$ being denoted $Z*$.

Let us add that in view of the very complex and untoward character of this optimisation task, we use the evolutionary algorithms, mostly of our own invention, like the two-level evolutionary method presented in [Sta03].

At this point let us mention, for instance, a multi-criteria allocation decision support system (MCADSS) used to assign students to groups, which is described in [WRJ92]. Because it is based on multiple criteria, the objective of MCADSS is to maximize the diversity of students within groups, while minimizing the average differences between groups. The problem can be, likewise, seen as a "reverse clustering" approach, although in this case the "reverse" is understood in quite a different manner (clustering with an objective opposite to the classical clustering objective of group homogeneity).

To close this section let us note that we use the "reverse clustering" approach in this particular case, because of the nature of data, composed of some 60 variables, of which many are binary, many are ordinal (Likert scale, mainly), some are quasi-cardinal, and some are qualitative, with distinct groups of "themes", composed of various numbers and kinds of partial variables. For such data, the standard approaches, like correlation analysis or factor analysis, are simply doomed to fail. At the same time, "reverse clustering" allows for both explicit formulation of hypotheses and for effective analysis, indicating the degree, to which a given ("hypothetical") partition can be actually reconstructed with the data one disposes of.

4 Some Preliminary Results and Their Interpretations

We shall now give a short report on the results, obtained with the use of the "reverse clustering' approach, for the data from the questionnaire-based survey, and the definite hypotheses, related to the possible cultural conditioning of the survey responses.

We start from an initial exercise, in which the given partition P_A was the one based on nationalities. According to the previously presented data, we dealt with, nominally, 4 nationalities, of quite varied cardinalities in the sample ($p_A = 4$). The optimum result, however, classified the observations differently. Namely, only two clusters were obtained in P_B ($p_B = 2$), as this is shown in Table 1.

The above result implied that there were exactly 3 "erroneously" classified observations ("Ukrainians" and "Other"), but, in fact, as this can be easily gathered from the general overview of the survey results, provided before, the partition into two clusters, as characterised above, was a perfect rendition of the division between the Polish and Romanian schools! Thus, we obtain a very strong indication of the influence, exerted by the "locational aspect", on the results of the survey, i.e. the response profiles.

Table 1. Results from the "reverse clustering" exercise based on the initial partition according to nationalities

Clusters assumed	Clusters obtained	
	Cluster 1	Cluster 2
Romanians	24	0
Poles	0	21
Ukrainians	0	2
Other	0	1

Then, an analogous exercise was performed, but with removal of the other features, which were obviously closely associated with nationality, namely, in this case, the school attended. The results are provided in Table 2.

Table 2. Results from the "reverse clustering" exercise based on the initial partition according to nationalities with exclusion of the variable indicating the school attended

Clusters assumed	Clusters obtained	
	1 "Romanians"	2 "non-Romanians"
Romanians	23	1
Poles	1	20
Ukrainians	0	2
Other (Turks)	0	1

Now, as this could be expected, the result is slightly less striking, but still very clear – only two additional "erroneous" assignments occurred, but, in fact, it is exactly these two observations that can be considered true "errors", the other three (as before – all classified in the non-Romanian cluster) representing just another perspective on the "proper partition of responses".

In order to verify this rather unexpectedly strong result, a similar exercise was performed, in which another initial partition, P_A, was assumed, namely that according to the schools considered. The results are provided in Tables 3 and 4 for two different cases analysed (one with a simple division between the Romanian and Polish schools, and another, where the Romanian schools were split into two categories).

It was again attempted to verify these results with the calculations, in which the closely associated variable of nationality was omitted. The result of this verification, analogous to that of Table 3, is provided in Table 5, while that, analogous to the one of Table 4, is given in Table 6.

Thus, again a very strong image appears of distinction between the almae matris of the particular respondents. Note that the "intermediate" character of

Table 3. Results from the "reverse clustering" exercise based on the initial partition according to higher schools, case 1

Clusters assumed	Clusters obtained	
	Cluster 1	Cluster 2
Romanian schools: BUES, UPB, NUPSPA, BBU	24	0
Polish school: WIT	0	24

Table 4. Results from the "reverse clustering" exercise based on the initial partition according to higher schools, case 2

Clusters assumed	Clusters obtained		
	Cluster 1	Cluster 2	Cluster 3
Romanian school: BUES	19	0	0
Romanian schools: UPB, NUPSPA, BBU	0	5	0
Polish school: WIT	0	0	24

Table 5. Results from the "reverse clustering" exercise based on the initial partition according to higher schools, case 1, without the nationality variable

Clusters assumed	Clusters obtained	
	Cluster 1	Cluster 2
Romanian schools: BUES, UPB, NUPSPA, BBU	20	4
Polish school: WIT	1	23

Table 6. Results from the "reverse clustering" exercise based on the initial partition according to higher schools, case 2; without the nationality variable

Clusters assumed	Clusters obtained		
	Cluster 1	Cluster 2	Cluster 3
Romanian school: BUES	19	0	0
Romanian schools: UPB, NUPSPA, BBU	1	2	2
Polish school: WIT	1	3	20

the second cluster from Table 6 is certainly largely associated with the relatively small number of respondents from the Romanian schools other than BUES. Definitely, omission of the nationality variable introduced definite "fuzziness" into the obtained image, but, certainly, the strength of the result is still striking.

Another aspect that is definitely of interest in terms of cultural differences is that of the gender-related distinctions. The result of the respective exercise is presented in Table 7.

Thus, it turns out that the gender aspect is also in a certain, but quite distinct way a dividing criterion with respect to the broad spectrum of the response items.

Table 7. Results from the "reverse clustering" exercise based on the initial partition according to gender

Clusters assumed	Clusters obtained	
	Cluster 1 (Females)	Cluster 2 (Males)
Females	20	3
Males	2	23

Altogether, the results here provided constitute, indeed, a demonstration of the very high importance of the "cultural background", understood in quite broad sense, involving a number of aspects, for the response profiles, and hence, hypothetically, for the potential behaviour in any crowdsourcing exercise.

5 A Comment on Interpretation

As indicated previously, the results from the "reverse clustering" demonstrate that it is possible to devise a regular partition of the given set of objects into subsets, by appropriately selecting the elements of the vector of parameters Z, characterising the clustering procedure, that is possibly close to a given partition. Thus, we do not deal with any "correlation-like" measure or any other statistical construct, but with the demonstration that certain manipulations can yield a partition very similar to the one we have hypothesised.

This, altogether, sounds to provide a very strong argument for the actual existence of a division of the set of objects or observations. It must be emphasised, though, that in all the exercises here illustrated, the vector Z included the variable weights, and, indeed, the values of these weights were also selected in searching for the partition most similar to the one assumed. This, definitely, somewhat weakens the strength of the conclusions drawn from the study, but it must be stressed that in this selection actually only few variables (response items) were eliminated. Hence, it can be maintained that the results have a substantive significance, and do not constitute just an academic data manipulation.

The educational system will change in the near future both in the way the lessons will be delivered and also from the point of view of students' expectations. The year 2020 made the video conferencing solutions, such as Zoom, Skype, Google Meet, Webex, Microsoft Teams, and yet a couple of others, widely used not just for personal purposes, but as the only solution available for the continuity of the educational and business process. In [AVR20] a comparison is presented between these video conferencing solutions, their advantages and disadvantages, because these virtual meeting solutions are not only used to have informal meeting with friends, but also are used nowadays to have formal meetings in business and educational sector.

Even if some video conferencing solutions have been relatively popular around the world, it is still surprising that some new solutions, such as Zoom and Google Meet, were rapidly adopted both in business and educational sector. Not only the costs for licensing are important when choosing a solution or another, but also the functionalities offered by each solution, how easy it is to install and to use both from desktop and mobile devices, and also other facilities which made these platforms so popular in a very short period of time, differ significantly and have to be properly assessed prior to actual application in a serious environment.

Another important criterion, potentially differentiating the use of the web platforms for video conferencing is the cultural profile of the participants. It would be very interesting to have an analysis of the use of current video conferencing solutions, based on gender, age, country and educational level of users. Certainly, the preferences regarding specific solutions differ among individuals, institutions, societies and cultures.

Other web conferencing benefits were described by authors of [SBF09], where it is mentioned that web conferencing offers a more efficient, flexible and dynamic approach to meetings, while retaining some of the aspects of the face-to-face. Through video conferencing solutions, people can save time and money related to travelling.

All in all, given the here exemplified strong evidence on the dependence of the cultural and societal backgrounds on the educational and technological perspectives of the young, an insight into the possible effectiveness of any dispersed decision-oriented or design-oriented effort from the point of view of adaptation to the backgrounds mentioned is of very high value.

6 Conclusions and Further Work

The "reverse clustering" approach is very useful in obtaining a complete representation of the data analysed. The clusters obtained demonstrate that people with different backgrounds that can be attributed to definite various characteristics have different expectations, according to their mind-set, local or community traditions and cultural profile of the country they come from.

Considering that in the near future many human activities will radically change (due to the coronavirus pandemic and the technological revolution), some jobs will disappear and new ones will be created, the educational system should be redesigned in order to satisfy the needs of currently generations of students. If the teaching and evaluation process will continue to be held online (using Zoom, Google Meet and other similar solutions), many aspects of the educational process should be reconsidered in order to preserve the quality and face the challenges of webinars, tutorials and other resources available on the internet.

References

[AVR20] Arul Vallarasi, S., Regi, S.: Effective virtual platforms for online meeting. High Technol. Lett. 26(7), 1034–1041 (2020). http://www.gjstx-e.cn/gallery/102-july2020.pdf

[CO19] Ciurea, C., Owsiński, J.W.: Consensus versus crowdsourcing in collaborative decision-making applied in cultural institutions. Inf. Technol. Quant. Manag. Procedia Comput. Sci. 162, 547–554 (2019)

[OCF20] Owsiński, J.W., Ciurea, C., Filip, F.: Mobilising students to forward thinking of their studies – preliminary results of a crowdsourcing effort. In: Proceedings of the IE 2020 International Conference (2020). www.conferenceie.ase.ro

[OKO17] Owsiński, J.W., Kacprzyk, J., Opara, K., Stańczak, J., Zadrożny, S.: Using a reverse engineering type paradigm in clustering: An evolutionary programming based approach. In: Torra, W.V., Dalbom, A., Narukawa, Y. (eds.) Fuzzy Sets, Rough Sets, Multisets and Clustering. Dedicated to Prof. Sadaaki Miyamoto. Studies in Computational Intelligence, p. 671, Springer (2017)

[OOS17] Owsiński, J.W., Opara, K., Stańczak, J., Kacprzyk, J., Zadrożny, S.: Reverse clustering. an outline for a concept and its use. Toxicol. Environ. Chem. **99**(2017)

[OSZ18] Owsiński, J.W., Stańczak, J., Zadrożny, S.: Designing the municipality typology for planning purposes: the use of reverse clustering and evolutionary algorithms. In: Daniele, P., Scrimali, L. (eds.) New Trends in Emerging Complex Real Life Problems. ASS, vol. 1, pp. 399–406. Springer, Cham (2018). https://doi.org/10.1007/978-3-030-00473-6_42

[OSO21] Owsiński, J.W., Stańczak, J., Opara, K., Zadrożny, S., Kacprzyk, J.: Reverse Clustering: Formulation, Interpretation and Case Studies. Studies in Computation Intelligence 957. Springer International Publishing, Cham (2021). https://doi.org/10.1007/978-3-030-69359-6

[Sta03] Stańczak, J.: Biologically inspired methods for control of evolutionary algorithms. Control Cybern. **32**(2), 411–433 (2003)

[SBF09] Suduc, A.M., Bîzoi, M., Filip, F.G.: Exploring multimedia web conferencing. Inform. Econ. **13**(3), 5–17 (2009)

[WRJ92] Weitz Rob, R., Jelassi, M.T.: Assigning students to groups: a multi-criteria decision support system approach. Decis. Sci. **23**(3), 1992, 746–757 (1992)

Generalized Net Model of Library Activities Using Intuitionistic Fuzzy Estimations

Gabriela Hristova and Veselina Bureva(✉)

"Prof. Asen Zlatarov" University, "Prof. Yakimov" Blvd, 8010 Burgas, Bulgaria
vesito_ka@abv.bg

Abstract. In the current research work a Generalized net (GN) model of the library processes is discussed. The simulation of the GN model presents the stages of the work activities in library. The proposed Generalized net model of library can be used for monitoring the working processes. In the last step of the GN model intuitionistic fuzzy evaluations for book and intuitionistic fuzzy estimations of the library are presented. They will help to analyze the trends in renting books in the library.

Keywords: Generalized net · Intuitionistic fuzzy estimation · Library

1 Introduction

The library is an organization providing many activities - collects, stores, organizes, completes, catalogs, serves readers and provides free use of its library and information resources. The main tasks of the library are: rendering assistance to the Ministry of Culture in the implementation of the state policy in the field of library and information services on the territory of the country; performing library and information services quality and interlibrary loan; creating and strengthening reading habits from an early age through the use of various forms of work; stimulating the imagination and creativity of children and young people through cultural events; support for self-education as well as formal learning at all levels through programs initiated in the library; increasing the information, computer and language literacy of library users through training courses and consultations; providing readers of the library with free access to the Internet; development of the processes of automation of the library activity, coordination of the construction of a regional automated library network; participation in projects of the Ministry of Culture; replenishment of the book fund; processing and arranging in the fund of the newly purchased library documents; arranging the book depository with many copies and more valuable editions for their better preservation; performs reference and bibliographic activity, information and publishing activity; creates electronic library databases and provides access to national information centers; provides access to its catalogs and provides remote services to its users; carries out editorial and publishing activity: develops, publishes and distributes bibliographic indexes, local lore and information editions, advertising editions, researches and articles; maintains partnerships with libraries, institutions and foundations in the country and abroad. Librarian

K. T. Atanassov et al. (Eds.): IWIFSGN 2020/BOS/SOR 2020, LNNS 338, pp. 381–396, 2022.
https://doi.org/10.1007/978-3-030-95929-6_29

science is a huge research area with very many approaches, hundreds or thousands of papers, and very many conferences [13].

2 Generalized Net Model of Library Evaluation Using Intuitionistic Fuzzy Estimations

Generalized nets are introduced in 1982 by Prof. Krassimir Atanassov. There are extensions of Petri nets as continuation of the notation of modeling the parallel processes. Generalized nets are defined in different and more complex form. Generalized net model is constructed by transitions and relations. The relations are presented by tokens in places which move its position every time when the transition is activated. The tokens enter the places of the transitions with initial characteristics. In the time of their moving between the transitions in the generalized net the tokens receive new characteristics. The movement of the tokens in the generalized net forms their history. This is the first difference between the Generalized nets and Petri nets. Each place of transition can have one input and one output arc. The places that have not entering arc are called input places. The places that not have leaving arc are presented as output places. In the graphical form of Generalized net the input places are added in the left of the transition and the output places are visualized in the right of the transition. The transition is potentially fireable when the token enters in its input place. The transition is fired when all the tokens from input places transfer towards output places. The second difference between Generalized Nets and Petri nets is the "place-transition" relation. A transition in Generalized net has more complex from and it may contain m input places and n output places where $m, n \geq 1$. The third difference between Generalized Nets and Petri nets is the index matrix which is used for describing the transition conditions. The time of functioning is the fourth difference between Generalized nets and all types of Petri nets. The time in Generalized nets can be assigned. The definition of Generalized nets is based on the discrete time scale: in each moment the user can see the status of Generaized net. The Generalized net theory contains algorithms for places movements and transition functioning [4–6].

GN models in different scientific fields are already published: medicine, artificial intelligence, transport, telecommunications, data mining, neural networls [1–3, 8, 12] and etc. The processes of university are modeled using Generalized nets in [9–11].The Generalized Net Model Generalized Net Model of the Library processes is made using GNDraw software [7]. It contains 13 transitions and 59 places (Fig. 1). The set of transitions A has the following form:

$$A = \{Z_1, Z_2, Z_3, Z_4, Z_5, Z_6, Z_7, Z_8, Z_9, Z_{10}, Z_{11}, Z_{12}, Z_{13}\},$$

where the transitions describe the following processes:

- Z_1 - "librarians";
- Z_2 - "readers";
- Z_3 - "returning a book";
- Z_4 - "reading a book in the library".
- Z_5 - "renting a book";
- Z_6 - "making a request for book";

- Z_7 - "getting a book from books warehouse";
- Z_8 - "scanning a book in the library";
- Z_9 - "recording readers activity";
- Z_{10} - "intuitionistic fuzzy estimations of the books in the library";
- Z_{11} - "inventory of documents";
- Z_{12} - "dropping of documents";
- Z_{13} - "registration of new documents";

Initially in the place l_8 there is one token. It will be in its own place during the entire time the GN is functioning. It has the following characteristic:

"Librarians".

The token in place L_8 generates new tokens at certain points in time which can move to places l_2, l_3, l_4, l_5, l_6, l_7, l_{14}, l_{47}, l_{52} and l_{56} with the characteristics:

"Selected librarian for returning a book"

in place l_2,

"Selected librarian for reading a book in the library"

in place l_3,

"Selected librarian for renting a book"

in place l_4,

"Selected librarian for making a request for book"

in place l_5,

"Selected librarian for scanning a book in the library"

in place l_6,

"Selected librarian for recording readers activity"

in place l_7,

"Selected librarian providing reader service"

in place l_{14},

"Selected librarian for making inventory of documents"

in place l_{47},

"Selected librarian for dropping of documents"

in place l_{52},

"Selected librarian for registration of new documents"

in place l_{56}.

Initially in the place l_{34} there is one token. It will be in its own place during the entire time the GN is functioning. It has the following characteristic:

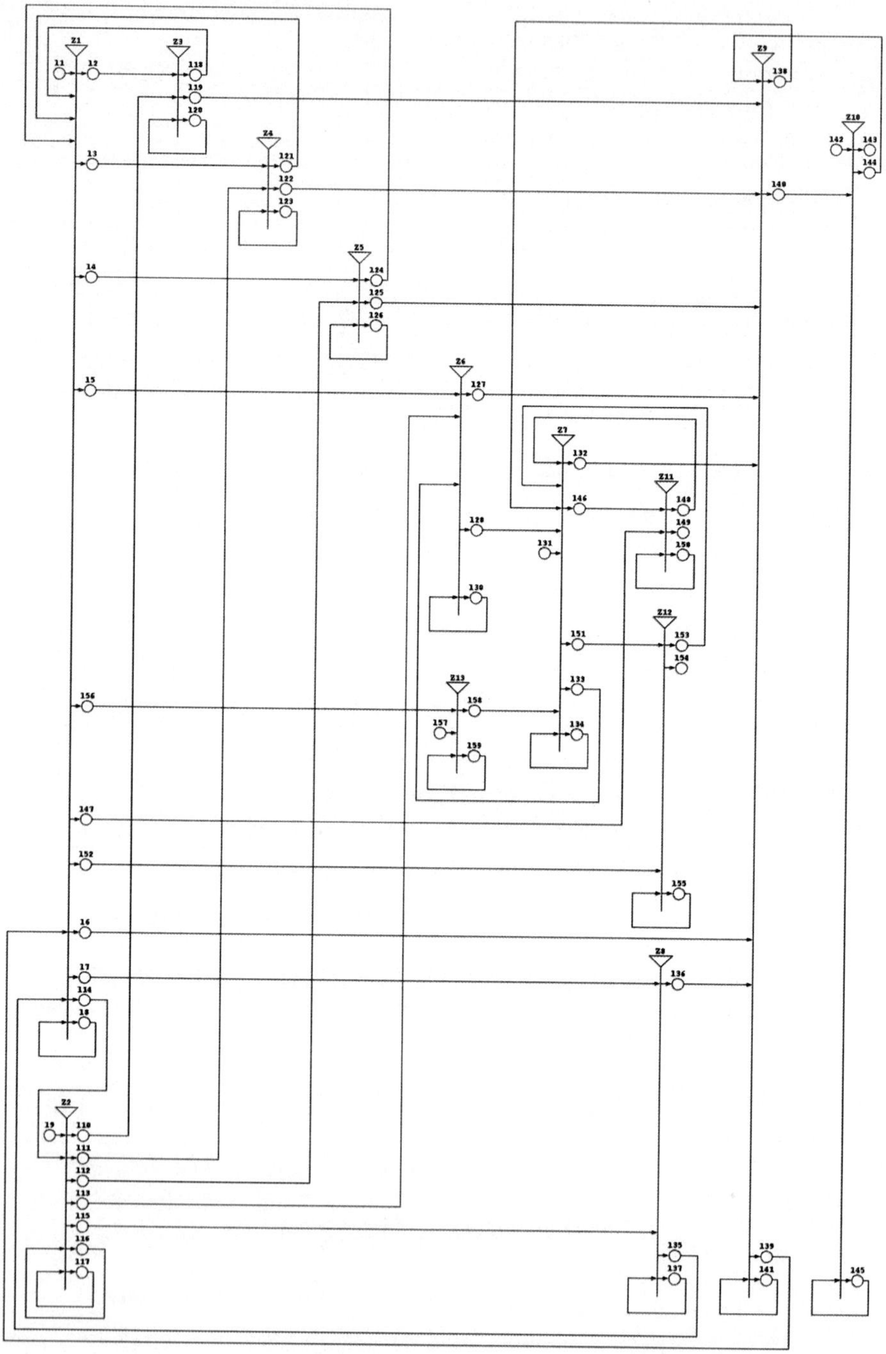

Fig. 1. Generalized net model of library evaluation using intuitionistic fuzzy estimations

"*Books warehouse*".

The token in place L_{34} generates new tokens at certain points in time which can move to places l_{32}, l_{46}, l_{51} and l_{33} with the characteristics:

"*Book from storage for reader request for recording*"
in place l_{32},
"*Book from storage for reader request*"
in place l_{33},
"*Books for inventory*"
in place l_{46},
"*Book for dropping*"
in place l_{51}.

Token enter the net via place l_1 with initial characteristic: "*librarian*".
The transition Z_1 has the form:

$$Z_1 = \langle \{l_1, l_8, l_{18}, l_{21}, l_{24}, l_{35}, l_{39}\}, \{l_2, l_3, l_4, l_5, l_6, l_7, l_8, l_{14}, l_{47}, l_{52}, l_{56}\}, R_1, \vee(l_1, l_8, l_{18}, l_{21}, l_{24}, l_{35}, l_{39})\rangle,$$

where

$R_1 =$

	l_2	l_3	l_4	l_5	l_6	l_7	l_8	l_{14}	l_{47}	l_{52}	l_{56}
l_1	*false*	*false*	*false*	*false*	*false*	*false*	*true*	*false*	*false*	*false*	*false*
l_8	$W_{8,2}$	$W_{8,3}$	$W_{8,4}$	$W_{8,5}$	$W_{8,6}$	$W_{8,7}$	$W_{8,8}$	$W_{8,14}$	$W_{8,47}$	$W_{8,52}$	$W_{8,56}$
l_{18}	*false*	*false*	*false*	*false*	*false*	*false*	*true*	*false*	*false*	*false*	*false*
l_{21}	*false*	*false*	*false*	*false*	*false*	*false*	*true*	*false*	*false*	*false*	*false*
l_{24}	*false*	*false*	*false*	*false*	*false*	*false*	*true*	*false*	*false*	*false*	*false*
l_{35}	*false*	*false*	*false*	*false*	*false*	*false*	*true*	*false*	*false*	*false*	*false*
l_{39}	*false*	*false*	*false*	*false*	*false*	*false*	*true*	*false*	*false*	*false*	*false*

and

- $W_{8,2}$ = "there is selected librarian for returning a book";
- $W_{8,3}$ = "there is selected librarian for reading a book in the library";
- $W_{8,4}$ = "there is selected librarian for making a request for book";
- $W_{8,5}$ = "there is selected librarian for renting a book";
- $W_{8,6}$ = "there is selected librarian for scanning a book in the library";
- $W_{8,7}$ = "there is selected librarian for recording readers activity";
- $W_{8,14}$ = "there is a librarian for reader card verification";
- $W_{8,47}$ = "there is a librarian for inventory of documents";
- $W_{8,52}$ = "there is a librarian for dropping of documents";
- $W_{8,56}$ = "there is a librarian for registration of new documents";

- $W_{8,8} = \neg(W_{8,2} \wedge W_{8,3} \wedge W_{8,4} \wedge W_{8,5} \wedge W_{8,6} \wedge W_{8,7} \wedge W_{8,14} \wedge W_{8,47} \wedge W_{8,52} \wedge W_{8,56})$.

The token in place L_8 generates new tokens at certain points in time which can move to places $l_2, l_3, l_4, l_5, l_6, l_7, l_{14}, l_{47}, l_{52}$ and l_{56} with the characteristics:

"*Selected librarian for returning a book*"
in place l_2,
"*Selected librarian for reading a book in the library*"
in place l_3,
"*Selected librarian for renting a book*"
in place l_4,
"*Selected librarian for making a request for book*"
in place l_5,
"*Selected librarian for scanning a book in the library*"
in place l_6,
"*Selected librarian for recording readers activity*"
in place l_7,
"*Selected librarian providing reader service*"
in place l_{14},
"*Selected librarian for making inventory of documents*"
in place l_{47},
"*Selected librarian for dropping of documents*"
in place l_{52},
"*Selected librarian for registration of new documents*"
in place l_{56}.

Token enter the net via place l_9 with initial characteristic: "*Reader*".
The transition Z_2 has the form:

$$Z_2 = \langle\{l_9, l_{14}, l_{16}, l_{17}\}, \{l_{10}, l_{11}, l_{12}, l_{13}, l_{15}, l_{16}, l_{17}\}, R_2, \vee(\wedge(l_9, l_{14}, l_{17}), l_{16})\rangle,$$

where

$R_2 =$

	l_{10}	l_{11}	l_{12}	l_{13}	l_{15}	l_{16}	l_{17}
l_9	*false*	*false*	*false*	*false*	*false*	*false*	*true*
l_{14}	*false*	*false*	*false*	*false*	*false*	*false*	*true*
l_{16}	$W_{16,10}$	$W_{16,11}$	$W_{16,12}$	$W_{16,13}$	$W_{16,15}$	$W_{16,16}$	*false*
l_{17}	$W_{17,10}$	$W_{17,11}$	$W_{17,12}$	$W_{17,13}$	$W_{17,15}$	$W_{17,16}$	$W_{17,17}$

and

- $W_{17,10} =$ "there is a reader for returning a book";
- $W_{17,11} =$ "there is a reader for reading a book in the library";
- $W_{17,12} =$ "there is a reader for renting a book";

- $W_{17,13}$ = "there is a reader for making a request for book";
- $W_{17,15}$ = "there is a reader for scanning a book in the library";
- $W_{17,16}$ = "there is a reader with card verification issues";
- $W_{17,17} = \neg(W_{17,10} \wedge W_{17,11} \wedge W_{17,12} \wedge W_{17,13} \wedge W_{17,15} \wedge W_{17,16})$;
- $W_{16,10}$ = "there is a reader with fixed card verification issues for returning a book";
- $W_{16,11}$ = "there is a reader with fixed card verification issues for reading a book in the library";
- $W_{16,12}$ = "there is a reader with fixed card verification issues for renting a book";
- $W_{16,13}$ = "there is a reader with fixed card verification issues for making a request for book";
- $W_{16,15}$ = "there is a reader with fixed card verification issues for scanning a book in the library";
- $W_{16,16} = \neg(W_{16,10} \wedge W_{16,11} \wedge W_{16,12} \wedge W_{16,13} \wedge W_{16,15})$.

The tokens in place L_{17} generate new tokens that enter in places $l_{10}, l_{11}, l_{12}, l_{13}, l_{15}$ and l_{16} with the characteristics:

"*Reader for returning a book*"

in place l_{10},

"*Reader for reading a book in the library*"

in place l_{11},

"*Reader for renting a book*"

in place l_{12},

"*Reader for making a request for book*"

in place l_{13},

"*Reader for scanning a book in the library*"

in place l_{15},

"*Reader with card verification issues*"

in place l_{16}.

At the second activation of the transition the token from place L_{16} generates new tokens that enter the places $l_{10}, l_{11}, l_{12}, l_{13}$ and l_{15} with the characteristics:

"*Reader with fixed card verification issues for returning a book*"

in place l_{10},

"*Reader with fixed card verification issues for reading a book in the library*"

in place l_{11},

"*Reader with fixed card verification issues for renting a book*"

in place l_{12},

"*Reader with fixed card verification issues for making a request for book*"

in place l_{13},

"*Reader with fixed card verification issues for scanning a book in the library*"

in place l_{15}.

The transition Z_3 has the form:

$$Z_3 = \langle\{l_2, l_{10}, l_{20}\}, \{l_{18}, l_{19}, l_{20}\}, R_3, \vee(\wedge(l_2, l_{10}), l_{20})\rangle,$$

where

$$R_3 = \begin{array}{c|ccc} & l_{18} & l_{19} & l_{20} \\ \hline l_2 & false & false & true \\ l_{10} & false & false & true \\ l_{20} & W_{20,18} & W_{20,19} & W_{20,20} \end{array}$$

and

- $W_{20,18}$ = "there is reader that returned book to the librarian";
- $W_{20,19}$ = "there is readers activity of returned book to the librarian for recording";
- $W_{20,20} = \neg(W_{20,18} \wedge W_{20,19})$.

The tokens in place L_{20} generate new tokens that enter in places l_{18} and l_{19} with the characteristics:

"*Returned book from the reader to the librarian*"

in place l_{18},

"*Readers activity of returned book from the reader to the librarian for recording*"

in place l_{19}.

The transition Z_4 has the form:

$$Z_4 = \langle\{l_3, l_{11}, l_{23}\}, \{l_{21}, l_{22}, l_{23}\}, R_4, \vee(\wedge(l_3, l_{11}), l_{23})\rangle,$$

where

$$R_4 = \begin{array}{c|ccc} & l_{21} & l_{22} & l_{23} \\ \hline l_3 & false & false & true \\ l_{11} & false & false & true \\ l_{23} & W_{23,21} & W_{23,22} & W_{23,23} \end{array}$$

and

- $W_{23,21}$ = "there is reader for reading book in library";
- $W_{23,22}$ = "there is the readers activity of reading a book in library for recording";
- $W_{23,23} = \neg(W_{23,21} \wedge W_{23,22})$.

The tokens in place L_{23} generate new tokens that enter in places l_{21} and l_{22} with the characteristics:

"*Serviced by the librarian reader for reading a book*"

in place l_{21},

"*Reader activity of reading a book for recording*"

in place l_{22}.

The transition Z_5 has the form:

$$Z_5 = \langle\{l_4, l_{12}, l_{26}\}, \{l_{24}, l_{25}, l_{26}\}, R_5, \vee(\wedge(l_4, l_{12}), l_{26})\rangle,$$

where

$$R_5 = \begin{array}{c|ccc} & l_{24} & l_{25} & l_{26} \\ \hline l_4 & false & false & true \\ l_{12} & false & false & true \\ l_{26} & W_{26,24} & W_{26,25} & W_{26,26} \end{array}$$

and

- $W_{26,24} =$ "there is serviced by the librarian renting a book reader";
- $W_{26,25} =$ "there is the readers activity of renting a book for recording";
- $W_{26,26} = \neg(W_{26,24} \wedge W_{26,25})$.

The tokens in place L_{26} generate new tokens that enter in places l_{24} and l_{25} with the characteristics:

"*Serviced by the librarian reader for renting a book*"

in place l_{24},

"*Record of the readers activity for renting a book in the library*"

in place l_{25}.

The transition Z_6 has the form:

$$Z_6 = \langle\{l_5, l_{13}, l_{30}, l_{33}\}, \{l_{27}, l_{28}, l_{30}\}, R_6, \vee(\wedge(\wedge(l_5, l_{13}), l_{33}), l_{30})\rangle,$$

where

$$R_6 = \begin{array}{c|ccc} & l_{27} & l_{28} & l_{30} \\ \hline l_5 & false & false & true \\ l_{13} & false & false & true \\ l_{30} & W_{30,27} & W_{30,28} & W_{30,30} \\ l_{33} & false & false & true \end{array}$$

and

- $W_{30,27}$ = "there is serviced by the librarian reader that makes request for book";
- $W_{30,28}$ = "there is the readers activity of making request for recording";
- $W_{30,30} = \neg(W_{30,27} \wedge W_{30,28})$.

The tokens in place L_{30} generate new tokens that enter in places l_{27} and l_{28} with the characteristics:

"*Serviced by the librarian reader making request for book*"

in place l_{27},

"*Readers activity of making request for recording*"

in place l_{28}.

Token enter the net via place l_{31} with initial characteristic: "*storage resources*".
The transition Z_7 has the form:

$$Z_7 = \langle\{l_{28}, l_{31}, l_{34}, l_{38}, l_{48}, l_{53}, l_{58}\}, \{l_{32}, l_{33}, l_{34}, l_{46}, l_{51}\}, R_7, \vee(l_{28}, l_{31}, l_{34}, l_{48}, l_{53}, l_{58})\rangle,$$

where

$R_7 =$	l_{32}	l_{33}	l_{34}	l_{46}	l_{51}
l_{28}	*false*	*false*	*true*	*false*	*false*
l_{31}	*false*	*false*	*true*	*false*	*false*
l_{34}	$W_{34,32}$	$W_{34,33}$	$W_{34,34}$	$W_{34,46}$	$W_{34,51}$
l_{38}	*false*	*false*	*true*	*false*	*false*
l_{48}	*false*	*false*	*true*	*false*	*false*
l_{53}	*false*	*false*	*true*	*false*	*false*
l_{58}	*false*	*false*	*true*	*false*	*false*

and

- $W_{34,32}$ = "there is the readers activity for getting a book from storage for recording";
- $W_{34,33}$ = "there is a book from storage for reader request";
- $W_{34,46}$ = "there are books for inventory";
- $W_{34,51}$ = "there are books for dropping";
- $W_{34,34} = \neg(W_{34,32} \wedge W_{34,33} \wedge W_{34,46} \wedge W_{34,51})$.

The tokens in place L_{34} generate new tokens that enter in places l_{32}, l_{33}, l_{46} and l_{51} with the characteristics:

"Book from storage for reader request for recording"

in place l_{32},

"Book from storage for reader request"

in place l_{33},

"Books for inventory"

in place l_{46},

"Books for dropping"

in place l_{51}.

The transition Z_8 has the form:

$$Z_8 = \langle\{l_7, l_{15}, l_{37}\}, \{l_{35}, l_{36}, l_{37}\}, R_8, \vee(\wedge(l_7, l_{15}), l_{37})\rangle,$$

where

$$R_8 = \begin{array}{c|ccc} & l_{35} & l_{36} & l_{37} \\ \hline l_7 & false & false & true \\ l_{15} & false & false & true \\ l_{37} & W_{37,35} & W_{37,36} & W_{37,37} \end{array}$$

and

- $W_{37,35}$ = "there is a librarian, serviced reader for scanning a book in the library";
- $W_{37,36}$ = "there is a reader that scan a book for recording";
- $W_{37,37} = \neg(W_{37,35} \wedge W_{37,36})$.

The tokens in place L_{37} generate new tokens that enter in places l_{35} and l_{36} with the characteristics:

"Librarian, serviced reader for scanning a book in the library"

in place l_{32},

"Reader that scan a book for recording"

in place l_{33}.

The transition Z_9 has the form:

$$Z_9 = \langle\{l_6, l_{19}, l_{22}, l_{25}, l_{27}, l_{32}, l_{36}, l_{41}, l_{44}\}, \{l_{38}, l_{39}, l_{40}, l_{41}\}, R_9, \vee(\wedge(l_6, l_{19}, l_{22}, l_{25}, l_{27}, l_{32}, l_{36}), l_{41}, l_{44})\rangle,$$

where

$$R_9 = \begin{array}{c|cccc} & l_{38} & l_{39} & l_{40} & l_{41} \\ \hline l_6 & false & false & false & true \\ l_{19} & false & false & false & true \\ l_{22} & false & false & false & true \\ l_{25} & false & false & false & true \\ l_{27} & false & false & false & true \\ l_{32} & false & false & false & true \\ l_{36} & false & false & false & true \\ l_{41} & W_{41,38} & W_{41,39} & W_{41,40} & W_{41,41} \\ l_{44} & false & false & false & true \end{array}$$

and

- $W_{41,38}$ = "there are books that are dropped out from the library catalogue according its intuitionistic fuzzy estimation";
- $W_{41,39}$ = "there is a librarian";
- $W_{41,40}$ = "there is recorded readers activity";
- $W_{41,41} = \neg(W_{41,38} \wedge W_{41,39} \wedge W_{41,40})$.

The tokens in place L_{41} generate new tokens that enter in places l_{38}, l_{39} and l_{40} with the characteristics:

"*Books that are dropped out from the library catalogue according its intuitionistic fuzzy estimation*"

in place l_{38},

"*Librarian*"

in place l_{39},

"*Recorded readers activity*"

in place l_{40}.

The transition Z_{10} has the form:

$$Z_{10} = \langle\{l_{40}, l_{42}, l_{45}\}, \{l_{43}, l_{44}, l_{45}\}, R_{10}, \vee(\wedge(l_{40}, l_{42}), l_{45})\rangle,$$

where

$$R_{10} = \begin{array}{c|ccc} & l_{43} & l_{44} & l_{45} \\ \hline l_{40} & false & false & true \\ l_{42} & false & false & true \\ l_{45} & W_{45,43} & W_{45,44} & W_{45,45} \end{array}$$

and

- $W_{45,43}$ = "there are intuitionistic fuzzy estimations";
- $W_{45,44}$ = "there are intuitionistic fuzzy estimations for storing";
- $W_{45,45} = \neg(W_{45,43} \wedge W_{45,44})$.

The tokens in place L_{45} generate new tokens that enter in places l_{43} and l_{44} with the characteristics:

"*Intuitionistic fuzzy estimations*"

in place l_{43},

"*Intuitionistic fuzzy estimations for storing*"

in place l_{44}.

The activity of the library is estimated by two types of intuitionistic fuzzy evaluations. The intuitionistic fuzzy evaluation has the form of $\langle a, b \rangle$ where $a + b \leq 1$. The first type of evaluations investigates a book in the library. The second type of estimations determines the library books activity. Let we have the following notations:

– *First case* – intuitionistic fuzzy estimation of book $\langle \mu_b^t, \nu_b^t \rangle$:

- Number of months in which the book is rented – n;
- Number of months in which the book is the library (the book is not rented) – l;
- Number of months in which the return of a book is overdue – r;
- Number of months in which the books belongs to the library – t.

The degree of membership and the degree of non-membership have the following form:

$$\mu_b^t = \frac{n}{t}, \nu_b^t = \frac{l}{t}. \tag{1}$$

The degree of uncertainty is:

$$\pi_b^t = \frac{r}{t}. \tag{2}$$

where T is non-empty set t is a given time moment: $t \in T$.

– *Second case* – intuitionistic fuzzy estimation of the library $\langle \mu_L^t, \nu_L^t \rangle$:

- Number of rented books – B;
- Number of the free books in the library – L;
- Number of books with overdue return – R;
- Total number of books in the library – T.

The degree of membership and the degree of non-membership have the following forms:

$$\mu_L^t = \frac{B}{T}, \nu_L^t = \frac{L}{T}. \tag{3}$$

The degree of uncertainty is:

$$\pi_L^t = \frac{R}{T}. \tag{4}$$

where T is non-empty set t is a given time moment: $t \in T$.

The transition Z_{11} has the form:

$$Z_{11} = \langle \{l_{46}, l_{47}, l_{50}\}, \{l_{48}, l_{49}, l_{50}\}, R_{11}, \vee(\wedge(l_{46}, l_{47}), l_{50})\rangle,$$

where

$R_{11} =$

	l_{48}	l_{49}	l_{50}
l_{46}	*false*	*false*	*true*
l_{47}	*false*	*false*	*true*
l_{50}	$W_{50,48}$	$W_{50,49}$	$W_{50,50}$

and

- $W_{50,48}$ = "there are documents and books for storing after inventory";
- $W_{50,49}$ = "there are dropped books after inventory";
- $W_{50,50} = \neg(W_{50,48} \wedge W_{50,49})$.

The tokens in place l_{50} generate new tokens that enter in places l_{48} and l_{49} with the characteristics:

"*Documents and books for storing after inventory*"

in place l_{48},

"*Dropped books after inventory*"

in place l_{49}.

The transition Z_{12} has the form:

$$Z_{12} = \langle \{l_{51}, l_{52}, l_{55}\}, \{l_{53}, l_{54}, l_{55}\}, R_{12}, \vee(\wedge(l_{51}, l_{52}), l_{55})\rangle,$$

where

$R_{12} =$

	l_{53}	l_{54}	l_{55}
l_{51}	*false*	*false*	*true*
l_{52}	*false*	*false*	*true*
l_{55}	$W_{55,53}$	$W_{55,54}$	$W_{55,55}$

and

- $W_{55,53}$ = "there are documents and books for storing after dropping stage";

- $W_{55,54}$ = "there are dropped books after dropping stage";
- $W_{55,55} = \neg(W_{55,53} \wedge W_{5,54})$.

The tokens in place L_{55} generate new tokens that enter in places l_{53} and l_{54} with the characteristics:

"*Documents and books for storing after dropping stage*"

in place l_{53},

"*Dropped books after dropped stage*"

in place l_{54}.

Token enter the net via place l_{27} with initial characteristic: "*new books and documents*".

The transition Z_{13} has the form:

$$Z_{13} = \langle\{l_{56}, l_{57}, l_{59}\}, \{l_{58}, l_{59}\}, R_{13}, \vee(\wedge(l_{56}, l_{57}), l_{59})\rangle,$$

where

$$R_{13} = \begin{array}{c|cc} & l_{58} & l_{59} \\ \hline l_{56} & false & true \\ l_{57} & false & true \\ l_{59} & W_{59,58} & W_{59,59} \end{array}$$

and

- $W_{59,58}$ = "there are documents and books for storing in the library";
- $W_{59,59} = \neg W_{59,58}$.

The tokens in place L_{59} generate new token that enters in place l_{58} with the characteristic:

"*Documents and books for storing in the library*"

in place l_{53}.

3 Conclusion

In the current research work a Generalized net model with intuitionistic fuzzy evaluations of the library activities is constructed. The processes performed in the library by the librarians and readers are discussed. The transitions for inventory, dropping books and registration of new books can be replaced with subnets.

Acknowledgments. The authors are grateful for the support provided by the project "*New Instruments for Knowledge Discovery from Data, and Their Modelling*" funded by the Bulgarian National Science Fund (no. DN-02-10/2016).

References

1. Atanassov, K.: Generalized nets as a tool for the modelling of data mining processes. In: Sgurev, V., Yager, R.R., Kacprzyk, J., Jotsov, V. (eds.) Innovative Issues in Intelligent Systems. SCI, vol. 623, pp. 161–215. Springer, Cham (2016). https://doi.org/10.1007/978-3-319-27267-2_6
2. Atanassov, K.T., et al. (eds.): Advances and New Developments in Fuzzy Logic and Technology, vol. 1308. Springer, Cham (2021). https://doi.org/10.1007/978-3-030-77716-6
3. Atanassov, K.: Generalized Nets and Intuitionistic Fuzziness in Data Mining. Prof. Marin Drinov Publishing House (2021)
4. Atanassov, K., Sotirova, E.: Generalized Nets. Prof. M. Drinov Academic Publishing House, Sofia (2017). (in Bulgarian)
5. Atanassov, K.: Generalized Nets, p. 1. World Scientific, Singapore/London (1991)
6. Atanassov, K.: On Generalized Nets Theory. "Prof. Marin Drinov" Academic Publishing House, Sofia (2007)
7. Ikonomov, N.: GNDraw – software application for creating generalized nets. Issues Intuitionistic Fuzzy Sets Generalized Nets **13**, 61–71 (2017)
8. Krawczak, M.: Multilayer Neural Networks: A Generalized Net Perspective. Studies in Computational Intelligence, vol. 478. Springer, Heidelberg (2013). https://doi.org/10.1007/978-3-319-00248-4
9. Shannon, A., et al.: Towards a model of the digital university: a generalized net model for producing course timetables and for evaluating the quality of subjects. In: Sgurev, V., Hadjiski, M., Kacprzyk, J. (eds.) Intelligent Systems: From Theory to Practice. SCI, vol. 299, pp. 373–381. Springer, Heidelberg (2010). https://doi.org/10.1007/978-3-642-13428-9_17
10. Shannon, A., et al.: Generalized Net Modelling of University Processes. KvB Institute of Technology (2005)
11. Shannon, A., et al.: Generalized net model of the process of applying to university. In: 12th International Workshop on Generalized Nets Burgas, pp. 27–31 (2012)
12. Zoteva, D., Angelova, N.: Generalized nets. An overview of the main results and applications. In: Atanassov, K.T. (ed.) Research in Computer Science in the Bulgarian Academy of Sciences. SCI, vol. 934, pp. 177–226. Springer, Cham (2021). https://doi.org/10.1007/978-3-030-72284-5_10
13. Wikipedia, Library science. https://en.wikipedia.org/wiki/Library_science. Accessed 5 Aug 2021

Generalized Net Model of Computer Based Registration and Rehabilitation of Cognitive Impairments in Multiple Sclerosis

Lyudmila Todorova[1](✉), Valentina Ignatova[2], Peter Vassilev[1], and Jivko Surchev[3]

[1] Department of Bioinformatics and Mathematical Modelling, Institute of Biophysics and Biomedical Engineering, Bulgarian Academy of Sciences, Acad. G. Bonchev Street, bl. 105, 1113 Sofia, Bulgaria
lpt@biomed.bas.bg

[2] Clinic of Neurology Multiprofile Hospital for Active Treatment, National Heart Hospital, Konyovitsa Street, 65, 1309 Sofia, Bulgaria
valyaig@abv.bg

[3] Department of Neurosurgery, St Ivan Rilski University Hospital, Medical University of Sofia, Akad. Ivan Geshov Blvd., 15, 1431 Sofia, Bulgaria
j_surchev@abv.bg

Abstract. Cognitive decline occurs along with physical disability in approximately two-thirds of persons suffering from multiple sclerosis (MS) and has devastating influence on quality of life at all stages and in all subtypes of the disease. The modern immunomodulatory therapy has a satisfactory effect on the relapse rate and disease progression, but does not improve the degree of cognitive impairment. Over the last decade, the number of studies on treatment and management of cognitive impairment in people with MS are rapidly growing. A generalized net model for registration, assessment and rehabilitation of cognitive impairment in persons with MS has been developed. The assessment of cognitive status was made on the basis of results obtained through computer-based neuropsychological tests. Based on comparison of the results from MS patients with those of healthy controls, the degree of cognitive deficit was determined. The input of the generalized net was formed by the data from patient's medical records, which were categorized according to disease duration, grade of disability, age, sex and employment. As a result of applying the InterCriteria Analysis, the output of the model provides information about the type of impaired cognitive domains and the degree of cognitive deterioration. This allows the therapist to build an appropriate rehabilitation strategy for each individual patient. The rehabilitation module contains a set of different rehabilitation games with several levels of difficulty. The frequency, as well as the degree of complexity of cognitive training is based on the patient's results and the general cognitive reserve registered at baseline. The system is implemented in a cloud infrastructure and is planned to be provided as an IT service for neurological departments and clinics. In this way, the system is available for testing and rehabilitation of unlimited number of people (both healthy controls and persons with MS) at any time. The developed GN model provides a mechanism for description and simulation of these parallel processes. In this way, the proposed model will be useful in determining the individual approach to each patient and will help optimizing the results of cognitive rehabilitation. The

K. T. Atanassov et al. (Eds.): IWIFSGN 2020/BOS/SOR 2020, LNNS 338, pp. 397–407, 2022.
https://doi.org/10.1007/978-3-030-95929-6_30

analysis of the model will allow outlining the future directions for therapeutic approaches on cognitive impairments in MS, which will lead to inclusion of cognitive rehabilitation and training programs in routine clinical practice within the next decade.

Keywords: Multiple sclerosis · Cognitive impairment · Cognitive rehabilitation · Computer training · Generalized Nets · InterCriteria analysis

1 Introduction

Multiple sclerosis (MS) affects individuals of working age, damaging numerous neurological modalities, including cortical functions. The course of the disease is characterized by progressive neurodegeneration and brain atrophy (both cerebral and spinal involvement), which further aggravates the physical deficit and worsens the cognitive status. Cognitive decline occurs along with physical impairment in approximately two-thirds of the individuals with MS [1–6], and has a devastating effect on the quality of life at all stages and in all subtypes of the disease [7]. However, cognitive decline, which is difficult to detect during routine neurological examinations, is not yet the focus of clinical practice [7, 8].

Modern immunomodulatory therapy is widely used in the treatment of the disease and has a satisfactory effect on the frequency of relapses and progression, but the same time, it does not improve the degree of cognitive impairment. Detection of early changes in cognition requires targeted neuropsychological examination, which unfortunately is often missed in routine practice [9]. Neuropsychological status is currently assessed by a neurologist/neuropsychologist based on the results of tests performed by the patient. A paper version of the tests is usually used. This form of neuropsychological examination is time consuming, requires the involvement of a qualified specialist during the study, makes it difficult to archive and compare the patient's results. Until recently, testing of the main neuropsychological domains took about 90 min using modern neuropsychological batteries designed for MS (Brief Repeatable Battery of Neuropsychological tests [10] and Minimal Assessment of Cognitive Function in Multiple Sclerosis [11]).

In 2012, Benedict and co-authors introduced the Brief International Cognitive Assessment for MS (BICAMS), the implementation of which largely overcomes the aforementioned restrictions [12]. It concerns a short battery of tests (Symbol digit modalities test - SDMT [13], California Verbal Learning Test - CVLT2 [14], Brief Visuospatial Memory Test - BVMTR [15]) assessing the cognitive ability of people with multiple sclerosis, recommended by the International Consensus Commission of Experts. The performance of classic version of BICAMS takes 15 min and re-quires only paper, pencil and stopwatch [16].

Computer based cognitive batteries are increasingly used in world practice, which facilitate the integration of neuropsychological assessments into routine clinical practice in patients with multiple sclerosis and more subtly detect the presence of cognitive impairment [17, 18]. Recent studies demonstrate a better perception and preferably computer cognitive tests as compared to classical variants [19], without taking into account a significant difference in the results [20]. They allow the collection and statistical processing of large amount of information. In addition, subjectivism is avoided as a result of research by different neuropsychologists, uniform administration, data recording and evaluation procedures are ensured [21].

In line with global trends, a bilingual computer system has been developed for assessment and rehabilitation of cognitive status in people with MS CogniSoft, including:

- Beck Depression Inventory (BDI-II) [22];
- Diagnostic tests to examine memory and executive functions, adapted by BICAMS;
- Tests for cognitive rehabilitation.

Generalized Nets (GN) [23, 24] are a tool for modeling of parallel processes and are a proven approach for use in a wide range of areas [25–30] which is relevant and evolving to this day. The GNs are a generalization of Petri nets and other extension of theirs. In GNs the tokens enter the net with initial characteristics and during their movement they receive new characteristic gathering their history. Every transition is associated with an index matrix [25], with elements predicates which determine the direction of transition for each token – from input to output places. As an evidence of the suitability of the GNs as a modelling tool for biological and medical processes one can refer, for instance, to the review and the bibliography of Jürden Will and Monika Heiner [31], where more than 30 GN-models of various such processes are cited.

That is why a Generalized Nets model for registration, assessment and rehabilitation of cognitive impairment in people with MS has been developed. The assessment of cognitive status was made on the basis of results obtained through computer-based neuropsychological tests of the computer system MS CogniSoft through application of InterCriteria Analysis [32, 33].

The system is implemented in a cloud infrastructure and is planned to be provided as an IT service for neurological departments and clinics. In this way, the system is available for testing and rehabilitation of unlimited number of people (both healthy controls and persons with MS) at any time. The developed GN model provides a mechanism for description and simulation of these parallel processes.

2 GN Model

The input of the Generalized Net is formed by the patient's data extracted from the medical records. The model is composed by 10 transitions and 27 places (Fig. 1).

The first characteristics of token α in place l_1 are:

"patient's initials, demographic data, data from electronic record of the subject"

Transition z_1 of the GN-model has the following form:

$$z_1 = \langle \{l_1, l_3, l_{27}\}, \{l_2\}, r_1, \wedge(\wedge(l_1), \vee(l_3, l_{27}))\rangle$$

where the index matrix has the form.

$r_1 =$

	l_2
l_1	true
l_3	true
l_{27}	true

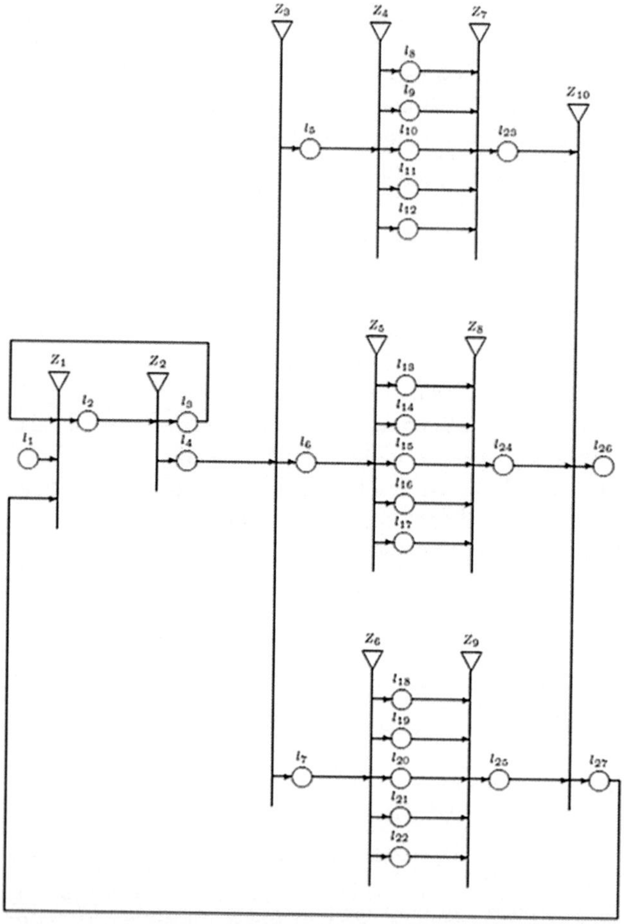

Fig. 1. The Generalized Net model for registration, assessment and rehabilitation of cognitive impairment in persons with MS.

Token α gets characteristic "results from BDI" in place l_2.
Transition z_2 of the GN-model has the following form:

$$z_2 = \langle\{l_2\}, \{l_3, l_4\}, r_2, \wedge(l_2)\rangle$$

where the index matrix has the form.

	l_3	l_4
$r_2 =$ l_2	$w_{3,2}$	$w_{4,2}$

$w_{2,3}$ = "the result from BDI test is ≥31".
$w_{2,4}$ = "the result from BDI test is <31".

Token α from place l_2 moves to one of the following places l_3 or l_4, where it obtains characteristic.

"Patient experience severe depression and the cognitive testing is not appropriate at the present moment" in place l_3;

or

"the patient is not depressed at the current moment and the cognitive testing could be performed" at place l_4.

Transition z_3 of the GN-model has the following form:

$$z_3 = \langle \{l_4\}, \{l_5, l_6, l_7\}, r_3, \wedge(l_4)\rangle$$

where the index matrix has the form.

	l_5	l_6	l_7
$r_3 =$ l_4	true	true	true

Token α from place l_4 splits into three tokens, which receive the following characteristics:

"result form SDMT test" at place l_5;
"result form BMVT test" at place l_6;
"result form CVMT test" at place l_7.

Transition z_4 of the GN-model has the following form:

$$z_4 = \langle \{l_5\}, \{l_8, l_9, l_{10}, l_{11}, l_{12}\}, r_4, \wedge(l_5)\rangle$$

where the index matrix has the form.

	l_8	l_9	l_{10}	l_{11}	l_{12}
$r_4 =$ l_5	$w_{5,8}$	$w_{5,9}$	$w_{5,10}$	$w_{5,11}$	$w_{5,12}$

$w_{5,8}$ = "the result from SDMT test is ≥ (norm)"
$w_{5,9}$ = "the result from SDMT test is (value of mild cognitive impairment)"
$w_{5,10}$ = "the result from SDMT test is (value of moderate cognitive impairment)".
$w_{5,11}$ = "the result from SDMT test is (value of severe cognitive impairment)"
$w_{5,12}$ = "the result from SDMT test is (value of very severe cognitive impairment)".

Token α from place l_5 splits into five tokens, which receive the following characteristics:

"the result from SDMT is within the normal range" at place l_8;
"the result from SDMT register mild cognitive impairment" at place l_9;
"the result from SDMT register moderate cognitive impairment" at place l_{10};
"the result from SDMT register severe cognitive impairment" at place l_{11};
"the result from SDMT register extremely severe cognitive impairment" at place l_{12}.

Transition z_5 of the GN-model has the following form:

$$z_5 = \langle \{l_6\}, \{l_{13}, l_{14}, l_{15}, l_{16}, l_{17}\}, r_5, \wedge(l_6) \rangle$$

where the index matrix has the form.

$$r_5 = \begin{array}{c|ccccc} & l_{13} & l_{14} & l_{15} & l_{16} & l_{17} \\ \hline l_6 & w_{6,13} & w_{6,14} & w_{6,15} & w_{6,16} & w_{6,17} \end{array}$$

$w_{6,13} =$ "the result from BMVT test is $\geq$ (norm)"
$w_{6,14} =$ "the result from BMVT test is (value of mild cognitive impairment)"
$w_{6,15} =$ "the result from BMVT test is (value of moderate cognitive impairment)".
$w_{6,16} =$ "the result from BMVT test is (value of severe cognitive impairment)".
$w_{6,17} =$ "the result from BMVT test is (value of very severe cognitive impairment)".

Token α from place l_6 splits into five tokens, which receive the following characteristics:

"the result from BMVT is within the normal range" at place l_{13};
"the result from BMVT register mild cognitive impairment" at place l_{14};
"the result from BMVT register moderate cognitive impairment" at place l_{15};
"the result from BMVT register severe cognitive impairment" at place l_{16};
"the result from BMVT register extremely severe cognitive impairment" at place l_{17}.

Transition z_6 of the GN-model has the following form:

$$z_6 = \langle \{l_7\}, \{l_{18}, l_{19}, l_{20}, l_{21}, l_{22}\}, r_6, \wedge(l_7) \rangle$$

where the index matrix has the form.

$$r_6 = \begin{array}{c|ccccc} & l_{18} & l_{19} & l_{20} & l_{21} & l_{22} \\ \hline l_7 & w_{7,18} & w_{7,19} & w_{7,20} & w_{7,21} & w_{7,22} \end{array}$$

$w_{7,18} =$ "the result from CVMT test is $\geq$ (norm)"
$w_{7,19} =$ "the result from CVMT test is (value of mild cognitive impairment)"
$w_{7,20} =$ "the result from CVMT test is (value of moderate cognitive impairment)".
$w_{7,21} =$ "the result from CVMT test is (value of severe cognitive impairment)".
$w_{7,22} =$ "the result from CVMT test is (value of very severe cognitive impairment)".

Token α from place l_7 splits into five tokens, which receive the following characteristics:

"the result from CVMT is within the normal range" at place l_{18};
"the result from CVMT register mild cognitive impairment" at place l_{19};
"the result from CVMT register moderate cognitive impairment" at place l_{20};
"the result from CVMT register severe cognitive impairment" at place l_{21};
"the result from CVMT register extremely severe cognitive impairment" at place l_{22}.

Transition z_7 of the GN-model has the following form:

$$z_7 = \langle\{l_8, l_9, l_{10}, l_{11}, l_{12}\}, \{l_{23}\}, r_7, \vee(l_8, l_9, l_{10}, l_{11}, l_{12})\rangle$$

where the index matrix has the form.

$r_7 =$

	l_{23}
l_8	true
l_9	true
l_{10}	true
l_{11}	true
l_{12}	true

Transition z_8 of the GN-model has the following form:

$$z_8 = \langle\{l_{13}, l_{14}, l_{15}, l_{16}, l_{17}\}, \{l_{24}\}, r_8, \vee(l_{13}, l_{14}, l_{15}, l_{16}, l_{17})\rangle$$

where the index matrix has the form.

$r_8 =$

	l_{24}
l_{13}	true
l_{14}	true
l_{15}	true
l_{16}	true
l_{17}	true

Transition z_9 of the GN-model has the following form:

$$z_9 = \langle\{l_{18}, l_{19}, l_{20}, l_{21}, l_{22}\}, \{l_{25}\}, r_9, \vee(l_{18}, l_{19}, l_{20}, l_{21}, l_{22})\rangle$$

where the index matrix has the form.

$r_9 =$	l_{25}
l_{18}	true
l_{19}	true
l_{20}	true
l_{21}	true
l_{22}	true

The characteristics of token α at places:

l_{23} - "list with the most commonly administered cognitive rehabilitation games at the appropriate cognitive level, registered through SDMT"

l_{24} - "list with the most commonly administered cognitive rehabilitation games at the appropriate cognitive level, registered through BMVT".

l_{25} - "list with the most commonly administered cognitive rehabilitation games at the appropriate cognitive level, registered through CVMT";

Transition z_{10} of the GN-model has the following form:

$$z_{10} = \langle\{l_{23}, l_{24}, l_{25}\}, \{l_{26}, l_{27}\}, r_{10}, \wedge(l_{23}, l_{24}, l_{25})\rangle$$

where the index matrix has the form.

$r_{10} =$	l_{26}	l_{27}
l_{23}	$w_{23,26}$	$w_{23,27}$
l_{24}	$w_{24,26}$	$w_{24,27}$
l_{25}	$w_{25,26}$	$w_{25,27}$

$w_{23,26} = w_{24,26} = w_{25,26}$ "all characteristics of token α at places l_{23},l_{24} и l_{25} mean that" "the result is within the normal range"

$w_{23,27} = w_{24,27} = w_{25,27} \neg w_{23,26}$

l_{27} - "administration of personal cognitive rehabilitation with individual level of difficulty and frequency of training";

l_{26} – "cognitive impairments are not registered".

InterCriteria Analysis was applied at the current study for search of correlations between the grade and type of the cognitive impairment (affected cognitive domains) and between:

- grade of physical disability (EDSS);
- duration of multiple sclerosis;
- age;
- education level;
- working status.

Using InterCriteria Analysis we tried to define trends in cognitive. Finding of such relationships will allow to predict the cognitive decline during the time according to individual patient's specifics-form of MS, age and duration of the disease. Outlining the individual trend of course of the disease, as well as its prognosis, will help to apply complex target therapy including individual cognitive rehabilitation, according to the specific needs.

3 Conclusion

Patient data are categorized according to disease duration, degree of disability, age, gender, and employment. Based on a comparison of the results of healthy individuals and patients, analyzed by InterCriteria Analysis, the result of the model contains information about the impaired cognitive areas and the degree of impairment. The outcome of the GN model provides a solution for the therapist's recommendation for building a rehabilitation strategy - rehabilitation programs or batteries from rehabilitation games, level of difficulty and frequency of cognitive training.

Thus, the proposed model will be useful in determining the individual approach to each patient and will help to optimize the results of cognitive rehabilitation. The analysis of the model will provide an opportunity to outline future guidelines for therapeutic approaches to cognitive impairment in MS, which will lead to the inclusion of programs for cognitive rehabilitation and training in routine clinical practice over the next decade.

Acknowledgments. This research is supported by the Bulgarian National Science Fund under Grant KP-06-N22/1/2018 "Theoretical research and applications of InterCriteria Analysis".

References

1. Cotter, J., et al.: Investigating domain-specific cognitive impairment among patients with multiple sclerosis using touchscreen cognitive testing in routine clinical care. Front. Neurol. **9,** 331 (2018)
2. Donaldson, E., Patel, V.P., Shammi, P., Feinstein, A.: Why sex matters: a cognitive study of people with multiple sclerosis. Cogni. Behav Neurol. **32**(1), 39–45 (2019)
3. Rao, S.M., Leo, G.J., Ellington, L., Nauertz, T., Bernardin, L., Unverzagt, F.: Cognitive dysfunction in multiple sclerosis: II. Impact on employment and social functioning. Neurology **41**, 692–696 (1991)
4. Strober, L., Englert, J., Munschauer, F., Weinstock-Guttman, B., Rao, S., Benedict, R.H.B.: Sensitivity of conventional memory tests in multiple sclerosis: comparing the Rao Brief Repeatable Neuropsychological Battery and the Minimal Assessment of Cognitive Function in MS. Mult. Scler. J. **15**(9), 1077–1084 (2009)
5. Figved, N., Klevan, G., Myhr, K.M., Omdal, R., Aarsland, D.: Neuropsychiatric symptoms in patients with multiple sclerosis. Acta Psychiat. Scand, **112**(6), 463–468 (2005)
6. Genova, H.M., Lengenfelder, J., Chiaravalloti, N.D., Moore, N.B., DeLuca, J.: Processing speed versus working memory: contributions to an information-processing task in multiple sclerosis. Appl. Neuropsychol. **19**(2), 132–140 (2012)
7. Langdon, D.W., et al.: Recommendations for a brief international cognitive assessment for multiple sclerosis (BICAMS). Mult. Scler. J. **18**(6), 891–898 (2012)

8. Dennison, L., Moss-Morr, R., Chalder, T.: A review of psychological correlates of adjustment in patients with multiple sclerosis. Clin. Psychol. Rev. **29**, 141–153 (2009)
9. Langdon, D.: Cognitive impairment in multiple sclerosis – recent advances and future prospects. Eur Neurol Rev **5**, 69–72 (2010)
10. Rao, S.M.: A Manual for the Brief, Repeatable Battery of Neuropsychological Tests in Multiple Sclerosis. National Multiple Sclerosis Society, New York (1991)
11. Benedict, R.H., et al.: Validity of the minimal assessment of cognitive function in multiple sclerosis (MACFIMS). J. Int. Neuropsychol. Soc. **12**(4), 549–558 (2006)
12. Benedict, R.H., et al.: Brief International Cognitive Assessment for MS (BICAMS): international standards for validation. BMC Neurol. **12**(1), 55 (2012)
13. Smith, A.: Symbol Digit Modalities Test: Manual. Western Psychological Services, Los Angeles (1982)
14. Delis, D.C., Kramer, J.H., Kaplan, E., Ober, B.A.: California Verbal Learning Test Manual, 2nd edn,, Adult version, Psychological Corporation, San Antonio (2000)
15. Benedict, R.: Brief Visuospatial Memory Test-Revised: Professional Manual. Psychological Assessment Resources, Odessa (1997)
16. Corfield, F.: Validation of the ipad brief international cognitive assessment for multiple sclerosis (BICAMS). Doctoral Thesis, Holloway, University of London (2018)
17. Golan, D., et al.: Validity of a multidomain computerized cognitive assessment battery for patients with multiple sclerosis. Mult. Scler. Relat. Disord. **30**, 154–162 (2019)
18. Edgar, C., et al.: Cognitive performance in relapsing remitting multiple sclerosis: a longitudinal study in daily practice using a brief computerized cognitive battery. BMC Neurol. **11**(1), 68 (2011)
19. De Meijer, L., et al.: Monitoring cognitive change in multiple sclerosis using a computerized cognitive battery. Mult. Scler. J. Exp., Transl. Clin. **4**(4), 2055217318815513 (2018)
20. Lapshin, H., Lanctôt, K.L., O'Connor, P., Feinstein, A.: Assessing the validity of a computer-generated cognitive screening instrument for patients with multiple sclerosis. Mult. Scler. J. **19**(14), 1905–1912 (2013)
21. Goretti, B., et al.: The brief international cognitive assessment for multiple sclerosis (BICAMS): normative values with gender, age and education corrections in the Italian population. BMC Neurol. **14**(1), 171 (2014)
22. Beck, A.T., Steer, R.A., Brown, G.: Beck depression inventory–II. Psychol. Assess. (1996)
23. Atanassov, K.: Generalized Nets. World Scientific, Singapore (1991)
24. Atanassov, K.: On Generalized Nets Theory. Prof. M. Drinov Academic Publishing House, Sofia (2007)
25. Atanassov, K.T.: Index Matrices: Towards an Augmented Matrix Calculus, pp. 1–110. Springer International Publishing, Cham (2014). https://doi.org/10.1007/978-3-319-10945-9
26. Atanassov, K., Sotirova, E., Andonov, V.: Generalized net model of multicriteria decision making procedure using intercriteria analysis. In Advances in Fuzzy Logic and Technology, Springer, Cham, pp. 99–111 (2017). https://doi.org/10.1007/978-3-319-66830-7
27. Pencheva, T., Roeva, O., Shannon, A.: Generalized net models of basic genetic algorithm operators. In: Imprecision and Uncertainty in Information Representation and Processing, Springer, Cham, pp. 305–325 (2016). https://doi.org/10.1007/978-3-319-26302-1
28. Ribagin, S., Roeva O., Pencheva T.: Generalized Net model of asymptomatic osteoporosis diagnosing. In: Proceedings of the 2016 IEEE 8th International Conference on Intelligent Systems IEEE, IS 2016, pp. 604–608 (2016)
29. Roeva, O., Shannon A., Pencheva T.: Description of simple genetic algorithm modifications using generalized nets. In: IS'2012 - 6th IEEE International Conference Intelligent Systems, Proceedings, pp. 178–183 (2012)

30. Atanassov, K.: Generalized net model of blood donation processes. uncertainty and imprecision in decision making and decision support: new challenges, solutions and perspectives. Selected Papers from BOS-2018, held on September 24–26, 2018, and IWIFSGN-2018, held on September 27–28, 2018 in Warsaw, Poland, 1081, p. 147 (2020)
31. Will, J., Heiner, M.: Petri nets in biology, chemistry, and medicine. Computer Science Reports 04/02, University of Technology at Cottbus (2002)
32. Atanassov, K., Mavrov, D., Atanassova, V.: Intercriteria decision making: a new approach for multicriteria decision making, based on index matrices and intuit Ionistic fuzzy sets. Iss. Intuit. Fuzzy Sets Gene. Nets **11**, 1–8 (2014)
33. Atanassov, K., Atanassova, V., Gluhchev, G.: InterCriteria analysis: ideas and problems. Notes Intuit. Fuzzy Sets **21**(1), 81–88 (2015)

A Generalized Net Model of the Normal Heart Functioning

Valentin Vassilev[1,2(✉)], Evdokia Sotirova[1,4], and Krassimir Atanassov[3,4]

[1] Faculty of Public Health and Health Care, Prof. Assen Zlatarov University, 1, Prof. Yakimov Street, 8010 Burgas, Bulgaria
esotirova@btu.bg

[2] Vascular Surgery Clinic, University Hospital for Active Treatment, 73, Stefan Stambolov Boul., 8000 Burgas, Bulgaria
vailvasilev@mail.bg

[3] Department of Bioinformatics and Mathematical Modelling, Institute of Biophysics and Biomedical Engineering, Bulgarian Academy of Sciences, Acad. G. Bonchev Street, Bl. 105, 1113 Sofia, Bulgaria
krat@bas.bg

[4] Intelligent Systems Laboratory, Prof. Asen Zlatarov University, 8000 Bourgas, Bulgaria

Abstract. Within the last 20 years a lot of Generalized Net (GN) models of the ways of functioning of the different systems and organs in human body and models, related to description of biomedical processes in the living organisms were constructed. In this paper a GN model of the normal heart functioning is developed. It gives possibilities to evaluate heart's parameters and to analyze its behavior and functioning. The GN model can be detailed and complicated and also it can be extended to accommodate some disorders of the cardiovascular system.

Keywords: Cardiovascular system · Diastole · Generalized net · Systole

1 Introduction: Basic Anatomy and Physiology of the Cardiovascular (Circulatory) System

The cardiovascular (circulatory) system consists of three main parts (see [6,7]):

1. A "pump", which makes blood flow through the blood vessels. The role of a "pump" is played by the heart, which contracts periodically and in this way pumps blood into the blood vessels.
2. Circulating fluid - blood.
3. A network of blood vessels, through which the blood in the body flows.

The lymphatic vessels are also a part of the circulatory system.

The circulatory system has two components, a systematic circulation and a pulmonary circulation.

K. T. Atanassov et al. (Eds.): IWIFSGN 2020/BOS/SOR 2020, LNNS 338, pp. 408–418, 2022.
https://doi.org/10.1007/978-3-030-95929-6_31

The systematic circulation begins at the left ventricle of the heart, which pumps arterial blood into the aorta, after which the blood flows consequently into the smaller caliber arteries, arterioles, capillaries, venules and veins, through which the venous blood flows back into the right atrium of the heart.

The pulmonary circulation begins at the right ventricle, which pumps venous blood into the pulmonary artery. The pulmonary artery divides into smaller arteries, arterioles, capillaries, venules and veins, which return the arterial blood back into the left atrium of the heart [4].

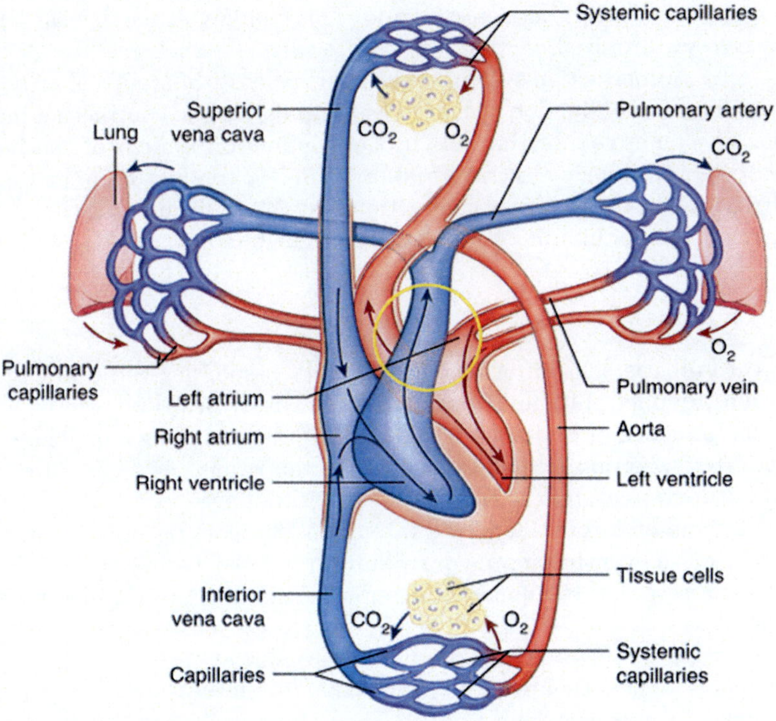

Fig. 1. The cardiovascular (circulatory) system

1.1 Anatomy of the Heart

The heart is a muscle, which plays the part of a double pump, responsible for the flowing of blood into the two components of the circulatory system. The heart consists of two parts (halves), separated from each other by a septum: a left half, through which flows arterial blood, and a right half, through which flows venous blood. There are two septums that separate the left and right atria, and the left and right ventricles. The septums are not fenestrated, and they prevent the mixing of the venous and arterial blood flows.

Each heart half consists of two chambers - an atrium and a ventricle. Each atrium is separated from the underlying ventricle by a valve. The whole heart has 4 chambers: two atria and two ventricles.

1.2 Anatomy of the Heart Muscle (Myocardium)

There are two types of muscle tissue in the myocardium: a contracting myocardium and myocardial cells, which play a part in the conduction of electrical impulses, which lead to the autonomic contraction of the heart.

The cardiomyocytes (myocardial cells), are split at the end, which enables them to connect to each other consequently. This enables all the myocardial cells to produce a synchronized contraction.

The electrical conduction system of the heart consists of groups of myocardial cells, specialized in conduction the electrical impulse to the contracting myocardial cells. The main role of the cells in the conduction system of the heart is to generate spontaneously electrical impulses, and to conduct them at the right speed. Structures of the electrical conduction system of the heart: sinoatrial node, atrioventricular node, bundle of His, left and right bundles, Purkinje fibers.

1.3 Physiology of the Myocardium

The myocardium, as a lot of other types muscle tissue, has the characteristics of excitation, conductibility and contractibility. The myocardium can be excited only by the influence of excitation electrical impulses, coming from the cells in the electrical conduction system, which are autonomous. All structures of the electrical conduction system of the heart have autonomy.

Normally, the heart contracts at a rate of 60–90 beats per minute, at rest.

The cells of the contracting myocardium do not have autonomy. They contract only as a response to an electrical impulse coming from the electrical conduction system.

The electrical impulses in the conduction system of the heart, arise at the sinoatrial node - then they excite the atrial musculature - then the impulse goes through the atrioventricular node - then the bundle of His - the left and right bundles - Purkinje fibers - ventricular musculature. The speed at which the impulses are transmitted is not the same in all parts of the heart. The slowest speed at which the impulses are transmitted is measured at the atrioventricular node (0.05 m/s), and the highest speed is measured at the Purkinje fibers. (1–4 m/s).

1.4 Cardiac Cycle

The main function of the heart is to pump blood into the blood vessels. This is accomplished by the periodic contraction (Systole) and relaxation (diastole) of the heart muscle. One systole and one diastole of the atria and ventricles makes one cardiac cycle. During one cardiac cycle, the blood flows only one way and does not regurgitate (normally) back - due to the valvular apparatus of the heart.

There are two types of valves in the heart: atroventricular valves and semilunar valves. Atrioventricular valves are the mitral (bicuspid) valve, which separates the left atrium from the left ventricle, and the tricuspid valve, which separates the right atrium from the right ventricle. The semilunar valves are the aortic valve and the pulmonary valve. The valves open and close passively under the influence of the gradient of pressures on the two sides of the valves - they open when the pressure in the chambers above them is higher than the pressure below them, and they close when the pressure in the chamber below them is higher than the pressure in the chambers above them.

Phases of the cardiac cycle

1. Passive ventricular filling - the atria and the ventricles are relaxed (diastole) and fill passively with blood from the large veins, which open into the atria. Venous blood from the superior vena cava and the inferior vena cava flows through the right atrium into the right ventricle. Arterial blood from the four/five pulmonary veins flows into the left atrium and then into the left ventricle. During this phase the atrioventricular valves are open, and the semilunar valves are closed.
2. Atrial contraction - this makes blood flow from the atria to the ventricles, which leads to the addition of a portion of blood into the ventricles. At the end of this phase the volume of the chambers is at its maximum, which is called an end diastolic volume. The end diastolic volume at rest is 120–140 ml. The contraction of the atria gives off 10–20% of the volume of blood in the ventricle. During phase 2, the state of the valves is the same as in phase 1.
3. Isovolumetric contraction of the ventricles - The contraction of the ventricular myocardium raises the pressure on the blood inside. Because of the higher pressure in the ventricles the atrioventricular valves close. During this phase the valves of the heart are closed. This period continues until the pressure in the ventricles becomes higher than the pressure in the large blood vessels (aorta and pulmonary artery). This is when the semilunar valves open and the next phase begins.
4. Pumping of blood out of the chambers - the most important phase of the cardiac cycle, as it is through this phase that blood is supplied to the organs. During the period of pumping, the semilunar valves are open, and the atrioventricular valves are closed, which does not permit blood from the ventricles to regurgitate back to the atria. When the ventricular musculature begins to relax and the pressure in the chambers goes below the pressure in the aorta and the pulmonary artery, the semilunar valves close, which means the end of the pumping cycle. In the end of the pumping cycle, there is still a certain amount of blood left in the ventricles. The volume that is left in the ventricles at the end of the systole is called an end systolic volume. At rest it amounts to around 60 ml. The volume, which is pumped during the systole, is called a stroke volume. At rest, the stroke volume of the heart is 70 ml.

5. Isovolumetric relaxation of the ventricles. During this phase the semilunar valves are closed, the pressure in the ventricles lowers further, but as it still is higher than the pressure in the atria, the atrioventricular valves are still closed, and this is why the volume of the ventricles remains the same. When the pressure in the ventricles goes below that in the atria, the atrioventricular valves open. With this, phase 4 ends, and phase 1 begins - and the cycle repeats. The duration of each cardiac cycle depends on the heart rate and it is calculated by the formula: Cardiac cycle (in seconds) = 60/HF (heart rate) in bpm. At rest, the average duration of one cardiac cycle is 0.8 s, but it can vary between 0.7 and 1 s. The duration of the diastole is greater than that of the systole - at cardiac cycle duration 0.8 s, the duration of the diastole is 0.5 s, and that of the systole –0.3 s. At higher heart rate, the duration of the diastole shortens compared to that of the systole.

Modelling the different medical processes, different organs and systems in the human body is a powerful tool to study the complex and orchestrated biological processes. This models can be used for simulation of real processes and objects in the human body with the aim of using the information obtained in the process of medical diagnosis and research.

The heart is a complex system working thanks to many fine mechanisms for regulation and adaptation according to the inner environment of the organism and the outer environment [13].

Some major aspects of multi-scale modeling of the heart are introduced in [14].

The modelling of the heart functioning can be used for simulations into precision medicine with the ultimate goals to facilitate clinical decision making, guide treatment planning, and accelerate device design [12,15]. In [10] an investigation in human heart modeling in electrophysiology, cardiac mechanics, and fluid dynamics are presented. The clinically relevant applications of these models for pacing lead failure, heart failure, drug development, ventricular assist devices, edge-to-edge repair, and annuloplasty are explained. In [11] computational models of cardiac function, based on the anatomy, biophysics, and biochemistry of the heart are discussed.

A computer model for recreation the electrical activity of the ventricles in a human heart was created in University of Oxford [16].

According [16] the developed new human heart model decides theoretical inconsistencies from the last 25 years.

These models can be used for simulation of the heart functioning of healthy persons or they can be used for investigation of the heart status in the case when there are some disorders of the circulatory system.

In this paper a GN model of the normal heart functioning is developed. Within the last 20 years a lot of GN models of the ways of functioning of the different systems and organs in human body were constructed. The description of the different systems and organs by a uniform mathematical approach, a meaningful comparison, can be made among them and they can be investigated from the point of view of speed and performance, resources used, accuracy of the result, and others. The GN models in medicine can be used for simulation of real processes with educational purpose. They can guide specialists in studying the logic of the processes related to diagnoses and help in acquiring knowledge and diagnostic skills [8]. The fact that GNs are appropriate as a modelling tool for biological and medical processes is made evident, for example, by the review and the bibliography [17], where more than 30 GN-models of various such processes are cited.

2 A Generalized Net Model

Generalized nets (GNs) [1–3,8], are one of the tools for modeling parallel processes running in real time. By this reason, the process of the sound heart functioning will be described through a GN and the benefit of such a model will be discussed.

The GN-model contains 7 transitions, 19 places and 6 types of tokens. The sense of these GN-components is the following (see Fig. 2).

The β-tokens represent the blood. For brevity, their indices are omitted.

Transition Z_1 means the human body. It contains the place l_2 in which permanently stays token χ with initial and current characteristic

"the quantity of the blood circulating in the human body (in the

current time-moment)".

Transition Z_2 and more exactly, its place l_4 means the right auricle.

Transition Z_3 and more exactly, its place l_6 means the right camera.

Transition Z_4 means the lungs. It contains the place l_{10} in which permanently stays token λ with initial and current characteristic

"the quantity of the entering blood (in the current time-moment)".

Transition Z_4 contains two places l_7 and l_8 representing a part of the respiratory system.

Transition Z_5 means the autonomous nervous system. It contains the place l_{15} in which permanently stays token ν (in the present model) without characteristics. It generates signals $\sigma_1, \sigma_2, \sigma_3, \sigma_4$ to both auricles and both cameras. These signals are generated in the following order: $\sigma_3, \sigma_1, \sigma_4, \sigma_2$. Each one of them has the characteristic

"power of the electrical impulse (in the current time-moment)".

Transition Z_6 and more exactly, its place l_{17} means the left auricle.

Transition Z_7 and more exactly, its place l_{19} means the left camera.

The transitions have the following forms (see Fig. 2).

$$Z_1 = \langle \{l_2, l_{18}\}, \{l_1, l_2\}, \begin{array}{c|cc} & l_1 & l_2 \\ \hline l_2 & true & true \\ l_{18} & false & true \end{array} \rangle.$$

The token β in place l_1 obtains the characteristic

"the quantity of the venous blood in the large circle of the blood circulation; the current time-moment".

$$Z_2 = \langle \{l_1, l_4, l_{13}\}, \{l_3, l_4\}, \begin{array}{c|cc} & l_3 & l_4 \\ \hline l_1 & false & true \\ l_4 & W_{4,3} & true \\ l_{13} & false & true \end{array} \rangle,$$

where
$W_{4,3}$ = "on the previous time-step, a σ_3-token united with the β-token in l_4".

The token β in place l_4 obtains the characteristic

"the quantity of the blood entering the right auricle; the current time-moment".

On the next time-step, the same token β obtains in place l_3 the characteristic

"the quantity of the blood from auricle in right camera; the current time-moment".

$$Z_3 = \langle \{l_3, l_6, l_{14}\}, \{l_5, l_6\}, \begin{array}{c|cc} & l_5 & l_6 \\ \hline l_3 & false & true \\ l_6 & W_{6,5} & true \\ l_{14} & false & true \end{array} \rangle,$$

where
$W_{6,5}$ = "on the previous time-step, a σ_4-token united with the β-token in l_6".

The token β does not obtain a new characteristic in place l_6, while on the next time-step, it obtains in place l_5 the characteristic

"the quantity of the blood from right camera pushed to the lung through the lung artery; the current time-moment".

At each time-moment, token α (for brevity, with an index) enters place l_7 with the characteristic

"the quantity of the inhaled oxygen".

$$Z_4 = \langle \{l_5, l_7, l_{10}\}, \{l_8, l_9, l_{10}\}, \begin{array}{c|ccc} & l_8 & l_9 & l_{10} \\ \hline l_5 & false & false & true \\ l_7 & true & false & false \\ l_{10} & false & true & false \end{array} \rangle.$$

The token α entering place l_8 obtains the characteristic

"the quantity of the exhaled oxygen".

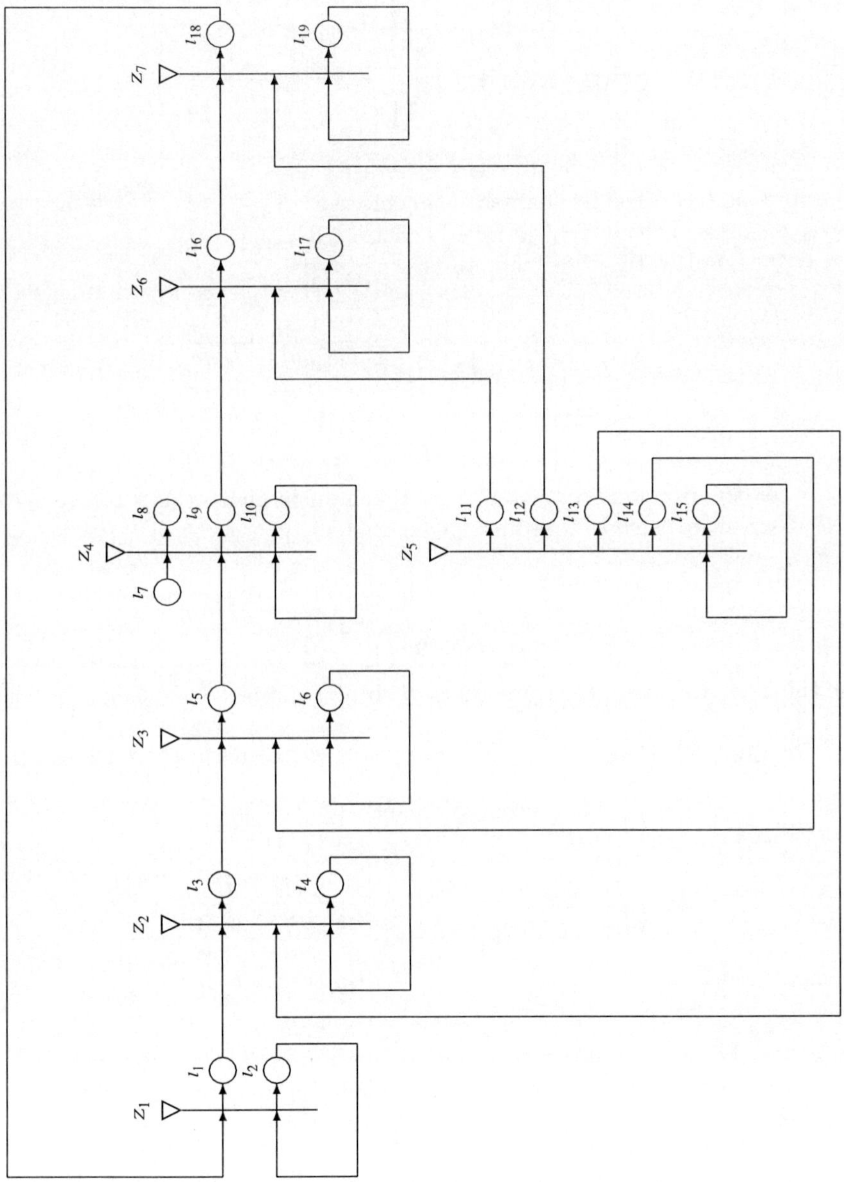

Fig. 2. The generalized net model

The token β in place l_9 obtains the characteristic

"the quantity of the blood leaving the lung; the current time-moment".

$$Z_5 = \langle \{l_{15}\}, \{l_{11}, l_{12}, l_{13}, l_{14}, l_{15}\}, \begin{array}{c|ccccc} & l_{11} & l_{12} & l_{13} & l_{14} & l_{15} \\ \hline l_{15} & W_{15,11} & W_{15,12} & W_{15,13} & W_{15,14} & true \end{array} \rangle,$$

where
$W_{15,11}$ = "a signal to the left auricle is generated",
$W_{15,12}$ = "a signal to the left camera is generated",
$W_{15,13}$ = "a signal to the right auricle is generated",
$W_{15,14}$ = "a signal to the right camera is generated".

$$Z_6 = \langle \{l_9, l_{11}, l_{17}\}, \{l_{16}, l_{17}\}, \begin{array}{c|cc} & l_{16} & l_{17} \\ \hline l_9 & false & true \\ l_{11} & false & true \\ l_{17} & W_{17,16} & true \end{array} \rangle,$$

where
$W_{17,16}$ = "on the previous time-step, a σ_1-token united with the β-token in l_{17}".
The token β in place l_{17} obtains the characteristic

"the quantity of the blood entering the left auricle; the current time-moment".

On the next time-step, the same token β obtains in place l_{16} the characteristic

"the quantity of the blood from auricle in left camera; the current time-moment".

$$Z_7 = \langle \{l_{12}, l_{16}, l_{19}\}, \{l_{18}, l_{19}\}, \begin{array}{c|cc} & l_{18} & l_{19} \\ \hline l_{12} & false & true \\ l_{16} & false & true \\ l_{19} & W_{19,18} & true \end{array} \rangle,$$

where
$W_{19,18}$ = "on the previous time-step, a σ_2-token united with the β-token in l_{19}".
The token β does not obtain a new characteristic in place l_{19}, while on the next time-step, it obtains in place l_{18} the characteristic

"the quantity of the blood from left camera pushed to the human body through the aorta; the current time-moment".

3 Conclusion

The so constructed GN-model describes the process of the normal heart functioning. On the one hand, it can be used for simulation of the heart functioning of healthy persons (e.g., sportsmen) and on the other - it can be a basis of GN-models of the heart status in the case when there are some disorders of the circulatory system.

The most common disorders that can lead to alterations in the normal physiology of the heart, can be separated into three groups:

1. Disorders of the heart:
   1.1. Valvular disorders - stenosis or regurgitation,
   1.2. Cardiomyopathies,
   1.3. Ischemic heart disease - atherosclerosis of the coronary vessels,
   1.4. Disorders of the electrical conduction system of the heart;
2. Disorders of the pulmonary circulatory system - disorders of the lungs, which can lead to reduction in lung tissue and as a result a higher vascular resistance in the pulmonary circulation;
3. Disorders of the systemic circulatory system - atherosclerosis of the large blood vessels and disorders of the internal organs, which can lead to a higher vascular resistance in the systematic circulation.

In future, we plan to construct separate GN-models for these disorders.

Acknowledgements. The present research has been supported by the Bulgarian National Science Fund under Grant Ref. No. DN-02/10 "New Instruments for Knowledge Discovery from Data and Their Modelling".

References

1. Alexieva, J., Choy, E., Koycheva, E.: Review and bibliography on generalized nets theory and applications. In: Choy, E., Krawczak, M., Shannon, E., Szmidt, V. (eds.) A Survey of Generalized Nets, Raffles KvB Monograph No. 10, pp. 207–301 (2007)
2. Atanassov, K.: Generalized Nets. World Scientific, Singapore, New Jersey, London (1991)
3. Atanassov, K.: On Generalized Nets Theory, "Prof. M. Drinov" Academic Publishing House, Sofia (2007)
4. Barrett, K.E., Barman, S., Boitano, S., Brooks, H.: Ganong's Review of Medical Physiology, 25th edn. McGraw-Hill, New York (2016)
5. Dzhambazova, E.: Human Physiology. Climent Ohridski University Publishing House, Soifa (2015).(in Bulgarian)
6. Piryov, B., Vassilev, V., Belchev, C.: Human Anatomy and Physiology. Mnemozina, Sofia (1995).(in Bulgarian)
7. Vitanova, L., Garchev, P.: Human Physiology. Arso, Sofia (2020).(in Bulgarian)
8. Atanassov, K.: Generalized nets as a tool for the modelling of data mining processes. In: Sgurev, V., Yager, R.R., Kacprzyk, J., Jotsov, V. (eds.) Innovative Issues in Intelligent Systems. SCI, vol. 623, pp. 161–215. Springer, Cham (2016). https://doi.org/10.1007/978-3-319-27267-2_6

9. Hunter, P., Pullan, A., Smaill, B.: Modeling total heart function. Ann. Rev, Biomed. Eng. **5**(1), 147–177 (2003)
10. Peirlinck, M., et al.: Precision medicine in human heart modeling. Biomech. Model Mechanobiol, 20, 803–831 (2001). https://doi.org/10.1007/s10237-021-01421-z
11. Hori, M., Suga, H., Baan, J., Yellin, E.: Cardiac Mechanics and Function in the Normal and Diseased Heart, pp. 155–245. Springer-Verlag, Tokyo, Chapter C (1989). https://doi.org/10.1007/978-4-431-67957-8
12. LeGrice, I., Hunter, P., Young, A., Smaill, B.: The architecture of the heart: a data-based model. Phil. Trans. R. Soc. Lond. Ser A Math. Phys. Eng. Sci. 359(1783), 1217–1232 (2001)
13. Glass, L., Hunter, P., McCulloch, A.: Theory of Heart: Biomechanics, Biophysics, and Nonlinear Dynamics of Cardiac Function. Springer-Verlag, New York (1991)
14. Mansi, T., Passerini, T., Comaniciu. D.: Artificial Intelligence for Computational Modeling of the Heart. Academic Press, New York (2019)
15. Trayanova, N.: Whole-heart modeling: applications to cardiac electrophysiology and electromechanics. Circul. Res. **108**(1), 113–128 (2011)
16. The Medical Sciences Division: New Human Heart Model set to Boost Future Cardiac Research and Therapies, University of Oxford, 3 February 2020. https://www.medsci.ox.ac.uk/news/new-human-heart-model-set-to-boost-future-cardiac-research-and-therapies
17. Will, J., Heiner, M.: Petri Nets in Biology, Chemistry, and Medicine. Computer Science Reports 04/02, University of Technology at Cottbus (2002)

Generalized Net Model of Biometric Multifactor Authentication System

Zhelyana Ivanova, Veselina Bureva(✉), and Sotir Sotirov

Laboratory of Intelligent Systems, "Prof. Dr. Assen Zlatarov" University, "Prof. Yakimov" Blvd., 8010 Burgas, Bulgaria
zhelyana.ivanova@abv.bg, {vbureva,ssotirov}@btu.bg

Abstract. The aim of the current research work is to investigate biometric multifactor system. The basics of the theories of multifactor systems and biometric systems are presented in the paper. Generalized net model of biometric multifactor authentication system is constructed. The presented model use hand (palm) geometry, palm vein and cardiac signal biometrics. The biometrics and factors are acquired in the same time. The biometric identification methods are applied. The features are compared to the stored information for individual. If it is no information for the user in the database the process of assignment have to be started. In the end rank level fusion is applied and the process of decision making is made.

Keywords: Cardiac signal · Generalized nets · Biometrics · Hand geometry · Multifactor authentication system · Palm geometry · Security · Mobile phone

1 Introduction

Multifactor system is a multifactor authentication solution that can contains the factors describing the following categories: "Something that the individual knows", "Something that the individual has", "Something that the individual is", "Somewhere the individual is". The individual knows a password or personal identification number (PIN). The user has a mobile phone, a bank card or a key. The biometrics of the individual that can be used in the systems are fingerprint, iris, voice, typing speed, pattern in key press intervals. The individual can have connection to a specific computing network or using a GPS signal to identify the location. These conditions describe multifactor systems [21]. In the other side multibiometric systems can be separated in different types: multi-sensor systems, multi-algorithm systems, multi-instance system, multi-sample systems, multimodal systems. Multi-sensor systems capture a single biometric trait using multiple sensors. In multi-algorithm systems the biometric data is processed using multiple algorithms. In multi-instance system multiple instances of the same body trait are captured. In multi-sample systems a sensor is used to acquire multiple samples of the same biometric trait. Multimodal systems combine different body traits to confirm identity [11, 14, 19, 20, 24, 30]. Depending of the type of the biometric system appropriate method of biometric fusion is used. Biometric fusion can be classified into fusion prior to matching phase

K. T. Atanassov et al. (Eds.): IWIFSGN 2020/BOS/SOR 2020, LNNS 338, pp. 419–435, 2022.
https://doi.org/10.1007/978-3-030-95929-6_32

and fusion after matching phase. Sensor-level fusion is applied in multi-sensor or multi-sample algorithms, where data have to be combined after its acquisition. Feature-level fusion is used in the biometric system containing multiple features extracted from the same or different input data [11]. Score-level fusion is applied to the algorithms where the match scores produced by several matchers are fused together. Rank-level fusion is performed after comparing the input patterns with the templates in the database [25, 28]. Decision-level fusion is used for algorithms where fusion is performed at the decision level (Fig. 1).

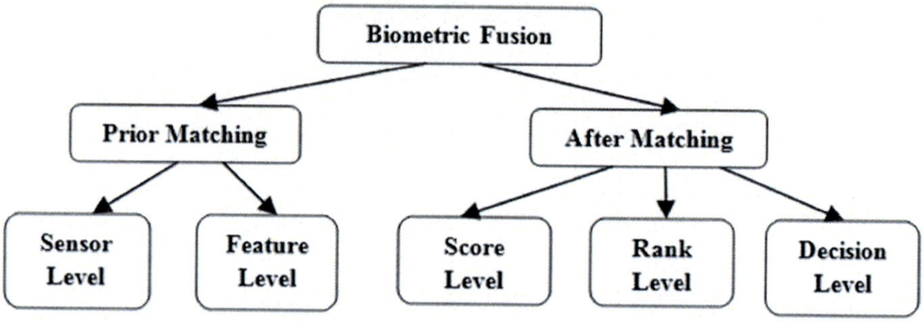

Fig. 1. Biometric fusion

In [10] the categories of biometric fusion are explained and Generalized net model of multimodal biometric system is constructed. In the current paper biometrics and factors are combined.

2 Generalized Net Model of Biometric Multifactor Authentication System

In the current research work a biometric multifactor authentication system is presented. The individual has to identify itself using device (phone). The factors as phone identity number, location, time, passwords are previously stored in database. The templates of user palm geometry, palm vein and cardiac signal are processed and inserted into the database for future comparisons (Fig. 2). In the proposed biometric multifactor system the processes of hand (palm) vein recognition, palm vein identification, cardiac signal detection and factors matching are executed in parallel.

Discrete event system consists of discrete events occurring asynchronously over time. Two key features of a discrete event system are known: its dynamics is event driven and its state variables belong to a discrete set [13, 27]. Petri nets are introduced in 1939 by Carl Petri. They are discrete event dynamic systems. Petri nets are constructed in the form of directed bipartite graph having two elements transitions and positions. The place contains tokens. The enabled transition is when the input place of the transition has token and all the other places are connected to the transition [12, 15, 25, 26, 31]. In the literature multiple Petri extensions are defined. Generalized nets are introduced in 1982 by Prof.

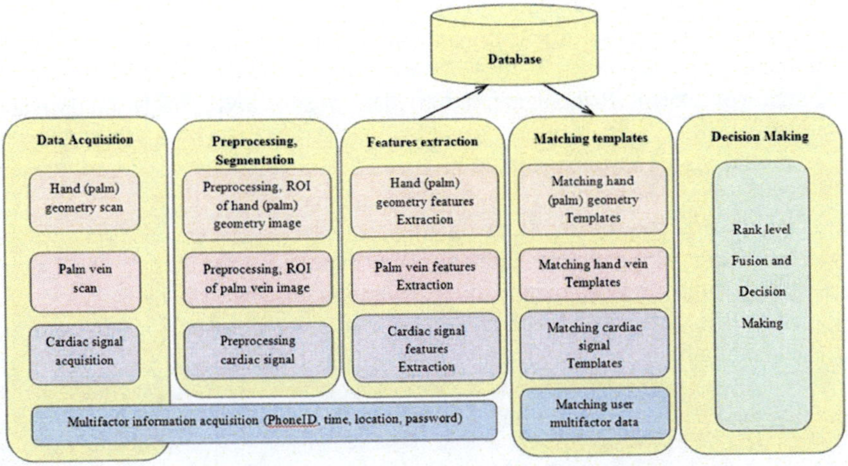

Fig. 2. Process of biometric multifactor authentication

Krassimir Atanassov as extension of Petri nets. Generalized nets are used for parallel modeling and have complex structure. Generalized nets and Petri nets have four main differences. The first difference is the history of the tokens: the movement of the tokens between the transitions. The second difference nets is the "place-transition" relation. The transition may contain m input places and n output places where m, n $\geq$ 1. The third difference the index matrix describing the transition conditions. The time of functioning is the fourth difference between Generalized nets and all types of Petri nets. The concepts of Generalized nets (GNs) and index matrices are introduced in [5–8]. GN models for pattern recognition of individual biometrics are presented in [2–4]. GN models for biometric access-control system and multimodal biometric system are constructed [1, 9]. The biometric methods as iris recognition [17, 29], face recognition [22], fingerprints recognition [10], signature verification [16] are already modeled using the apparatus of GNs. In the future works the investigation will be extended with constructing GN models for biometrics as gait recognition, speech recognition, ear recognition, retina recognition, key stroke recognition [11].

In the current paper GN model of the biometric multifactor authentication system using GNDraw software [18] is constructed (Fig. 3). It contains the following set of transitions:

$$A = \{Z_1, Z_2, Z_3, Z_4, Z_5, Z_6, Z_7, Z_8, Z_9, Z_{10}, Z_{11}, Z_{12}, Z_{13}, Z_{14}, Z_{15}, Z_{16}\},$$

where the transitions describe these process:

- Z_1 - "individual";
- Z_2 - "hand (palm) geometry scanner";
- Z_3 - "palm vein scanner";
- Z_4 - "cardiac signal acquisition";
- Z_5 - "hand (palm) geometry image quality assessment and preprocessing";
- Z_6 - "palm vein image quality assessment and preprocessing";

- Z_7 - "cardiac signal preprocessing";
- Z_8 - "region of interest (RoI, Segmentation) identification and palm vein features extraction";
- Z_9 - "region of interest (RoI, Segmentation) identification and palm geometry features extraction";
- Z_{10} - "cardiac signal features extraction";
- Z_{11} - "storing multifactor data for individuals in database";
- Z_{12} - "rank fusion and decision making";
- Z_{13} - "biometric palm geometry identification (matching templates)";
- Z_{14} - "biometric palm vein identification (matching templates)";
- Z_{15} - "biometric cardiac signal identification (matching templates)";
- Z_{16} - "matching multifactor input information with stored data".

Initially, there is one α_1-token that is located in place L_6 with initial characteristic:

"*biometric multifactor information of individuals*".

In the next time-moments this token is split into two or more. The original α_1-token will continue to stay in place L_{48}, while the other α-tokens will move to the next transitions.

Token enters the net via place l_1 with initial characteristics:

"*individual*".

The transition Z_1 has the form:

$$Z_1 = \langle\{l_1, l_{30}, l_{37}, l_{40}, l_{47}, l_{51}\}, \{l_2, l_3, l_4, l_6, l_7\}, R_1, \vee(l_1, l_{30}, l_{37}, l_{40}, l_{47}, l_{51})\rangle,$$

where

$R_1 =$

	l_2	l_3	l_4	l_6	l_7
l_1	*true*	*true*	*true*	*true*	*true*
l_{30}	*true*	*false*	*false*	*false*	*false*
l_{37}	*false*	*true*	*false*	*false*	*false*
l_{40}	*false*	*false*	*true*	*false*	*false*
l_{47}	*true*	*true*	*true*	*true*	*false*
l_{51}	*true*	*true*	*true*	*true*	*true*

Token enters the net via place l_8 with initial characteristics:

"*parameters for palm geometry image scanning*".

The transition Z_2 has the form:

$$Z_2 = \langle\{l_2, l_8, l_{10}\}, \{l_9, l_{10}\}, R_2, \vee(\wedge(l_2, l_8), l_{10})\rangle,$$

where

$$R_2 = \begin{array}{c|cc} & l_9 & l_{10} \\ \hline l_2 & false & true \\ l_8 & false & true \\ l_{10} & W_{10,9} & W_{10,10} \end{array}$$

and

- $W_{10,9}$ = "there is scanned hand geometry";
- $W_{10,10} = \neg W_{10,9}$.

The token that enters the place l_9 has the following characteristic:

"scanned hand geometry"

in place l_9.

The process of hand (palm) geometry scanning is described. The individual scan its hand (palm) geometry using special type of scanner. The parameters are used for setting the device.

Token enters the net via place l_{11} with initial characteristics:

"parameters for palm vein scanning".

The transition Z_3 has the form:

$$Z_3 = \langle \{l_3, l_{11}, l_{13}\}, \{l_{12}, l_{13}\}, R_3, \vee(\wedge(l_3, l_{11}), l_{13}) \rangle,$$

where

$$R_3 = \begin{array}{c|cc} & l_{12} & l_{13} \\ \hline l_3 & false & true \\ l_{11} & false & true \\ l_{13} & W_{13,12} & W_{13,13} \end{array}$$

and

- $W_{13,12}$ = "there is scanned palm vein";
- $W_{13,13} = \neg W_{13,12}$.

The token that enters the place l_{12} has the following characteristic:

"scanned palm vein"

in place l_{12}.

The palm veins of the individual are scanned using appropriate type of sensor integrated to the device. The parameters are used for setting the scanner.

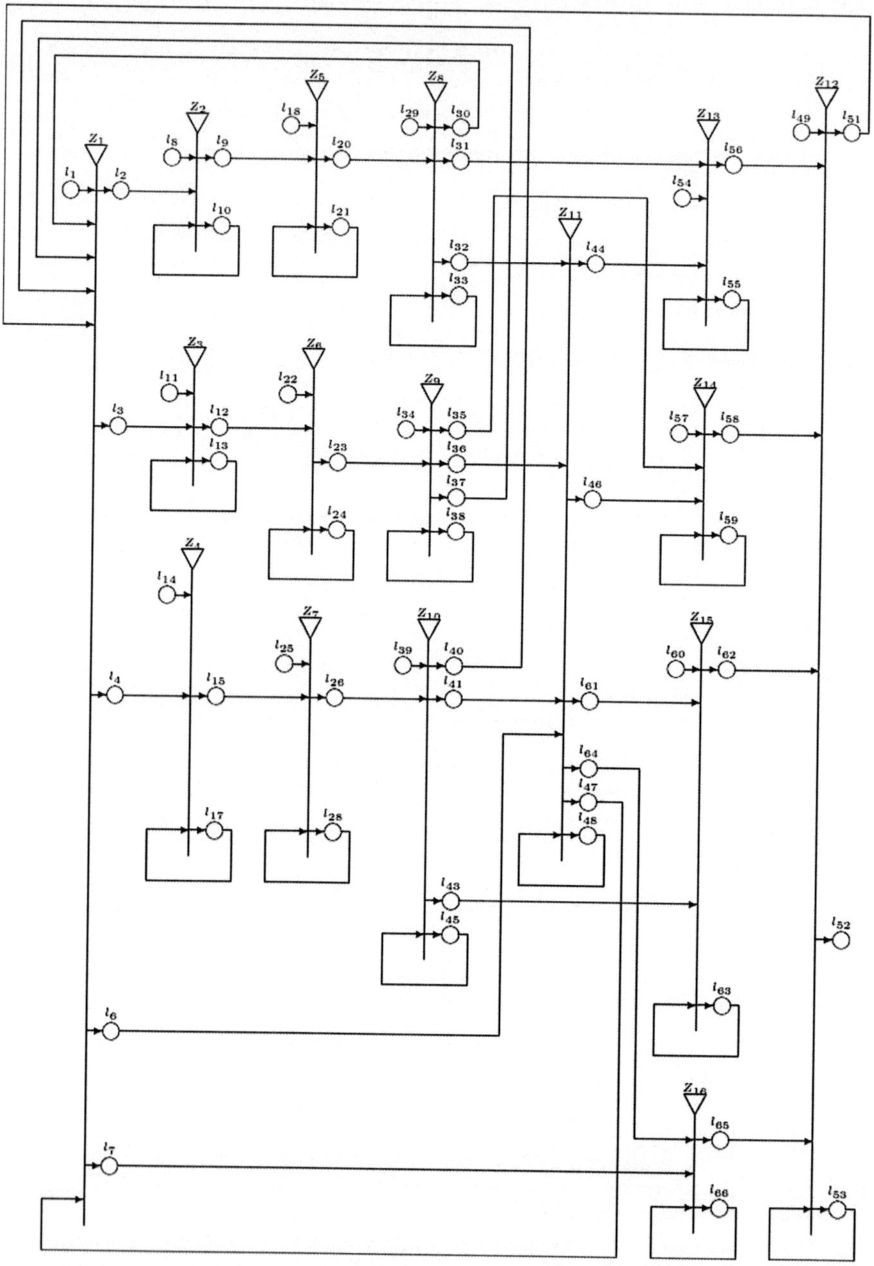

Fig. 3. Generalized net model of biometric multifactor authentication system

Token enters the net via place l_{14} with initial characteristics:

"parameters for cardiac signal acquisition".

The transition Z_4 has the form:

$$Z_4 = \langle\{l_4, l_{14}, l_{17}\}, \{l_{15}, l_{17}\}, R_4, \vee(\wedge(l_4, l_{14}), l_{17})\rangle,$$

where

$$R_4 = \begin{array}{c|cc} & l_{15} & l_{17} \\ \hline l_4 & false & true \\ l_{14} & false & true \\ l_{17} & W_{17,15} & W_{17,17} \end{array}$$

and

- $W_{17,15,1}$ = "there are acquired cardiac signals in calm and excited states";
- $W_{17,17} = \neg W_{17,15}$.

The token that enters the place l_{15} has the following characteristic:

"acquired cardiac signals in calm and excited states"

in place l_{15}.

Cardiac signal acquisition is made several times depending on the emotional states of the individual. The scanner is set by the proposed parameters. Special type of sensor is used.

Token enters the net via place l_{18} with initial characteristics:

"parameters for hand geometry image preprocessing".

The transition Z_5 has the form:

$$Z_5 = \langle\{l_9, l_{18}, l_{21}\}, \{l_{20}, l_{21}\}, R_5, \vee(\wedge(l_9, l_{18}), l_{21})\rangle,$$

where

$$R_5 = \begin{array}{c|cc} & l_{20} & l_{21} \\ \hline l_9 & false & true \\ l_{18} & false & true \\ l_{21} & W_{21,20} & W_{21,21} \end{array}$$

and

- $W_{21,20}$ = "there is preprocessed hand geometry image";
- $W_{21,21} = \neg W_{21,20}$.

The token that enters the place l_{12} has the following characteristic:

"*preprocessed hand geometry image*"

in place l_{20}.

The image of scanned hand (palm) geometry of the individual is preprocessed using methods for noise removal and techniques image quality improving. The segmentation of the region of interest (RoI) is made.

Token enters the net via place l_{22} with initial characteristics:

"*parameters for palm vein image preprocessing*".

The transition Z_6 has the form:

$$Z_6 = \langle\{l_{12}, l_{22}, l_{24}\}, \{l_{23}, l_{24}\}, R_6, \vee(\wedge(l_{12}, l_{22}), l_{24})\rangle,$$

where

$$R_6 = \begin{array}{c|cc} & l_{23} & l_{24} \\ \hline l_{12} & false & true \\ l_{22} & false & true \\ l_{24} & W_{24,23} & W_{24,24} \end{array}$$

and

- $W_{24,23}$ = "there is preprocessed palm vein image";
- $W_{24,24} = \neg W_{24,23}$.

The token that enters the place l_{23} has the following characteristic:

"*preprocessed palm vein image*"

in place l_{23}.

The scanned palm vein image is preprocessed using methods for noise removal and techniques for quality improving. The segmentation of the RoI is made.

Token enters the net via place l_{25} with initial characteristics:

"*parameters for cardiac signal preprocessing*".

The transition Z_7 has the form:

$$Z_7 = \langle\{l_{15}, l_{25}, l_{28}\}, \{l_{26,}, l_{28}\}, R_7, \vee(\wedge(l_{15}, l_{25}), l_{28})\rangle,$$

where

$R_7 =$	l_{26}	l_{28}
l_{15}	*false*	*true*
l_{25}	*false*	*true*
l_{28}	$W_{28,26}$	$W_{28,28}$

and

- $W_{28,26} =$ "there are preprocessed cardiac signals in calm and excited states";
- $W_{28,28} = \neg W_{28,26}$.

The token that enters the place l_{26} has the following characteristic:

"*preprocessed cardiac signals in calm and excited states*"

in place l_{26}.

The cardiac signals of individual depending on its emotions are preprocessed using methods for noise removal and signal alignment. Additional methods also can be used for processing the signal.

Token enters the net via place l_{29} with initial characteristics:

"*parameters for hand geometry features extraction*".

The transition Z_8 has the form:

$$Z_8 = \langle \{l_{20}, l_{29}, l_{33}\}, \{l_{30}, l_{31}, l_{32}, l_{33}\}, R_8, \vee(\wedge(l_{20}, l_{29}), l_{33})\rangle,$$

where

$R_8 =$	l_{30}	l_{31}	l_{32}	l_{33}
l_{20}	*false*	*false*	*false*	*true*
l_{29}	*false*	*false*	*false*	*true*
l_{33}	$W_{33,30}$	$W_{33,31}$	$W_{33,32}$	$W_{33,33}$

and

- $W_{33,30} =$ "it is necessary the new hand geometry scan";
- $W_{33,31} =$ "there is hand geometry template for biometric identification";
- $W_{33,32} =$ "there is hand geometry template for storing in database";
- $W_{33,33} = \neg (W_{33,30} \wedge W_{33,31} \wedge W_{33,32})$.

The tokens that enter the places l_{30}, l_{31} and l_{32} have the following characteristics:

"*need of new hand geometry scan*"

in place l_{30},

"hand geometry template for biometric identification"

in place l_{31},

"hand geometry template for storing in database"

in place l_{32}.

After hand (palm) geometry image preprocessing the futures extraction step is applied. The individual features of the user will be determined using appropriate algorithms and parameters.

Token enters the net via place l_{34} with initial characteristics:

"parameters for palm vein features extraction".

The transition Z_9 has the form:

$$Z_9 = \langle \{l_{25}, l_{34}, l_{38}\}, \{l_{35}, l_{36}, l_{37}, l_{38}\}, R_9, \vee(\wedge(l_{25}, l_{34}), l_{38}) \rangle,$$

where

$$R_9 = \begin{array}{c|cccc} & l_{35} & l_{36} & l_{37} & l_{38} \\ \hline l_{25} & false & false & false & true \\ l_{34} & false & false & false & true \\ l_{38} & W_{38,35} & W_{38,36} & W_{38,37} & W_{38,38} \end{array}$$

and

- $W_{38,35}$ = "there is palm vein template for biometric identification";
- $W_{38,36}$ = "there is palm vein template for storing in database";
- $W_{38,37}$ = "it is necessary new palm vein scan";
- $W_{38,38} = \neg (W_{38,35} \wedge W_{38,36} \wedge W_{38,37})$.

The tokens that enter the places l_{35}, l_{36} and l_{37} have the following characteristics:

"palm vein template for biometric identification"

in place l_{30},

"palm vein template for storing in database"

in place l_{31},

"need of the new palm vein scan"

in place l_{32}.

Palm vein features extraction is a step of pattern recognition process for individual features identification. In this step the unique scars of the people skin as veins are determined. The step of palm vein features extraction use algorithms and setting parameters as way to extract the important information.

Token enters the net via place l_{34} with initial characteristics:

"parameters for cardiac signal features extraction"

The transition Z_{10} has the form:

$$Z_{10} = \langle\{l_{26}, l_{39}, l_{45}\}, \{l_{40}, l_{41}, l_{43}, l_{45}\}, R_{10}, \vee(\wedge(l_{26}, l_{39}), l_{45})\rangle,$$

where

$$R_{10} = \begin{array}{c|cccc} & l_{40} & l_{41} & l_{43} & l_{45} \\ \hline l_{26} & false & false & false & true \\ l_{39} & false & false & false & true \\ l_{45} & W_{45,40} & W_{45,41} & W_{45,43} & W_{45,45} \end{array}$$

and

- $W_{45,40}$ = "it is necessary new cardiac signal acquisition";
- $W_{45,41}$ = "there are cardiac signal features of individual in calm and excited states for storing in database";
- $W_{45,43}$ = "there are cardiac signal features of individual in calm and excited states for authentication";
- $W_{45,45} = \neg(W_{45,40} \wedge W_{45,41} \wedge W_{45,43})$.

The tokens that enter the places l_{40}, l_{41} and l_{43} have the following characteristics:

"need of new cardiac signal acquisition"

in place l_{40},

"cardiac signal features of individual in calm and excited states for storing in database"

in place l_{41},

"cardiac signal features of individual in calm and excited states for authentication"

in place l_{43}.

Cardiac signal features extraction is a process getting the processed cardiac signal detecting the unique characteristics of the individual. Different parameters and algorithms can be used to extract the unique features from the signal.

The transition Z_{11} has the form:

$$Z_{11} = \langle\{l_6, l_{32}, l_{36}, l_{41}, l_{48}\}, \{l_{44}, l_{46}, l_{47}, l_{48}, l_{61}, l_{64}\}, R_{11}, \vee(\wedge(l_6, l_{32}, l_{36}, l_{41}), l_{48})\rangle,$$

where

$R_{11} =$

	l_{44}	l_{46}	l_{47}	l_{48}	l_{61}	l_{64}
l_6	*false*	*false*	*false*	*true*	*false*	*false*
l_{32}	*false*	*false*	*false*	*true*	*false*	*false*
l_{36}	*false*	*false*	*false*	*true*	*false*	*false*
l_{41}	*false*	*false*	*false*	*true*	*false*	*false*
l_{48}	$W_{48,44}$	$W_{48,46}$	$W_{48,47}$	$W_{48,48}$	$W_{48,61}$	$W_{48,64}$

and

- $W_{48,44}$ = "there are stored hand geometry data of individual for authentication";
- $W_{48,46}$ = "there are stored palm vein data of individual for authentication";
- $W_{48,47}$ = " it is necessary new characteristics of individual for storing to be provided";
- $W_{48,61}$ = "there are stored cardiac signal information of individual for authentication";
- $W_{48,64}$ = "there are stored passwords, location time and phone ID of the individual";
- $W_{48,48} = \neg (W_{48,44} \wedge W_{48,46} \wedge W_{48,47} \wedge W_{48,61} \wedge W_{48,64})$.

The tokens that enter the places l_{44}, l_{46}, l_{47}, l_{61} and l_{64} have the following characteristics:

"*stored hand geometry data of individual for authentication*"

in place l_{44},

"*stored palm vein data of individual for authentication*"

in place l_{46},

"*need of new characteristics of individual for storing*"

in place l_{47},

"*stored cardiac signal information of individual for authentication*"

in place l_{61},

"*stored passwords, location time and phone ID of the individual*"

in place l_{64}.

The database contains unique biometric templates and factors of the users assigned to the proposed system. If a new user has to be included in the data warehouse it has to apply a previous steps to extract its features. When the features of the person are stored once, they can be used for future access control activities.

Token enters the net via place l_{49} with initial characteristics:

"*parameters for rank fusion and decision making*".

The transition Z_{12} has the form:

$$Z_{12} = \langle \{l_{49}, l_{53}, l_{56}, l_{58}, l_{62}, l_{65}\}, \{l_{51}, l_{52}, l_{53}\}, R_{12}, \vee(\wedge(l_7, l_{49}, l_{56}, l_{58}, l_{62}), l_{53}) \rangle,$$

where

$R_{12} =$	l_{51}	l_{52}	l_{53}
l_{65}	*false*	*false*	*true*
l_{49}	*false*	*false*	*true*
l_{53}	$W_{53,51}$	$W_{53,52}$	$W_{53,53}$
l_{56}	*false*	*false*	*true*
l_{58}	*false*	*false*	*true*
l_{62}	*false*	*false*	*true*

and

- $W_{53,51}$ = "new individual characteristics for authentication have to be provided";
- $W_{53,52}$ = "there is a result of biometric multifactor authentication";
- $W_{53,53} = \neg(W_{53,51} \wedge W_{53,52})$.

The tokens that enter the places l_{51} and l_{52} have the following characteristics:

"*need of new individual characteristics for authentication*"

in place l_{51},

"*result of biometric multifactor authentication*"

in place l_{52}.

The results of matching phase of user factors and biometrics are used to determine its identity. The previously determined parameters as thresholds for rank fusion are used. According to the summarized result the system makes decision for the user identity.

Token enters the net via place l_{54} with initial characteristics:

"*parameters for hand geometry matching*".

The transition Z_{13} has the form:

$$Z_{13} = \langle \{l_{31}, l_{44}, l_{54}, l_{55}\}, \{l_{56}, l_{55}\}, R_{13}, \vee(\wedge(l_{31}, l_{44}, l_{54}), l_{55})\rangle,$$

where

$R_{13} =$	l_{56}	l_{55}
l_{31}	*false*	*true*
l_{44}	*false*	*true*
l_{54}	*false*	*true*
l_{55}	$W_{55,56}$	$W_{55,55}$

and

- $W_{55,56}$ = "there is a result from hand (palm) geometry matching";
- $W_{55,55} = \neg W_{55,56}$.

The token that enters the place l_{56} has the following characteristic:

"*result from hand (palm) geometry matching*"

in place l_{56}.

The step of hand (palm) geometry matching compares the input template to the stored template of the hand (palm) geometry of the user. The comparison is executed by algorithm and parameters for matching.

Token enters the net via place l_{57} with initial characteristics:

"*parameters for palm vein matching*".

The transition Z_{14} has the form:

$$Z_{14} = \langle\{l_{35}, l_{46}, l_{57}, l_{59}\}, \{l_{58}, l_{59}\}, R_{14}, \vee(\wedge(l_{35}, l_{46}, l_{57}), l_{59})\rangle,$$

where

$$R_{14} = \begin{array}{c|cc} & l_{58} & l_{59} \\ \hline l_{35} & false & true \\ l_{46} & false & true \\ l_{57} & false & true \\ l_{59} & W_{59,58} & W_{59,59} \end{array}$$

and

- $W_{59,58}$ = "there is a result from palm vein matching";
- $W_{59,59} = \neg W_{59,58}$.

The token that enters the place l_{58} has the following characteristic:

"*result from palm vein matching*"

in place l_{58}.

Palm vein matching is a step of juxtaposition of the input template of the user to the stored template of the palm vein in the database. The pattern matching algorithms and parameters are used.

Token enters the net via place l_{60} with initial characteristics:

"*parameters for cardiac signal matching*".

The transition Z_{15} has the form:

$$Z_{15} = \langle\{l_{43}, l_{60}, l_{61}, l_{63}\}, \{l_{62}, l_{63}\}, R_{15}, \vee(\wedge(l_{43}, l_{60}, l_{61}), l_{63})\rangle,$$

where

$$R_{15} = \begin{array}{c|cc} & l_{62} & l_{63} \\ \hline l_{43} & false & true \\ l_{60} & false & true \\ l_{61} & false & true \\ l_{63} & W_{63,62} & W_{63,63} \end{array}$$

and

- $W_{63,62}$ = "there is a result from cardiac signal matching";
- $W_{63,63} = \neg W_{63,62}$.

The token that enters the place l_{62} has the following characteristic:

"*result from cardiac signal matching*"

in place l_{62}.

Cardiac signal matching phase compares the stored templates for different emotional states of the user to the input templates. Pattern matching algorithms are applied to the features to extract their similarity.

The transition Z_{16} has the form:

$$Z_{16} = \langle\{l_7, l_{64}, l_{66}\}, \{l_{65}, l_{66}\}, R_{16}, \vee(\wedge(l_7, l_{64}), l_{66})\rangle,$$

where

$$R_{16} = \begin{array}{c|cc} & l_{65} & l_{66} \\ \hline l_7 & false & true \\ l_{64} & false & true \\ l_{66} & W_{66,65} & W_{66,66} \end{array}$$

and

- $W_{66,65}$ = "there is a result from multifactor matching of data";

The token that enters the place l_{65} has the following characteristic:

"*result from multifactor matching of data*"

in place l_{65}.

The input information of the user as unique device number, time of access request, location of the user device in the current moment of time and passwords are compared to the previously stored data in database to identify the individual.

3 Conclusion

Generalized net model of biometric multifactor authentication system is constructed. The proposed solution use multimodal biometric module and multifactor model. The user information is processed in parallel. According to the results of the biometric methods a decision making is made. The proposed model can be used for access control. In the future works a comparison of different types of biometrics in multifactor systems will be presented.

Acknowledgments. The authors are thankful for the support provided by the European Regional Development Fund and the Operational Program "Science and Education for Smart Growth" under contract UNITe No. BG05M2OP001-1.001-0004-C01 (2018–2023).

References

1. Atanassov, K., Boumbarov, O., Gluhchev, G., Hadjitodorov, S., Shannon, A., Vassilev, V.: A Generalized Net Model of Biometric Access-Control System. In: Proceedings of the 9th WSEAS International Conference on Automatic Control, Modeling & Simulation, Istanbul, Turkey, May 27–29, pp 78–81 (2007)
2. Atanassov, K., et al.: Generalized Nets Decision Making and Pattern Recognition. Warsaw School of Information Technology, Warszawa (2006)
3. Atanassov, K., Gluhchev, G., Hadjitodorov, S., Shannon, A., Vasilev, V.: Generalized Nets in Image Processing and Pattern Recognition, pp. 47–60. Sixth International Workshop on GNs, Sofia (2005)
4. Atanassov, K., Gluhchev, G., Hadjitodorov, S., Shannon, A., Vassilev, V.: "Generalized Nets and Pattern Recognition", KvB Visual Concepts Pty Ltd., Monograph No. 6, Sydney (2003)
5. Atanassov, K., Sotirova, E.: Generalized Nets. Prof. M. Drinov Academic Publishing House, Sofia (2017). (in Bulgarian)
6. Atanassov, K.: "Index Matrices: Towards an Augmented Matrix Calculus", Springer, Cham. (2014)
7. Atanassov, K.: "On Generalized Nets Theory". Prof. Marin Drinov Academic Publishing House, Sofia (2007)
8. Atanassov, K.: Generalized Nets. World Scientific, Singapore (1991)
9. Bureva, V., Sotirova, E., Bozov, H.: Generalized Net Model of Biometric Identification Process. In: 2018 20th International Symposium on Electrical Apparatus and Technologies (SIELA), Bourgas, pp. 1–4 (2018)
10. Bureva, V., Yovcheva, P., Sotirov, S.: Generalized net model of fingerprint recognition with intuitionistic fuzzy evaluations. In: Kacprzyk, J., Szmidt, E., Zadrożny, S., 11. Atanassov, K., Krawczak, M. (eds.) Advances in Fuzzy Logic and Technology 2017. IWIFSGN 2017, EUSFLAT 2017. AISC, vol. 641, pp. 286–294, Springer, Cham (2018). https://doi.org/10.1007/978-3-319-66830-7
11. Cantoni, V., Dimov, D., Tistarelli, M.: Biometric Authentication: First International Workshop, BIOMET 2014, Sofia, Bulgaria, June 23–24, 2014. Springer, Revised Selected Papers (2014)
12. Capek, J., Hub, M., Myskova, R.: Basic authentication procedure modelled by Petri nets. Int. J. Comput. Commun. **4**(4) (2010)
13. Cassandras C.: Lafortune, St.: Introduction to Discrete Event Systems, Springer-Verlag, Boston (2008). https://doi.org/10.1007/978-0-387-68612-7

14. Chen, Y., Fondeur J.C.: Biometric algorithms. In: Li, S.Z., Jain, A. (eds.) Encyclopedia of Biometrics. Springer, Boston (2009), https://doi.org/10.1007/978-0-387-73003-5_208
15. David, R., Alla, H.: Discrete, Continuous, and Hybrid Petri Nets, Springer (2010)
16. Gluhchev, G., Atanassov, K., Hadjitodorov, S., Szmidt, E.: A generalized net model for signature verification. In: Conference Proceedings of Tenth International Workshop on Generalized Nets, Sofia, 5 December 2009, pp. 27–30 (2009)
17. Gocheva, E., Sotirov S.: Modelling of the verification by iris scanning by generalized nets. In: Conference Proceedings of Ninth International Workshop on Generalized Nets, Sofia, pp 9–13 (2008)
18. Ikonomov, N.: GNDraw – software application for creating generalized nets. Iss. Intuit. Fuzzy Sets General. Nets **13**, 61–71 (2017)
19. Jain, A., Ross, A.: Nandakumar K.: Introduction to biometrics. In: Jain, A.K., Bolle, R., Pankanti, S. (eds.) Biometrics. Springer, Boston (2011). https://doi.org/10.1007/0-306-47044-6_1
20. Mason, J., Dave, R., Chatterjee, P., Graham-Allen, Esterline, A., Roy, K.: An Investigation of biometric authentication in the healthcare environment. Array **8** (2020)
21. Ometov, A., Bezzateev, S., Mäkitalo, N., Andreev, S., Mikkonen, T., Koucheryavy, Y.: Multifactor authentication: a survey. Cryptography **1** (2017)
22. Parvathi, R, Gluhchev, G,, Atanassov, K.: Generalized net model of face recognition. In: Conference proceedings of Ninth International Workshop on Generalized Nets, Sofia, pp 102–105 (2008)
23. Petrosyan, G., Ter-Vardanyan L., Gaboutchian A.: Modeling of biometric identification system using the colored petri nets. In: ISPRS - International Archives of the Photogrammetry, Remote Sensing and Spatial Information Sciences, pp. 37–42 (2015)
24. Ratha, N., Govindaraju, V.: Advances in Biometrics: Sensors, Algorithms and Systems, 3rd edn., Springer, London (2021). https://doi.org/10.1007/978-1-84628-921-7
25. Ratha, N.K., Govindaraju, V.: Advances in Biometrics, Sensors, Algorithms and Systems, Springer London (2008)
26. Reisig, W.: Understanding Petri Nets: Modeling Techniques, Analysis Methods, Case Studies, Springer, Berlin (2013). https://doi.org/10.1007/978-3-642-33278-4
27. Seatzu, C., Silva, M., Schuppen, J.: Control of Discrete-Event Systems, Springer (2013)
28. Singh, M., Singh, R., Ross, A.: A comprehensive overview of biometric fusion. Inf. Fusion **52**,187–205 (2019)
29. Sotirov, S.: Generalized net model of iris recognition using neural networks, annual of "informatics. In: Conference on Section Union of Scientists in Bulgaria, vol. 4, pp. 127–132 (2011)
30. Uhl, A., Busch, C., Marcel, S., Veldhuis, R.: Handbook of Vascular Biometrics, SpringerOpen (2020). https://doi.org/10.1007/978-3-030-27731-4
31. Wikipedia: Petri nets. https://en.wikipedia.org/wiki/Petri_net. Accessed 5 Aug 2021

Evolutionary k-Means Clustering Method with Controlled Number of Clusters Applied to Determine the Typology of Polish Municipalities

Jarosław Stańczak[1,2] and Jan W. Owsiński[1(✉)]

[1] Systems Research Institute, Newelska 6, Warsaw, Poland
{Jaroslaw.Stanczak,Jan.Owsinski}@ibspan.waw.pl
[2] Warsaw School of Information Technology, Newelska 6, Warsaw, Poland
J.Stanczak@wit.edu.pl

Abstract. Obtaining a method of grouping the data items in a number of categories, depending on the values of easy-to-interpret parameters, giving the ability to regulate to a certain degree the level of detail of data division (the number, size and location of detected clusters), became the impetus to develop the evolutionary k-means method and to examine its properties in conjunction with the test data, containing several important parameters of Polish municipalities.

The method presented is based on the widely known k-means approach, which, in combination with the evolutionary algorithm and some easy-to-obtain information on grouped data, allows for obtaining an adaptive method of data clustering. In this method, the degree of detail in the distribution of data can be easily adjusted using one parameter, and, depending on the value of this parameter, the method produces more detailed or more coarse data clustering, involving more or less of the detected clusters. This key parameter depends on some geometric characteristics of clustered data. It is, of course, possible to generate different divisions using classical k-means, by simply specifying different numbers of clusters to be detected, but in connection with evolutionary method and some information about clustered data it is possible to do it more efficiently and with an additional interpretation for the results.

The paper presents the evolutionary k-means method, the data considered and the obtained results, compared with the results generated by geographers.

Keywords: Clustering · Adaptive k-means · Evolutionary algorithm · Poland · Municipalities · Typology

1 Introduction

The k-means algorithm is a well-known grouping method, described in the middle and the second part of the 20th century in parallel by several researchers [Ste56,

K. T. Atanassov et al. (Eds.): IWIFSGN 2020/BOS/SOR 2020, LNNS 338, pp. 436–446, 2022.
https://doi.org/10.1007/978-3-030-95929-6_33

Llo82], while its present name was for the first time used by J. MacQueen in [Mac67].

In general, for the chosen number of clusters bigger than 1, the k-means algorithm is formally NP-hard [ADH09]. However, from the very start a simple and powerful heuristic has been proposed and used, providing fast results, even though without the assurance of correctness (global optimality of the solutions obtained). The latter issue, related to dependence of the results on the starting point of the procedure, is being overcome by application of numerous algorithm runs with different starting points, as well as special schemes for generating possibly good starting points. In any case, given that we have to deal with approximations, it seems justified that we use another heuristic method – the evolutionary algorithm (EA) in trying to solve the respective problem.

The k-means algorithm, mentioned above, is very simple, but, of course, has several drawbacks. The formal NP-hardness and the numerical problems, arising in connection with overcoming of the local character of solutions obtained, are among them, but in this article the problem of choice of the proper number of cluster centroids with their starting values will be the most important.

The evolutionary method is now almost a classical tool for solving difficult problems. It was also already used several times to improve the k-means algorithm, also for obtaining proper or optimal number of clusters (see, for instance [BaM02,LeC10]).

2 Basic Issues

2.1 Clustering

Formally, clustering, as understood here, is a disjunctive partitioning of set of data X, containing n-dimensional elements x, into p non-empty subsets called clusters, containing elements similar in some sense or measure, while the elements belonging to different clusters should be possibly dissimilar.

2.2 Basic Version of k-Means Algorithm

The basic algorithm of k-means can be described as follows:

```
1. Choose the number of sought clusters.
2. Generate starting positions of cluster centroids
3. Calculate distances of all clustered objects to all cluster
   centroids.
4. Assign objects to clusters with the closest centroids.
5. Update cluster centroids as geometric centers of their clusters.
6. If the assignment of objects to clusters in the subsequent two
   steps does not change, then go to 7, else go to 3.
7. End.
```

Algorithm 1. The k-means method.

The above algorithm can be perceived as effectively minimizing the following criterion:

$$C_D(P) = \sum_q \sum_{i \in A_q} d(x_i, x^q)^2 \quad (1)$$

where $d(.,.)$ – denotes the Euclidean distance; x_i – clustered data items; x^q – centroids of clusters A_q, $q = 1...p^{max}$.

Although the reference to centroids (means) implies the formulation of (1) with squared Euclidean distance, in many cases other Minkowski distances are used, defined as:

$$d(x_i, x_j) = (\sum_{ik} (x_{ik} - x_{jk})^h)^{\frac{1}{h}} \quad (2)$$

where k denotes the index of coordinates/attributes, describing the processed data items x_i, x_j, i, j – data indexes, h is an exponent, $h > 0$.

It is quite easy and, indeed, straightforward, to minimize the formula (1) for a given number of clusters p taking some starting point of the cluster centers x^q. It is owing to this simplicity that the k-means algorithms (their quite numerous varieties) are highly popular, as being very effective numerically (very fast convergence) to a local minimum of the objective function of type (1). However, for all these algorithms, choosing the right value for the number of clusters p is an important challenge, as this choice will affect the way the data set is ultimately divided. The solution to this fundamental deficiency of the k-means method is to be ensured here by the use of an evolutionary algorithm and a modified objective function of the problem, described in Sect. 3.

2.3 Standard Evolutionary Algorithm

The standard evolutionary algorithm works as this is shown below [Gol89].

```
1. Random initialization of the population of solutions.
2. Reproduction and modification of solutions using genetic
   operators.
3. Evaluation of obtained solutions.
4. Selection of individuals for the next generation.
5. If stop condition is not satisfied go to 2, else go to 6.
6. End.
```

Algorithm 2. The general evolutionary algorithm scheme.

This elementary scheme requires many problem specific definitions and improvements in order to work efficiently – [Mic96]. The adaptation of this simple version of EA to solve a given concrete problem requires the invention of a proper encoding of solutions, development of specialized genetic operators, appropriate for the accepted solution encoding, connected with the solved problem, and the fitness function to be optimized by the algorithm. The stop condition is usually constituted by a certain number of iterations (unless the problem specification provides a good "natural" stopping criterion, mainly related to the values of the fitness function). A short description of these EA adaptations can be found in the next section and a more detailed one in [Sta03].

3 Evolutionary k-Means Method with Controlled Number of Clusters

The modified evolutionary k-means method is a specialized EA with evaluation of the fitness function done according to the k-means paradigm. The combination of the EA with the k-means method is intended to ensure a better minimization of the modified quality criterion (3) – derived from the classic one (1), match better the selected clusters to the specificity of the data, and enable easy influence on the form of generated clustering using one parameter.

One solution of the problem, processed by EA, contains the chosen by EA variable number of clusters with values of centroid coordinates and optionally the exponent of Minkowski distance (2), but in the results, presented here in Sect. 4, the exponent value was set to constant value equal 2.

The EA in the here presented method is responsible for establishing the random starting values of the numbers of centroids and their coordinates and in subsequent iterations modifying them by genetic operators. Centroid coordinates are in each iteration recalculated by the k-means method during evaluation of solution for finding better clusterings, thus EA only proposes the starting points for them in each iteration, but the last word always belongs to the k-means method.

To avoid the potentially long calculations of the k-means algorithm (obtaining a solution that does not change in subsequent iterations), maximally five (or less, if possible) iterations of this algorithm are performed (which in most cases provides already quite satisfactory approximation of the respective local solution).

The evolutionary k-means method can be formulated as follows:

1. `Random initialization of the population of solutions.`
2. `Reproduction and modification of solutions using genetic operators.`
3. `Evaluation of obtained solutions:`
   a) `total minimized distance (3) is equal to infinity, the number of steps is equal to 0`
   b) `take the number and centers of sought clusters from evaluated solution,`
   c) `calculate distances (meant as in formula (2)) of all clustered objects to all cluster centroids,`
   d) `assign objects to clusters with the closest centroids,`

e) `update cluster centroids as geometric centers of their clusters,`
f) `if calculated total distance for new data clustering (3) is less than calculated in previous step and number of steps is less than 5, then go to b).`

4. `Selection of individuals for the next generation.`
5. `If stop condition of EA is not satisfied go to 2, else go to 6.`
6. `End.`

Algorithm 3. The evolutionary k-means algorithm scheme. The "solutions" are constituted by the set of locations of cluster representatives (assuming their variable number)

The minimized fitness function is similar to formula (1), namely:

$$C_{Dr}(P) = \sum_{q} \sum_{i \in A_q} d_r(x_i, x^q)^2 \tag{3}$$

where $d_r(.,.)$ denotes a modified Euclidean/Minkowski distance[1], x_i – clustered data, x^q – centroids of clusters A_q, $q = 1...p^{max}$.

The modified Euclidean/Minkowski distance $d_r(.,.)$, as used in (3), is calculated as follows:

$$d_r(x_i, x^q) = \begin{cases} R \textbf{ if } d(x_i, x^q) \le R \\ d(x_i, x^q) \textbf{ if } d(x_i, x^q) > R \end{cases} \tag{4}$$

The value of the threshold R is calculated on the basis of the properties of the grouped data and the given parameter r, $r \in [0, 1]$, which is meant to control the degree of detail of the clustering:

$$R = (1 - r) * 0.4 * d_{min}(x_i, x_j) + r * 0.8 * d_{max}(x_i, x_j), \tag{5}$$

where $d_{min}(x_i, x_j)$ is the minimum value (but bigger than zero[2]) of the Euclidean distance among grouped data items, while $d_{max}(x_i, x_j)$ is the maximum value of the Euclidean distance among grouped data items.

The R value is, in a specific sense, the "target radius" of generated clusters, although of course the actual clusters obtained may differ in size from those anticipated.

---

[1] In our case also other exponents (h) than 2 can be used in calculating distance, and they can be also tuned by the EA.

[2] For duplicate or identical objects among grouped data, repetitions are ignored, the nearest different data items are taken into account.

Values "0.4"[3] and "0.8"[4] come from the topological properties of possible distributions of data[5] and allow for obtaining from small clusters containing only one object (or only a few objects) for $r = 0$ to one big cluster containing all data for $r = 1$ with many different numbers of clusters[6] for $r \in [0, 1]$. The evolutionary algorithm can propose different data clusterings depending on the used value of r.

To sum up, the method operates on the following set of parameters, tuned by the EA: $Z = (x^q, r, p)$, in which we are looking for various "natural" data divisions into clusters, minimizing criterion (3) depending on the value of these parameters and the set of x^q cluster centers.

The specialized evolutionary algorithm is used to minimize the criterion (3).

The specialized evolutionary algorithm has in this case 5 mutation type operators that randomly modify the values of the components of Z, forming the solutions, such as:

- the number of clusters – p;
- initial values of cluster centers – x^q;
- weights of individual data attributes (in the experiments here reported this option is off, all weights are set to 1);
- exponent value when calculating the data distance - h (here not used and set to 2).
- value of r parameter (also can be set at predefined value).

The EA operates also two crossover operators:

- averaging crossover (weighted average parameter values from crossed solutions);
- uniform crossover (exchange of parameters between solutions).

The mechanism described in [Sta03] was used to manage the crossover and mutation operators, whose number is bigger and which are different than in the standard case. Similarly, the mixed selection method used to select solutions for the next iteration also comes from the work mentioned.

3 The radius value equal to 0.4 of the smallest distance between data, means that the method will generate as many clusters as there are different data items in the set (or sometimes a little less because also data farther away than R can be allocated to the cluster). This is not a very useful clustering, but it provides a natural upper limitation on the number of clusters.

4 The radius value equal to 0.8 of the largest distance between data means that the method should generate one cluster containing all data in the set. This is also not a very useful clustering, but it provides a natural lower limit on the number of clusters.

5 For a minimum value of the cluster radius equal to $0.4 * d_{min}$ ($r = 0$), we will obtain clusters with a radius that covers only one data item (for practical reasons, EA is focused on combining similar clusters, the effects of which can be seen in Table 3, where less clusters than the number of processed data is selected for $r = 0$). For the value of the cluster radius equal to $0.8 * d_{max}$ ($r = 1$) it is possible to include all grouped data in one cluster, given the appropriate selection of the center.

6 The value of r may be also tuned by EA to obtain the best value of formula (3), depending also on r, but it is likely that it would be $r = 0$..

4 The Data Describing the Typology of the Polish Municipalities

The data used for the experimental calculations come from the publications of the Central Statistical Office, describing 21 attributes of 2 478 Polish municipalities (the data used show the situation in 2015).

In the experiment here reported, also the typology of Polish municipalities is used, coming from [SlK16] and describing 10 functional types of municipalities. This typology was elaborated by the experts for purposes, related to planning at the local and regional levels.

Table 1. Attributes of data describing municipalities

No.	Attribute
1	Population
2	Built-up area
3	Share of transport areas
4	Density of population
5	Share of agricultural land
6	Share of built-up land
7	Share of forest areas
8	Share of population over 60 years of age
9	Share of the population below 20 years of age
10	Birth rate in the last 3 years
11	Migration balance in the last 3 years
12	Index of the average area of an agricultural holding
13	Share of registered working people
14	Registered economic activities per 1,000 inhabitants
15	Average employment rate in operating enterprises
16	The share of enterprises from industry and construction
17	Number of pupils and students per 1,000 inhabitants
18	Number of pupils and students of secondary schools per 1,000 inhabitants
19	Own income of the commune per capita
20	Share in PIT as part of the commune's budget
21	Share of expenses for social purposes in the communal budget

Table 2. Functional typology of Polish municipalities

Functional types	No. of units		Population		Area		Pop. density
	No.	%	$\times 10^3$	%	km^2	%	Pers./km^2
1. Functional urban areas of provincial capitals	33	1.3	9 557	24.8	4 720	1.5	2 025
2. External zones of provincial capitals	266	10.7	4 625	12.0	27 870	8.9	166
3. Functional urban areas of subregional centers	55	2.2	4 446	11.6	3 390	1.1	1 312
4. External zones of subregional centers	201	8.1	2 409	6.3	21 380	6.8	113
5. Multifunctional urban centers	147	5.9	3 938	10.2	10 390	3.3	379
6. Communes with developed transport functions	138	5.6	1 448	3.8	20 060	6.4	72
7. Communes with developed other non-agricultural functions	222	9.0	1 840	4.8	33 750	10.8	55
8. Communes with intensively developed agricultural functions	411	16.6	2 665	6.9	55 590	17.8	48
9. Communes with moderately developed agricultural functions	749	30.2	5 688	14.8	93 830	30.0	61
10. Communes featuring extensive development (with forests or nature protection areas)	257	10.4	1 878	4.9	41 590	13.3	45
Totals for Poland	2 479[a]	100	38 495	100	312 590	100	123

[a]Unfortunately, the number of municipalities (communes, gminas) is subject to some changes over time, thus different values of total number of municipalities are presented in this work.

The division of data, provided in [SlK16], and characterized in Table 2, will become the basis for comparing the properties of the here presented method with the division proposed by the experts.

5 Results of Computational Experiments

The first experiment shows how the clustering algorithm divides the data into clusters, depending on the imposed r value. Before calculations, the data attributes were normalized. The respective series of results is presented in Table 3.

As it can be seen, the number of obtained clusters decreases from a rather big value of 60 down to 1 with the increase of r, which is in line with the assumptions of the method. The fact that for $r = 0$ "only" 60 clusters are obtained instead of

2 478 (the number of classified data) is the result of AE and k-means functioning, also aimed at minimizing the number of very similar clusters, and not only at optimizing the objective function.

Table 3. Results showing the dependence of the number of obtained clusters on parameter r

r	$C_{Dr}(P)$	Number of detected clusters (p)
0.00	577.09	60
0.10	630.58	57
0.20	1038.59	46
0.30	1520.13	31
0.40	2004.66	15
0.45	2246.92	10
0.50	2489.17	7
0.60	2973.69	5
0.70	3458.20	3
0.80	3942.75	4
0.90	4427.23	2
1.00	4911.64	1

Then, Table 4 presents the distribution of experts' clustering (in rows) versus computed clustering (in columns) with the most possible mutual assignment of clusters for the apparently most similar of the obtained partitions $r = 0.45$ in Table 3, number of clusters = 10).

Table 4. Results for the division into 10 categories for Polish communes (the values in bold are an attempt to manually assign clusters to the typology, provided by the experts)

Categories: computed⟶ given↓	1	2	3	4	5	6	7	8	9	10	Totals
1	15	0	**14**	0	1	0	0	2	1	0	33
2	29	**98**	9	54	3	49	19	4	0	0	265
3	**26**	0	14	0	12	0	0	3	0	0	55
4	6	62	0	55	1	52	20	5	0	0	201
5	43	7	15	15	**36**	8	13	5	0	0	142
6	3	23	0	49	2	30	30	0	0	0	137
7	4	30	2	26	2	97	**44**	16	0	1	222
8	0	22	0	141	0	0	0	**333**	0	0	496
9	0	190	0	**258**	8	125	84	0	0	0	665
10	2	46	0	43	1	**141**	27	2	0	0	262
Totals	128	478	54	641	66	502	237	370	1	1	2 478

Figure 1 presents a map with the obtained clustering of Polish communes, characterized in Table 4. As it can be seen, we actually obtained 8 broad categories, and two additional categories, corresponding to outlier communes: Warsaw (the capital of Poland) and Kleszczów (home of the largest lignite mine and power plant in Europe) – the richest communes in Poland. However, categories indexed originally 4 (external zones of subregional centers) and 6 (communes with developed transport functions) from Table 2 did not appear in the partition obtained with the proposed method[7].

While the correspondence with the functional typology, proposed by experts (Table 2) is rather weak, at least in purely "arithmetic" terms, one can observe a degree of qualitative agreement between the two kinds of typologies, which makes the potential analysis, oriented at substantive, as well as formal aspects of the problem at hand, sensible.

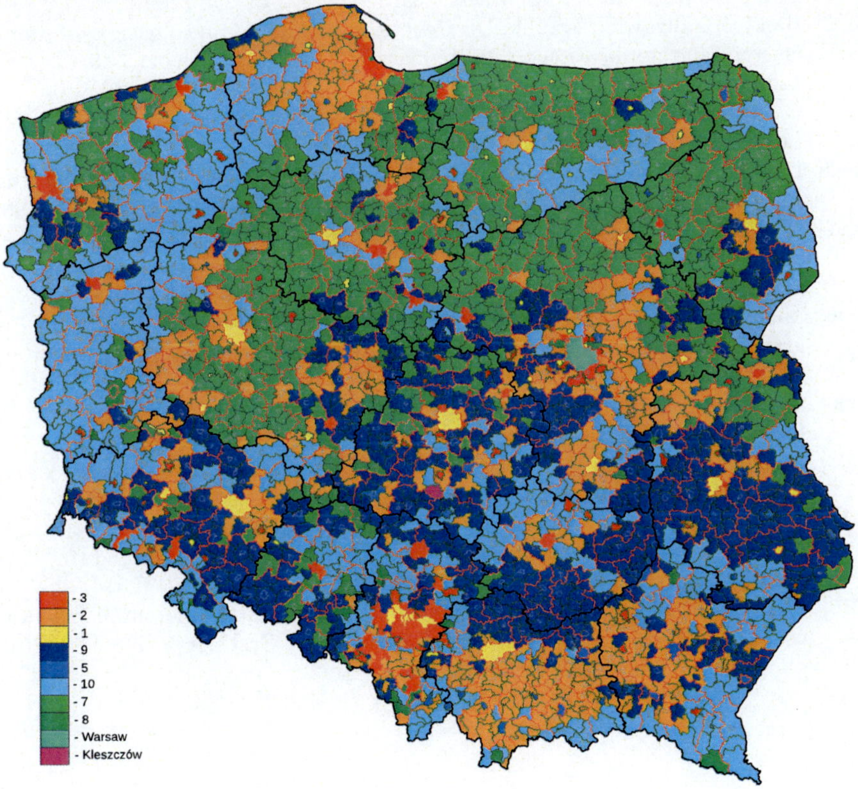

Fig. 1. The obtained division of communes into 10 categories using the k-means method for the entire Poland (with the interpretation according to Table 4 and the output data)

[7] Although this statement with respect to category 4 can be treated as debatable – see Table 4 – given that the assignment of the original and calculated clusters is by no means "absolute", it appears quite obvious that it is true for category 6.

6 Conclusions

The analysis of the data presented here shows that the proposed algorithm has interesting properties and thus has a great potential for use in pre-grouping of data. Further work should concentrate on the following three issues: checking the options, which have not been accounted for in the present series of experiments, more efficient linkage of the original k-means with the EA, and enhancement of the computational features (as of now, the computational features have not been the focus of interest, as primarily the substantive properties of the method were subject to verification).

References

[ADH09] Aloise, D., Deshpande, A., Hansen, P., Popat, P.: NP-hardness of Euclidean sum of-squares clustering. Mach. Learn. **75**(2), 245–248 (2009)

[BaM02] Bandyopadhyay, S., Maulik, U.: Genetic clustering for automatic evolution of clusters and application to image classification. Patt. Recogn. **35**(6), 1197–1208 (2002)

[Gol89] Goldberg, D.: Genetic Algorithms in Search Optimization and Machine Learning. Addison-Wesley, Cambridge (1989)

[Llo82] Least squares quantization in PCM. In: Bell Telephone Labs Memorandum, Murray Hill NJ, reprinted in: IEEE Trans. Inform. Theory **28**, 129–137 (1982)

[Mac67] MacQueen, J.: Some methods for classification and analysis of multivariate observations. In: Proceedings of the Fifth Berkeley Symposium on Mathematical Statistics and Probability, vol. 1, 281–297 (1967)

[Mic96] Michalewicz, Z.: Genetic Algorithms + Data Structures = Evolution Programs Springer Verlag, Berlin Heidelberg (1996). https://doi.org/10.1007/978-3-662-03315-9

[Ste56] Steinhaus, H.: Sur la division des corps materiels en parties. Bulletin de l'Academie Polonaise des Sciences Classe **3**(12), 801–804 (1956)

[Sta03] Stańczak, J.: Biologically inspired methods for control of evolutionary algorithms. Control Cybern **32**(2), 411–433 (2003)

[SlK16] Śleszyński, P., Komornicki, T.: Functional classification of Poland's communes (gminas) for the needs of the monitoring of spatial planning (in Polish with English summary). Przegląd Geograficzny **88**, 469–488 (2016)

[LeC10] Lee, W.-P., Chen, S.-W.: Automatic clustering with differential evolution using a cluster number oscillation method. Intell. Syst. Appl. 218–237 (2010)

Author Index

K. T. Atanassov et al. (Eds.): IWIFSGN 2020/BOS/SOR 2020, LNNS 338, pp. 447–448, 2022.
https://doi.org/10.1007/978-3-030-95929-6